ALL THE HELP, **RESOURCES**, AND PERSONAL **SUPPORT** YOU AND YOUR STUDENTS NEED!

2-Minute Tutorials and all of the resources you & your students need to get started
www.wileyplus.com/firstday

Student support from an experienced student user. Ask your local representative for details!

Collaborate with your colleagues, find a mentor, attend virtual and live events, and view resources
www.WhereFacultyConnect.com

Pre-loaded, ready-to-use assignments and presentations
www.wiley.com/college/quickstart

Technical Support 24/7 FAQs, online chat, and phone support
www.wileyplus.com/support

Your WileyPLUS Account Manager. Training and implementation support
www.wileyplus.com/accountmanager

www.wileyplus.com

MAKE IT YOURS!

9TH
EDITION

FUNDAMENTALS
OF PHYSICS

PART 2

9TH EDITION

HALLIDAY & RESNICK

FUNDAMENTALS OF PHYSICS

Jearl Walker

Cleveland State University

WILEY

John Wiley & Sons, Inc.

SPONSORING EDITOR   Geraldine Osnato

EXECUTIVE EDITOR   Stuart Johnson

ASSISTANT EDITOR   Aly Rentrop

ASSOCIATE MARKETING DIRECTOR   Christine Kushner

SENIOR PRODUCTION EDITOR   Elizabeth Swain

TEXT DESIGNER   Madelyn Lesure

COVER DESIGNER   M77 Design

DUMMY DESIGNER   Lee Goldstein

PHOTO EDITOR   Hilary Newman

EXECUTIVE MEDIA EDITOR   Thomas Kulesa

COVER IMAGE   ©Eric Heller/Photo Researchers, Inc.

This book was set in 10/12 Times Ten by Prepare and was printed and bound by R.R.Donnelley/Jefferson City. The cover was printed by R.R.Donnelley/Jefferson City.

This book is printed on acid free paper.

Library of Congress Cataloging-in-Publication Data

Halliday, David
  Fundamentals of physics / David Halliday, Robert Resnick, Jearl Walker.—9th ed.
     p. cm.
  Includes index.
Part 2: ISBN 978-0-470-54792-2 (pbk.)
Also catalogued as
Extended version: ISBN 978-0-470-46908-8
  1. Physics—Textbooks. I. Resnick, Robert II. Walker, Jearl III. Title.
   QC21.3.H35 2011
    530—dc22

                                        2009033774

Printed in the United States of America

10 9 8 7 6 5 4 3 2

CONTENTS

19 THE KINETIC THEORY OF GASES   507

20 ENTROPY AND THE SECOND LAW OF THERMODYNAMICS   536

APPENDICES

ANSWERS

INDEX   I-1

WHY I WROTE THIS BOOK

Fun with a big challenge. That is how I have regarded physics since the day when Sharon, one of the students in a class I taught as a graduate student, suddenly demanded of me, "What has any of this got to do with my life?" Of course I immediately responded, "Sharon, this has everything to do with your life—this is physics."

She asked me for an example. I thought and thought but could not come up with a single one. That night I began writing the book *The Flying Circus of Physics* (John Wiley & Sons Inc., 1975) for Sharon but also for me because I realized her complaint was mine. I had spent six years slugging my way through many dozens of physics textbooks that were carefully written with the best of pedagogical plans, but there was something missing. Physics is the most interesting subject in the world because it is about how the world works, and yet the textbooks had been thoroughly wrung of any connection with the real world. The fun was missing.

I have packed a lot of real-world physics into this HRW book, connecting it with the new edition of *The Flying Circus of Physics*. Much of the material comes from the HRW classes I teach, where I can judge from the faces and blunt comments what material and presentations work and what do not. The notes I make on my successes and failures there help form the basis of this book. My message here is the same as I had with every student I've met since Sharon so long ago: "Yes, you *can* reason from basic physics concepts all the way to valid conclusions about the real world, and that understanding of the real world is where the fun is."

I have many goals in writing this book but the overriding one is to provide instructors with tools by which they can teach students how to effectively read scientific material, identify fundamental concepts, reason through scientific questions, and solve quantitative problems. This process is not easy for either students or instructors. Indeed, the course associated with this book may be one of the most challenging of all the courses taken by a student. However, it can also be one of the most rewarding because it reveals the world's fundamental clockwork from which all scientific and engineering applications spring.

Many users of the eighth edition (both instructors and students) sent in comments and suggestions to improve the book. These improvements are now incorporated into the narrative and problems throughout the book. The publisher John Wiley & Sons and I regard the book as an ongoing project and encourage more input from users. You can send suggestions, corrections, and positive or negative comments to John Wiley & Sons or Jearl Walker (mail address: Physics Department, Cleveland State University, Cleveland, OH 44115 USA; or email address: physics@wiley.com; or the blog site at www.flyingcircusofphysics.com). We may not be able to respond to all suggestions, but we keep and study each of them.

LEARNINGS TOOLS

Animation

Because today's students have a wide range of learning styles, I have produced a wide range of learning tools, both in this new edition and online in *WileyPLUS*:

ANIMATIONS of one of the key figures in each chapter. Here in the book, those figures are flagged with the swirling icon. In the online chapter in *WileyPLUS*, a mouse click begins the animation. I have chosen the figures that are rich in information so that a student can see the physics in action and played out over a minute or two instead of just being flat on a printed page. Not only does this give life to the physics, but the animation can be repeated as many times as a student wants.

VIDEOS I have made well over 1000 instructional videos, with more coming each semester. Students can watch me draw or type on the screen as they hear me talk about a solution, tutorial, sample problem, or review, very much as they would experience were they sitting next to me in my office while I worked out something on a notepad. An instructor's lectures and tutoring will always be the most valuable learning tools, but my videos are available 24 hours a day, 7 days a week, and can be repeated indefinitely.

- **Video tutorials on subjects in the chapters**. I chose the subjects that challenge the students the most, the ones that my students scratch their heads about.

Video Review

GO Tutorial

- **Video reviews of high school math,** such as basic algebraic manipulations, trig functions, and simultaneous equations.

- **Video introductions to math,** such as vector multiplication, that will be new to the students.

- **Video presentations of every Sample Problem** in the textbook chapters (both 8e and 9e). My intent is to work out the physics, starting with the Key Ideas instead of just grabbing a formula. However, I also want to demonstrate how to read a sample problem, that is, how to read technical material to learn problem-solving procedures that can be transferred to other types of problems.

- **Video solutions to 20% of the end-of chapter problems.** The availability and timing of these solutions are controlled by the instructor. For example, they might be available after a homework deadline or a quiz. Each solution is not simply a plug-and-chug recipe. Rather I build a solution from the Key Ideas to the first step of reasoning and to a final solution. The student learns not just how to solve a particular problem but how to tackle any problem, even those that require *physics courage*.

- **Video examples of how to read data from graphs** (more than simply reading off a number with no comprehension of the physics).

WILEY PLUS   **READING MATERIAL** I have written a large number of reading resources for *WileyPLUS*.

- **Every sample problem in the textbook** (both 8e and 9e) is available online in both reading and video formats.

- **Hundreds of additional sample problems.** These are available as stand-alone resources but (at the discretion of the instructor) they are also linked out of the homework problems. So, if a homework problem deals with, say, forces on a block on a ramp, a link to a related sample problem is provided. However, the sample problem is not just a replica of the homework problem and thus does not provide a solution that can be merely duplicated without comprehension.

- **GO** **GO Tutorials** for 10% of the end-of-chapter homework problems. In multiple steps, I lead a student through a homework problem, starting with the Key Ideas and giving hints when wrong answers are submitted. However, I purposely leave the last step (for the final answer) to the student so that they are responsible at the end. Some online tutorial systems trap a student when wrong answers are given, which can generate a lot of frustration. My GO Tutorials are not traps, because at any step along the way, a student can return to the main problem.

- **Hints on every end-of-chapter homework problem** are available online (at the discretion of the instructor). I wrote these as true hints about the main ideas and the general procedure for a solution, not as recipes that provide an answer without any comprehension.

WILEY PLUS   **EVALUATION MATERIALS** Both self-evaluations and instructor evaluations are available.

- **Reading questions are available within each online section.** I wrote these so that they do not require analysis or any deep understanding; rather they simply test whether a student has read the section. When a student opens up a section, a randomly chosen reading question (from a bank of questions) appears at the end. The instructor can decide whether the question is part of the grading for that section or whether it is just for the benefit of the student.

- **Checkpoints are available within most sections.** I wrote these so that they require analysis and decisions about the physics in the section. *Answers to all checkpoints are in the back of the book.*

✓**CHECKPOINT 1**

The figure shows a circular path taken by a particle. If the instantaneous velocity of the particle is $\vec{v} = (2 \text{ m/s})\hat{i} - (2 \text{ m/s})\hat{j}$, through which quadrant is the particle moving at that instant if it is traveling (a) clockwise and (b) counterclockwise around the circle? For both cases, draw $\vec{v}$ on the figure.

Checkpoint

- **All end-of-chapter homework questions and problems** in the book (and many more problems) are available in *WileyPLUS*. The instructor can construct a homework assignment and control how it is graded when the answers are submitted online. For example, the instructor controls the deadline for submission and how many attempts a student is allowed on an answer. The

Expanded Figure

instructor also controls which, if any, learning aids are available with each homework problem. Such links can include hints, sample problems, in-chapter reading materials, video tutorials, video math reviews, and even video solutions (which can be made available to the students after, say, a homework deadline).

- **Symbolic notation problems** are available in every chapter and require algebraic answers.

WILEY PLUS DEMONSTRATIONS AND INTERACTIVE SIMULATIONS These have been produced by a number of instructors, to provide the experience of a computerized lab and lecture-room demonstrations.

ART PROGRAM

- Many of the figures in the book have been modified to make the physics ideas more pronounced.
- At least one key figure per chapter has been greatly expanded so that its message is conveyed in steps.

Fig. 4-9 The *projectile motion* of an object launched into the air at the origin of a coordinate system and with launch velocity $\vec{v}_0$ at angle θ_0. The motion is a combination of vertical motion (constant acceleration) and horizontal motion (constant velocity), as shown by the velocity components.

FLYING CIRCUS OF PHYSICS

- Flying Circus material has been incorporated into the text in several ways: Sample Problems, text examples, and end-of-chapter Problems. The purpose of this is two-fold: (1) make the subject more interesting and engaging, (2) show the student that the world around them can be examined and understood using the fundamental principles of physics.
- Links to *The Flying Circus of Physics* are shown throughout the text material and end-of-chapter problems with a biplane icon.   In the electronic version of this book, clicking on the icon takes you to the corresponding item in *Flying Circus*. The bibliography of *Flying Circus* (over 11 000 references to scientific and engineering journals) is located at www.flyingcircusofphysics.com.

SAMPLE PROBLEMS are chosen to demonstrate how problems can be solved with reasoned solutions rather than quick and simplistic plugging of numbers into an equation with no regard for what the equation means.

KEY IDEAS in the sample problems focus a student on the basic concepts at the root of the solution to a problem. In effect, these key ideas say, "We start our solution by using this basic concept, a procedure that prepares us for solving many other problems. We don't start by grabbing an equation for a quick plug-and-chug, a procedure that prepares us for nothing."

WHAT IS PHYSICS? The narrative of every chapter begins with this question, and with an answer that pertains to the subject of the chapter. (A plumber once asked me, "What do you do for a living?" I replied, "I teach physics." He thought for several minutes and then asked, "What is physics?" The plumber's career was entirely based on physics, yet he did not even know what physics is. Many students in introductory physics do not know what physics is but assume that it is irrelevant to their chosen career.)

ICONS FOR ADDITIONAL HELP. When worked-out solutions are provided either in print or electronically for certain of the odd-numbered problems, the statements for those problems include an icon to alert both student and instructor as to where the solutions are located. An icon guide is provided here and at the beginning of each set of problems

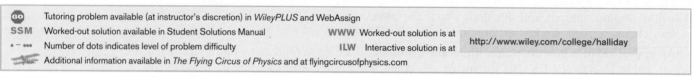

Icon Guide

VERSIONS OF THE TEXT

To accommodate the individual needs of instructors and students, the ninth edition of *Fundamentals of Physics* is available in a number of different versions.

The **Regular Edition** consists of Chapters 1 through 37 (ISBN 978-0-470-04472-8).

The **Extended Edition** contains seven additional chapters on quantum physics and cosmology, Chapters 1–44 (ISBN 978-0-471-75801-3).

Both editions are available as single, hardcover books, or in the following alternative versions:
Volume 1 - Chapters 1–20 (Mechanics and Thermodynamics), hardcover, ISBN 978-0-47004473-5
Volume 2 - Chapters 21–44 (E&M, Optics, and Quantum Physics), hardcover, ISBN 978-0-470-04474-2

INSTRUCTOR SUPPLEMENTS

INSTRUCTOR'S SOLUTIONS MANUAL by Sen-Ben Liao, Lawrence Livermore National Laboratory. This manual provides worked-out solutions for all problems found at the end of each chapter.

INSTRUCTOR COMPANION SITE http://www.wiley.com/college/halliday

- **Instructor's Manual** This resource contains lecture notes outlining the most important topics of each chapter; demonstration experiments; laboratory and computer projects; film and video sources; answers to all Questions, Exercises, Problems, and Checkpoints; and a correlation guide to the Questions, Exercises, and Problems in the previous edition. It also contains a complete list of all problems for which solutions are available to students (SSM, WWW, and ILW).

- **Lecture PowerPoint Slides** by Sudipa Kirtley of The Rose Hulman Institute. These PowerPoint slides serve as a helpful starter pack for instructors, outlining key concepts and incorporating figures and equations from the text.

- **Classroom Response Systems ("Clicker") Questions** by David Marx, Illinois State University. There are two sets of questions available: Reading Quiz questions and Interactive Lecture questions. The Reading Quiz questions are intended to be relatively straightforward for any student who reads the assigned material. The Interactive Lecture questions are intended for use in an interactive lecture setting.

- **Wiley Physics Simulations** by Andrew Duffy, Boston University. This is a collection of 50 interactive simulations (Java applets) that can be used for classroom demonstrations.

- **Wiley Physics Demonstrations** by David Maiullo, Rutgers University. This is a collection of digital videos of 80 standard physics demonstrations. They can be shown in class or accessed from the Student Companion site. There is an accompanying Instructor's Guide that includes "clicker" questions.

- **Test Bank** The Test Bank includes more than 2200 multiple-choice questions. These items are also available in the Computerized Test Bank which provides full editing features to help you customize tests (available in both IBM and Macintosh versions). The Computerized Test Bank is offered in both Diploma and Respondus.

- *Instructor's Solutions Manual*, in both MSWord and PDF files.

- All text illustrations, suitable for both classroom projection and printing.

ONLINE HOMEWORK AND QUIZZING. In addition to *WileyPLUS*, *Fundamentals of Physics*, ninth edition, also supports WebAssignPLUS and LON-CAPA, which are other programs that give instructors the ability to deliver and grade homework and quizzes online. WebAssign PLUS also offers students an online version of the text.

STUDENT SUPPLEMENTS

STUDENT COMPANION SITE. The web site http://www.wiley.com/college/halliday was developed specifically for *Fundamentals of Physics*, ninth edition, and is designed to further assist students in the study of physics. It includes solutions to selected end-of-chapter problems (which are identified with a www icon in the text); self-quizzes; simulation exercises; tips on how to make best use of a programmable calculator; and the Interactive LearningWare tutorials that are described below.

STUDENT STUDY GUIDE by Thomas Barrett of Ohio State University. The Student Study Guide consists of an overview of the chapter's important concepts, problem solving techniques and detailed examples.

STUDENT SOLUTIONS MANUAL by Sen-Ben Liao, Lawrence Livermore National Laboratory. This manual provides students with complete worked-out solutions to 15 percent of the problems found at the end of each chapter within the text. The Student Solutions Manual for the ninth edition is written using an innovative approach called TEAL which stands for Think, Express, Analyze, and Learn. This learning strategy was originally developed at the Massachusetts Institute of Technology and has proven to be an effective learning tool for students. These problems with TEAL solutions are indicated with an SSM icon in the text.

INTERACTIVE LEARNINGWARE. This software guides students through solutions to 200 of the end-of-chapter problems. These problems are indicated with an ILW icon in the text. The solutions process is developed interactively, with appropriate feedback and access to error-specific help for the most common mistakes.

INTRODUCTORY PHYSICS WITH CALCULUS AS A SECOND LANGUAGE: *Mastering Problem Solving* by Thomas Barrett of Ohio State University. This brief paperback teaches the student how to approach problems more efficiently and effectively. The student will learn how to recognize common patterns in physics problems, break problems down into manageable steps, and apply appropriate techniques. The book takes the student step by step through the solutions to numerous examples.

ACKNOWLEDGMENTS

A great many people have contributed to this book. J. Richard Christman, of the U.S. Coast Guard Academy, has once again created many fine supplements; his recommendations to this book have been invaluable. Sen-Ben Liao of Lawrence Livermore National Laboratory, James Whitenton of Southern Polytechnic State University, and Jerry Shi, of Pasadena City College, performed the Herculean task of working out solutions for every one of the homework problems in the book. At John Wiley publishers, the book received support from Stuart Johnson and Geraldine Osnato, the editors who oversaw the entire project from start to finish, and Tom Kulesa, who coordi-

nated the state-of-the-art media package. We thank Elizabeth Swain, the production editor, for pulling all the pieces together during the complex production process. We also thank Maddy Lesure for her design of the text and art direction of the cover; Lee Goldstein for her page make-up; and Lilian Brady for her proofreading. Hilary Newman was inspired in the search for unusual and interesting photographs. Both the publisher John Wiley & Sons, Inc. and Jearl Walker would like to thank the following for comments and ideas about the 8th edition: Jonathan Abramson, Portland State University; Omar Adawi, Parkland College; Edward Adelson, The Ohio State

University; Steven R. Baker, Naval Postgraduate School; George Caplan, Wellesley College; Richard Kass, The Ohio State University; M. R. Khoshbin-e-Khoshnazar, Research Institution for Curriculum Development & Educational Innovations (Tehran); Stuart Loucks, American River College; Laurence Lurio, Northern Illinois University; Ponn Maheswaranathan, Winthrop University; Joe McCullough, Cabrillo College; Don N. Page, University of Alberta; Elie Riachi, Fort Scott Community College; Andrew G. Rinzler, University of Florida; Dubravka Rupnik, Louisiana State University; Robert Schabinger, Rutgers University; Ruth

Schwartz, Milwaukee School of Engineering; Nora Thornber, Raritan Valley Community College; Frank Wang, LaGuardia Community College; Graham W. Wilson, University of Kansas; Roland Winkler, Northern Illinois University; Ulrich Zurcher, Cleveland State University. Finally, our external reviewers have been outstanding and we acknowledge here our debt to each member of that team.

Maris A. Abolins, *Michigan State University*

Edward Adelson, *Ohio State University*

Nural Akchurin, *Texas Tech*

Yildirim Aktas, *University of North Carolina- Charlotte*

Barbara Andereck, *Ohio Wesleyan University*

Tetyana Antimirova, *Ryerson University*

Mark Arnett, *Kirkwood Community College*

Arun Bansil, *Northeastern University*

Richard Barber, *Santa Clara University*

Neil Basecu, *Westchester Community College*

Anand Batra, *Howard University*

Richard Bone, *Florida International University*

Michael E. Browne, *University of Idaho*

Timothy J. Burns, *Leeward Community College*

Joseph Buschi, *Manhattan College*

Philip A. Casabella, *Rensselaer Polytechnic Institute*

Randall Caton, *Christopher Newport College*

Roger Clapp, *University of South Florida*

W. R. Conkie, *Queen's University*

Renate Crawford, *University of Massachusetts-Dartmouth*

Mike Crivello, *San Diego State University*

Robert N. Davie, Jr., *St. Petersburg Junior College*

Cheryl K. Dellai, *Glendale Community College*

Eric R. Dietz, *California State University at Chico*

N. John DiNardo, *Drexel University*

Eugene Dunnam, *University of Florida*

Robert Endorf, *University of Cincinnati*

F. Paul Esposito, *University of Cincinnati*

Jerry Finkelstein, *San Jose State University*

Robert H. Good, *California State University-Hayward*

Michael Gorman, *University of Houston*

Benjamin Grinstein, *University of California, San Diego*

John B. Gruber, *San Jose State University*

Ann Hanks, *American River College*

Randy Harris, *University of California-Davis*

Samuel Harris, *Purdue University*

Harold B. Hart, *Western Illinois University*

Rebecca Hartzler, *Seattle Central Community College*

John Hubisz, *North Carolina State University*

Joey Huston, *Michigan State University*

David Ingram, *Ohio University*

Shawn Jackson, *University of Tulsa*

Hector Jimenez, *University of Puerto Rico*

Sudhakar B. Joshi, *York University*

Leonard M. Kahn, *University of Rhode Island*

Sudipa Kirtley, *Rose-Hulman Institute*

Leonard Kleinman, *University of Texas at Austin*

Craig Kletzing, *University of Iowa*

Peter F. Koehler, *University of Pittsburgh*

Arthur Z. Kovacs, *Rochester Institute of Technology*

Kenneth Krane, *Oregon State University*

Priscilla Laws, *Dickinson College*

Edbertho Leal, *Polytechnic University of Puerto Rico*

Vern Lindberg, *Rochester Institute of Technology*

Peter Loly, *University of Manitoba*

James MacLaren, *Tulane University*

Andreas Mandelis, *University of Toronto*

Robert R. Marchini, *Memphis State University*

Andrea Markelz, *University at Buffalo, SUNY*

Paul Marquard, *Caspar College*

David Marx, *Illinois State University*

Dan Mazilu, *Washington and Lee University*

James H. McGuire, *Tulane University*

David M. McKinstry, *Eastern Washington University*

Jordon Morelli, *Queen's University*

Eugene Mosca, *United States Naval Academy*

Eric R. Murray, *Georgia Institute of Technology, School of Physics*

James Napolitano, *Rensselaer Polytechnic Institute*

Blaine Norum, *University of Virginia*

Michael O'Shea, *Kansas State University*

Patrick Papin, *San Diego State University*

Kiumars Parvin, *San Jose State University*

Robert Pelcovits, *Brown University*

Oren P. Quist, *South Dakota State University*

Joe Redish, *University of Maryland*

Timothy M. Ritter, *University of North Carolina at Pembroke*

Dan Styer, *Oberlin College*

Frank Wang, *LaGuardia Community College*

12

EQUILIBRIUM AND ELASTICITY

12-1 WHAT IS PHYSICS?

Human constructions are supposed to be stable in spite of the forces that act on them. A building, for example, should be stable in spite of the gravitational force and wind forces on it, and a bridge should be stable in spite of the gravitational force pulling it downward and the repeated jolting it receives from cars and trucks.

One focus of physics is on what allows an object to be stable in spite of any forces acting on it. In this chapter we examine the two main aspects of stability: the *equilibrium* of the forces and torques acting on rigid objects and the *elasticity* of nonrigid objects, a property that governs how such objects can deform. When this physics is done correctly, it is the subject of countless articles in physics and engineering journals; when it is done incorrectly, it is the subject of countless articles in newspapers and legal journals.

12-2 Equilibrium

Consider these objects: (1) a book resting on a table, (2) a hockey puck sliding with constant velocity across a frictionless surface, (3) the rotating blades of a ceiling fan, and (4) the wheel of a bicycle that is traveling along a straight path at constant speed. For each of these four objects,

1. The linear momentum $\vec{P}$ of its center of mass is constant.
2. Its angular momentum $\vec{L}$ about its center of mass, or about any other point, is also constant.

We say that such objects are in **equilibrium.** The two requirements for equilibrium are then

$$\vec{P} = \text{a constant} \quad \text{and} \quad \vec{L} = \text{a constant.} \tag{12-1}$$

Our concern in this chapter is with situations in which the constants in Eq. 12-1 are zero; that is, we are concerned largely with objects that are not moving in any way—either in translation or in rotation—in the reference frame from which we observe them. Such objects are in **static equilibrium.** Of the four objects mentioned at the beginning of this section, only one—the book resting on the table—is in static equilibrium.

The balancing rock of Fig. 12-1 is another example of an object that, for the present at least, is in static equilibrium. It shares this property with countless other structures, such as cathedrals, houses, filing cabinets, and taco stands, that remain stationary over time.

As we discussed in Section 8-6, if a body returns to a state of static equilibrium after having been displaced from that state by a force, the body is said to be in *stable* static equilibrium. A marble placed at the bottom of a hemispherical

Fig. 12-1 A balancing rock. Although its perch seems precarious, the rock is in static equilibrium. *(Symon Lobsang/Photis/Jupiter Images Corp.)*

Fig. 12-2 (*a*) A domino balanced on one edge, with its center of mass vertically above that edge. The gravitational force $\vec{F}_g$ on the domino is directed through the supporting edge. (*b*) If the domino is rotated even slightly from the balanced orientation, then $\vec{F}_g$ causes a torque that increases the rotation. (*c*) A domino upright on a narrow side is somewhat more stable than the domino in (*a*). (*d*) A square block is even more stable.

To tip the block, the center of mass must pass over the supporting edge.

com

$\vec{F}_g$ $\vec{F}_g$ $\vec{F}_g$ $\vec{F}_g$

Supporting edge

(*a*)   (*b*)   (*c*)   (*d*)

bowl is an example. However, if a small force can displace the body and end the equilibrium, the body is in *unstable* static equilibrium.

For example, suppose we balance a domino with the domino's center of mass vertically above the supporting edge, as in Fig. 12-2*a*. The torque about the supporting edge due to the gravitational force $\vec{F}_g$ on the domino is zero because the line of action of $\vec{F}_g$ is through that edge. Thus, the domino is in equilibrium. Of course, even a slight force on it due to some chance disturbance ends the equilibrium. As the line of action of $\vec{F}_g$ moves to one side of the supporting edge (as in Fig. 12-2*b*), the torque due to $\vec{F}_g$ increases the rotation of the domino. Therefore, the domino in Fig. 12-2*a* is in unstable static equilibrium.

The domino in Fig. 12-2*c* is not quite as unstable. To topple this domino, a force would have to rotate it through and then beyond the balance position of Fig. 12-2*a*, in which the center of mass is above a supporting edge. A slight force will not topple this domino, but a vigorous flick of the finger against the domino certainly will. (If we arrange a chain of such upright dominos, a finger flick against the first can cause the whole chain to fall.)

The child's square block in Fig. 12-2*d* is even more stable because its center of mass would have to be moved even farther to get it to pass above a supporting edge. A flick of the finger may not topple the block. (This is why you never see a chain of toppling square blocks.) The worker in Fig. 12-3 is like both the domino and the square block: Parallel to the beam, his stance is wide and he is stable; perpendicular to the beam, his stance is narrow and he is unstable (and at the mercy of a chance gust of wind).

The analysis of static equilibrium is very important in engineering practice. The design engineer must isolate and identify all the external forces and torques that may act on a structure and, by good design and wise choice of materials, ensure that the structure will remain stable under these loads. Such analysis is necessary to ensure, for example, that bridges do not collapse under their traffic and wind loads and that the landing gear of aircraft will function after the shock of rough landings.

12-3 The Requirements of Equilibrium

The translational motion of a body is governed by Newton's second law in its linear momentum form, given by Eq. 9-27 as

$$\vec{F}_{\text{net}} = \frac{d\vec{P}}{dt}.$$ (12-2)

If the body is in translational equilibrium—that is, if $\vec{P}$ is a constant—then $d\vec{P}/dt = 0$ and we must have

$$\vec{F}_{\text{net}} = 0 \quad \text{(balance of forces)}.$$ (12-3)

Fig. 12-3 A construction worker balanced on a steel beam is in static equilibrium but is more stable parallel to the beam than perpendicular to it.
(*Robert Brenner/PhotoEdit*)

The rotational motion of a body is governed by Newton's second law in its angular momentum form, given by Eq. 11-29 as

$$\vec{\tau}_{net} = \frac{d\vec{L}}{dt}.$$ (12-4)

If the body is in rotational equilibrium—that is, if $\vec{L}$ is a constant—then $d\vec{L}/dt = 0$ and we must have

$$\vec{\tau}_{net} = 0 \qquad \text{(balance of torques).}$$ (12-5)

Thus, the two requirements for a body to be in equilibrium are as follows:

1. The vector sum of all the external forces that act on the body must be zero.

2. The vector sum of all external torques that act on the body, measured about *any* possible point, must also be zero.

These requirements obviously hold for *static* equilibrium. They also hold for the more general equilibrium in which $\vec{P}$ and $\vec{L}$ are constant but not zero.

Equations 12-3 and 12-5, as vector equations, are each equivalent to three independent component equations, one for each direction of the coordinate axes:

Balance of forces	Balance of torques	
$F_{net,x} = 0$	$\tau_{net,x} = 0$	
$F_{net,y} = 0$	$\tau_{net,y} = 0$	(12-6)
$F_{net,z} = 0$	$\tau_{net,z} = 0$	

We shall simplify matters by considering only situations in which the forces that act on the body lie in the *xy* plane. This means that the only torques that can act on the body must tend to cause rotation around an axis parallel to the *z* axis. With this assumption, we eliminate one force equation and two torque equations from Eqs. 12-6, leaving

$$F_{net,x} = 0 \qquad \text{(balance of forces),}$$ (12-7)

$$F_{net,y} = 0 \qquad \text{(balance of forces),}$$ (12-8)

$$\tau_{net,z} = 0 \qquad \text{(balance of torques).}$$ (12-9)

Here, $\tau_{net,z}$ is the net torque that the external forces produce either about the *z* axis or about *any* axis parallel to it.

A hockey puck sliding at constant velocity over ice satisfies Eqs. 12-7, 12-8, and 12-9 and is thus in equilibrium *but not in static equilibrium.* For static equilibrium, the linear momentum $\vec{P}$ of the puck must be not only constant but also zero; the puck must be at rest on the ice. Thus, there is another requirement for static equilibrium:

3. The linear momentum $\vec{P}$ of the body must be zero.

✓CHECKPOINT 1

The figure gives six overhead views of a uniform rod on which two or more forces act perpendicularly to the rod. If the magnitudes of the forces are adjusted properly (but kept nonzero), in which situations can the rod be in static equilibrium?

(a)   (b)   (c)   (d)   (e)   (f)

12-4 The Center of Gravity

The gravitational force on an extended body is the vector sum of the gravitational forces acting on the individual elements (the atoms) of the body. Instead of considering all those individual elements, we can say that

 The gravitational force $\vec{F}_g$ on a body effectively acts at a single point, called the **center of gravity** (cog) of the body.

Here the word "effectively" means that if the forces on the individual elements were somehow turned off and force $\vec{F}_g$ at the center of gravity were turned on, the net force and the net torque (about any point) acting on the body would not change.

Until now, we have assumed that the gravitational force $\vec{F}_g$ acts at the center of mass (com) of the body. This is equivalent to assuming that the center of gravity is at the center of mass. Recall that, for a body of mass M, the force $\vec{F}_g$ is equal to $M\vec{g}$, where $\vec{g}$ is the acceleration that the force would produce if the body were to fall freely. In the proof that follows, we show that

If $\vec{g}$ is the same for all elements of a body, then the body's center of gravity (cog) is coincident with the body's center of mass (com).

This is approximately true for everyday objects because $\vec{g}$ varies only a little along Earth's surface and decreases in magnitude only slightly with altitude. Thus, for objects like a mouse or a moose, we have been justified in assuming that the gravitational force acts at the center of mass. After the following proof, we shall resume that assumption.

Proof

First, we consider the individual elements of the body. Figure 12-4a shows an extended body, of mass M, and one of its elements, of mass m_i. A gravitational force $\vec{F}_{gi}$ acts on each such element and is equal to $m_i\vec{g}_i$. The subscript on $\vec{g}_i$ means $\vec{g}_i$ is the gravitational acceleration *at the location of the element i* (it can be different for other elements).

In Fig. 12-4a, each force $\vec{F}_{gi}$ produces a torque τ_i on the element about the origin O, with moment arm x_i. Using Eq. 10-41 ($\tau = r_\perp F$), we can write torque τ_i as

$$\tau_i = x_i F_{gi}. \tag{12-10}$$

The net torque on all the elements of the body is then

$$\tau_{net} = \sum \tau_i = \sum x_i F_{gi}. \tag{12-11}$$

Next, we consider the body as a whole. Figure 12-4b shows the gravitational force $\vec{F}_g$ acting at the body's center of gravity. This force produces a torque τ on the body

Fig. 12-4 (a) An element of mass m_i in an extended body. The gravitational force $\vec{F}_{gi}$ on the element has moment arm x_i about the origin O of the coordinate system. (b) The gravitational force $\vec{F}_g$ on a body is said to act at the center of gravity (cog) of the body. Here $\vec{F}_g$ has moment arm x_{cog} about origin O.

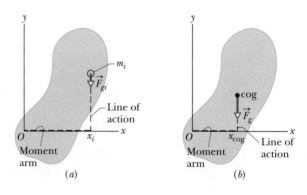

about O, with moment arm x_{cog}. Again using Eq. 10-41, we can write this torque as

$$\tau = x_{cog}F_g. \qquad (12\text{-}12)$$

The gravitational force $\vec{F}_g$ on the body is equal to the sum of the gravitational forces $\vec{F}_{gi}$ on all its elements, so we can substitute ΣF_{gi} for F_g in Eq. 12-12 to write

$$\tau = x_{cog}\Sigma F_{gi}. \qquad (12\text{-}13)$$

Now recall that the torque due to force $\vec{F}_g$ acting at the center of gravity is equal to the net torque due to all the forces $\vec{F}_{gi}$ acting on all the elements of the body. (That is how we defined the center of gravity.) Thus, τ in Eq. 12-13 is equal to τ_{net} in Eq. 12-11. Putting those two equations together, we can write

$$x_{cog}\Sigma F_{gi} = \Sigma x_i F_{gi}.$$

Substituting $m_i g_i$ for F_{gi} gives us

$$x_{cog}\Sigma m_i g_i = \Sigma x_i m_i g_i. \qquad (12\text{-}14)$$

Now here is a key idea: If the accelerations g_i at all the locations of the elements are the same, we can cancel g_i from this equation to write

$$x_{cog}\Sigma m_i = \Sigma x_i m_i. \qquad (12\text{-}15)$$

The sum Σm_i of the masses of all the elements is the mass M of the body. Therefore, we can rewrite Eq. 12-15 as

$$x_{cog} = \frac{1}{M}\Sigma x_i m_i. \qquad (12\text{-}16)$$

The right side of this equation gives the coordinate x_{com} of the body's center of mass (Eq. 9-4). We now have what we sought to prove:

$$x_{cog} = x_{com}. \qquad (12\text{-}17)$$

12-5 Some Examples of Static Equilibrium

In this section we examine several sample problems involving static equilibrium. In each, we select a system of one or more objects to which we apply the equations of equilibrium (Eqs. 12-7, 12-8, and 12-9). The forces involved in the equilibrium are all in the xy plane, which means that the torques involved are parallel to the z axis. Thus, in applying Eq. 12-9, the balance of torques, we select an axis parallel to the z axis about which to calculate the torques. Although Eq. 12-9 is satisfied for *any* such choice of axis, you will see that certain choices simplify the application of Eq. 12-9 by eliminating one or more unknown force terms.

✓CHECKPOINT 2

The figure gives an overhead view of a uniform rod in static equilibrium. (a) Can you find the magnitudes of unknown forces $\vec{F}_1$ and $\vec{F}_2$ by balancing the forces? (b) If you wish to find the magnitude of force $\vec{F}_2$ by using a balance of torques equation, where should you place a rotation axis to eliminate $\vec{F}_1$ from the equation? (c) The magnitude of $\vec{F}_2$ turns out to be 65 N. What then is the magnitude of $\vec{F}_1$?

Balancing a leaning ladder

In Fig. 12-5a, a ladder of length $L = 12$ m and mass $m = 45$ kg leans against a slick wall (that is, there is no friction between the ladder and the wall). The ladder's upper end is at height $h = 9.3$ m above the pavement on which the lower end is supported (the pavement is not frictionless). The ladder's center of mass is $L/3$ from the lower end, along the length of the ladder. A firefighter of mass $M = 72$ kg climbs the ladder until her center of mass is $L/2$ from the lower end. What then are the magnitudes of the forces on the ladder from the wall and the pavement?

KEY IDEAS

First, we choose our system as being the firefighter and ladder, together, and then we draw the free-body diagram of Fig. 12-5b to show the forces acting on the system. Because the system is in static equilibrium, the balancing equations for both forces and torques (Eqs. 12-7 through 12-9) can be applied to it.

Calculations: In Fig. 12-5b, the firefighter is represented with a dot within the boundary of the ladder. The gravitational force on her is represented with its equivalent expression $M\vec{g}$, and that vector has been shifted along its line of action (the line extending through the force vector), so that its tail is on the dot. (The shift does not alter a torque due to $M\vec{g}$ about any axis perpendicular to the figure. Thus, the shift does not affect the torque balancing equation that we shall be using.)

The only force on the ladder from the wall is the horizontal force $\vec{F}_w$ (there cannot be a frictional force along a frictionless wall, so there is no vertical force on the ladder from the wall). The force $\vec{F}_p$ on the ladder from the pavement has two components: a horizontal component $\vec{F}_{px}$ that is a static frictional force and a vertical component $\vec{F}_{py}$ that is a normal force.

To apply the balancing equations, let's start with the torque balancing of Eq. 12-9 ($\tau_{net,z} = 0$). To choose an axis about which to calculate the torques, note that we have unknown forces ($\vec{F}_w$ and $\vec{F}_p$) at the two ends of the ladder. To eliminate, say, $\vec{F}_p$ from the calculation, we place the axis at point O, perpendicular to the figure (Fig. 12-5b). We also place the origin of an xy coordinate system at O. We can find torques about O with any of Eqs. 10-39 through 10-41, but Eq. 10-41 ($\tau = r_\perp F$) is easiest to use here. *Making a wise choice about the placement of the origin can make our torque calculation much easier.*

To find the moment arm $r_\perp$ of the horizontal force $\vec{F}_w$ from the wall, we draw a line of action through that vector (it is the horizontal dashed line shown in Fig. 12-5c). Then $r_\perp$ is the perpendicular distance between O and the line of action. In Fig. 12-5c, $r_\perp$ extends along the y axis and is equal to the height h. We similarly draw lines of action for the gravitational force vectors $M\vec{g}$ and $m\vec{g}$ and see that their moment arms extend along the x axis. For the distance a shown in Fig. 12-5a, the moment arms are $a/2$ (the firefighter is halfway up the ladder) and $a/3$ (the ladder's center of mass is one-third of the way up the ladder), respectively. The moment arms for $\vec{F}_{px}$ and $\vec{F}_{py}$ are zero because the origin is located at the point where both of those forces act.

Now, with torques written in the form $r_\perp F$, the balancing equation $\tau_{net,z} = 0$ becomes

$$-(h)(F_w) + (a/2)(Mg) + (a/3)(mg)$$
$$+ (0)(F_{px}) + (0)(F_{py}) = 0. \quad (12\text{-}18)$$

(Recall our rule: A positive torque corresponds to counterclockwise rotation and a negative torque corresponds to clockwise rotation.)

Using the Pythagorean theorem for the right triangle made by the ladder in Fig. 11-5a, we find that

$$a = \sqrt{L^2 - h^2} = 7.58 \text{ m}.$$

Then Eq. 12-18 gives us

$$F_w = \frac{ga(M/2 + m/3)}{h}$$
$$= \frac{(9.8 \text{ m/s}^2)(7.58 \text{ m})(72/2 \text{ kg} + 45/3 \text{ kg})}{9.3 \text{ m}}$$
$$= 407 \text{ N} \approx 410 \text{ N}. \qquad \text{(Answer)}$$

Now we need to use the force balancing equations and Fig. 12-5d. The equation $F_{net,x} = 0$ gives us

$$F_w - F_{px} = 0,$$

so
$$F_{px} = F_w = 410 \text{ N}. \qquad \text{(Answer)}$$

The equation $F_{net,y} = 0$ gives us

$$F_{py} - Mg - mg = 0,$$

so $\quad F_{py} = (M + m)g = (72 \text{ kg} + 45 \text{ kg})(9.8 \text{ m/s}^2)$
$$= 1146.6 \text{ N} \approx 1100 \text{ N}. \qquad \text{(Answer)}$$

 Additional examples, video, and practice available at *WileyPLUS*

(a)

Frictionless
System
L
Fire-
fighter
com
Ladder
com
h
a

(b)

Here are all
the forces.

$\vec{F}_w$

Fire-
fighter

$\vec{F}_{py}$

$M\vec{g}$ Ladder

$m\vec{g}$

$\vec{F}_{px}$ O   x

a/3
a/2

(c)

$\vec{F}_{py}$

$\vec{F}_{px}$ O   x

Choosing the
rotation axis
here eliminates
the torques
due to these
forces.

$\vec{F}_w$

This moment
arm is
perpendicular
to the
line of action.

h

O   x

$M\vec{g}$

Here
too.

O   x

a/2

$m\vec{g}$

Here
too.

O   x

a/3

Fig. 12-5 (a) A firefighter climbs halfway up a
ladder that is leaning against a frictionless wall.
The pavement beneath the ladder is not frictionless. (b) A free-body diagram, showing the forces
that act on the *firefighter + ladder* system. The origin O of a coordinate system is placed at the point
of application of the unknown force $\vec{F}_p$ (whose
vector components $\vec{F}_{px}$ and $\vec{F}_{py}$ are shown). (c)
Calculating the torques. (d) Balancing the forces.

(d)

These horizontal
forces balance.

$\vec{F}_w$

$\vec{F}_{px}$ O   x

These
vertical
forces
balance.

$M\vec{g}$

$m\vec{g}$

$\vec{F}_{py}$

O   x

Sample Problem

Balancing a leaning boom

Figure 12-6a shows a safe (mass $M = 430$ kg) hanging by a rope (negligible mass) from a boom ($a = 1.9$ m and $b = 2.5$ m) that consists of a uniform hinged beam ($m = 85$ kg) and horizontal cable (negligible mass).

(a) What is the tension T_c in the cable? In other words, what is the magnitude of the force $\vec{T}_c$ on the beam from the cable?

KEY IDEAS

The system here is the beam alone, and the forces on it are shown in the free-body diagram of Fig. 12-6b. The force from the cable is $\vec{T}_c$. The gravitational force on the beam acts at the beam's center of mass (at the beam's center) and is represented by its equivalent $m\vec{g}$. The vertical component of the force on the beam from the hinge is $\vec{F}_v$, and the horizontal component of the force from the hinge is $\vec{F}_h$. The force from the rope supporting the safe is $\vec{T}_r$. Because beam, rope, and safe are stationary, the magnitude of $\vec{T}_r$ is equal to the weight of the safe: $T_r = Mg$. We place the origin O of an xy coordinate system at the hinge. Because the system is in static equilibrium, the balancing equations apply to it.

Calculations: Let us start with Eq. 12-9 ($\tau_{\text{net},z} = 0$). Note that we are asked for the magnitude of force $\vec{T}_c$ and not of forces $\vec{F}_h$ and $\vec{F}_v$ acting at the hinge, at point O. To eliminate $\vec{F}_h$ and $\vec{F}_v$ from the torque calculation, we should calculate torques about an axis that is perpendicular to the figure at point O. Then $\vec{F}_h$ and $\vec{F}_v$ will have moment arms of zero. The lines of action for $\vec{T}_c$, $\vec{T}_r$, and $m\vec{g}$ are dashed in Fig. 12-6b. The corresponding moment arms are a, b, and $b/2$.

Writing torques in the form of $r_\perp F$ and using our rule about signs for torques, the balancing equation $\tau_{\text{net},z} = 0$ becomes

$$(a)(T_c) - (b)(T_r) - (\tfrac{1}{2}b)(mg) = 0. \quad (12\text{-}19)$$

Substituting Mg for T_r and solving for T_c, we find that

$$T_c = \frac{gb(M + \tfrac{1}{2}m)}{a}$$

$$= \frac{(9.8 \text{ m/s}^2)(2.5 \text{ m})(430 \text{ kg} + 85/2 \text{ kg})}{1.9 \text{ m}}$$

$$= 6093 \text{ N} \approx 6100 \text{ N}. \quad \text{(Answer)}$$

(b) Find the magnitude F of the net force on the beam from the hinge.

KEY IDEA

Now we want the horizontal component F_h and vertical component F_v so that we can combine them to get the mag-

Fig. 12-6 (a) A heavy safe is hung from a boom consisting of a horizontal steel cable and a uniform beam. (b) A free-body diagram for the beam.

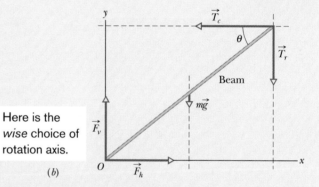

Here is the *wise* choice of rotation axis.

nitude F of the net force. Because we know T_c, we apply the force balancing equations to the beam.

Calculations: For the horizontal balance, we write $F_{\text{net},x} = 0$ as

$$F_h - T_c = 0, \quad (12\text{-}20)$$

and so

$$F_h = T_c = 6093 \text{ N}.$$

For the vertical balance, we write $F_{\text{net},y} = 0$ as

$$F_v - mg - T_r = 0.$$

Substituting Mg for T_r and solving for F_v, we find that

$$F_v = (m + M)g = (85 \text{ kg} + 430 \text{ kg})(9.8 \text{ m/s}^2)$$

$$= 5047 \text{ N}.$$

From the Pythagorean theorem, we now have

$$F = \sqrt{F_h^2 + F_v^2}$$

$$= \sqrt{(6093 \text{ N})^2 + (5047 \text{ N})^2} \approx 7900 \text{ N}. \quad \text{(Answer)}$$

Note that F is substantially greater than either the combined weights of the safe and the beam, 5000 N, or the tension in the horizontal wire, 6100 N.

 Additional examples, video, and practice available at *WileyPLUS*

Sample Problem

Balancing a horizantal beam

In Fig. 12-7a, a uniform beam, of length L and mass $m = 1.8$ kg, is at rest on two scales. A uniform block, with mass $M = 2.7$ kg, is at rest on the beam, with its center a distance $L/4$ from the beam's left end. What do the scales read?

The first steps in the solution of *any* problem about static equilibrium are these: Clearly define the system to be analyzed and then draw a free-body diagram of it, indicating all the forces on the system. Here, let us choose the system as the beam and block taken together. Then the forces on the system are shown in the free-body diagram of Fig. 12-7b. (Choosing the system takes experience, and often there can be more than one good choice.) Because the system is in static equilibrium, we can apply the balance of forces equations (Eqs. 12-7 and 12-8) and the balance of torques equation (Eq. 12-9) to it.

Calculations: The normal forces on the beam from the scales are $\vec{F}_l$ on the left and $\vec{F}_r$ on the right. The scale readings that we want are equal to the magnitudes of those forces. The

(a)

The vertical forces balance but that is not enough.

We must also balance torques, with a *wise* choice of rotation axis.

(b)

Fig. 12-7 (*a*) A beam of mass m supports a block of mass M. (*b*) A free-body diagram, showing the forces that act on the system *beam + block*.

gravitational force $\vec{F}_{g,beam}$ on the beam acts at the beam's center of mass and is equal to $m\vec{g}$. Similarly, the gravitational force $\vec{F}_{g,block}$ on the block acts at the block's center of mass and is equal to $M\vec{g}$. However, to simplify Fig. 12-7b, the block is represented by a dot within the boundary of the beam and $\vec{F}_{g,block}$ is drawn with its tail on that dot. (This shift of the vector $\vec{F}_{g,block}$ along its line of action does not alter the torque due to $\vec{F}_{g,block}$ about any axis perpendicular to the figure.)

The forces have no x components, so Eq. 12-7 ($F_{net,x} = 0$) provides no information. For the y components, Eq. 12-8 ($F_{net,y} = 0$) gives us

$$F_l + F_r - Mg - mg = 0. \qquad (12\text{-}21)$$

This equation contains two unknowns, the forces F_l and F_r, so we also need to use Eq. 12-9, the balance of torques equation. We can apply it to *any* rotation axis perpendicular to the plane of Fig. 12-7. Let us choose a rotation axis through the left end of the beam. We shall also use our general rule for assigning signs to torques: If a torque would cause an initially stationary body to rotate clockwise about the rotation axis, the torque is negative. If the rotation would be counterclockwise, the torque is positive. Finally, we shall write the torques in the form $r_\perp F$, where the moment arm $r_\perp$ is 0 for $\vec{F}_l$, $L/4$ for $M\vec{g}$, $L/2$ for $m\vec{g}$, and L for $\vec{F}_r$.

We now can write the balancing equation ($\tau_{net,z} = 0$) as

$$(0)(F_l) - (L/4)(Mg) - (L/2)(mg) + (L)(F_r) = 0,$$

which gives us

$$F_r = \tfrac{1}{4}Mg + \tfrac{1}{2}mg$$
$$= \tfrac{1}{4}(2.7 \text{ kg})(9.8 \text{ m/s}^2) + \tfrac{1}{2}(1.8 \text{ kg})(9.8 \text{ m/s}^2)$$
$$= 15.44 \text{ N} \approx 15 \text{ N}. \qquad \text{(Answer)}$$

Now, solving Eq. 12-21 for F_l and substituting this result, we find

$$F_l = (M + m)g - F_r$$
$$= (2.7 \text{ kg} + 1.8 \text{ kg})(9.8 \text{ m/s}^2) - 15.44 \text{ N}$$
$$= 28.66 \text{ N} \approx 29 \text{ N}. \qquad \text{(Answer)}$$

Notice the strategy in the solution: When we wrote an equation for the balance of force components, we got stuck with two unknowns. If we had written an equation for the balance of torques around some *arbitrary* axis, we would have again gotten stuck with those two unknowns. However, because we chose the axis to pass through the point of application of one of the unknown forces, here $\vec{F}_l$, we did not get stuck. Our choice neatly eliminated that force from the torque equation, allowing us to solve for the other unknown force magnitude F_r. Then we returned to the equation for the balance of force components to find the remaining unknown force magnitude.

Fig. 12-8 The table is an indeterminate structure. The four forces on the table legs differ from one another in magnitude and cannot be found from the laws of static equilibrium alone.

12-6 Indeterminate Structures

For the problems of this chapter, we have only three independent equations at our disposal, usually two balance of forces equations and one balance of torques equation about a given rotation axis. Thus, if a problem has more than three unknowns, we cannot solve it.

Consider an unsymmetrically loaded car. What are the forces—all different—on the four tires? Again, we cannot find them because we have only three independent equations. Similarly, we can solve an equilibrium problem for a table with three legs but not for one with four legs. Problems like these, in which there are more unknowns than equations, are called **indeterminate.**

Yet solutions to indeterminate problems exist in the real world. If you rest the tires of the car on four platform scales, each scale will register a definite reading, the sum of the readings being the weight of the car. What is eluding us in our efforts to find the individual forces by solving equations?

The problem is that we have assumed—without making a great point of it—that the bodies to which we apply the equations of static equilibrium are perfectly rigid. By this we mean that they do not deform when forces are applied to them. Strictly, there are no such bodies. The tires of the car, for example, deform easily under load until the car settles into a position of static equilibrium.

We have all had experience with a wobbly restaurant table, which we usually level by putting folded paper under one of the legs. If a big enough elephant sat on such a table, however, you may be sure that if the table did not collapse, it would deform just like the tires of a car. Its legs would all touch the floor, the forces acting upward on the table legs would all assume definite (and different) values as in Fig. 12-8, and the table would no longer wobble. Of course, we (and the elephant) would be thrown out onto the street but, in principle, how do we find the individual values of those forces acting on the legs in this or similar situations where there is deformation?

To solve such indeterminate equilibrium problems, we must supplement equilibrium equations with some knowledge of *elasticity,* the branch of physics and engineering that describes how real bodies deform when forces are applied to them. The next section provides an introduction to this subject.

✓ **CHECKPOINT 3**

A horizontal uniform bar of weight 10 N is to hang from a ceiling by two wires that exert upward forces $\vec{F}_1$ and $\vec{F}_2$ on the bar. The figure shows four arrangements for the wires. Which arrangements, if any, are indeterminate (so that we cannot solve for numerical values of $\vec{F}_1$ and $\vec{F}_2$)?

12-7 Elasticity

When a large number of atoms come together to form a metallic solid, such as an iron nail, they settle into equilibrium positions in a three-dimensional *lattice,* a repetitive arrangement in which each atom is a well-defined equilibrium distance from its nearest neighbors. The atoms are held together by interatomic forces that are modeled as tiny springs in Fig. 12-9. The lattice is remarkably rigid, which is another way of saying that the "interatomic springs" are extremely stiff. It is for this reason that we perceive many ordinary objects, such as metal ladders, tables, and spoons, as perfectly rigid. Of course, some ordinary objects, such as garden hoses or rubber gloves, do not strike us as rigid at all. The atoms that make up these objects *do not* form a rigid lattice like that of Fig. 12-9 but are aligned in long, flexible molecular chains, each chain being only loosely bound to its neighbors.

All real "rigid" bodies are to some extent **elastic,** which means that we can change their dimensions slightly by pulling, pushing, twisting, or compressing them. To get a feeling for the orders of magnitude involved, consider a vertical steel rod 1 m long and 1 cm in diameter attached to a factory ceiling. If you hang a subcompact car from the free end of such a rod, the rod will stretch but only by about 0.5 mm, or 0.05%. Furthermore, the rod will return to its original length when the car is removed.

If you hang two cars from the rod, the rod will be permanently stretched and will not recover its original length when you remove the load. If you hang three cars from the rod, the rod will break. Just before rupture, the elongation of the rod will be less than 0.2%. Although deformations of this size seem small, they are important in engineering practice. (Whether a wing under load will stay on an airplane is obviously important.)

Figure 12-10 shows three ways in which a solid might change its dimensions when forces act on it. In Fig. 12-10*a*, a cylinder is stretched. In Fig. 12-10*b*, a cylinder is deformed by a force perpendicular to its long axis, much as we might deform a pack of cards or a book. In Fig. 12-10*c*, a solid object placed in a fluid under high pressure is compressed uniformly on all sides. What the three deformation types have in common is that a **stress,** or deforming force per unit area, produces a **strain,** or unit deformation. In Fig. 12-10, *tensile stress* (associated with stretching) is illustrated in (*a*), *shearing stress* in (*b*), and *hydraulic stress* in (*c*).

The stresses and the strains take different forms in the three situations of Fig. 12-10, but—over the range of engineering usefulness—stress and strain are proportional to each other. The constant of proportionality is called a **modulus of elasticity,** so that

$$\text{stress} = \text{modulus} \times \text{strain}. \tag{12-22}$$

Fig. 12-9 The atoms of a metallic solid are distributed on a repetitive three-dimensional lattice. The springs represent interatomic forces.

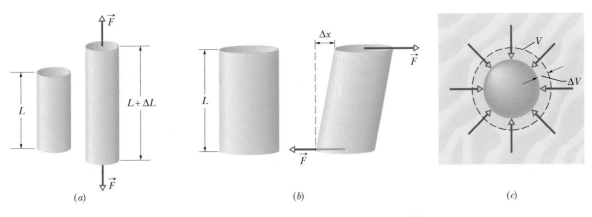

(*a*)   (*b*)   (*c*)

Fig. 12-10 (*a*) A cylinder subject to *tensile stress* stretches by an amount ΔL. (*b*) A cylinder subject to *shearing stress* deforms by an amount Δx, somewhat like a pack of playing cards would. (*c*) A solid sphere subject to uniform *hydraulic stress* from a fluid shrinks in volume by an amount ΔV. All the deformations shown are greatly exaggerated.

Fig. 12-11   A test specimen used to determine a stress–strain curve such as that of Fig. 12-12. The change ΔL that occurs in a certain length L is measured in a tensile stress–strain test.

Fig. 12-12   A stress–strain curve for a steel test specimen such as that of Fig. 12-11. The specimen deforms permanently when the stress is equal to the *yield strength* of the specimen's material. It ruptures when the stress is equal to the *ultimate strength* of the material.

In a standard test of tensile properties, the tensile stress on a test cylinder (like that in Fig. 12-11) is slowly increased from zero to the point at which the cylinder fractures, and the strain is carefully measured and plotted. The result is a graph of stress versus strain like that in Fig. 12-12. For a substantial range of applied stresses, the stress–strain relation is linear, and the specimen recovers its original dimensions when the stress is removed; it is here that Eq. 12-22 applies. If the stress is increased beyond the **yield strength** S_y of the specimen, the specimen becomes permanently deformed. If the stress continues to increase, the specimen eventually ruptures, at a stress called the **ultimate strength** S_u.

Tension and Compression

For simple tension or compression, the stress on the object is defined as F/A, where F is the magnitude of the force applied perpendicularly to an area A on the object. The strain, or unit deformation, is then the dimensionless quantity $\Delta L/L$, the fractional (or sometimes percentage) change in a length of the specimen. If the specimen is a long rod and the stress does not exceed the yield strength, then not only the entire rod but also every section of it experiences the same strain when a given stress is applied. Because the strain is dimensionless, the modulus in Eq. 12-22 has the same dimensions as the stress—namely, force per unit area.

The modulus for tensile and compressive stresses is called the **Young's modulus** and is represented in engineering practice by the symbol E. Equation 12-22 becomes

$$\frac{F}{A} = E\,\frac{\Delta L}{L}. \tag{12-23}$$

The strain $\Delta L/L$ in a specimen can often be measured conveniently with a *strain gage* (Fig. 12-13), which can be attached directly to operating machinery with an adhesive. Its electrical properties are dependent on the strain it undergoes.

Although the Young's modulus for an object may be almost the same for tension and compression, the object's ultimate strength may well be different for the two types of stress. Concrete, for example, is very strong in compression but is so weak in tension that it is almost never used in that manner. Table 12-1 shows the Young's modulus and other elastic properties for some materials of engineering interest.

Shearing

In the case of shearing, the stress is also a force per unit area, but the force vector lies in the plane of the area rather than perpendicular to it. The strain is the dimensionless ratio $\Delta x/L$, with the quantities defined as shown in Fig. 12-10b. The corresponding modulus, which is given the symbol G in engineering practice, is called the **shear modulus.** For shearing, Eq. 12-22 is written as

$$\frac{F}{A} = G\,\frac{\Delta x}{L}. \tag{12-24}$$

Shearing stresses play a critical role in the buckling of shafts that rotate under load and in bone fractures caused by bending.

Hydraulic Stress

In Fig. 12-10c, the stress is the fluid pressure p on the object, and, as you will see in Chapter 14, pressure is a force per unit area. The strain is $\Delta V/V$, where V is the

Fig. 12-13   A strain gage of overall dimensions 9.8 mm by 4.6 mm. The gage is fastened with adhesive to the object whose strain is to be measured; it experiences the same strain as the object. The electrical resistance of the gage varies with the strain, permitting strains up to 3% to be measured. *(Courtesy Vishay Micro-Measurements Group, Raleigh, NC)*

Table 12-1

Some Elastic Properties of Selected Materials of Engineering Interest

Material	Density ρ (kg/m³)	Young's Modulus E (10^9 N/m²)	Ultimate Strength S_u (10^6 N/m²)	Yield Strength S_y (10^6 N/m²)
Steel[a]	7860	200	400	250
Aluminum	2710	70	110	95
Glass	2190	65	50[b]	—
Concrete[c]	2320	30	40[b]	—
Wood[d]	525	13	50[b]	—
Bone	1900	9[b]	170[b]	—
Polystyrene	1050	3	48	—

[a]Structural steel (ASTM-A36).   [b]In compression.
[c]High strength   [d]Douglas fir.

original volume of the specimen and ΔV is the absolute value of the change in volume. The corresponding modulus, with symbol B, is called the **bulk modulus** of the material. The object is said to be under *hydraulic compression,* and the pressure can be called the *hydraulic stress.* For this situation, we write Eq. 12-22 as

$$p = B\,\frac{\Delta V}{V}. \qquad (12\text{-}25)$$

The bulk modulus is 2.2×10^9 N/m² for water and 1.6×10^{11} N/m² for steel. The pressure at the bottom of the Pacific Ocean, at its average depth of about 4000 m, is 4.0×10^7 N/m². The fractional compression $\Delta V/V$ of a volume of water due to this pressure is 1.8%; that for a steel object is only about 0.025%. In general, solids—with their rigid atomic lattices—are less compressible than liquids, in which the atoms or molecules are less tightly coupled to their neighbors.

Sample Problem

Stress and strain of elongated rod

One end of a steel rod of radius $R = 9.5$ mm and length $L = 81$ cm is held in a vise. A force of magnitude $F = 62$ kN is then applied perpendicularly to the end face (uniformly across the area) at the other end, pulling directly away from the vise. What are the stress on the rod and the elongation ΔL and strain of the rod?

KEY IDEAS

(1) Because the force is perpendicular to the end face and uniform, the stress is the ratio of the magnitude F of the force to the area A. The ratio is the left side of Eq. 12-23. (2) The elongation ΔL is related to the stress and Young's modulus E by Eq. 12-23 ($F/A = E\,\Delta L/L$). (3) Strain is the ratio of the elongation to the initial length L.

Calculations: To find the stress, we write

$$\text{stress} = \frac{F}{A} = \frac{F}{\pi R^2} = \frac{6.2 \times 10^4 \text{ N}}{(\pi)(9.5 \times 10^{-3}\text{ m})^2}$$
$$= 2.2 \times 10^8 \text{ N/m}^2. \qquad \text{(Answer)}$$

The yield strength for structural steel is 2.5×10^8 N/m², so this rod is dangerously close to its yield strength.

We find the value of Young's modulus for steel in Table 12-1. Then from Eq. 12-23 we find the elongation:

$$\Delta L = \frac{(F/A)L}{E} = \frac{(2.2 \times 10^8 \text{ N/m}^2)(0.81\text{ m})}{2.0 \times 10^{11} \text{ N/m}^2}$$
$$= 8.9 \times 10^{-4} \text{ m} = 0.89 \text{ mm}. \qquad \text{(Answer)}$$

For the strain, we have

$$\frac{\Delta L}{L} = \frac{8.9 \times 10^{-4} \text{ m}}{0.81 \text{ m}}$$
$$= 1.1 \times 10^{-3} = 0.11\%. \qquad \text{(Answer)}$$

 Additional examples, video, and practice available at *WileyPLUS*

Balancing a wobbly table

A table has three legs that are 1.00 m in length and a fourth leg that is longer by $d = 0.50$ mm, so that the table wobbles slightly. A steel cylinder with mass $M = 290$ kg is placed on the table (which has a mass much less than M) so that all four legs are compressed but unbuckled and the table is level but no longer wobbles. The legs are wooden cylinders with cross-sectional area $A = 1.0$ cm^2; Young's modulus is $E = 1.3 \times 10^{10}$ N/m^2. What are the magnitudes of the forces on the legs from the floor?

We take the table plus steel cylinder as our system. The situation is like that in Fig. 12-8, except now we have a steel cylinder on the table. If the tabletop remains level, the legs must be compressed in the following ways: Each of the short legs must be compressed by the same amount (call it ΔL_3) and thus by the same force of magnitude F_3. The single long leg must be compressed by a larger amount ΔL_4 and thus by a force with a larger magnitude F_4. In other words, for a level tabletop, we must have

$$\Delta L_4 = \Delta L_3 + d. \qquad (12\text{-}26)$$

From Eq. 12-23, we can relate a change in length to the force causing the change with $\Delta L = FL/AE$, where L is the original length of a leg. We can use this relation to replace ΔL_4 and ΔL_3 in Eq. 12-26. However, note that we can approximate the original length L as being the same for all four legs.

Calculations: Making those replacements and that approximation gives us

$$\frac{F_4 L}{AE} = \frac{F_3 L}{AE} + d. \qquad (12\text{-}27)$$

We cannot solve this equation because it has two unknowns, F_4 and F_3.

To get a second equation containing F_4 and F_3, we can use a vertical y axis and then write the balance of vertical forces ($F_{\text{net},y} = 0$) as

$$3F_3 + F_4 - Mg = 0, \qquad (12\text{-}28)$$

where Mg is equal to the magnitude of the gravitational force on the system. (*Three* legs have force $\vec{F}_3$ on them.) To solve the simultaneous equations 12-27 and 12-28 for, say, F_3, we first use Eq. 12-28 to find that $F_4 = Mg - 3F_3$. Substituting that into Eq. 12-27 then yields, after some algebra,

$$F_3 = \frac{Mg}{4} - \frac{dAE}{4L}$$

$$= \frac{(290 \text{ kg})(9.8 \text{ m/s}^2)}{4}$$

$$- \frac{(5.0 \times 10^{-4} \text{ m})(10^{-4} \text{ m}^2)(1.3 \times 10^{10} \text{ N/m}^2)}{(4)(1.00 \text{ m})}$$

$$= 548 \text{ N} \approx 5.5 \times 10^2 \text{ N}. \qquad \text{(Answer)}$$

From Eq. 12-28 we then find

$$F_4 = Mg - 3F_3 = (290 \text{ kg})(9.8 \text{ m/s}^2) - 3(548 \text{ N})$$

$$\approx 1.2 \text{ kN}. \qquad \text{(Answer)}$$

You can show that to reach their equilibrium configuration, the three short legs are each compressed by 0.42 mm and the single long leg by 0.92 mm.

 Additional examples, video, and practice available at *WileyPLUS*

Static Equilibrium A rigid body at rest is said to be in **static equilibrium.** For such a body, the vector sum of the external forces acting on it is zero:

$$\vec{F}_{\text{net}} = 0 \qquad \text{(balance of forces).} \qquad (12\text{-}3)$$

If all the forces lie in the xy plane, this vector equation is equivalent to two component equations:

$$F_{\text{net},x} = 0 \quad \text{and} \quad F_{\text{net},y} = 0 \qquad \text{(balance of forces).} \qquad (12\text{-}7, 12\text{-}8)$$

Static equilibrium also implies that the vector sum of the external torques acting on the body about *any* point is zero, or

$$\vec{\tau}_{\text{net}} = 0 \qquad \text{(balance of torques).} \qquad (12\text{-}5)$$

If the forces lie in the xy plane, all torque vectors are parallel to the

z axis, and Eq. 12-5 is equivalent to the single component equation

$$\tau_{\text{net},z} = 0 \qquad \text{(balance of torques).} \qquad (12\text{-}9)$$

Center of Gravity The gravitational force acts individually on each element of a body. The net effect of all individual actions may be found by imagining an equivalent total gravitational force $\vec{F}_g$ acting at the **center of gravity.** If the gravitational acceleration $\vec{g}$ is the same for all the elements of the body, the center of gravity is at the center of mass.

Elastic Moduli Three **elastic moduli** are used to describe the elastic behavior (deformations) of objects as they respond to forces that act on them. The **strain** (fractional change in length) is linearly related to the applied **stress** (force per unit area) by the proper modulus, according to the general relation

$$\text{stress} = \text{modulus} \times \text{strain}. \qquad (12\text{-}22)$$

Tension and Compression When an object is under tension or compression, Eq. 12-22 is written as

$$\frac{F}{A} = E\frac{\Delta L}{L}, \qquad (12\text{-}23)$$

where $\Delta L/L$ is the tensile or compressive strain of the object, F is the magnitude of the applied force $\vec{F}$ causing the strain, A is the cross-sectional area over which $\vec{F}$ is applied (perpendicular to A, as in Fig. 12-10a), and E is the **Young's modulus** for the object. The stress is F/A.

Shearing When an object is under a shearing stress, Eq. 12-22 is written as

$$\frac{F}{A} = G\frac{\Delta x}{L}, \qquad (12\text{-}24)$$

where $\Delta x/L$ is the shearing strain of the object, Δx is the displacement of one end of the object in the direction of the applied force $\vec{F}$ (as in Fig. 12-10b), and G is the **shear modulus** of the object. The stress is F/A.

Hydraulic Stress When an object undergoes *hydraulic compression* due to a stress exerted by a surrounding fluid, Eq. 12-22 is written as

$$p = B\frac{\Delta V}{V}, \qquad (12\text{-}25)$$

where p is the pressure (*hydraulic stress*) on the object due to the fluid, $\Delta V/V$ (the strain) is the absolute value of the fractional change in the object's volume due to that pressure, and B is the **bulk modulus** of the object.

QUESTIONS

1 Figure 12-14 shows three situations in which the same horizontal rod is supported by a hinge on a wall at one end and a cord at its other end. Without written calculation, rank the situations according to the magnitudes of (a) the force on the rod from the cord, (b) the vertical force on the rod from the hinge, and (c) the horizontal force on the rod from the hinge, greatest first.

Fig. 12-14 Question 1.

2 In Fig. 12-15, a rigid beam is attached to two posts that are fastened to a floor. A small but heavy safe is placed at the six positions indicated, in turn. Assume that the mass of the beam is negligible compared to that of the safe. (a) Rank the positions according to the force on post A due to the safe, greatest compression first, greatest tension last, and indicate where, if anywhere, the force is zero. (b) Rank them according to the force on post B.

Fig. 12-15 Question 2.

3 Figure 12-16 shows four overhead views of rotating uniform disks that are sliding across a frictionless floor. Three forces, of

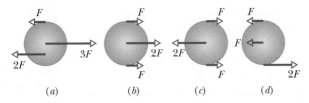

Fig. 12-16 Question 3.

magnitude F, $2F$, or $3F$, act on each disk, either at the rim, at the center, or halfway between rim and center. The force vectors rotate along with the disks, and, in the "snapshots" of Fig. 12-16, point left or right. Which disks are in equilibrium?

4 A ladder leans against a frictionless wall but is prevented from falling because of friction between it and the ground. Suppose you shift the base of the ladder toward the wall. Determine whether the following become larger, smaller, or stay the same (in magnitude): (a) the normal force on the ladder from the ground, (b) the force on the ladder from the wall, (c) the static frictional force on the ladder from the ground, and (d) the maximum value $f_{s,\text{max}}$ of the static frictional force.

5 Figure 12-17 shows a mobile of toy penguins hanging from a ceiling. Each crossbar is horizontal, has negligible mass, and extends three times as far to the right of the wire supporting it as to the left. Penguin 1 has mass $m_1 = 48$ kg. What are the masses of (a) penguin 2, (b) penguin 3, and (c) penguin 4?

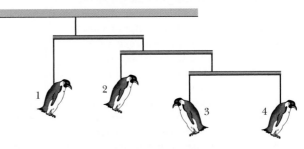

Fig. 12-17 Question 5.

6 Figure 12-18 shows an overhead view of a uniform stick on which four forces act. Suppose we choose a rotation axis through point O, calculate the torques about that axis due to the forces, and find that these torques balance. Will the torques balance if, instead, the rotation axis is chosen to be at (a) point A (on the stick), (b) point B (on line with the stick), or (c) point C (off to

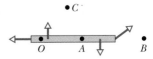

Fig. 12-18 Question 6.

one side of the stick)? (d) Suppose, instead, that we find that the torques about point O do not balance. Is there another point about which the torques will balance?

7 In Fig. 12-19, a stationary 5 kg rod AC is held against a wall by a rope and friction between rod and wall. The uniform rod is 1 m long, and angle $\theta = 30°$. (a) If you are to find the magnitude of the force $\vec{T}$ on the rod from the rope with a single equation, at what labeled point should a rotation axis be placed? With that choice of axis and counterclockwise torques positive, what is the sign of (b) the torque τ_w due to the rod's weight and (c) the torque τ_r due to the pull on the rod by the rope? (d) Is the magnitude of τ_r greater than, less than, or equal to the magnitude of τ_w?

Fig. 12-19
Question 7.

8 Three piñatas hang from the (stationary) assembly of massless pulleys and cords seen in Fig. 12-20. One long cord runs from the ceiling at the right to the lower pulley at the left, looping halfway around all the pulleys. Several shorter cords suspend pulleys from

Fig. 12-20 Question 8.

the ceiling or piñatas from the pulleys. The weights (in newtons) of two piñatas are given. (a) What is the weight of the third piñata? (*Hint:* A cord that loops halfway around a pulley pulls on the pulley with a net force that is twice the tension in the cord.) (b) What is the tension in the short cord labeled with T?

9 In Fig. 12-21, a vertical rod is hinged at its lower end and attached to a cable at its upper end. A horizontal force $\vec{F}_a$ is to be applied to the rod as shown. If the point at which the force is applied is moved up the rod, does the tension in the cable increase, decrease, or remain the same?

Fig. 12-21 Question 9.

10 Figure 12-22 shows a horizontal block that is suspended by two wires, A and B, which are identical except for their original lengths. The center of mass of the block is closer to wire B than to wire A. (a) Measuring torques about the block's center of mass, state whether the magnitude of the torque due to wire A is greater than, less than, or equal to the magnitude of the torque due to wire B. (b) Which wire exerts more force on the block? (c) If the wires are now equal in length, which one was originally shorter (before the block was suspended)?

Fig. 12-22 Question 10.

PROBLEMS

GO	Tutoring problem available (at instructor's discretion) in *WileyPLUS* and WebAssign
SSM	Worked-out solution available in Student Solutions Manual
•–•••	Number of dots indicates level of problem difficulty
	Additional information available in *The Flying Circus of Physics* and at flyingcircusofphysics.com

WWW Worked-out solution is at
ILW Interactive solution is at http://www.wiley.com/college/halliday

sec. 12-4 The Center of Gravity

•1 Because g varies so little over the extent of most structures, any structure's center of gravity effectively coincides with its center of mass. Here is a fictitious example where g varies more significantly. Figure 12-23 shows an array of six particles, each with mass m, fixed to the edge of a rigid structure of negligible mass. The distance between adjacent particles along the edge is 2.00 m. The following table gives the value of g (m/s²) at each particle's location. Using the coordinate system shown, find (a) the x coordinate x_{com} and (b) the y coordinate y_{com} of the center of mass of the six-particle system. Then find (c) the x coordinate x_{cog} and (d) the y coordinate y_{cog} of the center of gravity of the six-particle system.

Fig. 12-23
Problem 1.

Particle	g	Particle	g
1	8.00	4	7.40
2	7.80	5	7.60
3	7.60	6	7.80

sec. 12-5 Some Examples of Static Equilibrium

•2 An automobile with a mass of 1360 kg has 3.05 m between the front and rear axles. Its center of gravity is located 1.78 m behind the front axle. With the automobile on level ground, determine the magnitude of the force from the ground on (a) each front wheel (assuming equal forces on the front wheels) and (b) each rear wheel (assuming equal forces on the rear wheels).

•3 **SSM** **WWW** In Fig. 12-24, a uniform sphere of mass $m = 0.85$ kg and radius $r = 4.2$ cm is held in place by a massless rope attached to a frictionless wall a distance $L = 8.0$ cm above the center of the sphere. Find (a) the tension in the rope and (b) the force on the sphere from the wall.

•4 An archer's bow is drawn at its midpoint until the tension in the string is equal to the force exerted by the archer. What is the angle between the two halves of the string?

Fig. 12-24
Problem 3.

•5 **ILW** A rope of negligible mass is stretched horizontally between two supports that are 3.44 m apart. When an object of weight 3160 N is hung at the center of the rope, the rope is observed to sag by 35.0 cm. What is the tension in the rope?

•6   A scaffold of mass 60 kg and length 5.0 m is supported in a horizontal position by a vertical cable at each end. A window washer of mass 80 kg stands at a point 1.5 m from one end. What is the tension in (a) the nearer cable and (b) the farther cable?

•7   A 75 kg window cleaner uses a 10 kg ladder that is 5.0 m long. He places one end on the ground 2.5 m from a wall, rests the upper end against a cracked window, and climbs the ladder. He is 3.0 m up along the ladder when the window breaks. Neglect friction between the ladder and window and assume that the base of the ladder does not slip. When the window is on the verge of breaking, what are (a) the magnitude of the force on the window from the ladder, (b) the magnitude of the force on the ladder from the ground, and (c) the angle (relative to the horizontal) of that force on the ladder?

•8   A physics Brady Bunch, whose weights in newtons are indicated in Fig. 12-25, is balanced on a seesaw. What is the number of the person who causes the largest torque about the rotation axis at *fulcrum f* directed (a) out of the page and (b) into the page?

Fig. 12-25   Problem 8.

•9   SSM   A meter stick balances horizontally on a knife-edge at the 50.0 cm mark. With two 5.00 g coins stacked over the 12.0 cm mark, the stick is found to balance at the 45.5 cm mark. What is the mass of the meter stick?

•10   GO   The system in Fig. 12-26 is in equilibrium, with the string in the center exactly horizontal. Block A weighs 40 N, block B weighs 50 N, and angle ϕ is 35°. Find (a) tension T_1, (b) tension T_2, (c) tension T_3, and (d) angle θ.

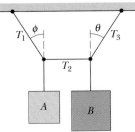

Fig. 12-26   Problem 10.

•11   SSM   Figure 12-27 shows a diver of weight 580 N standing at the end of a diving board with a length of $L = 4.5$ m and negligible mass. The board is fixed to two pedestals (supports) that are separated by distance $d = 1.5$ m. Of the forces acting on the board, what are the (a) magnitude and (b) direction (up or down) of the force from the left pedestal and the (c) magnitude and (d) direction (up or down) of the force from the right pedestal? (e) Which pedestal (left or right) is being stretched, and (f) which pedestal is being compressed?

Fig. 12-27   Problem 11.

•12   In Fig. 12-28, trying to get his car out of mud, a man ties one end of a rope around the front bumper and the other end tightly around a utility pole 18 m away. He then pushes sideways on the rope at its midpoint with a force of 550 N, displacing the center of the rope 0.30 m, but the car barely moves. What is the magnitude of the force on the car from the rope? (The rope stretches somewhat.)

Fig. 12-28   Problem 12.

•13   Figure 12-29 shows the anatomical structures in the lower leg and foot that are involved in standing on tiptoe, with the heel raised slightly off the floor so that the foot effectively contacts the floor only at point P. Assume distance $a = 5.0$ cm, distance $b = 15$ cm, and the person's weight $W = 900$ N. Of the forces acting on the foot, what are the (a) magnitude and (b) direction (up or down) of the force at point A from the calf muscle and the (c) magnitude and (d) direction (up or down) of the force at point B from the lower leg bones?

Fig. 12-29   Problem 13.

•14   In Fig. 12-30, a horizontal scaffold, of length 2.00 m and uniform mass 50.0 kg, is suspended from a building by two cables. The

Fig. 12-30   Problem 14.

scaffold has dozens of paint cans stacked on it at various points. The total mass of the paint cans is 75.0 kg. The tension in the cable at the right is 722 N. How far horizontally from *that* cable is the center of mass of the system of paint cans?

•15   ILW   Forces $\vec{F}_1$, $\vec{F}_2$, and $\vec{F}_3$ act on the structure of Fig. 12-31, shown in an overhead view. We wish to put the structure in equilibrium by applying a fourth force, at a point such as P. The fourth force has vector components $\vec{F}_h$ and $\vec{F}_v$. We are given that $a = 2.0$ m, $b = 3.0$ m, $c = 1.0$ m, $F_1 = 20$ N, $F_2 = 10$ N, and $F_3 = 5.0$ N. Find (a) F_h, (b) F_v, and (c) d.

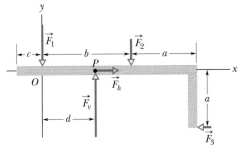

Fig. 12-31   Problem 15.

•16   A uniform cubical crate is 0.750 m on each side and weighs 500 N. It rests on a floor with one edge against a very small, fixed obstruction. At what least height above the floor must a horizontal force of magnitude 350 N be applied to the crate to tip it?

•**17** In Fig. 12-32, a uniform beam of weight 500 N and length 3.0 m is suspended horizontally. On the left it is hinged to a wall; on the right it is supported by a cable bolted to the wall at distance D above the beam. The least tension that will snap the cable is 1200 N. (a) What value of D corresponds to that tension? (b) To prevent the cable from snapping, should D be increased or decreased from that value?

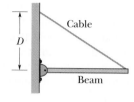
Fig. 12-32 Problem 17.

•**18** GO In Fig. 12-33, horizontal scaffold 2, with uniform mass $m_2 = 30.0$ kg and length $L_2 = 2.00$ m, hangs from horizontal scaffold 1, with uniform mass $m_1 = 50.0$ kg. A 20.0 kg box of nails lies on scaffold 2, centered at distance $d = 0.500$ m from the left end. What is the tension T in the cable indicated?

Fig. 12-33 Problem 18.

•**19** To crack a certain nut in a nutcracker, forces with magnitudes of at least 40 N must act on its shell from both sides. For the nutcracker of Fig. 12-34, with distances $L = 12$ cm and $d = 2.6$ cm, what are the force components $F_\perp$ (perpendicular to the handles) corresponding to that 40 N?

Fig. 12-34 Problem 19.

•**20** A bowler holds a bowling ball ($M = 7.2$ kg) in the palm of his hand (Fig. 12-35). His upper arm is vertical; his lower arm (1.8 kg) is horizontal. What is the magnitude of (a) the force of the biceps muscle on the lower arm and (b) the force between the bony structures at the elbow contact point?

Biceps

Elbow contact point

←4.0 cm
← 15 cm →
33 cm

M

Lower arm (forearm plus hand) center of mass

Fig. 12-35 Problem 20.

••**21** ILW The system in Fig. 12-36 is in equilibrium. A concrete block of mass 225 kg hangs from the end of the uniform strut of mass 45.0 kg. For angles $\phi = 30.0°$ and $\theta = 45.0°$, find (a) the tension T in the cable and the (b) horizontal and (c) vertical components of the force on the strut from the hinge.

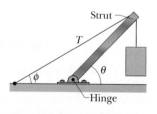
Fig. 12-36 Problem 21.

••**22** GO In Fig. 12-37, a 55 kg rock climber is in a lie-back climb along a fissure, with hands pulling on one side of the fissure and feet pressed against the opposite side. The fissure has width $w = 0.20$ m, and the center of mass of the climber is a horizontal distance $d = 0.40$ m from the fissure. The coefficient of static friction between hands and rock is $\mu_1 = 0.40$, and between boots and rock it is $\mu_2 = 1.2$. (a) What is the least horizontal pull by the hands and push by the feet that will keep the climber stable? (b) For the horizontal pull of (a), what must be the vertical distance h between hands and feet? If the climber encounters wet rock, so that μ_1 and μ_2 are reduced, what happens to (c) the answer to (a) and (d) the answer to (b)?

Fig. 12-37 Problem 22.

••**23** In Fig. 12-38, one end of a uniform beam of weight 222 N is hinged to a wall; the other end is supported by a wire that makes angles $\theta = 30.0°$ with both wall and beam. Find (a) the tension in the wire and the (b) horizontal and (c) vertical components of the force of the hinge on the beam.

Fig. 12-38 Problem 23.

••**24** GO In Fig. 12-39, a climber with a weight of 533.8 N is held by a belay rope connected to her climbing harness and belay device; the force of the rope on her has a line of action through her center of mass. The indicated angles are $\theta = 40.0°$ and $\phi = 30.0°$. If her feet are on the verge of sliding on the vertical wall, what is the coefficient of static friction between her climbing shoes and the wall?

Fig. 12-39 Problem 24.

••**25** SSM WWW In Fig. 12-40, what magnitude of (constant) force $\vec{F}$ applied horizontally at the axle of the wheel is necessary to raise the wheel over an obstacle of height $h = 3.00$ cm? The wheel's radius is $r = 6.00$ cm, and its mass is $m = 0.800$ kg.

Fig. 12-40 Problem 25.

••26 GO 🛩 In Fig. 12-41, a climber leans out against a vertical ice wall that has negligible friction. Distance a is 0.914 m and distance L is 2.10 m. His center of mass is distance $d = 0.940$ m from the feet–ground contact point. If he is on the verge of sliding, what is the coefficient of static friction between feet and ground?

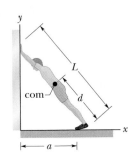

Fig. 12-41   Problem 26.

••27 In Fig. 12-42, a 15 kg block is held in place via a pulley system. The person's upper arm is vertical; the forearm is at angle $\theta = 30°$ with the horizontal. Forearm and hand together have a mass of 2.0 kg, with a center of mass at distance $d_1 = 15$ cm from the contact point of the forearm bone and the upper-arm bone (humerus). The triceps muscle pulls vertically upward on the forearm at distance $d_2 = 2.5$ cm behind that contact point. Distance d_3 is 35 cm. What are the (a) magnitude and (b) direction (up or down) of the force on the forearm from the triceps muscle and the (c) magnitude and (d) direction (up or down) of the force on the forearm from the humerus?

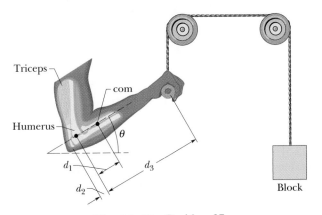

Fig. 12-42   Problem 27.

••28 GO In Fig. 12-43, suppose the length L of the uniform bar is 3.00 m and its weight is 200 N. Also, let the block's weight $W = 300$ N and the angle $\theta = 30.0°$. The wire can withstand a maximum tension of 500 N. (a) What is the maximum possible distance x before the wire breaks? With the block placed at this maximum x, what are the (b) horizontal and (c) vertical components of the force on the bar from the hinge at A?

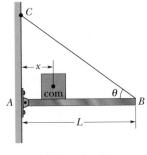

Fig. 12-43
Problems 28 and 34.

••29 A door has a height of 2.1 m along a y axis that extends vertically upward and a width of 0.91 m along an x axis that extends outward from the hinged edge of the door. A hinge 0.30 m from the top and a hinge 0.30 m from the bottom each support half the door's mass, which is 27 kg. In unit-vector notation, what are the forces on the door at (a) the top hinge and (b) the bottom hinge?

••30 In Fig. 12-44, a 50.0 kg uniform square sign, of edge length $L = 2.00$ m, is hung from a horizontal rod of length $d_h = 3.00$ m and negligible mass. A cable is attached to the end of the rod and to a point on the wall at distance $d_v = 4.00$ m above the point where the rod is hinged to the wall. (a) What is the tension in the cable? What are the (b) magnitude and (c) direction (left or right) of the horizontal component of the force on the rod from the wall, and the (d) magnitude and (e) direction (up or down) of the vertical component of this force?

Fig. 12-44
Problem 30.

••31 GO In Fig. 12-45, a nonuniform bar is suspended at rest in a horizontal position by two massless cords. One cord makes the angle $\theta = 36.9°$ with the vertical; the other makes the angle $\phi = 53.1°$ with the vertical. If the length L of the bar is 6.10 m, compute the distance x from the left end of the bar to its center of mass.

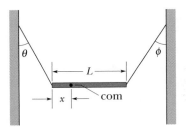

Fig. 12-45   Problem 31.

••32 In Fig. 12-46, the driver of a car on a horizontal road makes an emergency stop by applying the brakes so that all four wheels lock and skid along the road. The coefficient of kinetic friction between tires and road is 0.40. The separation between the front and rear axles is $L = 4.2$ m, and the center of mass of the car is located at distance $d = 1.8$ m behind the front axle and distance $h = 0.75$ m above the road. The car weighs 11 kN. Find the magnitude of (a) the braking acceleration of the car, (b) the normal force on each rear wheel, (c) the normal force on each front wheel, (d) the braking force on each rear wheel, and (e) the braking force on each front wheel. (*Hint:* Although the car is not in translational equilibrium, it *is* in rotational equilibrium.)

Fig. 12-46   Problem 32.

••33 Figure 12-47a shows a vertical uniform beam of length L that is hinged at its lower end. A horizontal force $\vec{F}_a$ is applied to the beam at distance y from the lower end. The beam remains vertical because of a cable attached at the upper end, at angle θ with the horizontal. Figure 12-47b gives the tension T in the cable as a function of the position of the applied force given as a fraction

y/L of the beam length. The scale of the T axis is set by $T_s = 600$ N. Figure 12-47c gives the magnitude F_h of the horizontal force on the beam from the hinge, also as a function of y/L. Evaluate (a) angle θ and (b) the magnitude of $\vec{F}_a$.

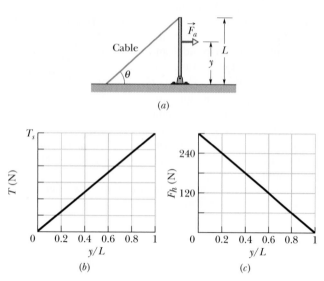

(a)

(b)

(c)

Fig. 12-47   Problem 33.

••34   In Fig. 12-43, a thin horizontal bar AB of negligible weight and length L is hinged to a vertical wall at A and supported at B by a thin wire BC that makes an angle θ with the horizontal. A block of weight W can be moved anywhere along the bar; its position is defined by the distance x from the wall to its center of mass. As a function of x, find (a) the tension in the wire, and the (b) horizontal and (c) vertical components of the force on the bar from the hinge at A.

••35   SSM   WWW   A cubical box is filled with sand and weighs 890 N. We wish to "roll" the box by pushing horizontally on one of the upper edges. (a) What minimum force is required? (b) What minimum coefficient of static friction between box and floor is required? (c) If there is a more efficient way to roll the box, find the smallest possible force that would have to be applied directly to the box to roll it. (*Hint:* At the onset of tipping, where is the normal force located?)

••36   Figure 12-48 shows a 70 kg climber hanging by only the *crimp hold* of one hand on the edge of a shallow horizontal ledge in a rock wall. (The fingers are pressed down to gain purchase.) Her feet touch the rock wall at distance $H = 2.0$ m directly below her crimped fingers but do not provide any support. Her center of mass is distance $a = 0.20$ m from the wall. Assume that the force from the ledge supporting her fingers is equally shared by the four fingers. What are the values of the (a) horizontal component F_h and (b) vertical component F_v of the force on *each* fingertip?

••37   In Fig. 12-49, a uniform plank, with a length L of 6.10 m and a weight of 445 N, rests on the ground and against a frictionless roller

Fig. 12-48 Problem 36.

at the top of a wall of height $h = 3.05$ m. The plank remains in equilibrium for any value of $\theta \geq 70°$ but slips if $\theta < 70°$. Find the coefficient of static friction between the plank and the ground.

••38   In Fig. 12-50, uniform beams A and B are attached to a wall with hinges and loosely bolted together (there is no torque of one on the other). Beam A has length $L_A = 2.40$ m and mass 54.0 kg; beam B has mass 68.0 kg. The two hinge points are separated by distance $d = 1.80$ m. In unit-vector notation, what is the force on (a) beam A due to its hinge, (b) beam A due to the bolt, (c) beam B due to its hinge, and (d) beam B due to the bolt?

Fig. 12-49   Problem 37.

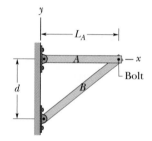

Fig. 12-50   Problem 38.

•••39   For the stepladder shown in Fig. 12-51, sides AC and CE are each 2.44 m long and hinged at C. Bar BD is a tie-rod 0.762 m long, halfway up. A man weighing 854 N climbs 1.80 m along the ladder. Assuming that the floor is frictionless and neglecting the mass of the ladder, find (a) the tension in the tie-rod and the magnitudes of the forces on the ladder from the floor at (b) A and (c) E. (*Hint:* Isolate parts of the ladder in applying the equilibrium conditions.)

•••40   Figure 12-52a shows a horizontal uniform beam of mass m_b and length L that is supported on the left by a hinge attached to a wall and on the

Fig. 12-51   Problem 39.

(a)

(b)

Fig. 12-52   Problem 40.

right by a cable at angle θ with the horizontal. A package of mass m_p is positioned on the beam at a distance x from the left end. The total mass is $m_b + m_p = 61.22$ kg. Figure 12-52b gives the tension T in the cable as a function of the package's position given as a fraction x/L of the beam length. The scale of the T axis is set by $T_a = 500$ N and $T_b = 700$ N. Evaluate (a) angle θ, (b) mass m_b, and (c) mass m_p.

•••41 A crate, in the form of a cube with edge lengths of 1.2 m, contains a piece of machinery; the center of mass of the crate and its contents is located 0.30 m above the crate's geometrical center. The crate rests on a ramp that makes an angle θ with the horizontal. As θ is increased from zero, an angle will be reached at which the crate will either tip over or start to slide down the ramp. If the coefficient of static friction μ_s between ramp and crate is 0.60, (a) does the crate tip or slide and (b) at what angle θ does this occur? If $\mu_s = 0.70$, (c) does the crate tip or slide and (d) at what angle θ does this occur? (*Hint:* At the onset of tipping, where is the normal force located?)

•••42 In Fig.12-5 and the associated sample problem, let the coefficient of static friction μ_s between the ladder and the pavement be 0.53. How far (in percent) up the ladder must the firefighter go to put the ladder on the verge of sliding?

sec. 12-7 Elasticity

•43 SSM ILW A horizontal aluminum rod 4.8 cm in diameter projects 5.3 cm from a wall. A 1200 kg object is suspended from the end of the rod. The shear modulus of aluminum is 3.0×10^{10} N/m². Neglecting the rod's mass, find (a) the shear stress on the rod and (b) the vertical deflection of the end of the rod.

•44 Figure 12-53 shows the stress–strain curve for a material. The scale of the stress axis is set by $s = 300$, in units of 10^6 N/m². What are (a) the Young's modulus and (b) the approximate yield strength for this material?

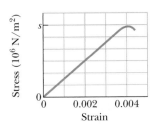

Fig. 12-53 Problem 44.

•45 In Fig. 12-54, a lead brick rests horizontally on cylinders A and B. The areas of the top faces of the cylinders are related by $A_A = 2A_B$; the Young's moduli of the cylinders are related by $E_A = 2E_B$. The cylinders had identical lengths before the brick was placed on them. What fraction of the brick's mass is supported (a) by cylinder A and (b) by cylinder B? The horizontal distances between the center of mass of the brick and the centerlines of the cylinders are d_A for cylinder A and d_B for cylinder B. (c) What is the ratio d_A/d_B?

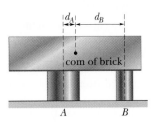

Fig. 12-54 Problem 45.

•46 ✈ Figure 12-55 shows an approximate plot of stress versus strain for a spider-web thread, out to the point of breaking at a strain of 2.00. The vertical axis scale is set by values $a = 0.12$ GN/m², $b = 0.30$ GN/m², and $c = 0.80$ GN/m². Assume that the thread has an ini-

Fig. 12-55 Problem 46.

tial length of 0.80 cm, an initial cross-sectional area of 8.0×10^{-12} m², and (during stretching) a constant volume. Assume also that when the single thread snares a flying insect, the insect's kinetic energy is transferred to the stretching of the thread. (a) How much kinetic energy would put the thread on the verge of breaking? What is the kinetic energy of (b) a fruit fly of mass 6.00 mg and speed 1.70 m/s and (c) a bumble bee of mass 0.388 g and speed 0.420 m/s? Would (d) the fruit fly and (e) the bumble bee break the thread?

•47 A tunnel of length $L = 150$ m, height $H = 7.2$ m, and width 5.8 m (with a flat roof) is to be constructed at distance $d = 60$ m beneath the ground. (See Fig. 12-56.) The tunnel roof is to be supported entirely by square steel columns, each with a cross-sectional area of 960 cm². The mass of 1.0 cm³ of the ground material is 2.8 g. (a) What is the total weight of the ground material the columns must support? (b) How many columns are needed to keep the compressive stress on each column at one-half its ultimate strength?

Fig. 12-56 Problem 47.

•48 Figure 12-57 shows the stress versus strain plot for an aluminum wire that is stretched by a machine pulling in opposite directions at the two ends of the wire. The scale of the stress axis is set by $s = 7.0$, in units of 10^7 N/m². The wire has an initial length of 0.800 m and an initial cross-sectional area of 2.00×10^{-6} m². How much work does the force from the machine do on the wire to produce a strain of 1.00×10^{-3}?

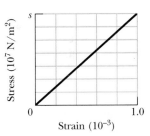

Fig. 12-57 Problem 48.

•49 GO In Fig. 12-58, a 103 kg uniform log hangs by two steel wires, A and B, both of radius 1.20 mm. Initially, wire A was 2.50 m long and 2.00 mm shorter than wire B. The log is now horizontal. What are the magnitudes of the forces on it from (a) wire A and (b) wire B? (c) What is the ratio d_A/d_B?

•••50 ✈ GO Figure 12-59 represents an insect caught at the midpoint of a spider-web thread. The thread breaks under a stress of 8.20×10^8 N/m² and a strain of 2.00.

Fig. 12-58 Problem 49.

Fig. 12-59 Problem 50.

Initially, it was horizontal and had a length of 2.00 cm and a cross-sectional area of 8.00×10^{-12} m². As the thread was stretched under the weight of the insect, its volume remained constant. If the weight of the insect puts the thread on the verge of breaking, what is the insect's mass? (A spider's web is built to break if a potentially harmful insect, such as a bumble bee, becomes snared in the web.)

•••**51**   Figure 12-60 is an overhead view of a rigid rod that turns about a vertical axle until the identical rubber stoppers A and B are forced against rigid walls at distances $r_A = 7.0$ cm and $r_B = 4.0$ cm from the axle. Initially the stoppers touch the walls without being compressed. Then force $\vec{F}$ of magnitude 220 N is applied perpendicular to the rod at a distance $R = 5.0$ cm from the axle. Find the magnitude of the force compressing (a) stopper A and (b) stopper B.

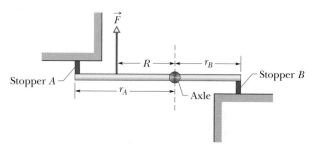

Fig. 12-60   Problem 51.

Additional Problems

52   After a fall, a 95 kg rock climber finds himself dangling from the end of a rope that had been 15 m long and 9.6 mm in diameter but has stretched by 2.8 cm. For the rope, calculate (a) the strain, (b) the stress, and (c) the Young's modulus.

53 SSM In Fig. 12-61, a rectangular slab of slate rests on a bedrock surface inclined at angle $\theta = 26°$. The slab has length $L = 43$ m, thickness $T = 2.5$ m, and width $W = 12$ m, and 1.0 cm³ of it has a mass of 3.2 g. The coefficient of static friction between slab and bedrock is 0.39. (a) Calculate the component of the gravitational force on the slab parallel to the bedrock surface. (b) Calculate the magnitude of the static frictional force on the slab. By comparing (a) and (b), you can see that the slab is in danger of sliding. This is prevented only by chance protrusions of bedrock. (c) To stabilize the slab, bolts are to be driven perpendicular to the bedrock surface (two bolts are shown). If each bolt has a cross-sectional area of 6.4 cm² and will snap under a shearing stress of 3.6×10^8 N/m², what is the minimum number of bolts needed? Assume that the bolts do not affect the normal force.

Fig. 12-61   Problem 53.

54   A uniform ladder whose length is 5.0 m and whose weight is 400 N leans against a frictionless vertical wall. The coefficient of static friction between the level ground and the foot of the ladder is 0.46. What is the greatest distance the foot of the ladder can be placed from the base of the wall without the ladder immediately slipping?

55 SSM In Fig. 12-62, block A (mass 10 kg) is in equilibrium, but it would slip if block B (mass 5.0 kg) were any heavier. For

angle $\theta = 30°$, what is the coefficient of static friction between block A and the surface below it?

56   Figure 12-63a shows a uniform ramp between two buildings that allows for motion between the buildings due to strong winds. At its left end, it is hinged to the building wall; at its right end, it has a roller that can roll along the building wall. There is no vertical force on the roller from the building, only a horizontal force with magnitude F_h. The horizontal distance between the buildings is $D = 4.00$ m.

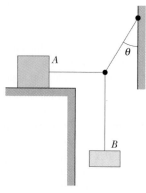

Fig. 12-62   Problem 55.

The rise of the ramp is $h = 0.490$ m. A man walks across the ramp from the left. Figure 12-63b gives F_h as a function of the horizontal distance x of the man from the building at the left. The scale of the F_h axis is set by $a = 20$ kN and $b = 25$ kN. What are the masses of (a) the ramp and (b) the man?

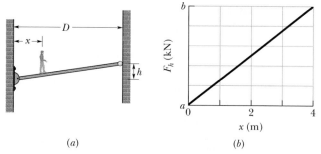

(a)   *(b)*

Fig. 12-63   Problem 56.

57   In Fig. 12-64, a 10 kg sphere is supported on a frictionless plane inclined at angle $\theta = 45°$ from the horizontal. Angle ϕ is 25°. Calculate the tension in the cable.

58   In Fig. 12-65a, a uniform 40.0 kg beam is centered over two rollers. Vertical lines across the beam mark off equal lengths. Two of the lines are centered over the rollers; a 10.0 kg package of tamales is centered over roller B. What are the magnitudes of the forces on the beam from (a) roller A and (b) roller B? The beam is then rolled to the left until the right-hand end is centered over roller B (Fig. 12-65b). What now are the magnitudes of the forces on the beam from (c) roller A and (d) roller B? Next, the beam is rolled to the right. Assume that it has a length of 0.800 m. (e) What horizontal distance between the package and roller B puts the beam on the verge of losing contact with roller A?

Fig. 12-64   Problem 57.

Fig. 12-65   Problem 58.

59 SSM In Fig. 12-66, an 817 kg construction bucket is suspended by a cable A that is attached at O to two other cables B and C, making angles $\theta_1 = 51.0°$ and $\theta_2 = 66.0°$ with the horizontal. Find the tensions in (a) cable A, (b) cable B, and (c) cable C. (*Hint:* To avoid solving two equations in two unknowns, position the axes as shown in the figure.)

Fig. 12-66 Problem 59.

60 In Fig. 12-67, a package of mass m hangs from a short cord that is tied to the wall via cord 1 and to the ceiling via cord 2. Cord 1 is at angle $\phi = 40°$ with the horizontal; cord 2 is at angle θ. (a) For what value of θ is the tension in cord 2 minimized? (b) In terms of mg, what is the minimum tension in cord 2?

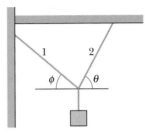

Fig. 12-67 Problem 60.

61 ILW The force $\vec{F}$ in Fig. 12-68 keeps the 6.40 kg block and the pulleys in equilibrium. The pulleys have negligible mass and friction. Calculate the tension T in the upper cable. (*Hint:* When a cable wraps halfway around a pulley as here, the magnitude of its net force on the pulley is twice the tension in the cable.)

62 A mine elevator is supported by a single steel cable 2.5 cm in diameter. The total mass of the elevator cage and occupants is 670 kg. By how much does the cable stretch when the elevator hangs by (a) 12 m of cable and (b) 362 m of cable? (Neglect the mass of the cable.)

Fig. 12-68
Problem 61.

63 Four bricks of length L, identical and uniform, are stacked on top of one another (Fig. 12-69) in such a way that part of each extends beyond the one beneath. Find, in terms of L, the maximum values of (a) a_1, (b) a_2, (c) a_3, (d) a_4, and (e) h, such that the stack is in equilibrium.

Fig. 12-69 Problem 63.

64 In Fig. 12-70, two identical, uniform, and frictionless spheres, each of mass m, rest in a rigid rectangular container. A line connecting their centers is at 45° to the horizontal. Find the magni-

tudes of the forces on the spheres from (a) the bottom of the container, (b) the left side of the container, (c) the right side of the container, and (d) each other. (*Hint:* The force of one sphere on the other is directed along the center–center line.)

Fig. 12-70 Problem 64.

65 In Fig. 12-71, a uniform beam with a weight of 60 N and a length of 3.2 m is hinged at its lower end, and a horizontal force $\vec{F}$ of magnitude 50 N acts at its upper end. The beam is held vertical by a cable that makes angle $\theta = 25°$ with the ground and is attached to the beam at height $h = 2.0$ m. What are (a) the tension in the cable and (b) the force on the beam from the hinge in unit-vector notation?

Fig. 12-71 Problem 65.

66 A uniform beam is 5.0 m long and has a mass of 53 kg. In Fig. 12-72, the beam is supported in a horizontal position by a hinge and a cable, with angle $\theta = 60°$. In unit-vector notation, what is the force on the beam from the hinge?

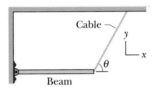

Fig. 12-72 Problem 66.

67 A solid copper cube has an edge length of 85.5 cm. How much stress must be applied to the cube to reduce the edge length to 85.0 cm? The bulk modulus of copper is 1.4×10^{11} N/m^2.

68 A construction worker attempts to lift a uniform beam off the floor and raise it to a vertical position. The beam is 2.50 m long and weighs 500 N. At a certain instant the worker holds the beam momentarily at rest with one end at distance $d = 1.50$ m above the floor, as shown in Fig. 12-73, by exerting a force $\vec{P}$ on the beam, perpendicular to the beam. (a) What is the magnitude P? (b) What is the magnitude of the (net) force of the floor on the beam? (c) What is the minimum value the coefficient of static friction between beam and floor can have in order for the beam not to slip at this instant?

Fig. 12-73 Problem 68.

69 SSM In Fig. 12-74, a uniform rod of mass m is hinged to a building at its lower end, while its upper end is held in place by a rope attached to the wall. If angle $\theta_1 = 60°$, what value must angle θ_2 have so that the tension in the rope is equal to $mg/2$?

Fig. 12-74
Problem 69.

70 A 73 kg man stands on a level bridge of length L. He is at distance $L/4$ from one end. The bridge is uniform and weighs 2.7 kN. What are the magnitudes of the vertical forces on the bridge from its supports at (a) the end farther from him and (b) the nearer end?

71 SSM A uniform cube of side length 8.0 cm rests on a horizontal floor. The coefficient of static friction between cube and floor is μ. A horizontal pull $\vec{P}$ is applied perpendicular to one of the vertical faces of the cube, at a distance 7.0 cm above the floor on the vertical midline of the cube face. The magnitude of $\vec{P}$ is gradually increased. During that increase, for what values of μ will the cube eventually (a) begin to slide and (b) begin to tip? (*Hint:* At the onset of tipping, where is the normal force located?)

72 The system in Fig. 12-75 is in equilibrium. The angles are $\theta_1 = 60°$ and $\theta_2 = 20°$, and the ball has mass $M = 2.0$ kg. What is the tension in (a) string ab and (b) string bc?

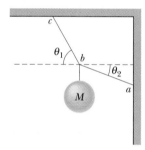

Fig. 12-75 Problem 72.

73 SSM A uniform ladder is 10 m long and weighs 200 N. In Fig. 12-76, the ladder leans against a vertical, frictionless wall at height $h = 8.0$ m above the ground. A horizontal force $\vec{F}$ is applied to the ladder at distance $d = 2.0$ m from its base (measured along the ladder). (a) If force magnitude $F = 50$ N, what is the force of the ground on the ladder, in unit-vector notation? (b) If $F = 150$ N, what is the force of the ground on the ladder, also in unit-vector notation? (c) Suppose the coefficient of

Fig. 12-76 Problem 73.

static friction between the ladder and the ground is 0.38; for what minimum value of the force magnitude F will the base of the ladder just barely start to move toward the wall?

74 A pan balance is made up of a rigid, massless rod with a hanging pan attached at each end. The rod is supported at and free to rotate about a point not at its center. It is balanced by unequal masses placed in the two pans. When an unknown mass m is placed in the left pan, it is balanced by a mass m_1 placed in the right pan; when the mass m is placed in the right pan, it is balanced by a mass m_2 in the left pan. Show that $m = \sqrt{m_1 m_2}$.

75 The rigid square frame in Fig. 12-77 consists of the four side bars AB, BC, CD, and DA plus two diagonal bars AC and BD, which pass each other freely at E. By means of the turnbuckle G, bar AB is put under tension, as if its ends were subject to horizontal, outward forces $\vec{T}$ of magnitude 535 N. (a) Which of the other bars are in tension? What are the magnitudes of (b) the forces causing the tension in those bars and (c) the forces causing compression in the other bars? (*Hint:* Symmetry considerations can lead to considerable simplification in this problem.)

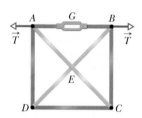

Fig. 12-77 Problem 75.

76 A gymnast with mass 46.0 kg stands on the end of a uniform balance beam as shown in Fig. 12-78. The beam is 5.00 m long and has a mass of 250 kg (excluding the mass of the two supports). Each support is 0.540 m from its end of the beam. In unit-vector notation, what are the forces on the beam due to (a) support 1 and (b) support 2?

Fig. 12-78 Problem 76.

77 Figure 12-79 shows a 300 kg cylinder that is horizontal. Three steel wires support the cylinder from a ceiling. Wires 1 and 3 are attached at the ends of the cylinder, and wire 2 is attached at the center. The wires each have a cross-sectional area of 2.00×10^{-6} m^2. Initially (before the cylinder was put in place) wires 1 and 3 were 2.0000 m long and wire 2 was 6.00 mm longer than that. Now (with the cylinder in place) all three wires have been stretched. What is the tension in (a) wire 1 and (b) wire 2?

Fig. 12-79 Problem 77.

78 In Fig. 12-80, a uniform beam of length 12.0 m is supported by a horizontal cable and a hinge at angle $\theta = 50.0°$. The tension in the cable is 400 N. In unit-vector notation, what are (a) the gravitational force on the beam and (b) the force on the beam from the hinge?

Fig. 12-80 Problem 78.

79 Four bricks of length L, identical and uniform, are stacked on a table in two ways, as shown in Fig. 12-81 (compare with Problem 63). We seek to maximize the overhang distance h in both arrangements. Find the optimum distances a_1, a_2, b_1, and b_2, and calculate h for the two arrangements.

(*a*)

(*b*)

Fig. 12-81   Problem 79.

80 A cylindrical aluminum rod, with an initial length of 0.8000 m and radius 1000.0 μm, is clamped in place at one end and then stretched by a machine pulling parallel to its length at its other end. Assuming that the rod's density (mass per unit volume) does not change, find the force magnitude that is required of the machine to decrease the radius to 999.9 μm. (The yield strength is not exceeded.)

81 A beam of length L is carried by three men, one man at one end and the other two supporting the beam between them on a crosspiece placed so that the load of the beam is equally divided among the three men. How far from the beam's free end is the crosspiece placed? (Neglect the mass of the crosspiece.)

82 If the (square) beam in Fig. 12-6a and the associated sample problem is of Douglas fir, what must be its thickness to keep the compressive stress on it to $\frac{1}{6}$ of its ultimate strength?

83 Figure 12-82 shows a stationary arrangement of two crayon boxes and three cords. Box A has a mass of 11.0 kg and is on a ramp at angle $\theta = 30.0°$; box B has a mass of 7.00 kg and hangs on a cord. The cord connected to box A is parallel to the ramp, which is frictionless. (a) What is the tension in the upper cord, and (b) what angle does that cord make with the horizontal?

Fig. 12-82   Problem 83.

13

GRAVITATION

13-1 WHAT IS PHYSICS?

One of the long-standing goals of physics is to understand the gravitational force—the force that holds you to Earth, holds the Moon in orbit around Earth, and holds Earth in orbit around the Sun. It also reaches out through the whole of our Milky Way galaxy, holding together the billions and billions of stars in the Galaxy and the countless molecules and dust particles between stars. We are located somewhat near the edge of this disk-shaped collection of stars and other matter, 2.6×10^4 light-years (2.5×10^{20} m) from the galactic center, around which we slowly revolve.

The gravitational force also reaches across intergalactic space, holding together the Local Group of galaxies, which includes, in addition to the Milky Way, the Andromeda Galaxy (Fig. 13-1) at a distance of 2.3×10^6 light-years away from Earth, plus several closer dwarf galaxies, such as the Large Magellanic Cloud. The Local Group is part of the Local Supercluster of galaxies that is being drawn by the gravitational force toward an exceptionally massive region of space called the Great Attractor. This region appears to be about 3.0×10^8 light-years from Earth, on the opposite side of the Milky Way. And the gravitational force is even more far-reaching because it attempts to hold together the entire universe, which is expanding.

This force is also responsible for some of the most mysterious structures in the universe: *black holes*. When a star considerably larger than our Sun burns out, the gravitational force between all its particles can cause the star to collapse in on itself and thereby to form a black hole. The gravitational force at the surface of such a collapsed star is so strong that neither particles nor light can escape from the surface (thus the term "black hole"). Any star coming too near a black hole can be ripped apart by the strong gravitational force and pulled into the hole. Enough captures like this yields a *supermassive black hole*. Such mysterious monsters appear to be common in the universe.

Although the gravitational force is still not fully understood, the starting point in our understanding of it lies in the *law of gravitation* of Isaac Newton.

13-2 Newton's Law of Gravitation

Physicists like to study seemingly unrelated phenomena to show that a relationship can be found if the phenomena are examined closely enough. This search for unification has been going on for centuries. In 1665, the 23-year-old Isaac Newton made a basic contribution to physics when he showed that the force that holds the Moon in its orbit is the same force that makes an apple fall. We take this knowledge so much for granted now that it is not easy for us to comprehend the ancient belief that the motions of earthbound bodies and heavenly bodies were different in kind and were governed by different laws.

Newton concluded not only that Earth attracts both apples and the Moon but also that every body in the universe attracts every other body; this tendency of bodies to move toward one another is called **gravitation.** Newton's conclusion takes a little getting used to, because the familiar attraction of Earth for earth-

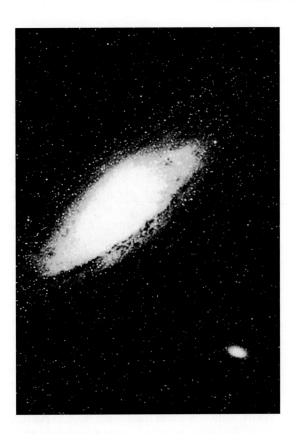

Fig. 13-1   The Andromeda Galaxy. Located 2.3×10^6 light-years from us, and faintly visible to the naked eye, it is very similar to our home galaxy, the Milky Way. *(Courtesy NASA)*

bound bodies is so great that it overwhelms the attraction that earthbound bodies have for each other. For example, Earth attracts an apple with a force magnitude of about 0.8 N. You also attract a nearby apple (and it attracts you), but the force of attraction has less magnitude than the weight of a speck of dust.

Newton proposed a *force law* that we call **Newton's law of gravitation:** Every particle attracts any other particle with a **gravitational force** of magnitude

$$F = G \frac{m_1 m_2}{r^2} \qquad \text{(Newton's law of gravitation).} \qquad (13\text{-}1)$$

Here m_1 and m_2 are the masses of the particles, r is the distance between them, and G is the **gravitational constant,** with a value that is now known to be

$$G = 6.67 \times 10^{-11} \ \text{N} \cdot \text{m}^2/\text{kg}^2$$
$$= 6.67 \times 10^{-11} \ \text{m}^3/\text{kg} \cdot \text{s}^2. \qquad (13\text{-}2)$$

In Fig. 13-2a, $\vec{F}$ is the gravitational force acting on particle 1 (mass m_1) due to particle 2 (mass m_2). The force is directed toward particle 2 and is said to be an *attractive force* because particle 1 is attracted toward particle 2. The magnitude of the force is given by Eq. 13-1.

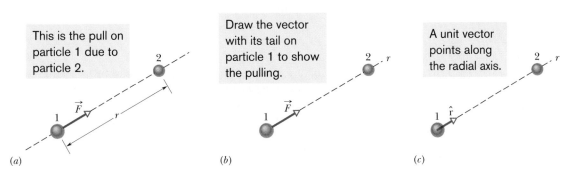

This is the pull on particle 1 due to particle 2.

Draw the vector with its tail on particle 1 to show the pulling.

A unit vector points along the radial axis.

(a)   (b)   (c)

Fig. 13-2   (a) The gravitational force $\vec{F}$ on particle 1 due to particle 2 is an attractive force because particle 1 is attracted to particle 2. (b) Force $\vec{F}$ is directed along a radial coordinate axis r extending from particle 1 through particle 2. (c) $\vec{F}$ is in the direction of a unit vector $\hat{\mathrm{r}}$ along the r axis.

Fig. 13-3 The apple pulls up on Earth just as hard as Earth pulls down on the apple.

We can describe $\vec{F}$ as being in the positive direction of an r axis extending radially from particle 1 through particle 2 (Fig. 13-2b). We can also describe $\vec{F}$ by using a radial unit vector $\hat{r}$ (a dimensionless vector of magnitude 1) that is directed away from particle 1 along the r axis (Fig. 13-2c). From Eq. 13-1, the force on particle 1 is then

$$\vec{F} = G\frac{m_1 m_2}{r^2}\hat{r}. \tag{13-3}$$

The gravitational force on particle 2 due to particle 1 has the same magnitude as the force on particle 1 but the opposite direction. These two forces form a third-law force pair, and we can speak of the gravitational force *between* the two particles as having a magnitude given by Eq. 13-1. This force between two particles is not altered by other objects, even if they are located between the particles. Put another way, no object can shield either particle from the gravitational force due to the other particle.

The strength of the gravitational force—that is, how strongly two particles with given masses at a given separation attract each other—depends on the value of the gravitational constant G. If G—by some miracle—were suddenly multiplied by a factor of 10, you would be crushed to the floor by Earth's attraction. If G were divided by this factor, Earth's attraction would be so weak that you could jump over a building.

Although Newton's law of gravitation applies strictly to particles, we can also apply it to real objects as long as the sizes of the objects are small relative to the distance between them. The Moon and Earth are far enough apart so that, to a good approximation, we can treat them both as particles—but what about an apple and Earth? From the point of view of the apple, the broad and level Earth, stretching out to the horizon beneath the apple, certainly does not look like a particle.

Newton solved the apple–Earth problem by proving an important theorem called the *shell theorem:*

> A uniform spherical shell of matter attracts a particle that is outside the shell as if all the shell's mass were concentrated at its center.

Earth can be thought of as a nest of such shells, one within another and each shell attracting a particle outside Earth's surface as if the mass of that shell were at the center of the shell. Thus, from the apple's point of view, Earth *does* behave like a particle, one that is located at the center of Earth and has a mass equal to that of Earth.

Suppose that, as in Fig. 13-3, Earth pulls down on an apple with a force of magnitude 0.80 N. The apple must then pull up on Earth with a force of magnitude 0.80 N, which we take to act at the center of Earth. Although the forces are matched in magnitude, they produce different accelerations when the apple is released. The acceleration of the apple is about 9.8 m/s², the familiar acceleration of a falling body near Earth's surface. The acceleration of Earth, however, measured in a reference frame attached to the center of mass of the apple–Earth system, is only about 1×10^{-25} m/s².

✔CHECKPOINT 1

A particle is to be placed, in turn, outside four objects, each of mass m: (1) a large uniform solid sphere, (2) a large uniform spherical shell, (3) a small uniform solid sphere, and (4) a small uniform shell. In each situation, the distance between the particle and the center of the object is d. Rank the objects according to the magnitude of the gravitational force they exert on the particle, greatest first.

13-3 Gravitation and the Principle of Superposition

Given a group of particles, we find the net (or resultant) gravitational force on any one of them from the others by using the **principle of superposition.** This is a general principle that says a net effect is the sum of the individual effects. Here, the principle means that we first compute the individual gravitational forces that act on our selected particle due to each of the other particles. We then find the net force by adding these forces vectorially, as usual.

For n interacting particles, we can write the principle of superposition for the gravitational forces on particle 1 as

$$\vec{F}_{1,\text{net}} = \vec{F}_{12} + \vec{F}_{13} + \vec{F}_{14} + \vec{F}_{15} + \cdots + \vec{F}_{1n}. \tag{13-4}$$

Here $\vec{F}_{1,\text{net}}$ is the net force on particle 1 due to the other particles and, for example, $\vec{F}_{13}$ is the force on particle 1 from particle 3. We can express this equation more compactly as a vector sum:

$$\vec{F}_{1,\text{net}} = \sum_{i=2}^{n} \vec{F}_{1i}. \tag{13-5}$$

What about the gravitational force on a particle from a real (extended) object? This force is found by dividing the object into parts small enough to treat as particles and then using Eq. 13-5 to find the vector sum of the forces on the particle from all the parts. In the limiting case, we can divide the extended object into differential parts each of mass dm and each producing a differential force $d\vec{F}$ on the particle. In this limit, the sum of Eq. 13-5 becomes an integral and we have

$$\vec{F}_1 = \int d\vec{F}, \tag{13-6}$$

in which the integral is taken over the entire extended object and we drop the subscript "net." If the extended object is a uniform sphere or a spherical shell, we can avoid the integration of Eq. 13-6 by assuming that the object's mass is concentrated at the object's center and using Eq. 13-1.

CHECKPOINT 2

The figure shows four arrangements of three particles of equal masses. (a) Rank the arrangements according to the magnitude of the net gravitational force on the particle labeled m, greatest first. (b) In arrangement 2, is the direction of the net force closer to the line of length d or to the line of length D?

Net gravitational force, 2D, 3 particles

Figure 13-4a shows an arrangement of three particles, particle 1 of mass $m_1 = 6.0$ kg and particles 2 and 3 of mass $m_2 = m_3 = 4.0$ kg, and distance $a = 2.0$ cm. What is the net gravitational force $\vec{F}_{1,\text{net}}$ on particle 1 due to the other particles?

KEY IDEAS

(1) Because we have particles, the magnitude of the gravitational force on particle 1 due to either of the other particles is given by Eq. 13-1 ($F = Gm_1m_2/r^2$). (2) The direction of either gravitational force on particle 1 is toward the particle responsible for it. (3) Because the forces are not along a single axis, we *cannot* simply add or subtract their magnitudes or their components to get the net force. Instead, we must add them as vectors.

Calculations: From Eq. 13-1, the magnitude of the force $\vec{F}_{12}$ on particle 1 from particle 2 is

$$F_{12} = \frac{Gm_1m_2}{a^2} \qquad (13\text{-}7)$$

$$= \frac{(6.67 \times 10^{-11}\,\text{m}^3/\text{kg}\cdot\text{s}^2)(6.0\,\text{kg})(4.0\,\text{kg})}{(0.020\,\text{m})^2}$$

$$= 4.00 \times 10^{-6}\,\text{N}.$$

Similarly, the magnitude of force $\vec{F}_{13}$ on particle 1 from particle 3 is

$$F_{13} = \frac{Gm_1m_3}{(2a)^2} \qquad (13\text{-}8)$$

$$= \frac{(6.67 \times 10^{-11}\,\text{m}^3/\text{kg}\cdot\text{s}^2)(6.0\,\text{kg})(4.0\,\text{kg})}{(0.040\,\text{m})^2}$$

$$= 1.00 \times 10^{-6}\,\text{N}.$$

Force $\vec{F}_{12}$ is directed in the positive direction of the y axis (Fig. 13-4b) and has only the y component F_{12}. Similarly, $\vec{F}_{13}$ is directed in the negative direction of the x axis and has only the x component $-F_{13}$ (Fig. 13-4c).

To find the net force $\vec{F}_{1,\text{net}}$ on particle 1, we must add the two forces as vectors (Figs. 13-4d and e). We can do so on a vector-capable calculator. However, here we note that $-F_{13}$ and F_{12} are actually the x and y components of $\vec{F}_{1,\text{net}}$. Therefore, we can use Eq. 3-6 to find first the magnitude and then the direction of $\vec{F}_{1,\text{net}}$. The magnitude is

$$F_{1,\text{net}} = \sqrt{(F_{12})^2 + (-F_{13})^2}$$

$$= \sqrt{(4.00 \times 10^{-6}\,\text{N})^2 + (-1.00 \times 10^{-6}\,\text{N})^2}$$

$$= 4.1 \times 10^{-6}\,\text{N}. \qquad \text{(Answer)}$$

Relative to the positive direction of the x axis, Eq. 3-6 gives the direction of $\vec{F}_{1,\text{net}}$ as

$$\theta = \tan^{-1}\frac{F_{12}}{-F_{13}} = \tan^{-1}\frac{4.00 \times 10^{-6}\,\text{N}}{-1.00 \times 10^{-6}\,\text{N}} = -76°.$$

Is this a reasonable direction (Fig. 13-4f)? No, because the direction of $\vec{F}_{1,\text{net}}$ must be between the directions of $\vec{F}_{12}$ and $\vec{F}_{13}$. Recall from Chapter 3 that a calculator displays only one of the two possible answers to a $\tan^{-1}$ function. We find the other answer by adding 180°:

$$-76° + 180° = 104°, \qquad \text{(Answer)}$$

which *is* a reasonable direction for $\vec{F}_{1,\text{net}}$ (Fig. 13-4g).

 Additional examples, video, and practice available at *WileyPLUS*

Table 13-1

Variation of a_g with Altitude

Altitude (km)	a_g (m/s²)	Altitude Example
0	9.83	Mean Earth surface
8.8	9.80	Mt. Everest
36.6	9.71	Highest crewed balloon
400	8.70	Space shuttle orbit
35 700	0.225	Communications satellite

13-4 Gravitation Near Earth's Surface

Let us assume that Earth is a uniform sphere of mass M. The magnitude of the gravitational force from Earth on a particle of mass m, located outside Earth a distance r from Earth's center, is then given by Eq. 13-1 as

$$F = G\frac{Mm}{r^2}. \qquad (13\text{-}9)$$

If the particle is released, it will fall toward the center of Earth, as a result of the gravitational force $\vec{F}$, with an acceleration we shall call the **gravitational acceleration** $\vec{a}_g$. Newton's second law tells us that magnitudes F and a_g are related by

$$F = ma_g. \qquad (13\text{-}10)$$

Now, substituting F from Eq. 13-9 into Eq. 13-10 and solving for a_g, we find

$$a_g = \frac{GM}{r^2}. \qquad (13\text{-}11)$$

Table 13-1 shows values of a_g computed for various altitudes above Earth's surface. Notice a_g is significant even at 400 km.

We want the forces (pulls) on particle 1, *not* the forces on the other particles.

This is the force (pull) on particle 1 due to particle 2.

This is the force (pull) on particle 1 due to particle 3.

This is one way to show the net force on particle 1. Note the head-to-tail arrangement.

This is another way, also a head-to-tail arrangement.

A calculator's inverse tangent can give this for the angle.

But this is the correct angle.

Fig. 13-4   (*a*) An arrangement of three particles. The force on particle 1 due to (*b*) particle 2 and (*c*) particle 3. (*d*)–(*g*) Ways to combine the forces to get the net force magnitude and orientation.

Since Section 5-4, we have assumed that Earth is an inertial frame by neglecting its rotation. This simplification has allowed us to assume that the free-fall acceleration g of a particle is the same as the particle's gravitational acceleration (which we now call a_g). Furthermore, we assumed that g has the constant value 9.8 m/s^2 any place on Earth's surface. However, any g value measured at a given location will differ from the a_g value calculated with Eq. 13-11 for that location for three reasons: (1) Earth's mass is not distributed uniformly, (2) Earth is not a perfect sphere, and (3) Earth rotates. Moreover, because g differs from a_g, the same three reasons mean that the measured weight mg of a particle differs from the magnitude of the gravitational force on the particle as given by Eq. 13-9. Let us now examine those reasons.

1. **Earth's mass is not uniformly distributed.** The density (mass per unit volume) of Earth varies radially as shown in Fig. 13-5, and the density of the crust (outer section) varies from region to region over Earth's surface. Thus, g varies from region to region over the surface.

2. **Earth is not a sphere.** Earth is approximately an ellipsoid, flattened at the poles and bulging at the equator. Its equatorial radius (from its center point out to the equator) is greater than its polar radius (from its center point out

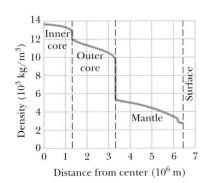

Fig. 13-5   The density of Earth as a function of distance from the center. The limits of the solid inner core, the largely liquid outer core, and the solid mantle are shown, but the crust of Earth is too thin to show clearly on this plot.

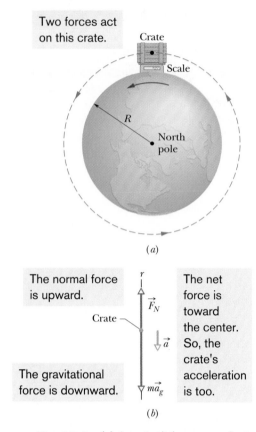

Two forces act on this crate.

Crate

Scale

R

North pole

(a)

The normal force is upward.

Crate

$\vec{F}_N$

$\vec{a}$

$m\vec{a}_g$

The gravitational force is downward.

r

The net force is toward the center. So, the crate's acceleration is too.

(b)

Fig. 13-6 (a) A crate sitting on a scale at Earth's equator, as seen by an observer positioned on Earth's rotation axis at some point above the north pole. (b) A free-body diagram for the crate, with a radial r axis extending from Earth's center. The gravitational force on the crate is represented with its equivalent $m\vec{a}_g$. The normal force on the crate from the scale is $\vec{F}_N$. Because of Earth's rotation, the crate has a centripetal acceleration $\vec{a}$ that is directed toward Earth's center.

to either north or south pole) by 21 km. Thus, a point at the poles is closer to the dense core of Earth than is a point on the equator. This is one reason the free-fall acceleration g increases if you were to measure it while moving at sea level from the equator toward the north or south pole. As you move, you are actually getting closer to the center of Earth and thus, by Newton's law of gravitation, g increases.

3. **Earth is rotating.** The rotation axis runs through the north and south poles of Earth. An object located on Earth's surface anywhere except at those poles must rotate in a circle about the rotation axis and thus must have a centripetal acceleration directed toward the center of the circle. This centripetal acceleration requires a centripetal net force that is also directed toward that center.

To see how Earth's rotation causes g to differ from a_g, let us analyze a simple situation in which a crate of mass m is on a scale at the equator. Figure 13-6a shows this situation as viewed from a point in space above the north pole.

Figure 13-6b, a free-body diagram for the crate, shows the two forces on the crate, both acting along a radial r axis that extends from Earth's center. The normal force $\vec{F}_N$ on the crate from the scale is directed outward, in the positive direction of the r axis. The gravitational force, represented with its equivalent $m\vec{a}_g$, is directed inward. Because it travels in a circle about the center of Earth as Earth turns, the crate has a centripetal acceleration $\vec{a}$ directed toward Earth's center. From Eq. 10-23 ($a_r = \omega^2 r$), we know this acceleration is equal to $\omega^2 R$, where ω is Earth's angular speed and R is the circle's radius (approximately Earth's radius). Thus, we can write Newton's second law for forces along the r axis ($F_{\text{net},r} = ma_r$) as

$$F_N - ma_g = m(-\omega^2 R). \qquad (13\text{-}12)$$

The magnitude F_N of the normal force is equal to the weight mg read on the scale. With mg substituted for F_N, Eq. 13-12 gives us

$$mg = ma_g - m(\omega^2 R), \qquad (13\text{-}13)$$

which says

$$\begin{pmatrix} \text{measured} \\ \text{weight} \end{pmatrix} = \begin{pmatrix} \text{magnitude of} \\ \text{gravitational force} \end{pmatrix} - \begin{pmatrix} \text{mass times} \\ \text{centripetal acceleration} \end{pmatrix}.$$

Thus, the measured weight is less than the magnitude of the gravitational force on the crate, because of Earth's rotation.

To find a corresponding expression for g and a_g, we cancel m from Eq. 13-13 to write

$$g = a_g - \omega^2 R, \qquad (13\text{-}14)$$

which says

$$\begin{pmatrix} \text{free-fall} \\ \text{acceleration} \end{pmatrix} = \begin{pmatrix} \text{gravitational} \\ \text{acceleration} \end{pmatrix} - \begin{pmatrix} \text{centripetal} \\ \text{acceleration} \end{pmatrix}.$$

Thus, the measured free-fall acceleration is less than the gravitational acceleration because of Earth's rotation.

The difference between accelerations g and a_g is equal to $\omega^2 R$ and is greatest on the equator (for one reason, the radius of the circle traveled by the crate is greatest there). To find the difference, we can use Eq. 10-5 ($\omega = \Delta\theta/\Delta t$) and Earth's radius $R = 6.37 \times 10^6$ m. For one rotation of Earth, θ is 2π rad and the time period Δt is about 24 h. Using these values (and converting hours to seconds), we find that g is less than a_g by only about 0.034 m/s² (small compared to 9.8 m/s²). Therefore, neglecting the difference in accelerations g and a_g is often justified. Similarly, neglecting the difference between weight and the magnitude of the gravitational force is also often justified.

Sample Problem

Difference in acceleration at head and feet

(a) An astronaut whose height h is 1.70 m floats "feet down" in an orbiting space shuttle at distance $r = 6.77 \times 10^6$ m away from the center of Earth. What is the difference between the gravitational acceleration at her feet and at her head?

KEY IDEAS

We can approximate Earth as a uniform sphere of mass M_E. Then, from Eq. 13-11, the gravitational acceleration at any distance r from the center of Earth is

$$a_g = \frac{GM_E}{r^2}. \quad (13\text{-}15)$$

We might simply apply this equation twice, first with $r = 6.77 \times 10^6$ m for the location of the feet and then with $r = 6.77 \times 10^6$ m + 1.70 m for the location of the head. However, a calculator may give us the same value for a_g twice, and thus a difference of zero, because h is so much smaller than r. Here's a more promising approach: Because we have a differential change dr in r between the astronaut's feet and head, we should differentiate Eq. 13-15 with respect to r.

Calculations: The differentiation gives us

$$da_g = -2\frac{GM_E}{r^3}\,dr, \quad (13\text{-}16)$$

where da_g is the differential change in the gravitational acceleration due to the differential change dr in r. For the astronaut, $dr = h$ and $r = 6.77 \times 10^6$ m. Substituting data into Eq. 13-16, we find

$$da_g = -2\frac{(6.67 \times 10^{-11}\,\text{m}^3/\text{kg}\cdot\text{s}^2)(5.98 \times 10^{24}\,\text{kg})}{(6.77 \times 10^6\,\text{m})^3}(1.70\,\text{m})$$

$$= -4.37 \times 10^{-6}\,\text{m/s}^2, \quad \text{(Answer)}$$

where the M_E value is taken from Appendix C. This result means that the gravitational acceleration of the astronaut's feet toward Earth is slightly greater than the gravitational acceleration of her head toward Earth. This difference in acceleration (often called a *tidal effect*) tends to stretch her body, but the difference is so small that she would never even sense the stretching, much less suffer pain from it.

(b) If the astronaut is now "feet down" at the same orbital radius $r = 6.77 \times 10^6$ m about a black hole of mass $M_h = 1.99 \times 10^{31}$ kg (10 times our Sun's mass), what is the difference between the gravitational acceleration at her feet and at her head? The black hole has a mathematical surface (*event horizon*) of radius $R_h = 2.95 \times 10^4$ m. Nothing, not even light, can escape from that surface or anywhere inside it. Note that the astronaut is well outside the surface (at $r = 229R_h$).

Calculations: We again have a differential change dr in r between the astronaut's feet and head, so we can again use Eq. 13-16. However, now we substitute $M_h = 1.99 \times 10^{31}$ kg for M_E. We find

$$da_g = -2\frac{(6.67 \times 10^{-11}\,\text{m}^3/\text{kg}\cdot\text{s}^2)(1.99 \times 10^{31}\,\text{kg})}{(6.77 \times 10^6\,\text{m})^3}(1.70\,\text{m})$$

$$= -14.5\,\text{m/s}^2. \quad \text{(Answer)}$$

This means that the gravitational acceleration of the astronaut's feet toward the black hole is noticeably larger than that of her head. The resulting tendency to stretch her body would be bearable but quite painful. If she drifted closer to the black hole, the stretching tendency would increase drastically.

 Additional examples, video, and practice available at *WileyPLUS*

13-5 Gravitation Inside Earth

Newton's shell theorem can also be applied to a situation in which a particle is located *inside* a uniform shell, to show the following:

 A uniform shell of matter exerts no net gravitational force on a particle located inside it.

Caution: This statement does *not* mean that the gravitational forces on the particle from the various elements of the shell magically disappear. Rather, it means that the *sum* of the force vectors on the particle from all the elements is zero.

If Earth's mass were uniformly distributed, the gravitational force acting on a particle would be a maximum at Earth's surface and would decrease as the particle moved outward, away from the planet. If the particle were to move inward, perhaps down a deep mine shaft, the gravitational force would change for two reasons. (1) It would tend to increase because the particle would be moving closer to the center of Earth. (2) It would tend to decrease because the thickening shell of material lying outside the particle's radial position would not exert any net force on the particle.

For a uniform Earth, the second influence would prevail and the force on the particle would steadily decrease to zero as the particle approached the center of Earth. However, for the real (nonuniform) Earth, the force on the particle actually increases as the particle begins to descend. The force reaches a maximum at a certain depth and then decreases as the particle descends farther.

Sample Problem

Tunnel through Earth's center, gravitation

In *Pole to Pole,* an early science fiction story by George Griffith, three explorers attempt to travel by capsule through a naturally formed (and, of course, fictional) tunnel directly from the south pole to the north pole (Fig. 13-7). According to the story, as the capsule approaches Earth's center, the gravitational force on the explorers becomes alarmingly large and then, exactly at the center, it suddenly but only momentarily disappears. Then the capsule travels through the second half of the tunnel, to the north pole.

Check Griffith's description by finding the gravitational force on the capsule of mass m when it reaches a distance r from Earth's center. Assume that Earth is a sphere of uniform density ρ (mass per unit volume).

KEY IDEAS

Newton's shell theorem gives us three ideas:

1. When the capsule is at radius r from Earth's center, the portion of Earth that lies outside a sphere of radius r does *not* produce a net gravitational force on the capsule.

2. The portion of Earth that lies inside that sphere *does* produce a net gravitational force on the capsule.

3. We can treat the mass M_{ins} of that inside portion of Earth as being the mass of a particle located at Earth's center.

Calculations: All three ideas tell us that we can write Eq. 13-1, for the magnitude of the gravitational force on the capsule, as

$$F = \frac{GmM_{ins}}{r^2}. \qquad (13\text{-}17)$$

To write the mass M_{ins} in terms of the radius r, we note that the volume V_{ins} containing this mass is $\frac{4}{3}\pi r^3$. Also, because we're assuming an Earth of uniform density, the density $\rho_{ins} = M_{ins}/V_{ins}$ is Earth's density ρ. Thus, we have

$$M_{ins} = \rho V_{ins} = \rho\,\frac{4\pi r^3}{3}. \qquad (13\text{-}18)$$

Then, after substituting this expression into Eq. 13-17 and

canceling, we have

$$F = \frac{4\pi Gm\rho}{3}\,r. \quad \text{(Answer)} \qquad (13\text{-}19)$$

This equation tells us that the force magnitude F depends linearly on the capsule's distance r from Earth's center. Thus, as r decreases, F also decreases (opposite of Griffith's description), until it is zero at Earth's center. At least Griffith got the zero-at-the-center detail correct.

Equation 13-19 can also be written in terms of the force vector $\vec{F}$ and the capsule's position vector $\vec{r}$ along a radial axis extending from Earth's center. Let K represent the collection of constants $4\pi Gm\rho/3$. Then, Eq. 13-19 becomes

$$\vec{F} = -K\vec{r}, \qquad (13\text{-}20)$$

in which we have inserted a minus sign to indicate that $\vec{F}$ and $\vec{r}$ have opposite directions. Equation 13-20 has the form of Hooke's law (Eq. 7-20, $\vec{F} = -k\vec{d}$). Thus, under the idealized conditions of the story, the capsule would oscillate like a block on a spring, with the center of the oscillation at Earth's center. After the capsule had fallen from the south pole to Earth's center, it would travel from the center to the north pole (as Griffith said) and then back again, repeating the cycle forever.

Fig. 13-7 A capsule of mass m falls from rest through a tunnel that connects Earth's south and north poles. When the capsule is at distance r from Earth's center, the portion of Earth's mass that is contained in a sphere of that radius is M_{ins}.

WILEY PLUS Additional examples, video, and practice available at *WileyPLUS*

13-6 Gravitational Potential Energy

In Section 8-4, we discussed the gravitational potential energy of a particle–Earth system. We were careful to keep the particle near Earth's surface, so that we could regard the gravitational force as constant. We then chose some reference configuration of the system as having a gravitational potential energy of zero. Often, in this configuration the particle was on Earth's surface. For particles not on Earth's surface, the gravitational potential energy decreased when the separation between the particle and Earth decreased.

Here, we broaden our view and consider the gravitational potential energy U of two particles, of masses m and M, separated by a distance r. We again choose a reference configuration with U equal to zero. However, to simplify the equations, the separation distance r in the reference configuration is now large enough to be approximated as *infinite*. As before, the gravitational potential energy decreases when the separation decreases. Since $U = 0$ for $r = \infty$, the potential energy is negative for any finite separation and becomes progressively more negative as the particles move closer together.

With these facts in mind and as we shall justify next, we take the gravitational potential energy of the two-particle system to be

$$U = -\frac{GMm}{r} \quad \text{(gravitational potential energy)}. \quad (13\text{-}21)$$

Note that $U(r)$ approaches zero as r approaches infinity and that for any finite value of r, the value of $U(r)$ is negative.

The potential energy given by Eq. 13-21 is a property of the system of two particles rather than of either particle alone. There is no way to divide this energy and say that so much belongs to one particle and so much to the other. However, if $M \gg m$, as is true for Earth (mass M) and a baseball (mass m), we often speak of "the potential energy of the baseball." We can get away with this because, when a baseball moves in the vicinity of Earth, changes in the potential energy of the baseball–Earth system appear almost entirely as changes in the kinetic energy of the baseball, since changes in the kinetic energy of Earth are too small to be measured. Similarly, in Section 13-8 we shall speak of "the potential energy of an artificial satellite" orbiting Earth, because the satellite's mass is so much smaller than Earth's mass. When we speak of the potential energy of bodies of comparable mass, however, we have to be careful to treat them as a system.

If our system contains more than two particles, we consider each pair of particles in turn, calculate the gravitational potential energy of that pair with Eq. 13-21 as if the other particles were not there, and then algebraically sum the results. Applying Eq. 13-21 to each of the three pairs of Fig. 13-8, for example, gives the potential energy of the system as

$$U = -\left(\frac{Gm_1m_2}{r_{12}} + \frac{Gm_1m_3}{r_{13}} + \frac{Gm_2m_3}{r_{23}}\right). \quad (13\text{-}22)$$

Proof of Equation 13-21

Let us shoot a baseball directly away from Earth along the path in Fig. 13-9. We want to find an expression for the gravitational potential energy U of the ball at point P along its path, at radial distance R from Earth's center. To do so, we first find the work W done on the ball by the gravitational force as the ball travels from point P to a great (infinite) distance from Earth. Because the gravitational force $\vec{F}(r)$ is a variable force (its magnitude depends on r), we must use the techniques of Section 7-8 to find the work. In vector notation, we can write

$$W = \int_R^\infty \vec{F}(r) \cdot d\vec{r}. \quad (13\text{-}23)$$

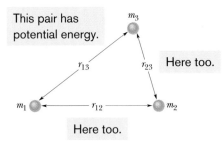

This pair has potential energy.

Here too.

Here too.

Fig. 13-8 A system consisting of three particles. The gravitational potential energy *of the system* is the sum of the gravitational potential energies of all three pairs of particles.

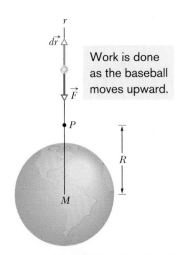

Work is done as the baseball moves upward.

Fig. 13-9 A baseball is shot directly away from Earth, through point P at radial distance R from Earth's center. The gravitational force $\vec{F}$ on the ball and a differential displacement vector $d\vec{r}$ are shown, both directed along a radial r axis.

The integral contains the scalar (or dot) product of the force $\vec{F}(r)$ and the differential displacement vector $d\vec{r}$ along the ball's path. We can expand that product as

$$\vec{F}(r) \cdot d\vec{r} = F(r)\, dr \cos \phi, \tag{13-24}$$

where ϕ is the angle between the directions of $\vec{F}(r)$ and $d\vec{r}$. When we substitute $180°$ for ϕ and Eq. 13-1 for $F(r)$, Eq. 13-24 becomes

$$\vec{F}(r) \cdot d\vec{r} = -\frac{GMm}{r^2}\, dr,$$

where M is Earth's mass and m is the mass of the ball.

Substituting this into Eq. 13-23 and integrating give us

$$W = -GMm \int_R^\infty \frac{1}{r^2}\, dr = \left[\frac{GMm}{r} \right]_R^\infty$$

$$= 0 - \frac{GMm}{R} = -\frac{GMm}{R}, \tag{13-25}$$

where W is the work required to move the ball from point P (at distance R) to infinity. Equation 8-1 ($\Delta U = -W$) tells us that we can also write that work in terms of potential energies as

$$U_\infty - U = -W.$$

Because the potential energy U_∞ at infinity is zero, U is the potential energy at P, and W is given by Eq. 13-25, this equation becomes

$$U = W = -\frac{GMm}{R}.$$

Switching R to r gives us Eq. 13-21, which we set out to prove.

Path Independence

In Fig. 13-10, we move a baseball from point A to point G along a path consisting of three radial lengths and three circular arcs (centered on Earth). We are interested in the total work W done by Earth's gravitational force $\vec{F}$ on the ball as it moves from A to G. The work done along each circular arc is zero, because the direction of $\vec{F}$ is perpendicular to the arc at every point. Thus, W is the sum of only the works done by $\vec{F}$ along the three radial lengths.

Now, suppose we mentally shrink the arcs to zero. We would then be moving the ball directly from A to G along a single radial length. Does that change W? No. Because no work was done along the arcs, eliminating them does not change the work. The path taken from A to G now is clearly different, but the work done by $\vec{F}$ is the same.

We discussed such a result in a general way in Section 8-3. Here is the point: The gravitational force is a conservative force. Thus, the work done by the gravitational force on a particle moving from an initial point i to a final point f is independent of the path taken between the points. From Eq. 8-1, the change ΔU in the gravitational potential energy from point i to point f is given by

$$\Delta U = U_f - U_i = -W. \tag{13-26}$$

Since the work W done by a conservative force is independent of the actual path taken, the change ΔU in gravitational potential energy is *also independent* of the path taken.

Potential Energy and Force

In the proof of Eq. 13-21, we derived the potential energy function $U(r)$ from the force function $\vec{F}(r)$. We should be able to go the other way—that is, to start from the potential energy function and derive the force function. Guided by Eq. 8-22

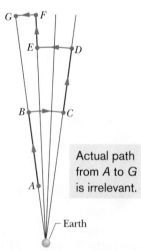

Actual path from A to G is irrelevant.

Fig. 13-10 Near Earth, a baseball is moved from point A to point G along a path consisting of radial lengths and circular arcs.

$(F(x) = -dU(x)/dx)$, we can write

$$F = -\frac{dU}{dr} = -\frac{d}{dr}\left(-\frac{GMm}{r}\right)$$

$$= -\frac{GMm}{r^2}.$$ (13-27)

This is Newton's law of gravitation (Eq. 13-1). The minus sign indicates that the force on mass m points radially inward, toward mass M.

Escape Speed

If you fire a projectile upward, usually it will slow, stop momentarily, and return to Earth. There is, however, a certain minimum initial speed that will cause it to move upward forever, theoretically coming to rest only at infinity. This minimum initial speed is called the (Earth) **escape speed.**

Consider a projectile of mass m, leaving the surface of a planet (or some other astronomical body or system) with escape speed v. The projectile has a kinetic energy K given by $\frac{1}{2}mv^2$ and a potential energy U given by Eq. 13-21:

$$U = -\frac{GMm}{R},$$

in which M is the mass of the planet and R is its radius.

When the projectile reaches infinity, it stops and thus has no kinetic energy. It also has no potential energy because an infinite separation between two bodies is our zero-potential-energy configuration. Its total energy at infinity is therefore zero. From the principle of conservation of energy, its total energy at the planet's surface must also have been zero, and so

$$K + U = \frac{1}{2}mv^2 + \left(-\frac{GMm}{R}\right) = 0.$$

This yields

$$v = \sqrt{\frac{2GM}{R}}.$$ (13-28)

Note that v does not depend on the direction in which a projectile is fired from a planet. However, attaining that speed is easier if the projectile is fired in the direction the launch site is moving as the planet rotates about its axis. For example, rockets are launched eastward at Cape Canaveral to take advantage of the Cape's eastward speed of 1500 km/h due to Earth's rotation.

Equation 13-28 can be applied to find the escape speed of a projectile from any astronomical body, provided we substitute the mass of the body for M and the radius of the body for R. Table 13-2 shows some escape speeds.

✓ CHECKPOINT 3

You move a ball of mass m away from a sphere of mass M. (a) Does the gravitational potential energy of the system of ball and sphere increase or decrease? (b) Is positive work or negative work done by the gravitational force between the ball and the sphere?

Table 13-2

Some Escape Speeds

Body	Mass (kg)	Radius (m)	Escape Speed (km/s)
Ceres[a]	1.17×10^{21}	3.8×10^5	0.64
Earth's moon[a]	7.36×10^{22}	1.74×10^6	2.38
Earth	5.98×10^{24}	6.37×10^6	11.2
Jupiter	1.90×10^{27}	7.15×10^7	59.5
Sun	1.99×10^{30}	6.96×10^8	618
Sirius B[b]	2×10^{30}	1×10^7	5200
Neutron star[c]	2×10^{30}	1×10^4	2×10^5

[a]The most massive of the asteroids.
[b]A *white dwarf* (a star in a final stage of evolution) that is a companion of the bright star Sirius.
[c]The collapsed core of a star that remains after that star has exploded in a *supernova* event.

Sample Problem

Asteroid falling from space, mechanical energy

An asteroid, headed directly toward Earth, has a speed of 12 km/s relative to the planet when the asteroid is 10 Earth radii from Earth's center. Neglecting the effects of Earth's atmosphere on the asteroid, find the asteroid's speed v_f when it reaches Earth's surface.

KEY IDEAS

Because we are to neglect the effects of the atmosphere on the asteroid, the mechanical energy of the asteroid–Earth system is conserved during the fall. Thus, the final mechanical energy (when the asteroid reaches Earth's surface) is equal to the initial mechanical energy. With kinetic energy K and gravitational potential energy U, we can write this as

$$K_f + U_f = K_i + U_i. \qquad (13\text{-}29)$$

Also, if we assume the system is isolated, the system's linear momentum must be conserved during the fall. Therefore, the momentum change of the asteroid and that of Earth must be equal in magnitude and opposite in sign. However, because Earth's mass is so much greater than the asteroid's mass, the change in Earth's speed is negligible relative to the change in the asteroid's speed. So, the change in Earth's kinetic energy is also negligible. Thus, we can assume that the kinetic energies in Eq. 13-29 are those of the asteroid alone.

Calculations: Let m represent the asteroid's mass and M represent Earth's mass (5.98×10^{24} kg). The asteroid is initially at distance $10R_E$ and finally at distance R_E, where R_E is

Earth's radius (6.37×10^6 m). Substituting Eq. 13-21 for U and $\frac{1}{2}mv^2$ for K, we rewrite Eq. 13-29 as

$$\frac{1}{2}mv_f^2 - \frac{GMm}{R_E} = \frac{1}{2}mv_i^2 - \frac{GMm}{10R_E}.$$

Rearranging and substituting known values, we find

$$v_f^2 = v_i^2 + \frac{2GM}{R_E}\left(1 - \frac{1}{10}\right)$$

$$= (12 \times 10^3 \text{ m/s})^2$$

$$+ \frac{2(6.67 \times 10^{-11} \text{ m}^3/\text{kg}\cdot\text{s}^2)(5.98 \times 10^{24} \text{ kg})}{6.37 \times 10^6 \text{ m}} 0.9$$

$$= 2.567 \times 10^8 \text{ m}^2/\text{s}^2,$$

and

$$v_f = 1.60 \times 10^4 \text{ m/s} = 16 \text{ km/s}. \qquad \text{(Answer)}$$

At this speed, the asteroid would not have to be particularly large to do considerable damage at impact. If it were only 5 m across, the impact could release about as much energy as the nuclear explosion at Hiroshima. Alarmingly, about 500 million asteroids of this size are near Earth's orbit, and in 1994 one of them apparently penetrated Earth's atmosphere and exploded 20 km above the South Pacific (setting off nuclear-explosion warnings on six military satellites). The impact of an asteroid 500 m across (there may be a million of them near Earth's orbit) could end modern civilization and almost eliminate humans worldwide.

 Additional examples, video, and practice available at *WileyPLUS*

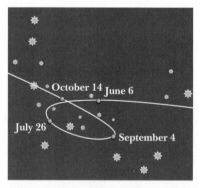

Fig. 13-11 The path seen from Earth for the planet Mars as it moved against a background of the constellation Capricorn during 1971. The planet's position on four days is marked. Both Mars and Earth are moving in orbits around the Sun so that we see the position of Mars relative to us; this relative motion sometimes results in an apparent loop in the path of Mars.

13-7 Planets and Satellites: Kepler's Laws

The motions of the planets, as they seemingly wander against the background of the stars, have been a puzzle since the dawn of history. The "loop-the-loop" motion of Mars, shown in Fig. 13-11, was particularly baffling. Johannes Kepler (1571–1630), after a lifetime of study, worked out the empirical laws that govern these motions. Tycho Brahe (1546–1601), the last of the great astronomers to make observations without the help of a telescope, compiled the extensive data from which Kepler was able to derive the three laws of planetary motion that now bear Kepler's name. Later, Newton (1642–1727) showed that his law of gravitation leads to Kepler's laws.

In this section we discuss each of Kepler's three laws. Although here we apply the laws to planets orbiting the Sun, they hold equally well for satellites, either natural or artificial, orbiting Earth or any other massive central body.

 1. THE LAW OF ORBITS: All planets move in elliptical orbits, with the Sun at one focus.

Figure 13-12 shows a planet of mass m moving in such an orbit around the Sun, whose mass is M. We assume that $M \gg m$, so that the center of mass of the planet–Sun system is approximately at the center of the Sun.

The orbit in Fig. 13-12 is described by giving its **semimajor axis** a and its **eccentricity** e, the latter defined so that ea is the distance from the center of the ellipse to either focus F or F'. *An eccentricity of zero corresponds to a circle,* in which the two foci merge to a single central point. The eccentricities of the planetary orbits are not large; so if the orbits are drawn to scale, they look circular. The eccentricity of the ellipse of Fig. 13-12, which has been exaggerated for clarity, is 0.74. The eccentricity of Earth's orbit is only 0.0167.

2. THE LAW OF AREAS: A line that connects a planet to the Sun sweeps out equal areas in the plane of the planet's orbit in equal time intervals; that is, the rate dA/dt at which it sweeps out area A is constant.

Qualitatively, this second law tells us that the planet will move most slowly when it is farthest from the Sun and most rapidly when it is nearest to the Sun. As it turns out, Kepler's second law is totally equivalent to the law of conservation of angular momentum. Let us prove it.

The area of the shaded wedge in Fig. 13-13a closely approximates the area swept out in time Δt by a line connecting the Sun and the planet, which are separated by distance r. The area ΔA of the wedge is approximately the area of a triangle with base $r\Delta\theta$ and height r. Since the area of a triangle is one-half of the base times the height, $\Delta A \approx \frac{1}{2}r^2\,\Delta\theta$. This expression for ΔA becomes more exact as Δt (hence $\Delta\theta$) approaches zero. The instantaneous rate at which area is being swept out is then

$$\frac{dA}{dt} = \tfrac{1}{2}r^2\frac{d\theta}{dt} = \tfrac{1}{2}r^2\omega, \qquad (13\text{-}30)$$

in which ω is the angular speed of the rotating line connecting Sun and planet.

Figure 13-13b shows the linear momentum $\vec{p}$ of the planet, along with the radial and perpendicular components of $\vec{p}$. From Eq. 11-20 ($L = rp_\perp$), the magnitude of the angular momentum $\vec{L}$ of the planet about the Sun is given by the product of r and $p_\perp$, the component of $\vec{p}$ perpendicular to r. Here, for a planet of mass m,

$$L = rp_\perp = (r)(mv_\perp) = (r)(m\omega r)$$
$$= mr^2\omega, \qquad (13\text{-}31)$$

where we have replaced $v_\perp$ with its equivalent ωr (Eq. 10-18). Eliminating $r^2\omega$ between Eqs. 13-30 and 13-31 leads to

$$\frac{dA}{dt} = \frac{L}{2m}. \qquad (13\text{-}32)$$

If dA/dt is constant, as Kepler said it is, then Eq. 13-32 means that L must also be constant—angular momentum is conserved. Kepler's second law is indeed equivalent to the law of conservation of angular momentum.

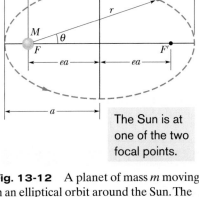

The Sun is at one of the two focal points.

Fig. 13-12 A planet of mass m moving in an elliptical orbit around the Sun. The Sun, of mass M, is at one focus F of the ellipse. The other focus is F', which is located in empty space. Each focus is a distance ea from the ellipse's center, with e being the eccentricity of the ellipse. The semimajor axis a of the ellipse, the perihelion (nearest the Sun) distance R_p, and the aphelion (farthest from the Sun) distance R_a are also shown.

The planet sweeps out this area.

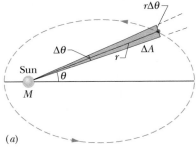

These are the two momentum components.

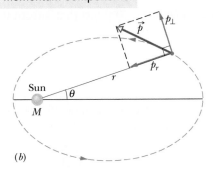

Fig. 13-13 (a) In time Δt, the line r connecting the planet to the Sun moves through an angle $\Delta\theta$, sweeping out an area ΔA (shaded). (b) The linear momentum $\vec{p}$ of the planet and the components of $\vec{p}$.

(a)

(b)

Fig. 13-14 A planet of mass m moving around the Sun in a circular orbit of radius r.

Table 13-3

Kepler's Law of Periods for the Solar System

Planet	Semimajor Axis a (10^{10} m)	Period T (y)	T^2/a^3 (10^{-34} y²/m³)
Mercury	5.79	0.241	2.99
Venus	10.8	0.615	3.00
Earth	15.0	1.00	2.96
Mars	22.8	1.88	2.98
Jupiter	77.8	11.9	3.01
Saturn	143	29.5	2.98
Uranus	287	84.0	2.98
Neptune	450	165	2.99
Pluto	590	248	2.99

3. THE LAW OF PERIODS: The square of the period of any planet is proportional to the cube of the semimajor axis of its orbit.

To see this, consider the circular orbit of Fig. 13-14, with radius r (the radius of a circle is equivalent to the semimajor axis of an ellipse). Applying Newton's second law ($F = ma$) to the orbiting planet in Fig. 13-14 yields

$$\frac{GMm}{r^2} = (m)(\omega^2 r). \qquad (13\text{-}33)$$

Here we have substituted from Eq. 13-1 for the force magnitude F and used Eq. 10-23 to substitute $\omega^2 r$ for the centripetal acceleration. If we now use Eq. 10-20 to replace ω with $2\pi/T$, where T is the period of the motion, we obtain Kepler's third law:

$$T^2 = \left(\frac{4\pi^2}{GM}\right) r^3 \qquad \text{(law of periods)}. \qquad (13\text{-}34)$$

The quantity in parentheses is a constant that depends only on the mass M of the central body about which the planet orbits.

Equation 13-34 holds also for elliptical orbits, provided we replace r with a, the semimajor axis of the ellipse. This law predicts that the ratio T^2/a^3 has essentially the same value for every planetary orbit around a given massive body. Table 13-3 shows how well it holds for the orbits of the planets of the solar system.

✔ CHECKPOINT 4

Satellite 1 is in a certain circular orbit around a planet, while satellite 2 is in a larger circular orbit. Which satellite has (a) the longer period and (b) the greater speed?

Sample Problem

Kepler's law of periods, Comet Halley

Comet Halley orbits the Sun with a period of 76 years and, in 1986, had a distance of closest approach to the Sun, its *perihelion distance* R_p, of 8.9×10^{10} m. Table 13-3 shows that this is between the orbits of Mercury and Venus.

(a) What is the comet's farthest distance from the Sun, which is called its *aphelion distance* R_a?

KEY IDEAS

From Fig. 13-12, we see that $R_a + R_p = 2a$, where a is the semimajor axis of the orbit. Thus, we can find R_a if we first find a. We can relate a to the given period via the law of periods (Eq. 13-34) if we simply substitute the semimajor axis a for r.

Calculations: Making that substitution and then solving for a, we have

$$a = \left(\frac{GMT^2}{4\pi^2}\right)^{1/3}. \qquad (13\text{-}35)$$

If we substitute the mass M of the Sun, 1.99×10^{30} kg, and the period T of the comet, 76 years or 2.4×10^9 s, into Eq. 13-35, we find that $a = 2.7 \times 10^{12}$ m. Now we have

$$R_a = 2a - R_p$$
$$= (2)(2.7 \times 10^{12} \text{ m}) - 8.9 \times 10^{10} \text{ m}$$
$$= 5.3 \times 10^{12} \text{ m}. \qquad \text{(Answer)}$$

Table 13-3 shows that this is a little less than the semimajor axis of the orbit of Pluto. Thus, the comet does not get farther from the Sun than Pluto.

(b) What is the eccentricity e of the orbit of comet Halley?

KEY IDEA

We can relate e, a, and R_p via Fig. 13-12, in which we see that $ea = a - R_p$.

Calculation: We have

$$e = \frac{a - R_p}{a} = 1 - \frac{R_p}{a} \qquad (13\text{-}36)$$

$$= 1 - \frac{8.9 \times 10^{10} \text{ m}}{2.7 \times 10^{12} \text{ m}} = 0.97. \qquad \text{(Answer)}$$

This tells us that, with an eccentricity approaching unity, this orbit must be a long thin ellipse.

 Additional examples, video, and practice available at *WileyPLUS*

13-8 Satellites: Orbits and Energy

As a satellite orbits Earth in an elliptical path, both its speed, which fixes its kinetic energy K, and its distance from the center of Earth, which fixes its gravitational potential energy U, fluctuate with fixed periods. However, the mechanical energy E of the satellite remains constant. (Since the satellite's mass is so much smaller than Earth's mass, we assign U and E for the Earth–satellite system to the satellite alone.)

The potential energy of the system is given by Eq. 13-21:

$$U = -\frac{GMm}{r}$$

(with $U = 0$ for infinite separation). Here r is the radius of the satellite's orbit, assumed for the time being to be circular, and M and m are the masses of Earth and the satellite, respectively.

To find the kinetic energy of a satellite in a circular orbit, we write Newton's second law ($F = ma$) as

$$\frac{GMm}{r^2} = m\frac{v^2}{r}, \tag{13-37}$$

where v^2/r is the centripetal acceleration of the satellite. Then, from Eq. 13-37, the kinetic energy is

$$K = \tfrac{1}{2}mv^2 = \frac{GMm}{2r}, \tag{13-38}$$

which shows us that for a satellite in a circular orbit,

$$K = -\frac{U}{2} \qquad \text{(circular orbit).} \tag{13-39}$$

The total mechanical energy of the orbiting satellite is

$$E = K + U = \frac{GMm}{2r} - \frac{GMm}{r}$$

or

$$E = -\frac{GMm}{2r} \qquad \text{(circular orbit).} \tag{13-40}$$

This tells us that for a satellite in a circular orbit, the total energy E is the negative of the kinetic energy K:

$$E = -K \qquad \text{(circular orbit).} \tag{13-41}$$

For a satellite in an elliptical orbit of semimajor axis a, we can substitute a for r in Eq. 13-40 to find the mechanical energy:

$$E = -\frac{GMm}{2a} \qquad \text{(elliptical orbit).} \tag{13-42}$$

Equation 13-42 tells us that the total energy of an orbiting satellite depends only on the semimajor axis of its orbit and not on its eccentricity e. For example, four orbits with the same semimajor axis are shown in Fig. 13-15; the same satellite would have the same total mechanical energy E in all four orbits. Figure 13-16 shows the variation of K, U, and E with r for a satellite moving in a circular orbit about a massive central body.

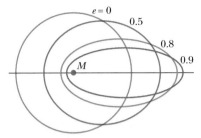

Fig. 13-15 Four orbits with different eccentricities e about an object of mass M. All four orbits have the same semimajor axis a and thus correspond to the same total mechanical energy E.

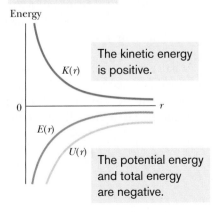

This is a plot of a satellite's energies versus orbit radius.

The kinetic energy is positive.

The potential energy and total energy are negative.

Fig. 13-16 The variation of kinetic energy K, potential energy U, and total energy E with radius r for a satellite in a circular orbit. For any value of r, the values of U and E are negative, the value of K is positive, and $E = -K$. As $r \to \infty$, all three energy curves approach a value of zero.

✔ CHECKPOINT 5

In the figure here, a space shuttle is initially in a circular orbit of radius r about Earth. At point P, the pilot briefly fires a forward-pointing thruster to decrease the shuttle's kinetic energy K and mechanical energy E. (a) Which of the dashed elliptical orbits shown in the figure will the shuttle then take? (b) Is the orbital period T of the shuttle (the time to return to P) then greater than, less than, or the same as in the circular orbit?

Sample Problem

Mechanical energy of orbiting bowling ball

A playful astronaut releases a bowling ball, of mass $m = 7.20$ kg, into circular orbit about Earth at an altitude h of 350 km.

(a) What is the mechanical energy E of the ball in its orbit?

KEY IDEA

We can get E from the orbital energy, given by Eq. 13-40 ($E = -GMm/2r$), if we first find the orbital radius r. (It is *not* simply the given altitude.)

Calculations: The orbital radius must be

$$r = R + h = 6370 \text{ km} + 350 \text{ km} = 6.72 \times 10^6 \text{ m},$$

in which R is the radius of Earth. Then, from Eq. 13-40, the mechanical energy is

$$
\begin{aligned}
E &= -\frac{GMm}{2r} \\
&= -\frac{(6.67 \times 10^{-11} \text{ N} \cdot \text{m}^2/\text{kg}^2)(5.98 \times 10^{24} \text{ kg})(7.20 \text{ kg})}{(2)(6.72 \times 10^6 \text{ m})} \\
&= -2.14 \times 10^8 \text{ J} = -214 \text{ MJ}. \quad \text{(Answer)}
\end{aligned}
$$

(b) What is the mechanical energy E_0 of the ball on the launchpad at Cape Canaveral (before it, the astronaut, and the spacecraft are launched)? From there to the orbit, what is the change ΔE in the ball's mechanical energy?

KEY IDEA

On the launchpad, the ball is *not* in orbit and thus Eq. 13-40 does *not* apply. Instead, we must find $E_0 = K_0 + U_0$, where K_0 is the ball's kinetic energy and U_0 is the gravitational potential energy of the ball–Earth system.

Calculations: To find U_0, we use Eq. 13-21 to write

$$
\begin{aligned}
U_0 &= -\frac{GMm}{R} \\
&= -\frac{(6.67 \times 10^{-11} \text{ N} \cdot \text{m}^2/\text{kg}^2)(5.98 \times 10^{24} \text{ kg})(7.20 \text{ kg})}{6.37 \times 10^6 \text{ m}} \\
&= -4.51 \times 10^8 \text{ J} = -451 \text{ MJ}.
\end{aligned}
$$

The kinetic energy K_0 of the ball is due to the ball's motion with Earth's rotation. You can show that K_0 is less than 1 MJ, which is negligible relative to U_0. Thus, the mechanical energy of the ball on the launchpad is

$$E_0 = K_0 + U_0 \approx 0 - 451 \text{ MJ} = -451 \text{ MJ}. \quad \text{(Answer)}$$

The *increase* in the mechanical energy of the ball from launchpad to orbit is

$$
\begin{aligned}
\Delta E &= E - E_0 = (-214 \text{ MJ}) - (-451 \text{ MJ}) \\
&= 237 \text{ MJ}. \quad \text{(Answer)}
\end{aligned}
$$

This is worth a few dollars at your utility company. Obviously the high cost of placing objects into orbit is not due to their required mechanical energy.

 Additional examples, video, and practice available at *WileyPLUS*

13-9 **Einstein and Gravitation**

Principle of Equivalence

Albert Einstein once said: "I was . . . in the patent office at Bern when all of a sudden a thought occurred to me: 'If a person falls freely, he will not feel his own weight.' I was startled. This simple thought made a deep impression on me. It impelled me toward a theory of gravitation."

Thus Einstein tells us how he began to form his **general theory of relativity.** The fundamental postulate of this theory about gravitation (the gravitating of objects toward each other) is called the **principle of equivalence,** which says that gravitation and acceleration are equivalent. If a physicist were locked up in a small box as in Fig. 13-17, he would not be able to tell whether the box was at rest on Earth (and subject only to Earth's gravitational force), as in Fig. 13-17a, or accelerating through interstellar space at 9.8 m/s^2 (and subject only to the force producing that acceleration), as in Fig. 13-17b. In both situations he would feel the same and would read the same value for his weight on a scale. Moreover, if he watched an object fall past him, the object would have the same acceleration relative to him in both situations.

Curvature of Space

We have thus far explained gravitation as due to a force between masses. Einstein showed that, instead, gravitation is due to a curvature of space that is caused by the masses. (As is discussed later in this book, space and time are entangled, so the curvature of which Einstein spoke is really a curvature of *spacetime,* the combined four dimensions of our universe.)

Picturing how space (such as vacuum) can have curvature is difficult. An analogy might help: Suppose that from orbit we watch a race in which two boats begin on Earth's equator with a separation of 20 km and head due south (Fig. 13-18a). To the sailors, the boats travel along flat, parallel paths. However, with time the boats draw together until, nearer the south pole, they touch. The sailors in the boats can interpret this drawing together in terms of a force acting on the boats. Looking on from space, however, we can see that the boats draw together simply because of the curvature of Earth's surface. We can see this because we are viewing the race from "outside" that surface.

(a)   (b)

Fig. 13-17 (a) A physicist in a box resting on Earth sees a cantaloupe falling with acceleration $a = 9.8$ m/s^2. (b) If he and the box accelerate in deep space at 9.8 m/s^2, the cantaloupe has the same acceleration relative to him. It is not possible, by doing experiments within the box, for the physicist to tell which situation he is in. For example, the platform scale on which he stands reads the same weight in both situations.

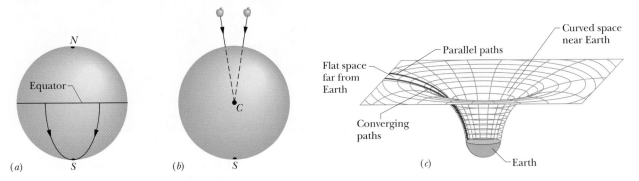

Fig. 13-18 (a) Two objects moving along lines of longitude toward the south pole converge because of the curvature of Earth's surface. (b) Two objects falling freely near Earth move along lines that converge toward the center of Earth because of the curvature of space near Earth. (c) Far from Earth (and other masses), space is flat and parallel paths remain parallel. Close to Earth, the parallel paths begin to converge because space is curved by Earth's mass.

Figure 13-18*b* shows a similar race: Two horizontally separated apples are dropped from the same height above Earth. Although the apples may appear to travel along parallel paths, they actually move toward each other because they both fall toward Earth's center. We can interpret the motion of the apples in terms of the gravitational force on the apples from Earth. We can also interpret the motion in terms of a curvature of the space near Earth, a curvature due to the presence of Earth's mass. This time we cannot see the curvature because we cannot get "outside" the curved space, as we got "outside" the curved Earth in the boat example. However, we can depict the curvature with a drawing like Fig. 13-18*c*; there the apples would move along a surface that curves toward Earth because of Earth's mass.

When light passes near Earth, the path of the light bends slightly because of the curvature of space there, an effect called *gravitational lensing*. When light passes a more massive structure, like a galaxy or a black hole having large mass, its path can be bent more. If such a massive structure is between us and a quasar (an extremely bright, extremely distant source of light), the light from the quasar can bend around the massive structure and toward us (Fig. 13-19*a*). Then, because the light seems to be coming to us from a number of slightly different directions in the sky, we see the same quasar in all those different directions. In some situations, the quasars we see blend together to form a giant luminous arc, which is called an *Einstein ring* (Fig. 13-19*b*).

Should we attribute gravitation to the curvature of spacetime due to the presence of masses or to a force between masses? Or should we attribute it to the actions of a type of fundamental particle called a *graviton,* as conjectured in some modern physics theories? Although our theories about gravitation have been enormously successful in describing everything from falling apples to planetary and stellar motions, we still do not fully understand it on either the cosmological scale or the quantum physics scale.

(a)   (b)

Fig. 13-19   (*a*) Light from a distant quasar follows curved paths around a galaxy or a large black hole because the mass of the galaxy or black hole has curved the adjacent space. If the light is detected, it appears to have originated along the backward extensions of the final paths (dashed lines). (*b*) The Einstein ring known as MG1131+0456 on the computer screen of a telescope. The source of the light (actually, radio waves, which are a form of invisible light) is far behind the large, unseen galaxy that produces the ring; a portion of the source appears as the two bright spots seen along the ring. (*Courtesy National Radio Astronomy Observatory*)

The Law of Gravitation Any particle in the universe attracts any other particle with a **gravitational force** whose magnitude is

$$F = G\frac{m_1 m_2}{r^2} \qquad \text{(Newton's law of gravitation),} \qquad (13\text{-}1)$$

where m_1 and m_2 are the masses of the particles, r is their separation, and G ($= 6.67 \times 10^{-11}$ N·m²/kg²) is the *gravitational constant*.

Gravitational Behavior of Uniform Spherical Shells
The gravitational force between extended bodies is found by adding (integrating) the individual forces on individual particles within the bodies. However, if either of the bodies is a uniform spherical shell or a spherically symmetric solid, the net gravitational force it exerts on an *external* object may be computed as if all the mass of the shell or body were located at its center.

Superposition Gravitational forces obey the **principle of superposition;** that is, if n particles interact, the net force $\vec{F}_{1,\text{net}}$ on a particle labeled particle 1 is the sum of the forces on it from all the other particles taken one at a time:

$$\vec{F}_{1,\text{net}} = \sum_{i=2}^{n} \vec{F}_{1i}, \qquad (13\text{-}5)$$

in which the sum is a vector sum of the forces $\vec{F}_{1i}$ on particle 1 from particles 2, 3, ..., n. The gravitational force $\vec{F}_1$ on a particle from an extended body is found by dividing the body into units of differential mass dm, each of which produces a differential force $d\vec{F}$ on the particle, and then integrating to find the sum of those forces:

$$\vec{F}_1 = \int d\vec{F}. \qquad (13\text{-}6)$$

Gravitational Acceleration The *gravitational acceleration* a_g of a particle (of mass m) is due solely to the gravitational force acting on it. When the particle is at distance r from the center of a uniform, spherical body of mass M, the magnitude F of the gravitational force on the particle is given by Eq. 13-1. Thus, by Newton's second law,

$$F = ma_g, \qquad (13\text{-}10)$$

which gives

$$a_g = \frac{GM}{r^2}. \qquad (13\text{-}11)$$

Free-Fall Acceleration and Weight Because Earth's mass is not distributed uniformly, because the planet is not perfectly spherical, and because it rotates, the actual free-fall acceleration $\vec{g}$ of a particle near Earth differs slightly from the gravitational acceleration $\vec{a}_g$, and the particle's weight (equal to mg) differs from the magnitude of the gravitational force on it (Eq. 13-1).

Gravitation Within a Spherical Shell A uniform shell of matter exerts no net gravitational force on a particle located inside it. This means that if a particle is located inside a uniform solid sphere at distance r from its center, the gravitational force exerted on the particle is due only to the mass M_{ins} that lies inside a sphere of radius r. This mass is given by

$$M_{\text{ins}} = \rho\,\frac{4\pi r^3}{3}, \qquad (13\text{-}18)$$

where ρ is the density of the sphere.

Gravitational Potential Energy The gravitational potential energy $U(r)$ of a system of two particles, with masses M and m and separated by a distance r, is the negative of the work that would be done by the gravitational force of either particle acting on the other if the separation between the particles were changed from infinite (very large) to r. This energy is

$$U = -\frac{GMm}{r} \qquad \text{(gravitational potential energy).} \qquad (13\text{-}21)$$

Potential Energy of a System If a system contains more than two particles, its total gravitational potential energy U is the sum of terms representing the potential energies of all the pairs. As an example, for three particles, of masses m_1, m_2, and m_3,

$$U = -\left(\frac{Gm_1 m_2}{r_{12}} + \frac{Gm_1 m_3}{r_{13}} + \frac{Gm_2 m_3}{r_{23}}\right). \qquad (13\text{-}22)$$

Escape Speed An object will escape the gravitational pull of an astronomical body of mass M and radius R (that is, it will reach an infinite distance) if the object's speed near the body's surface is at least equal to the **escape speed,** given by

$$v = \sqrt{\frac{2GM}{R}}. \qquad (13\text{-}28)$$

Kepler's Laws The motion of satellites, both natural and artificial, is governed by these laws:

1. *The law of orbits.* All planets move in elliptical orbits with the Sun at one focus.

2. *The law of areas.* A line joining any planet to the Sun sweeps out equal areas in equal time intervals. (This statement is equivalent to conservation of angular momentum.)

3. *The law of periods.* The square of the period T of any planet is proportional to the cube of the semimajor axis a of its orbit. For circular orbits with radius r,

$$T^2 = \left(\frac{4\pi^2}{GM}\right)r^3 \qquad \text{(law of periods),} \qquad (13\text{-}34)$$

where M is the mass of the attracting body—the Sun in the case of the solar system. For elliptical planetary orbits, the semimajor axis a is substituted for r.

Energy in Planetary Motion When a planet or satellite with mass m moves in a circular orbit with radius r, its potential energy U and kinetic energy K are given by

$$U = -\frac{GMm}{r} \quad \text{and} \quad K = \frac{GMm}{2r}. \qquad (13\text{-}21, 13\text{-}38)$$

The mechanical energy $E = K + U$ is then

$$E = -\frac{GMm}{2r}. \qquad (13\text{-}40)$$

For an elliptical orbit of semimajor axis a,

$$E = -\frac{GMm}{2a}. \qquad (13\text{-}42)$$

Einstein's View of Gravitation Einstein pointed out that gravitation and acceleration are equivalent. This **principle of equivalence** led him to a theory of gravitation (the **general theory of relativity**) that explains gravitational effects in terms of a curvature of space.

1 In Fig. 13-20, a central particle of mass M is surrounded by a square array of other particles, separated by either distance d or distance $d/2$ along the perimeter of the square. What are the magnitude and direction of the net gravitational force on the central particle due to the other particles?

Fig. 13-20 Question 1.

2 Figure 13-21 shows three arrangements of the same identical particles, with three of them placed on a circle of radius 0.20 m and the fourth one placed at the center of the circle. (a) Rank the arrangements according to the magnitude of the net gravitational force on the central particle due to the other three particles, greatest first. (b) Rank them according to the gravitational potential energy of the four-particle system, least negative first.

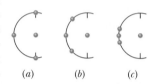

Fig. 13-21 Question 2.

3 In Fig. 13-22, a central particle is surrounded by two circular rings of particles, at radii r and R, with $R > r$. All the particles have mass m. What are the magnitude and direction of the net gravitational force on the central particle due to the particles in the rings?

Fig. 13-22 Question 3.

4 In Fig. 13-23, two particles, of masses m and $2m$, are fixed in place on an axis. (a) Where on the axis can a third particle of mass $3m$ be placed (other than at infinity) so that the net gravitational force on it from the first two particles is zero: to the left of the first two particles, to their right, between them but closer to the more massive particle, or between them but closer to the less massive particle? (b) Does the answer change if the third particle has, instead, a mass of $16m$? (c) Is there a point off the axis (other than infinity) at which the net force on the third particle would be zero?

Fig. 13-23 Question 4.

5 Figure 13-24 shows three situations involving a point particle P with mass m and a spherical shell with a uniformly distributed mass M. The radii of the shells are given. Rank the situations ac-

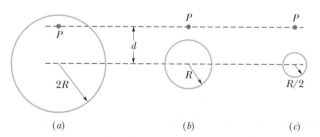

Fig. 13-24 Question 5.

cording to the magnitude of the gravitational force on particle P due to the shell, greatest first.

6 In Fig. 13-25, three particles are fixed in place. The mass of B is greater than the mass of C. Can a fourth particle (particle D) be placed somewhere so that the net gravitational force on particle A from particles B, C, and D is zero? If so, in which quadrant should it be placed and which axis should it be near?

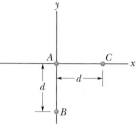

Fig. 13-25 Question 6.

7 Rank the four systems of equal-mass particles shown in Checkpoint 2 according to the absolute value of the gravitational potential energy of the system, greatest first.

8 Figure 13-26 gives the gravitational acceleration a_g for four planets as a function of the radial distance r from the center of the planet, starting at the surface of the planet (at radius R_1, R_2, R_3, or R_4). Plots 1 and 2 coincide for $r \geq R_2$; plots 3 and 4 coincide for $r \geq R_4$. Rank the four planets according to (a) mass and (b) mass per unit volume, greatest first.

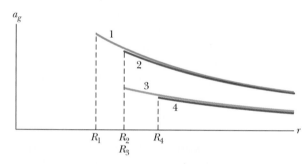

Fig. 13-26 Question 8.

9 Figure 13-27 shows three particles initially fixed in place, with B and C identical and positioned symmetrically about the y axis, at distance d from A. (a) In what direction is the net gravitational force $\vec{F}_{net}$ on A? (b) If we move C directly away from the origin, does $\vec{F}_{net}$ change in direction? If so, how and what is the limit of the change?

Fig. 13-27 Question 9.

10 Figure 13-28 shows six paths by which a rocket orbiting a moon might move from point a to point b. Rank the paths according to (a) the corresponding change in the gravitational potential energy of the rocket–moon system and (b) the net work done on the rocket by the gravitational force from the moon, greatest first.

Fig. 13-28 Question 10.

11 Figure 13-29 shows three uniform spherical planets that are identical in size and mass. The periods of rotation T for the planets are given, and six lettered points are indicated—three points are on the equators of the planets and three points are on the north poles. Rank the points according to the value of the free-fall acceleration g at them, greatest first.

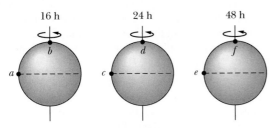

16 h   24 h   48 h

Fig. 13-29   Question 11.

12   In Fig. 13-30, a particle of mass m (which is not shown) is to be moved from an infinite distance to one of the three possible locations a, b, and c. Two other particles, of masses m and $2m$, are already fixed in place on the axis, as shown. Rank the three possible locations according to the work done by the net gravitational force on the moving particle due to the fixed particles, greatest first.

a $2m$ b m c

Fig. 13-30   Question 12.

PROBLEMS

WILEY
PLUS

GO   Tutoring problem available (at instructor's discretion) in *WileyPLUS* and WebAssign

SSM   Worked-out solution available in Student Solutions Manual

•–•••   Number of dots indicates level of problem difficulty

WWW   Worked-out solution is at

ILW   Interactive solution is at   http://www.wiley.com/college/halliday

Additional information available in *The Flying Circus of Physics* and at flyingcircusofphysics.com

sec. 13-2   Newton's Law of Gravitation

•1   ILW A mass M is split into two parts, m and $M - m$, which are then separated by a certain distance. What ratio m/M maximizes the magnitude of the gravitational force between the parts?

•2   *Moon effect.* Some people believe that the Moon controls their activities. If the Moon moves from being directly on the opposite side of Earth from you to being directly overhead, by what percent does (a) the Moon's gravitational pull on you increase and (b) your weight (as measured on a scale) decrease? Assume that the Earth–Moon (center-to-center) distance is 3.82×10^8 m and Earth's radius is 6.37×10^6 m.

•3   SSM What must the separation be between a 5.2 kg particle and a 2.4 kg particle for their gravitational attraction to have a magnitude of 2.3×10^{-12} N?

•4   The Sun and Earth each exert a gravitational force on the Moon. What is the ratio F_{Sun}/F_{Earth} of these two forces? (The average Sun–Moon distance is equal to the Sun–Earth distance.)

sec. 13-3   Gravitation and the Principle of Superposition

•5   *Miniature black holes.* Left over from the big-bang beginning of the universe, tiny black holes might still wander through the universe. If one with a mass of 1×10^{11} kg (and a radius of only 1×10^{-16} m) reached Earth, at what distance from your head would its gravitational pull on you match that of Earth's?

•6   GO In Fig. 13-31, a square of edge length 20.0 cm is formed by four spheres of masses $m_1 = 5.00$ g, $m_2 = 3.00$ g, $m_3 = 1.00$ g, and $m_4 = 5.00$ g. In unit-vector notation, what is the net gravitational force from them on a central sphere with mass $m_5 = 2.50$ g?

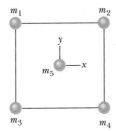

Fig. 13-31
Problem 6.

•7   *One dimension.* In Fig. 13-32, two point particles are fixed on an x axis separated by distance d. Particle A has mass m_A and par-

ticle B has mass $3.00m_A$. A third particle C, of mass $75.0m_A$, is to be placed on the x axis and near particles A and B. In terms of distance d, at what x coordinate should C be placed so that the net gravitational force on particle A from particles B and C is zero?

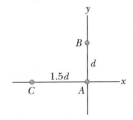

Fig. 13-32
Problem 7.

•8   In Fig. 13-33, three 5.00 kg spheres are located at distances $d_1 = 0.300$ m and $d_2 = 0.400$ m. What are the (a) magnitude and (b) direction (relative to the positive direction of the x axis) of the net gravitational force on sphere B due to spheres A and C?

Fig. 13-33   Problem 8.

•9   SSM   WWW We want to position a space probe along a line that extends directly toward the Sun in order to monitor solar flares. How far from Earth's center is the point on the line where the Sun's gravitational pull on the probe balances Earth's pull?

••10   GO *Two dimensions.* In Fig. 13-34, three point particles are fixed in place in an xy plane. Particle A has mass m_A, particle B has mass $2.00m_A$, and particle C has mass $3.00m_A$. A fourth particle D, with mass $4.00m_A$, is to be placed near the other three particles. In terms of distance d, at what (a) x coordinate and (b) y coordinate

Fig. 13-34   Problem 10.

should particle D be placed so that the net gravitational force on particle A from particles B, C, and D is zero?

••11 As seen in Fig. 13-35, two spheres of mass m and a third sphere of mass M form an equilateral triangle, and a fourth sphere of mass m_4 is at the center of the triangle. The net gravitational force on that central sphere from the three other spheres is zero. (a) What is M in terms of m? (b) If we double the value of m_4, what then is the magnitude of the net gravitational force on the central sphere?

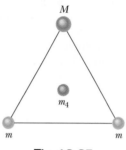

Fig. 13-35
Problem 11.

••12 In Fig. 13-36a, particle A is fixed in place at $x = -0.20$ m on the x axis and particle B, with a mass of 1.0 kg, is fixed in place at the origin. Particle C (not shown) can be moved along the x axis, between particle B and $x = \infty$. Figure 13-36b shows the x component $F_{net,x}$ of the net gravitational force on particle B due to particles A and C, as a function of position x of particle C. The plot actually extends to the right, approaching an asymptote of -4.17×10^{-10} N as $x \to \infty$. What are the masses of (a) particle A and (b) particle C?

 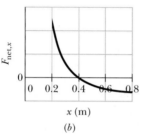

(a)   (b)

Fig. 13-36 Problem 12.

••13 Figure 13-37 shows a spherical hollow inside a lead sphere of radius $R = 4.00$ cm; the surface of the hollow passes through the center of the sphere and "touches" the right side of the sphere. The mass of the sphere before hollowing was $M = 2.95$ kg. With what gravitational force does the hollowed-out lead sphere attract a small sphere of mass $m = 0.431$ kg that lies at a distance $d = 9.00$ cm from the center of the lead sphere, on the straight line connecting the centers of the spheres and of the hollow?

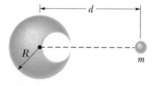

Fig. 13-37 Problem 13.

••14 〔GO〕 Three point particles are fixed in position in an xy plane. Two of them, particle A of mass 6.00 g and particle B of mass 12.0 g, are shown in Fig. 13-38, with a separation of $d_{AB} = 0.500$ m at angle $\theta = 30°$. Particle C, with mass 8.00 g, is not shown. The net gravitational force acting on particle A due to particles B and C is 2.77×10^{-14} N at an angle of $-163.8°$ from the positive direction of the x axis. What are (a) the x coordinate and (b) the y coordinate of particle C?

Fig. 13-38 Problem 14.

•••15 *Three dimensions.* Three point particles are fixed in place in an xyz coordinate system. Particle A, at the origin, has mass m_A. Particle B, at xyz coordinates $(2.00d, 1.00d, 2.00d)$, has mass $2.00m_A$, and particle C, at coordinates $(-1.00d, 2.00d, -3.00d)$, has mass $3.00m_A$. A fourth particle D, with mass $4.00m_A$, is to be placed near the other particles. In terms of distance d, at what (a) x, (b) y, and (c) z coordinate should D be placed so that the net gravitational force on A from B, C, and D is zero?

•••16 In Fig. 13-39, a particle of mass $m_1 = 0.67$ kg is a distance $d = 23$ cm from one end of a uniform rod with length $L = 3.0$ m and mass $M = 5.0$ kg. What is the magnitude of the gravitational force $\vec{F}$ on the particle from the rod?

Fig. 13-39 Problem 16.

sec. 13-4 Gravitation Near Earth's Surface

•17 (a) What will an object weigh on the Moon's surface if it weighs 100 N on Earth's surface? (b) How many Earth radii must this same object be from the center of Earth if it is to weigh the same as it does on the Moon?

•18 〔✈〕 *Mountain pull.* A large mountain can slightly affect the direction of "down" as determined by a plumb line. Assume that we can model a mountain as a sphere of radius $R = 2.00$ km and density (mass per unit volume) 2.6×10^3 kg/m³. Assume also that we hang a 0.50 m plumb line at a distance of $3R$ from the sphere's center and such that the sphere pulls horizontally on the lower end. How far would the lower end move toward the sphere?

•19 SSM At what altitude above Earth's surface would the gravitational acceleration be 4.9 m/s²?

•20 *Mile-high building.* In 1956, Frank Lloyd Wright proposed the construction of a mile-high building in Chicago. Suppose the building had been constructed. Ignoring Earth's rotation, find the change in your weight if you were to ride an elevator from the street level, where you weigh 600 N, to the top of the building.

••21 ILW Certain neutron stars (extremely dense stars) are believed to be rotating at about 1 rev/s. If such a star has a radius of 20 km, what must be its minimum mass so that material on its surface remains in place during the rapid rotation?

••22 The radius R_h and mass M_h of a black hole are related by $R_h = 2GM_h/c^2$, where c is the speed of light. Assume that the gravitational acceleration a_g of an object at a distance $r_o = 1.001R_h$ from the center of a black hole is given by Eq. 13-11 (it is, for large black holes). (a) In terms of M_h, find a_g at r_o. (b) Does a_g at r_o increase or decrease as M_h increases? (c) What is a_g at r_o for a very large black hole whose mass is 1.55×10^{12} times the solar mass of 1.99×10^{30} kg? (d) If an astronaut of height 1.70 m is at r_o with her feet down, what is the difference in gravitational acceleration between her head and feet? (e) Is the tendency to stretch the astronaut severe?

••23 One model for a certain planet has a core of radius R and mass M surrounded by an outer shell of inner radius R, outer radius $2R$, and mass $4M$. If $M = 4.1 \times 10^{24}$ kg and $R = 6.0 \times 10^6$ m, what is the gravitational acceleration of a particle at points (a) R and (b) $3R$ from the center of the planet?

sec. 13-5 Gravitation Inside Earth

•**24** Two concentric spherical shells with uniformly distributed masses M_1 and M_2 are situated as shown in Fig. 13-40. Find the magnitude of the net gravitational force on a particle of mass m, due to the shells, when the particle is located at radial distance (a) a, (b) b, and (c) c.

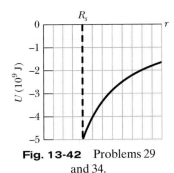

Fig. 13-40 Problem 24.

••**25** A solid uniform sphere has a mass of 1.0×10^4 kg and a radius of 1.0 m. What is the magnitude of the gravitational force due to the sphere on a particle of mass m located at a distance of (a) 1.5 m and (b) 0.50 m from the center of the sphere? (c) Write a general expression for the magnitude of the gravitational force on the particle at a distance $r \le 1.0$ m from the center of the sphere.

••**26** Consider a pulsar, a collapsed star of extremely high density, with a mass M equal to that of the Sun (1.98×10^{30} kg), a radius R of only 12 km, and a rotational period T of 0.041 s. By what percentage does the free-fall acceleration g differ from the gravitational acceleration a_g at the equator of this spherical star?

••**27** Figure 13-41 shows, not to scale, a cross section through the interior of Earth. Rather than being uniform throughout, Earth is divided into three zones: an outer *crust*, a *mantle*, and an inner *core*. The dimensions of these zones and the masses contained within them are shown on the figure. Earth has a total mass of 5.98×10^{24} kg and a radius of 6370 km. Ignore rotation and assume that Earth is spherical. (a) Calculate a_g at the surface. (b) Suppose that a bore hole (the *Mohole*) is driven to the crust–mantle interface at a depth of 25.0 km; what would be the value of a_g at the bottom of the hole? (c) Suppose that Earth were a uniform sphere with the same total mass and size. What would be the value of a_g at a depth of 25.0 km? (Precise measurements of a_g are sensitive probes of the interior structure of Earth, although results can be clouded by local variations in mass distribution.)

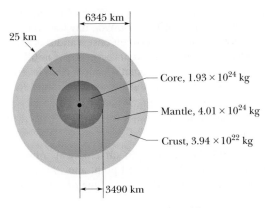

Fig. 13-41 Problem 27.

••**28** GO Assume a planet is a uniform sphere of radius R that (somehow) has a narrow radial tunnel through its center (Fig. 13-7). Also assume we can position an apple anywhere along the tunnel or outside the sphere. Let F_R be the magnitude of the gravitational force on the apple when it is located at the planet's surface. How far from the surface is there a point where the magnitude is $\frac{1}{2}F_R$ if we move the apple (a) away from the planet and (b) into the tunnel?

sec. 13-6 Gravitational Potential Energy

•**29** Figure 13-42 gives the potential energy function $U(r)$ of a projectile, plotted outward from the surface of a planet of radius R_s. What least kinetic energy is required of a projectile launched at the surface if the projectile is to "escape" the planet?

Fig. 13-42 Problems 29 and 34.

•**30** In Problem 1, what ratio m/M gives the least gravitational potential energy for the system?

•**31** SSM The mean diameters of Mars and Earth are 6.9×10^3 km and 1.3×10^4 km, respectively. The mass of Mars is 0.11 times Earth's mass. (a) What is the ratio of the mean density (mass per unit volume) of Mars to that of Earth? (b) What is the value of the gravitational acceleration on Mars? (c) What is the escape speed on Mars?

•**32** (a) What is the gravitational potential energy of the two-particle system in Problem 3? If you triple the separation between the particles, how much work is done (b) by the gravitational force between the particles and (c) by you?

•**33** What multiple of the energy needed to escape from Earth gives the energy needed to escape from (a) the Moon and (b) Jupiter?

•**34** Figure 13-42 gives the potential energy function $U(r)$ of a projectile, plotted outward from the surface of a planet of radius R_s. If the projectile is launched radially outward from the surface with a mechanical energy of -2.0×10^9 J, what are (a) its kinetic energy at radius $r = 1.25R_s$ and (b) its *turning point* (see Section 8-6) in terms of R_s?

••**35** GO Figure 13-43 shows four particles, each of mass 20.0 g, that form a square with an edge length of $d = 0.600$ m. If d is reduced to 0.200 m, what is the change in the gravitational potential energy of the four-particle system?

Fig. 13-43 Problem 35.

••**36** GO Zero, a hypothetical planet, has a mass of 5.0×10^{23} kg, a radius of 3.0×10^6 m, and no atmosphere. A 10 kg space probe is to be launched vertically from its surface. (a) If the probe is launched with an initial energy of 5.0×10^7 J, what will be its kinetic energy when it is 4.0×10^6 m from the center of Zero? (b) If the probe is to achieve a maximum distance of 8.0×10^6 m from the center of Zero, with what initial kinetic energy must it be launched from the surface of Zero?

••37 GO The three spheres in Fig. 13-44, with masses $m_A = 80$ g, $m_B = 10$ g, and $m_C = 20$ g, have their centers on a common line, with $L = 12$ cm and $d = 4.0$ cm. You move sphere B along the line until its center-to-center separation from C is $d = 4.0$ cm. How much work is done on sphere B (a) by you and (b) by the net gravitational force on B due to spheres A and C?

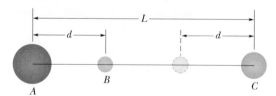

Fig. 13-44   Problem 37.

••38 In deep space, sphere A of mass 20 kg is located at the origin of an x axis and sphere B of mass 10 kg is located on the axis at $x = 0.80$ m. Sphere B is released from rest while sphere A is held at the origin. (a) What is the gravitational potential energy of the two-sphere system just as B is released? (b) What is the kinetic energy of B when it has moved 0.20 m toward A?

••39 SSM (a) What is the escape speed on a spherical asteroid whose radius is 500 km and whose gravitational acceleration at the surface is 3.0 m/s²? (b) How far from the surface will a particle go if it leaves the asteroid's surface with a radial speed of 1000 m/s? (c) With what speed will an object hit the asteroid if it is dropped from 1000 km above the surface?

••40 A projectile is shot directly away from Earth's surface. Neglect the rotation of Earth. What multiple of Earth's radius R_E gives the radial distance a projectile reaches if (a) its initial speed is 0.500 of the escape speed from Earth and (b) its initial kinetic energy is 0.500 of the kinetic energy required to escape Earth? (c) What is the least initial mechanical energy required at launch if the projectile is to escape Earth?

••41 SSM Two neutron stars are separated by a distance of 1.0×10^{10} m. They each have a mass of 1.0×10^{30} kg and a radius of 1.0×10^5 m. They are initially at rest with respect to each other. As measured from that rest frame, how fast are they moving when (a) their separation has decreased to one-half its initial value and (b) they are about to collide?

••42 GO Figure 13-45a shows a particle A that can be moved along a y axis from an infinite distance to the origin. That origin lies at the midpoint between particles B and C, which have identical

masses, and the y axis is a perpendicular bisector between them. Distance D is 0.3057 m. Figure 13-45b shows the potential energy U of the three-particle system as a function of the position of particle A along the y axis. The curve actually extends rightward and approaches an asymptote of -2.7×10^{-11} J as $y \to \infty$. What are the masses of (a) particles B and C and (b) particle A?

sec. 13-7 Planets and Satellites: Kepler's Laws

•43 (a) What linear speed must an Earth satellite have to be in a circular orbit at an altitude of 160 km above Earth's surface? (b) What is the period of revolution?

•44 A satellite is put in a circular orbit about Earth with a radius equal to one-half the radius of the Moon's orbit. What is its period of revolution in lunar months? (A lunar month is the period of revolution of the Moon.)

•45 The Martian satellite Phobos travels in an approximately circular orbit of radius 9.4×10^6 m with a period of 7 h 39 min. Calculate the mass of Mars from this information.

•46 The first known collision between space debris and a functioning satellite occurred in 1996: At an altitude of 700 km, a year-old French spy satellite was hit by a piece of an Ariane rocket. A stabilizing boom on the satellite was demolished, and the satellite was sent spinning out of control. Just before the collision and in kilometers per hour, what was the speed of the rocket piece relative to the satellite if both were in circular orbits and the collision was (a) head-on and (b) along perpendicular paths?

•47 SSM WWW The Sun, which is 2.2×10^{20} m from the center of the Milky Way galaxy, revolves around that center once every 2.5×10^8 years. Assuming each star in the Galaxy has a mass equal to the Sun's mass of 2.0×10^{30} kg, the stars are distributed uniformly in a sphere about the galactic center, and the Sun is at the edge of that sphere, estimate the number of stars in the Galaxy.

•48 The mean distance of Mars from the Sun is 1.52 times that of Earth from the Sun. From Kepler's law of periods, calculate the number of years required for Mars to make one revolution around the Sun; compare your answer with the value given in Appendix C.

•49 A comet that was seen in April 574 by Chinese astronomers on a day known by them as the Woo Woo day was spotted again in May 1994. Assume the time between observations is the period of the Woo Woo day comet and take its eccentricity as 0.11. What are (a) the semimajor axis of the comet's orbit and (b) its greatest distance from the Sun in terms of the mean orbital radius R_P of Pluto?

•50 An orbiting satellite stays over a certain spot on the equator of (rotating) Earth. What is the altitude of the orbit (called a *geosynchronous orbit*)?

•51 SSM A satellite, moving in an elliptical orbit, is 360 km above Earth's surface at its farthest point and 180 km above at its closest point. Calculate (a) the semimajor axis and (b) the eccentricity of the orbit.

•52 The Sun's center is at one focus of Earth's orbit. How far from this focus is the other focus, (a) in meters and (b) in terms of the solar radius, 6.96×10^8 m? The eccentricity is 0.0167, and the semimajor axis is 1.50×10^{11} m.

••53 A 20 kg satellite has a circular orbit with a period of 2.4 h and a radius of 8.0×10^6 m around a planet of unknown mass. If the magnitude of the gravitational acceleration on the surface of the planet is 8.0 m/s², what is the radius of the planet?

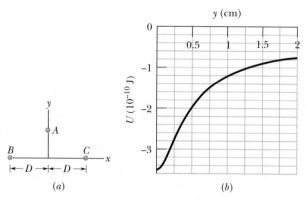

Fig. 13-45   Problem 42.

••**54** *Hunting a black hole.* Observations of the light from a certain star indicate that it is part of a binary (two-star) system. This visible star has orbital speed $v = 270$ km/s, orbital period $T = 1.70$ days, and approximate mass $m_1 = 6M_s$, where M_s is the Sun's mass, 1.99×10^{30} kg. Assume that the visible star and its companion star, which is dark and unseen, are both in circular orbits (Fig. 13-46). What multiple of M_s gives the approximate mass m_2 of the dark star?

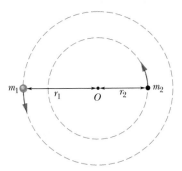

Fig. 13-46   Problem 54.

••**55**   In 1610, Galileo used his telescope to discover four prominent moons around Jupiter. Their mean orbital radii a and periods T are as follows:

Name	$a\ (10^8\ \text{m})$	$T\ (\text{days})$
Io	4.22	1.77
Europa	6.71	3.55
Ganymede	10.7	7.16
Callisto	18.8	16.7

(a) Plot log a (y axis) against log T (x axis) and show that you get a straight line. (b) Measure the slope of the line and compare it with the value that you expect from Kepler's third law. (c) Find the mass of Jupiter from the intercept of this line with the y axis.

••**56**   In 1993 the spacecraft *Galileo* sent home an image (Fig. 13-47) of asteroid 243 Ida and a tiny orbiting moon (now known as Dactyl), the first confirmed example of an asteroid–moon system. In the image, the moon, which is 1.5 km wide, is 100 km from the center of the asteroid, which is 55 km long. The shape of the moon's orbit is not well known; assume it is circular with a period

Fig. 13-47   Problem 56. A tiny moon (at right) orbits asteroid 243 Ida. *(Courtesy NASA)*

of 27 h. (a) What is the mass of the asteroid? (b) The volume of the asteroid, measured from the *Galileo* images, is 14 100 km³. What is the density (mass per unit volume) of the asteroid?

••**57**   ILW   In a certain binary-star system, each star has the same mass as our Sun, and they revolve about their center of mass. The distance between them is the same as the distance between Earth and the Sun. What is their period of revolution in years?

•••**58**   The presence of an unseen planet orbiting a distant star can sometimes be inferred from the motion of the star as we see it. As the star and planet orbit the center of mass of the star–planet system, the star moves toward and away from us with what is called the *line of sight velocity,* a motion that can be detected. Figure 13-48 shows a graph of the line of sight velocity versus time for the star 14 Herculis. The star's mass is believed to be 0.90 of the mass of our Sun. Assume that only one planet orbits the star and that our view is along the plane of the orbit. Then approximate (a) the planet's mass in terms of Jupiter's mass m_J and (b) the planet's orbital radius in terms of Earth's orbital radius r_E.

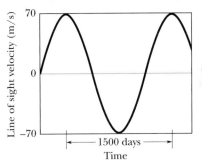

Fig. 13-48   Problem 58.

•••**59**   Three identical stars of mass M form an equilateral triangle that rotates around the triangle's center as the stars move in a common circle about that center. The triangle has edge length L. What is the speed of the stars?

sec. 13-8   Satellites: Orbits and Energy

•**60**   In Fig. 13-49, two satellites, A and B, both of mass $m = 125$ kg, move in the same circular orbit of radius $r = 7.87 \times 10^6$ m around Earth but in opposite senses of rotation and therefore on a collision course. (a) Find the total mechanical energy $E_A + E_B$ of the *two satellites + Earth* system before the collision. (b) If the collision is completely inelastic so that the wreckage remains as one piece of tangled material (mass = $2m$), find the total mechanical energy immediately after the collision. (c) Just after the collision, is the wreckage falling directly toward Earth's center or orbiting around Earth?

Fig. 13-49
Problem 60.

•**61**   (a) At what height above Earth's surface is the energy required to lift a satellite to that height equal to the kinetic energy required for the satellite to be in orbit at that height? (b) For greater heights, which is greater, the energy for lifting or the kinetic energy for orbiting?

•**62**   Two Earth satellites, A and B, each of mass m, are to be launched into circular orbits about Earth's center. Satellite A is to orbit at an altitude of 6370 km. Satellite *B* is to orbit at an altitude of 19 110 km. The radius of Earth R_E is 6370 km. (a) What is the ratio of the potential energy of satellite *B* to that of satellite *A*, in orbit? (b) What is the ratio of the kinetic energy of satellite *B* to that of satellite *A*, in orbit? (c) Which satellite has the greater total energy if each has a mass of 14.6 kg? (d) By how much?

•63 **SSM** **WWW** An asteroid, whose mass is 2.0×10^{-4} times the mass of Earth, revolves in a circular orbit around the Sun at a distance that is twice Earth's distance from the Sun. (a) Calculate the period of revolution of the asteroid in years. (b) What is the ratio of the kinetic energy of the asteroid to the kinetic energy of Earth?

•64 A satellite orbits a planet of unknown mass in a circle of radius 2.0×10^7 m. The magnitude of the gravitational force on the satellite from the planet is $F = 80$ N. (a) What is the kinetic energy of the satellite in this orbit? (b) What would *F* be if the orbit radius were increased to 3.0×10^7 m?

••65 A satellite is in a circular Earth orbit of radius *r*. The area *A* enclosed by the orbit depends on r^2 because $A = \pi r^2$. Determine how the following properties of the satellite depend on *r*: (a) period, (b) kinetic energy, (c) angular momentum, and (d) speed.

••66 One way to attack a satellite in Earth orbit is to launch a swarm of pellets in the same orbit as the satellite but in the opposite direction. Suppose a satellite in a circular orbit 500 km above Earth's surface collides with a pellet having mass 4.0 g. (a) What is the kinetic energy of the pellet in the reference frame of the satellite just before the collision? (b) What is the ratio of this kinetic energy to the kinetic energy of a 4.0 g bullet from a modern army rifle with a muzzle speed of 950 m/s?

•••67 What are (a) the speed and (b) the period of a 220 kg satellite in an approximately circular orbit 640 km above the surface of Earth? Suppose the satellite loses mechanical energy at the average rate of 1.4×10^5 J per orbital revolution. Adopting the reasonable approximation that the satellite's orbit becomes a "circle of slowly diminishing radius," determine the satellite's (c) altitude, (d) speed, and (e) period at the end of its 1500th revolution. (f) What is the magnitude of the average retarding force on the satellite? Is angular momentum around Earth's center conserved for (g) the satellite and (h) the satellite–Earth system (assuming that system is isolated)?

•••68 **GO** Two small spaceships, each with mass *m* = 2000 kg, are in the circular Earth orbit of Fig. 13-50, at an altitude *h* of 400 km. Igor, the commander of one of the ships, arrives at any fixed point in the orbit 90 s ahead of Picard, the commander of the other ship. What are the (a) period T_0 and (b) speed v_0 of the ships? At point *P* in Fig. 13-50, Picard fires an instantaneous burst in the forward direction, *reducing* his ship's speed by 1.00%. After this burst, he follows the elliptical orbit shown dashed in the figure. What are the (c) kinetic energy and (d) potential energy of his ship immediately after the burst? In Picard's new elliptical orbit, what are the (e) total energy *E*, (f) the semimajor axis *a*, and (g) the orbital period *T*? (h) How much earlier than Igor will Picard return to *P*?

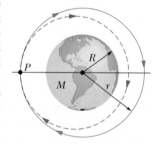

Fig. 13-50 Problem 68.

•69 In Fig. 13-17*b*, the scale on which the 60 kg physicist stands reads 220 N. How long will the cantaloupe take to reach the floor if the physicist drops it (from rest relative to himself) at a height of 2.1 m above the floor?

Additional Problems

70 The radius R_h of a black hole is the radius of a mathematical sphere, called the event horizon, that is centered on the black hole. Information from events inside the event horizon cannot reach the outside world. According to Einstein's general theory of relativity, $R_h = 2GM/c^2$, where *M* is the mass of the black hole and *c* is the speed of light.

Suppose that you wish to study a black hole near it, at a radial distance of $50R_h$. However, you do not want the difference in gravitational acceleration between your feet and your head to exceed 10 m/s² when you are feet down (or head down) toward the black hole. (a) As a multiple of our Sun's mass M_S, approximately what is the limit to the mass of the black hole you can tolerate at the given radial distance? (You need to estimate your height.) (b) Is the limit an upper limit (you can tolerate smaller masses) or a lower limit (you can tolerate larger masses)?

71 Several planets (Jupiter, Saturn, Uranus) are encircled by rings, perhaps composed of material that failed to form a satellite. In addition, many galaxies contain ring-like structures. Consider a homogeneous thin ring of mass *M* and outer radius *R* (Fig. 13-51). (a) What gravitational attraction does it exert on a particle of mass *m* located on the ring's central axis a distance *x* from the ring center? (b) Suppose the particle falls from rest as a result of the attraction of the ring of matter. What is the speed with which it passes through the center of the ring?

Fig. 13-51 Problem 71.

72 A typical neutron star may have a mass equal to that of the Sun but a radius of only 10 km. (a) What is the gravitational acceleration at the surface of such a star? (b) How fast would an object be moving if it fell from rest through a distance of 1.0 m on such a star? (Assume the star does not rotate.)

73 Figure 13-52 is a graph of the kinetic energy *K* of an asteroid versus its distance *r* from Earth's center, as the asteroid falls directly in toward that center. (a) What is the (approximate) mass of the asteroid? (b) What is its speed at $r = 1.945 \times 10^7$ m?

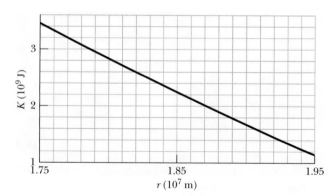

Fig. 13-52 Problem 73.

74 ✈ The mysterious visitor that appears in the enchanting story *The Little Prince* was said to come from a planet that "was scarcely any larger than a house!" Assume that the mass per unit volume of the planet is about that of Earth and that the planet does not appreciably spin. Approximate (a) the free-fall acceleration on the planet's surface and (b) the escape speed from the planet.

75 ILW The masses and coordinates of three spheres are as follows: 20 kg, $x = 0.50$ m, $y = 1.0$ m; 40 kg, $x = -1.0$ m, $y = -1.0$ m; 60 kg, $x = 0$ m, $y = -0.50$ m. What is the magnitude of the gravitational force on a 20 kg sphere located at the origin due to these three spheres?

76 SSM A very early, simple satellite consisted of an inflated spherical aluminum balloon 30 m in diameter and of mass 20 kg. Suppose a meteor having a mass of 7.0 kg passes within 3.0 m of the surface of the satellite. What is the magnitude of the gravitational force on the meteor from the satellite at the closest approach?

77 GO Four uniform spheres, with masses $m_A = 40$ kg, $m_B = 35$ kg, $m_C = 200$ kg, and $m_D = 50$ kg, have (x, y) coordinates of $(0, 50$ cm), $(0, 0)$, $(-80$ cm, $0)$, and $(40$ cm, $0)$, respectively. In unit-vector notation, what is the net gravitational force on sphere B due to the other spheres?

78 (a) In Problem 77, remove sphere A and calculate the gravitational potential energy of the remaining three-particle system. (b) If A is then put back in place, is the potential energy of the four-particle system more or less than that of the system in (a)? (c) In (a), is the work done by you to remove A positive or negative? (d) In (b), is the work done by you to replace A positive or negative?

79 SSM A certain triple-star system consists of two stars, each of mass m, revolving in the same circular orbit of radius r around a central star of mass M (Fig. 13-53). The two orbiting stars are always at opposite ends of a diameter of the orbit. Derive an expression for the period of revolution of the stars.

Fig. 13-53
Problem 79.

80 The fastest possible rate of rotation of a planet is that for which the gravitational force on material at the equator just barely provides the centripetal force needed for the rotation. (Why?) (a) Show that the corresponding shortest period of rotation is

$$T = \sqrt{\frac{3\pi}{G\rho}},$$

where ρ is the uniform density (mass per unit volume) of the spherical planet. (b) Calculate the rotation period assuming a density of 3.0 g/cm³, typical of many planets, satellites, and asteroids. No astronomical object has ever been found to be spinning with a period shorter than that determined by this analysis.

81 SSM In a double-star system, two stars of mass 3.0×10^{30} kg each rotate about the system's center of mass at radius 1.0×10^{11} m. (a) What is their common angular speed? (b) If a meteoroid passes through the system's center of mass perpendicular to their orbital plane, what minimum speed must it have at the center of mass if it is to escape to "infinity" from the two-star system?

82 A satellite is in elliptical orbit with a period of 8.00×10^4 s about a planet of mass 7.00×10^{24} kg. At aphelion, at radius 4.5×10^7 m, the satellite's angular speed is 7.158×10^{-5} rad/s. What is its angular speed at perihelion?

83 SSM In a shuttle craft of mass $m = 3000$ kg, Captain Janeway orbits a planet of mass $M = 9.50 \times 10^{25}$ kg, in a circular orbit of radius $r = 4.20 \times 10^7$ m. What are (a) the period of the orbit and (b) the speed of the shuttle craft? Janeway briefly fires a forward-pointing thruster, reducing her speed by 2.00%. Just then, what are (c) the speed, (d) the kinetic energy, (e) the gravitational potential energy, and (f) the mechanical energy of the shuttle craft? (g) What is the semimajor axis of the elliptical orbit now taken by the craft? (h) What is the difference between the period of the original circular orbit and that of the new elliptical orbit? (i) Which orbit has the smaller period?

84 A uniform solid sphere of radius R produces a gravitational acceleration of a_g on its surface. At what distance from the sphere's center are there points (a) inside and (b) outside the sphere where the gravitational acceleration is $a_g/3$?

85 ILW A projectile is fired vertically from Earth's surface with an initial speed of 10 km/s. Neglecting air drag, how far above the surface of Earth will it go?

86 An object lying on Earth's equator is accelerated (a) toward the center of Earth because Earth rotates, (b) toward the Sun because Earth revolves around the Sun in an almost circular orbit, and (c) toward the center of our galaxy because the Sun moves around the galactic center. For the latter, the period is 2.5×10^8 y and the radius is 2.2×10^{20} m. Calculate these three accelerations as multiples of $g = 9.8$ m/s².

87 (a) If the legendary apple of Newton could be released from rest at a height of 2 m from the surface of a neutron star with a mass 1.5 times that of our Sun and a radius of 20 km, what would be the apple's speed when it reached the surface of the star? (b) If the apple could rest on the surface of the star, what would be the approximate difference between the gravitational acceleration at the top and at the bottom of the apple? (Choose a reasonable size for an apple; the answer indicates that an apple would never survive near a neutron star.)

88 With what speed would mail pass through the center of Earth if falling in a tunnel through the center?

89 SSM The orbit of Earth around the Sun is *almost* circular: The closest and farthest distances are 1.47×10^8 km and 1.52×10^8 km respectively. Determine the corresponding variations in (a) total energy, (b) gravitational potential energy, (c) kinetic energy, and (d) orbital speed. (*Hint:* Use conservation of energy and conservation of angular momentum.)

90 A 50 kg satellite circles planet Cruton every 6.0 h. The magnitude of the gravitational force exerted on the satellite by Cruton is 80 N. (a) What is the radius of the orbit? (b) What is the kinetic energy of the satellite? (c) What is the mass of planet Cruton?

91 We watch two identical astronomical bodies A and B, each of mass m, fall toward each other from rest because of the gravitational force on each from the other. Their initial center-to-center separation is R_i. Assume that we are in an inertial reference frame that is stationary with respect to the center of mass of this two-body system. Use the principle of conservation of mechanical energy ($K_f + U_f = K_i + U_i$) to find the following when the center-to-center separation is $0.5R_i$: (a) the total kinetic energy of the system, (b) the kinetic energy of each body, (c) the speed of each body relative to us, and (d) the speed of body B relative to body A.

Next assume that we are in a reference frame attached to body A (we ride on the body). Now we see body B fall from rest toward us. From this reference frame, again use $K_f + U_f = K_i + U_i$ to find the following when the center-to-center separation is $0.5R_i$: (e) the kinetic energy of body B and (f) the speed of body B relative to body A. (g) Why are the answers to (d) and (f) different? Which answer is correct?

92   A 150.0 kg rocket moving radially outward from Earth has a speed of 3.70 km/s when its engine shuts off 200 km above Earth's surface. (a) Assuming negligible air drag, find the rocket's kinetic energy when the rocket is 1000 km above Earth's surface. (b) What maximum height above the surface is reached by the rocket?

93   Planet Roton, with a mass of 7.0×10^{24} kg and a radius of 1600 km, gravitationally attracts a meteorite that is initially at rest relative to the planet, at a distance great enough to take as infinite. The meteorite falls toward the planet. Assuming the planet is airless, find the speed of the meteorite when it reaches the planet's surface.

94   Two 20 kg spheres are fixed in place on a y axis, one at $y = 0.40$ m and the other at $y = -0.40$ m. A 10 kg ball is then released from rest at a point on the x axis that is at a great distance (effectively infinite) from the spheres. If the only forces acting on the ball are the gravitational forces from the spheres, then when the ball reaches the (x, y) point (0.30 m, 0), what are (a) its kinetic energy and (b) the net force on it from the spheres, in unit-vector notation?

95   Sphere A with mass 80 kg is located at the origin of an xy coordinate system; sphere B with mass 60 kg is located at coordinates (0.25 m, 0); sphere C with mass 0.20 kg is located in the first quadrant 0.20 m from A and 0.15 m from B. In unit-vector notation, what is the gravitational force on C due to A and B?

96   In his 1865 science fiction novel *From the Earth to the Moon,* Jules Verne described how three astronauts are shot to the Moon by means of a huge gun. According to Verne, the aluminum capsule containing the astronauts is accelerated by ignition of nitrocellulose to a speed of 11 km/s along the gun barrel's length of 220 m. (a) In g units, what is the average acceleration of the capsule and astronauts in the gun barrel? (b) Is that acceleration tolerable or deadly to the astronauts?

A modern version of such gun-launched spacecraft (although without passengers) has been proposed. In this modern version, called the SHARP (Super High Altitude Research Project) gun, ignition of methane and air shoves a piston along the gun's tube, compressing hydrogen gas that then launches a rocket. During this launch, the rocket moves 3.5 km and reaches a speed of 7.0 km/s. Once launched, the rocket can be fired to gain additional speed. (c) In g units, what would be the average acceleration of the rocket within the launcher? (d) How much additional speed is needed (via the rocket engine) if the rocket is to orbit Earth at an altitude of 700 km?

97   An object of mass m is initially held in place at radial distance $r = 3R_E$ from the center of Earth, where R_E is the radius of Earth. Let M_E be the mass of Earth. A force is applied to the object to move it to a radial distance $r = 4R_E$, where it again is held in place. Calculate the work done by the applied force during the move by integrating the force magnitude.

FLUIDS

14-1 WHAT IS PHYSICS?

The physics of fluids is the basis of hydraulic engineering, a branch of engineering that is applied in a great many fields. A nuclear engineer might study the fluid flow in the hydraulic system of an aging nuclear reactor, while a medical engineer might study the blood flow in the arteries of an aging patient. An environmental engineer might be concerned about the drainage from waste sites or the efficient irrigation of farmlands. A naval engineer might be concerned with the dangers faced by a deep-sea diver or with the possibility of a crew escaping from a downed submarine. An aeronautical engineer might design the hydraulic systems controlling the wing flaps that allow a jet airplane to land. Hydraulic engineering is also applied in many Broadway and Las Vegas shows, where huge sets are quickly put up and brought down by hydraulic systems.

Before we can study any such application of the physics of fluids, we must first answer the question "What is a fluid?"

14-2 What Is a Fluid?

A **fluid,** in contrast to a solid, is a substance that can flow. Fluids conform to the boundaries of any container in which we put them. They do so because a fluid cannot sustain a force that is tangential to its surface. (In the more formal language of Section 12-7, a fluid is a substance that flows because it cannot withstand a shearing stress. It can, however, exert a force in the direction perpendicular to its surface.) Some materials, such as pitch, take a long time to conform to the boundaries of a container, but they do so eventually; thus, we classify even those materials as fluids.

You may wonder why we lump liquids and gases together and call them fluids. After all (you may say), liquid water is as different from steam as it is from ice. Actually, it is not. Ice, like other crystalline solids, has its constituent atoms organized in a fairly rigid three-dimensional array called a crystalline lattice. In neither steam nor liquid water, however, is there any such orderly long-range arrangement.

14-3 Density and Pressure

When we discuss rigid bodies, we are concerned with particular lumps of matter, such as wooden blocks, baseballs, or metal rods. Physical quantities that we find useful, and in whose terms we express Newton's laws, are mass and force. We might speak, for example, of a 3.6 kg block acted on by a 25 N force.

With fluids, we are more interested in the extended substance and in properties that can vary from point to point in that substance. It is more useful to speak of **density** and **pressure** than of mass and force.

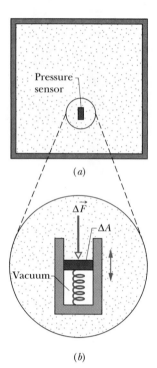

Fig. 14-1 (a) A fluid-filled vessel containing a small pressure sensor, shown in (b). The pressure is measured by the relative position of the movable piston in the sensor.

Density

To find the density ρ of a fluid at any point, we isolate a small volume element ΔV around that point and measure the mass Δm of the fluid contained within that element. The **density** is then

$$\rho = \frac{\Delta m}{\Delta V}. \qquad (14\text{-}1)$$

In theory, the density at any point in a fluid is the limit of this ratio as the volume element ΔV at that point is made smaller and smaller. In practice, we assume that a fluid sample is large relative to atomic dimensions and thus is "smooth" (with uniform density), rather than "lumpy" with atoms. This assumption allows us to write Eq. 14-1 as

$$\rho = \frac{m}{V} \qquad \text{(uniform density)}, \qquad (14\text{-}2)$$

where m and V are the mass and volume of the sample.

Density is a scalar property; its SI unit is the kilogram per cubic meter. Table 14-1 shows the densities of some substances and the average densities of some objects. Note that the density of a gas (see Air in the table) varies considerably with pressure, but the density of a liquid (see Water) does not; that is, gases are readily *compressible* but liquids are not.

Pressure

Let a small pressure-sensing device be suspended inside a fluid-filled vessel, as in Fig. 14-1a. The sensor (Fig. 14-1b) consists of a piston of surface area ΔA riding in a close-fitting cylinder and resting against a spring. A readout arrangement allows us to record the amount by which the (calibrated) spring is compressed by the surrounding fluid, thus indicating the magnitude ΔF of the force that acts normal to the piston. We define the **pressure** on the piston from the fluid as

$$p = \frac{\Delta F}{\Delta A}. \qquad (14\text{-}3)$$

In theory, the pressure at any point in the fluid is the limit of this ratio as the surface area ΔA of the piston, centered on that point, is made smaller and smaller. However, if the force is uniform over a flat area A, we can write Eq. 14-3 as

$$p = \frac{F}{A} \qquad \text{(pressure of uniform force on flat area)}, \qquad (14\text{-}4)$$

where F is the magnitude of the normal force on area A. (When we say a force is

Table 14-1			
Some Densities			
Material or Object	Density (kg/m³)	Material or Object	Density (kg/m³)
Interstellar space	10^{-20}	Iron	7.9×10^3
Best laboratory vacuum	10^{-17}	Mercury (the metal, not the planet)	13.6×10^3
Air: 20°C and 1 atm pressure	1.21	Earth: average	5.5×10^3
20°C and 50 atm	60.5	core	9.5×10^3
Styrofoam	1×10^2	crust	2.8×10^3
Ice	0.917×10^3	Sun: average	1.4×10^3
Water: 20°C and 1 atm	0.998×10^3	core	1.6×10^5
20°C and 50 atm	1.000×10^3	White dwarf star (core)	10^{10}
Seawater: 20°C and 1 atm	1.024×10^3	Uranium nucleus	3×10^{17}
Whole blood	1.060×10^3	Neutron star (core)	10^{18}

Table 14-2

Some Pressures

	Pressure (Pa)		Pressure (Pa)
Center of the Sun	2×10^{16}	Automobile tire[a]	2×10^5
Center of Earth	4×10^{11}	Atmosphere at sea level	1.0×10^5
Highest sustained laboratory pressure	1.5×10^{10}	Normal blood systolic pressure[a,b]	1.6×10^4
Deepest ocean trench (bottom)	1.1×10^8	Best laboratory vacuum	10^{-12}
Spike heels on a dance floor	10^6		

[a]Pressure in excess of atmospheric pressure.   [b]Equivalent to 120 torr on the physician's pressure gauge.

uniform over an area, we mean that the force is evenly distributed over every point of the area.)

We find by experiment that at a given point in a fluid at rest, the pressure p defined by Eq. 14-4 has the same value no matter how the pressure sensor is oriented. Pressure is a scalar, having no directional properties. It is true that the force acting on the piston of our pressure sensor is a vector quantity, but Eq. 14-4 involves only the *magnitude* of that force, a scalar quantity.

The SI unit of pressure is the newton per square meter, which is given a special name, the **pascal** (Pa). In metric countries, tire pressure gauges are calibrated in kilopascals. The pascal is related to some other common (non-SI) pressure units as follows:

$$1 \text{ atm} = 1.01 \times 10^5 \text{ Pa} = 760 \text{ torr} = 14.7 \text{ lb/in.}^2.$$

The *atmosphere* (atm) is, as the name suggests, the approximate average pressure of the atmosphere at sea level. The *torr* (named for Evangelista Torricelli, who invented the mercury barometer in 1674) was formerly called the *millimeter of mercury* (mm Hg). The pound per square inch is often abbreviated psi. Table 14-2 shows some pressures.

Sample Problem

Atmospheric pressure and force

A living room has floor dimensions of 3.5 m and 4.2 m and a height of 2.4 m.

(a) What does the air in the room weigh when the air pressure is 1.0 atm?

KEY IDEAS

(1) The air's weight is equal to mg, where m is its mass. (2) Mass m is related to the air density ρ and the air volume V by Eq. 14-2 ($\rho = m/V$).

Calculation: Putting the two ideas together and taking the density of air at 1.0 atm from Table 14-1, we find

$$mg = (\rho V)g$$
$$= (1.21 \text{ kg/m}^3)(3.5 \text{ m} \times 4.2 \text{ m} \times 2.4 \text{ m})(9.8 \text{ m/s}^2)$$
$$= 418 \text{ N} \approx 420 \text{ N}. \qquad \text{(Answer)}$$

This is the weight of about 110 cans of Pepsi.

(b) What is the magnitude of the atmosphere's downward force on the top of your head, which we take to have an area of 0.040 m²?

KEY IDEA

When the fluid pressure p on a surface of area A is uniform, the fluid force on the surface can be obtained from Eq. 14-4 ($p = F/A$).

Calculation: Although air pressure varies daily, we can approximate that $p = 1.0$ atm. Then Eq. 14-4 gives

$$F = pA = (1.0 \text{ atm})\left(\frac{1.01 \times 10^5 \text{ N/m}^2}{1.0 \text{ atm}}\right)(0.040 \text{ m}^2)$$
$$= 4.0 \times 10^3 \text{ N}. \qquad \text{(Answer)}$$

This large force is equal to the weight of the air column from the top of your head to the top of the atmosphere.

 Additional examples, video, and practice available at *WileyPLUS*

14-4 **Fluids at Rest**

Figure 14-2*a* shows a tank of water—or other liquid—open to the atmosphere. As every diver knows, the pressure *increases* with depth below the air–water interface. The diver's depth gauge, in fact, is a pressure sensor much like that of Fig. 14-1*b*. As every mountaineer knows, the pressure *decreases* with altitude as one ascends into the atmosphere. The pressures encountered by the diver and the mountaineer are usually called *hydrostatic pressures,* because they are due to fluids that are static (at rest). Here we want to find an expression for hydrostatic pressure as a function of depth or altitude.

Let us look first at the increase in pressure with depth below the water's surface. We set up a vertical *y* axis in the tank, with its origin at the air–water interface and the positive direction upward. We next consider a water sample contained in an imaginary right circular cylinder of horizontal base (or face) area *A*, such that y_1 and y_2 (both of which are *negative* numbers) are the depths below the surface of the upper and lower cylinder faces, respectively.

Figure 14-2*e* shows a free-body diagram for the water in the cylinder. The water is in *static equilibrium;* that is, it is stationary and the forces on it balance. Three forces act on it vertically: Force $\vec{F}_1$ acts at the top surface of the cylinder and is due to the water above the cylinder (Fig. 14-2*b*). Similarly, force $\vec{F}_2$ acts at the bottom surface of the cylinder and is due to the water just below the cylinder (Fig. 14-2*c*). The gravitational force on the water in the cylinder is represented by $m\vec{g}$, where *m* is the mass

Three forces act on this sample of water.

Fig. 14-2 (*a*) A tank of water in which a sample of water is contained in an imaginary cylinder of horizontal base area *A*. (*b*)–(*d*) Force $\vec{F}_1$ acts at the top surface of the cylinder; force $\vec{F}_2$ acts at the bottom surface of the cylinder; the gravitational force on the water in the cylinder is represented by $m\vec{g}$. (*e*) A free-body diagram of the water sample.

This downward force is due to the water pressure pushing on the *top* surface.

This upward force is due to the water pressure pushing on the *bottom* surface.

Gravity pulls downward on the sample.

The three forces balance.

of the water in the cylinder (Fig. 14-2d). The balance of these forces is written as

$$F_2 = F_1 + mg. \qquad (14\text{-}5)$$

We want to transform Eq. 14-5 into an equation involving pressures. From Eq. 14-4, we know that

$$F_1 = p_1 A \quad \text{and} \quad F_2 = p_2 A. \qquad (14\text{-}6)$$

The mass m of the water in the cylinder is, from Eq. 14-2, $m = \rho V$, where the cylinder's volume V is the product of its face area A and its height $y_1 - y_2$. Thus, m is equal to $\rho A(y_1 - y_2)$. Substituting this and Eq. 14-6 into Eq. 14-5, we find

$$p_2 A = p_1 A + \rho A g(y_1 - y_2)$$

or
$$p_2 = p_1 + \rho g(y_1 - y_2). \qquad (14\text{-}7)$$

This equation can be used to find pressure both in a liquid (as a function of depth) and in the atmosphere (as a function of altitude or height). For the former, suppose we seek the pressure p at a depth h below the liquid surface. Then we choose level 1 to be the surface, level 2 to be a distance h below it (as in Fig. 14-3), and p_0 to represent the atmospheric pressure on the surface. We then substitute

$$y_1 = 0, \quad p_1 = p_0 \quad \text{and} \quad y_2 = -h, \quad p_2 = p$$

into Eq. 14-7, which becomes

$$p = p_0 + \rho g h \qquad \text{(pressure at depth } h\text{)}. \qquad (14\text{-}8)$$

Note that the pressure at a given depth in the liquid depends on that depth but not on any horizontal dimension.

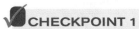 The pressure at a point in a fluid in static equilibrium depends on the depth of that point but not on any horizontal dimension of the fluid or its container.

Thus, Eq. 14-8 holds no matter what the shape of the container. If the bottom surface of the container is at depth h, then Eq. 14-8 gives the pressure p there.

In Eq. 14-8, p is said to be the total pressure, or **absolute pressure,** at level 2. To see why, note in Fig. 14-3 that the pressure p at level 2 consists of two contributions: (1) p_0, the pressure due to the atmosphere, which bears down on the liquid, and (2) $\rho g h$, the pressure due to the liquid above level 2, which bears down on level 2. In general, the difference between an absolute pressure and an atmospheric pressure is called the **gauge pressure.** (The name comes from the use of a gauge to measure this difference in pressures.) For the situation of Fig. 14-3, the gauge pressure is $\rho g h$.

Equation 14-7 also holds above the liquid surface: It gives the atmospheric pressure at a given distance above level 1 in terms of the atmospheric pressure p_1 at level 1 (*assuming* that the atmospheric density is uniform over that distance). For example, to find the atmospheric pressure at a distance d above level 1 in Fig. 14-3, we substitute

$$y_1 = 0, \quad p_1 = p_0 \quad \text{and} \quad y_2 = d, \quad p_2 = p.$$

Then with $\rho = \rho_{\text{air}}$, we obtain

$$p = p_0 - \rho_{\text{air}} g d.$$

Fig. 14-3 The pressure p increases with depth h below the liquid surface according to Eq. 14-8.

✓CHECKPOINT 1

The figure shows four containers of olive oil. Rank them according to the pressure at depth h, greatest first.

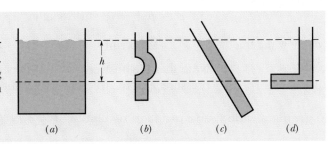

Sample Problem

Gauge pressure on a scuba diver

A novice scuba diver practicing in a swimming pool takes enough air from his tank to fully expand his lungs before abandoning the tank at depth L and swimming to the surface. He ignores instructions and fails to exhale during his ascent. When he reaches the surface, the difference between the external pressure on him and the air pressure in his lungs is 9.3 kPa. From what depth does he start? What potentially lethal danger does he face?

KEY IDEA

The pressure at depth h in a liquid of density ρ is given by Eq. 14-8 ($p = p_0 + \rho g h$), where the gauge pressure $\rho g h$ is added to the atmospheric pressure p_0.

Calculations: Here, when the diver fills his lungs at depth L, the external pressure on him (and thus the air pressure within his lungs) is greater than normal and given by Eq. 14-8 as

$$p = p_0 + \rho g L,$$

where p_0 is atmospheric pressure and ρ is the water's density

(998 kg/m³, from Table 14-1). As he ascends, the external pressure on him decreases, until it is atmospheric pressure p_0 at the surface. His blood pressure also decreases, until it is normal. However, because he does not exhale, the air pressure in his lungs remains at the value it had at depth L. At the surface, the pressure difference between the higher pressure in his lungs and the lower pressure on his chest is

$$\Delta p = p - p_0 = \rho g L,$$

from which we find

$$L = \frac{\Delta p}{\rho g} = \frac{9300 \text{ Pa}}{(998 \text{ kg/m}^3)(9.8 \text{ m/s}^2)}$$

$$= 0.95 \text{ m.} \qquad \text{(Answer)}$$

This is not deep! Yet, the pressure difference of 9.3 kPa (about 9% of atmospheric pressure) is sufficient to rupture the diver's lungs and force air from them into the depressurized blood, which then carries the air to the heart, killing the diver. If the diver follows instructions and gradually exhales as he ascends, he allows the pressure in his lungs to equalize with the external pressure, and then there is no danger.

Sample Problem

Balancing of pressure in a U-tube

The U-tube in Fig. 14-4 contains two liquids in static equilibrium: Water of density ρ_w (= 998 kg/m³) is in the right arm, and oil of unknown density ρ_x is in the left. Measurement gives l = 135 mm and d = 12.3 mm. What is the density of the oil?

KEY IDEAS

(1) The pressure p_{int} at the level of the oil–water interface in the left arm depends on the density ρ_x and height of the oil above the interface. (2) The water in the right arm *at the same level* must be at the same pressure p_{int}. The reason is that, because the water is in static equilibrium, pressures at points in the water at the same level must be the same even if the points are separated horizontally.

Calculations: In the right arm, the interface is a distance l below the free surface of the *water,* and we have, from Eq. 14-8,

$$p_{\text{int}} = p_0 + \rho_w g l \quad \text{(right arm).}$$

In the left arm, the interface is a distance $l + d$ below the free surface of the *oil,* and we have, again from Eq. 14-8,

$$p_{\text{int}} = p_0 + \rho_x g (l + d) \quad \text{(left arm).}$$

Fig. 14-4 The oil in the left arm stands higher than the water in the right arm because the oil is less dense than the water. Both fluid columns produce the same pressure p_{int} at the level of the interface.

Equating these two expressions and solving for the unknown density yield

$$\rho_x = \rho_w \frac{l}{l + d} = (998 \text{ kg/m}^3) \frac{135 \text{ mm}}{135 \text{ mm} + 12.3 \text{ mm}}$$

$$= 915 \text{ kg/m}^3. \qquad \text{(Answer)}$$

Note that the answer does not depend on the atmospheric pressure p_0 or the free-fall acceleration g.

14-5 Measuring Pressure

The Mercury Barometer

Figure 14-5a shows a very basic *mercury barometer,* a device used to measure the pressure of the atmosphere. The long glass tube is filled with mercury and inverted with its open end in a dish of mercury, as the figure shows. The space above the mercury column contains only mercury vapor, whose pressure is so small at ordinary temperatures that it can be neglected.

We can use Eq. 14-7 to find the atmospheric pressure p_0 in terms of the height h of the mercury column. We choose level 1 of Fig. 14-2 to be that of the air–mercury interface and level 2 to be that of the top of the mercury column, as labeled in Fig. 14-5a. We then substitute

$$y_1 = 0, \quad p_1 = p_0 \quad \text{and} \quad y_2 = h, \quad p_2 = 0$$

into Eq. 14-7, finding that

$$p_0 = \rho g h, \tag{14-9}$$

where ρ is the density of the mercury.

Fig. 14-5 (a) A mercury barometer. (b) Another mercury barometer. The distance h is the same in both cases.

(a)   (b)

For a given pressure, the height h of the mercury column does not depend on the cross-sectional area of the vertical tube. The fanciful mercury barometer of Fig. 14-5b gives the same reading as that of Fig. 14-5a; all that counts is the vertical distance h between the mercury levels.

Equation 14-9 shows that, for a given pressure, the height of the column of mercury depends on the value of g at the location of the barometer and on the density of mercury, which varies with temperature. The height of the column (in millimeters) is numerically equal to the pressure (in torr) *only* if the barometer is at a place where g has its accepted standard value of 9.80665 m/s^2 *and* the temperature of the mercury is 0°C. If these conditions do not prevail (and they rarely do), small corrections must be made before the height of the mercury column can be transformed into a pressure.

The Open-Tube Manometer

An *open-tube manometer* (Fig. 14-6) measures the gauge pressure p_g of a gas. It consists of a U-tube containing a liquid, with one end of the tube connected to the vessel whose gauge pressure we wish to measure and the other end open to the atmosphere. We can use Eq. 14-7 to find the gauge pressure in terms of the height h shown in Fig. 14-6. Let us choose levels 1 and 2 as shown in Fig. 14-6. We then substitute

$$y_1 = 0, \quad p_1 = p_0 \quad \text{and} \quad y_2 = -h, \quad p_2 = p$$

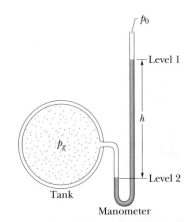

Fig. 14-6 An open-tube manometer, connected to measure the gauge pressure of the gas in the tank on the left. The right arm of the U-tube is open to the atmosphere.

Fig. 14-7 Lead shot (small balls of lead) loaded onto the piston create a pressure p_{ext} at the top of the enclosed (incompressible) liquid. If p_{ext} is increased, by adding more lead shot, the pressure increases by the same amount at all points within the liquid.

into Eq. 14-7, finding that

$$p_g = p - p_0 = \rho g h, \tag{14-10}$$

where ρ is the density of the liquid in the tube. The gauge pressure p_g is directly proportional to h.

The gauge pressure can be positive or negative, depending on whether $p > p_0$ or $p < p_0$. In inflated tires or the human circulatory system, the (absolute) pressure is greater than atmospheric pressure, so the gauge pressure is a positive quantity, sometimes called the *overpressure*. If you suck on a straw to pull fluid up the straw, the (absolute) pressure in your lungs is actually less than atmospheric pressure. The gauge pressure in your lungs is then a negative quantity.

14-6 Pascal's Principle

When you squeeze one end of a tube to get toothpaste out the other end, you are watching **Pascal's principle** in action. This principle is also the basis for the Heimlich maneuver, in which a sharp pressure increase properly applied to the abdomen is transmitted to the throat, forcefully ejecting food lodged there. The principle was first stated clearly in 1652 by Blaise Pascal (for whom the unit of pressure is named):

> A change in the pressure applied to an enclosed incompressible fluid is transmitted undiminished to every portion of the fluid and to the walls of its container.

Demonstrating Pascal's Principle

Consider the case in which the incompressible fluid is a liquid contained in a tall cylinder, as in Fig. 14-7. The cylinder is fitted with a piston on which a container of lead shot rests. The atmosphere, container, and shot exert pressure p_{ext} on the piston and thus on the liquid. The pressure p at any point P in the liquid is then

$$p = p_{ext} + \rho g h. \tag{14-11}$$

Let us add a little more lead shot to the container to increase p_{ext} by an amount Δp_{ext}. The quantities ρ, g, and h in Eq. 14-11 are unchanged, so the pressure change at P is

$$\Delta p = \Delta p_{ext}. \tag{14-12}$$

This pressure change is independent of h, so it must hold for all points within the liquid, as Pascal's principle states.

Pascal's Principle and the Hydraulic Lever

Figure 14-8 shows how Pascal's principle can be made the basis of a hydraulic lever. In operation, let an external force of magnitude F_i be directed downward on the left-hand (or input) piston, whose surface area is A_i. An incompressible liquid in the device then produces an upward force of magnitude F_o on the right-hand (or output) piston, whose surface area is A_o. To keep the system in equilibrium, there must be a downward force of magnitude F_o on the output piston from an external load (not shown). The force $\vec{F_i}$ applied on the left and the downward force $\vec{F_o}$ from the load on the right produce a change Δp in the pressure of the liquid that is given by

$$\Delta p = \frac{F_i}{A_i} = \frac{F_o}{A_o},$$

so

$$F_o = F_i \frac{A_o}{A_i}. \tag{14-13}$$

Fig. 14-8 A hydraulic arrangement that can be used to magnify a force $\vec{F_i}$. The work done is, however, not magnified and is the same for both the input and output forces.

Equation 14-13 shows that the output force F_o on the load must be greater than the input force F_i if $A_o > A_i$, as is the case in Fig. 14-8.

If we move the input piston downward a distance d_i, the output piston moves upward a distance d_o, such that the same volume V of the incompressible liquid is displaced at both pistons. Then

$$V = A_i d_i = A_o d_o,$$

which we can write as

$$d_o = d_i \frac{A_i}{A_o}. \qquad (14\text{-}14)$$

This shows that, if $A_o > A_i$ (as in Fig. 14-8), the output piston moves a smaller distance than the input piston moves.

From Eqs. 14-13 and 14-14 we can write the output work as

$$W = F_o d_o = \left(F_i \frac{A_o}{A_i} \right) \left(d_i \frac{A_i}{A_o} \right) = F_i d_i, \qquad (14\text{-}15)$$

which shows that the work W done *on* the input piston by the applied force is equal to the work W done *by* the output piston in lifting the load placed on it.

The advantage of a hydraulic lever is this:

> With a hydraulic lever, a given force applied over a given distance can be transformed to a greater force applied over a smaller distance.

The product of force and distance remains unchanged so that the same work is done. However, there is often tremendous advantage in being able to exert the larger force. Most of us, for example, cannot lift an automobile directly but can with a hydraulic jack, even though we have to pump the handle farther than the automobile rises and in a series of small strokes.

14-7 Archimedes' Principle

Figure 14-9 shows a student in a swimming pool, manipulating a very thin plastic sack (of negligible mass) that is filled with water. She finds that the sack and its contained water are in static equilibrium, tending neither to rise nor to sink. The downward gravitational force $\vec{F}_g$ on the contained water must be balanced by a net upward force from the water surrounding the sack.

This net upward force is a **buoyant force** $\vec{F}_b$. It exists because the pressure in the surrounding water increases with depth below the surface. Thus, the pressure near the bottom of the sack is greater than the pressure near the top, which means the forces on the sack due to this pressure are greater in magnitude near the bot-

The upward buoyant force on this sack of water equals the weight of the water.

Fig. 14-9 A thin-walled plastic sack of water is in static equilibrium in the pool. The gravitational force on the sack must be balanced by a net upward force on it from the surrounding water.

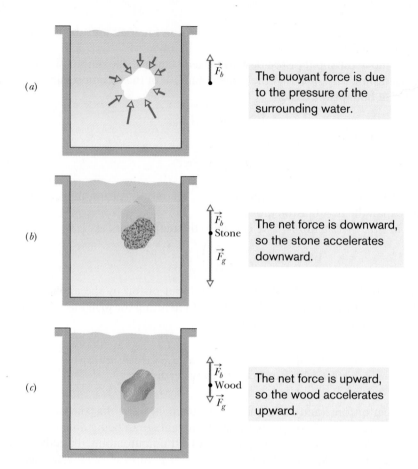

(a)

$\vec{F_b}$

The buoyant force is due to the pressure of the surrounding water.

(b)

$\vec{F_b}$
Stone
$\vec{F_g}$

The net force is downward, so the stone accelerates downward.

(c)

$\vec{F_b}$
Wood
$\vec{F_g}$

The net force is upward, so the wood accelerates upward.

Fig. 14-10 (a) The water surrounding the hole in the water produces a net upward buoyant force on whatever fills the hole. (b) For a stone of the same volume as the hole, the gravitational force exceeds the buoyant force in magnitude. (c) For a lump of wood of the same volume, the gravitational force is less than the buoyant force in magnitude.

tom of the sack than near the top. Some of the forces are represented in Fig. 14-10a, where the space occupied by the sack has been left empty. Note that the force vectors drawn near the bottom of that space (with upward components) have longer lengths than those drawn near the top of the sack (with downward components). If we vectorially add all the forces on the sack from the water, the horizontal components cancel and the vertical components add to yield the upward buoyant force $\vec{F_b}$ on the sack. (Force $\vec{F_b}$ is shown to the right of the pool in Fig. 14-10a.)

Because the sack of water is in static equilibrium, the magnitude of $\vec{F_b}$ is equal to the magnitude $m_f g$ of the gravitational force $\vec{F_g}$ on the sack of water: $F_b = m_f g$. (Subscript f refers to *fluid*, here the water.) In words, the magnitude of the buoyant force is equal to the weight of the water in the sack.

In Fig. 14-10b, we have replaced the sack of water with a stone that exactly fills the hole in Fig. 14-10a. The stone is said to *displace* the water, meaning that the stone occupies space that would otherwise be occupied by water. We have changed nothing about the shape of the hole, so the forces at the hole's surface must be the same as when the water-filled sack was in place. Thus, the same upward buoyant force that acted on the water-filled sack now acts on the stone; that is, the magnitude F_b of the buoyant force is equal to $m_f g$, the weight of the water displaced by the stone.

Unlike the water-filled sack, the stone is not in static equilibrium. The downward gravitational force $\vec{F_g}$ on the stone is greater in magnitude than the upward buoyant force, as is shown in the free-body diagram in Fig. 14-10b. The stone thus accelerates downward, sinking to the bottom of the pool.

Let us next exactly fill the hole in Fig. 14-10a with a block of lightweight wood, as in Fig. 14-10c. Again, nothing has changed about the forces at the hole's surface, so the magnitude F_b of the buoyant force is still equal to $m_f g$, the weight of the displaced water. Like the stone, the block is not in static equilibrium.

However, this time the gravitational force $\vec{F}_g$ is lesser in magnitude than the buoyant force (as shown to the right of the pool), and so the block accelerates upward, rising to the top surface of the water.

Our results with the sack, stone, and block apply to all fluids and are summarized in **Archimedes' principle:**

 When a body is fully or partially submerged in a fluid, a buoyant force $\vec{F}_b$ from the surrounding fluid acts on the body. The force is directed upward and has a magnitude equal to the weight $m_f g$ of the fluid that has been displaced by the body.

The buoyant force on a body in a fluid has the magnitude

$$F_b = m_f g \qquad \text{(buoyant force)}, \qquad (14\text{-}16)$$

where m_f is the mass of the fluid that is displaced by the body.

Floating

When we release a block of lightweight wood just above the water in a pool, the block moves into the water because the gravitational force on it pulls it downward. As the block displaces more and more water, the magnitude F_b of the upward buoyant force acting on it increases. Eventually, F_b is large enough to equal the magnitude F_g of the downward gravitational force on the block, and the block comes to rest. The block is then in static equilibrium and is said to be *floating* in the water. In general,

 When a body floats in a fluid, the magnitude F_b of the buoyant force on the body is equal to the magnitude F_g of the gravitational force on the body.

We can write this statement as

$$F_b = F_g \qquad \text{(floating)}. \qquad (14\text{-}17)$$

From Eq. 14-16, we know that $F_b = m_f g$. Thus,

 When a body floats in a fluid, the magnitude F_g of the gravitational force on the body is equal to the weight $m_f g$ of the fluid that has been displaced by the body.

We can write this statement as

$$F_g = m_f g \qquad \text{(floating)}. \qquad (14\text{-}18)$$

In other words, a floating body displaces its own weight of fluid.

Apparent Weight in a Fluid

If we place a stone on a scale that is calibrated to measure weight, then the reading on the scale is the stone's weight. However, if we do this underwater, the upward buoyant force on the stone from the water decreases the reading. That reading is then an apparent weight. In general, an **apparent weight** is related to the actual weight of a body and the buoyant force on the body by

$$\begin{pmatrix} \text{apparent} \\ \text{weight} \end{pmatrix} = \begin{pmatrix} \text{actual} \\ \text{weight} \end{pmatrix} - \begin{pmatrix} \text{magnitude of} \\ \text{buoyant force} \end{pmatrix},$$

which we can write as

$$\text{weight}_{\text{app}} = \text{weight} - F_b \qquad \text{(apparent weight)}. \qquad (14\text{-}19)$$

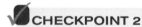**CHECKPOINT 2**

A penguin floats first in a fluid of density ρ_0, then in a fluid of density $0.95\rho_0$, and then in a fluid of density $1.1\rho_0$. (a) Rank the densities according to the magnitude of the buoyant force on the penguin, greatest first. (b) Rank the densities according to the amount of fluid displaced by the penguin, greatest first.

If, in some test of strength, you had to lift a heavy stone, you could do it more easily with the stone underwater. Then your applied force would need to exceed only the stone's apparent weight, not its larger actual weight, because the upward buoyant force would help you lift the stone.

The magnitude of the buoyant force on a floating body is equal to the body's weight. Equation 14-19 thus tells us that a floating body has an apparent weight of zero—the body would produce a reading of zero on a scale. (When astronauts prepare to perform a complex task in space, they practice the task floating underwater, where their apparent weight is zero as it is in space.)

Sample Problem

Floating, buoyancy, and density

In Fig. 14-11, a block of density $\rho = 800$ kg/m³ floats face down in a fluid of density $\rho_f = 1200$ kg/m³. The block has height $H = 6.0$ cm.

(a) By what depth h is the block submerged?

KEY IDEAS

(1) Floating requires that the upward buoyant force on the block match the downward gravitational force on the block. (2) The buoyant force is equal to the weight $m_f g$ of the fluid displaced by the submerged portion of the block.

Calculations: From Eq. 14-16, we know that the buoyant force has the magnitude $F_b = m_f g$, where m_f is the mass of the fluid displaced by the block's submerged volume V_f. From Eq. 14-2 ($\rho = m/V$), we know that the mass of the displaced fluid is $m_f = \rho_f V_f$. We don't know V_f but if we symbolize the block's face length as L and its width as W, then from Fig. 14-11 we see that the submerged volume must be $V_f = LWh$. If we now combine our three expressions, we find that the upward buoyant force has magnitude

$$F_b = m_f g = \rho_f V_f g = \rho_f LWhg. \qquad (14\text{-}20)$$

Similarly, we can write the magnitude F_g of the gravitational force on the block, first in terms of the block's mass

Floating means that the buoyant force matches the gravitational force.

Fig. 14-11 Block of height H floats in a fluid, to a depth of h.

m, then in terms of the block's density ρ and (full) volume V, and then in terms of the block's dimensions L, W, and H (the full height):

$$F_g = mg = \rho Vg = \rho_f LWHg. \qquad (14\text{-}21)$$

The floating block is stationary. Thus, writing Newton's second law for components along a vertical y axis with the positive direction upward ($F_{net,y} = ma_y$), we have

$$F_b - F_g = m(0),$$

or from Eqs. 14-20 and 14-21,

$$\rho_f LWhg - \rho LWHg = 0,$$

which gives us

$$h = \frac{\rho}{\rho_f} H = \frac{800 \text{ kg/m}^3}{1200 \text{ kg/m}^3} (6.0 \text{ cm})$$

$$= 4.0 \text{ cm.} \qquad \text{(Answer)}$$

(b) If the block is held fully submerged and then released, what is the magnitude of its acceleration?

Calculations: The gravitational force on the block is the same but now, with the block fully submerged, the volume of the displaced water is $V = LWH$. (The full height of the block is used.) This means that the value of F_b is now larger, and the block will no longer be stationary but will accelerate upward. Now Newton's second law yields

$$F_b - F_g = ma,$$

or

$$\rho_f LWHg - \rho LWHg = \rho LWHa,$$

where we inserted ρLWH for the mass m of the block. Solving for a leads to

$$a = \left(\frac{\rho_f}{\rho} - 1\right)g = \left(\frac{1200 \text{ kg/m}^3}{800 \text{ kg/m}^3} - 1\right)(9.8 \text{ m/s}^2)$$

$$= 4.9 \text{ m/s}^2. \qquad \text{(Answer)}$$

  Additional examples, video, and practice available at *WileyPLUS*

14-8 Ideal Fluids in Motion

The motion of *real fluids* is very complicated and not yet fully understood. Instead, we shall discuss the motion of an **ideal fluid**, which is simpler to handle mathematically and yet provides useful results. Here are four assumptions that we make about our ideal fluid; they all are concerned with *flow:*

1. **Steady flow**   In *steady* (or *laminar*) *flow,* the velocity of the moving fluid at any fixed point does not change with time. The gentle flow of water near the center of a quiet stream is steady; the flow in a chain of rapids is not. Figure 14-12 shows a transition from steady flow to *nonsteady* (or *nonlaminar* or *turbulent*) *flow* for a rising stream of smoke. The speed of the smoke particles increases as they rise and, at a certain critical speed, the flow changes from steady to nonsteady.

2. **Incompressible flow**   We assume, as for fluids at rest, that our ideal fluid is incompressible; that is, its density has a constant, uniform value.

3. **Nonviscous flow**   Roughly speaking, the viscosity of a fluid is a measure of how resistive the fluid is to flow. For example, thick honey is more resistive to flow than water, and so honey is said to be more viscous than water. Viscosity is the fluid analog of friction between solids; both are mechanisms by which the kinetic energy of moving objects can be transferred to thermal energy. In the absence of friction, a block could glide at constant speed along a horizontal surface. In the same way, an object moving through a nonviscous fluid would experience no *viscous drag force*—that is, no resistive force due to viscosity; it could move at constant speed through the fluid. The British scientist Lord Rayleigh noted that in an ideal fluid a ship's propeller would not work, but, on the other hand, in an ideal fluid a ship (once set into motion) would not need a propeller!

4. **Irrotational flow**   Although it need not concern us further, we also assume that the flow is *irrotational*. To test for this property, let a tiny grain of dust move with the fluid. Although this test body may (or may not) move in a circular path, in irrotational flow the test body will not rotate about an axis through its own center of mass. For a loose analogy, the motion of a Ferris wheel is rotational; that of its passengers is irrotational.

We can make the flow of a fluid visible by adding a *tracer*. This might be a dye injected into many points across a liquid stream (Fig. 14-13) or smoke particles added to a gas flow (Fig. 14-12). Each bit of a tracer follows a *streamline*, which is the path that a tiny element of the fluid would take as the fluid flows. Recall from Chapter 4 that the velocity of a particle is always tangent to the path taken by the particle. Here the particle is the fluid element, and its velocity $\vec{v}$ is always tangent to a streamline (Fig. 14-14). For this reason, two streamlines can never intersect; if they did, then an element arriving at their intersection would have two different velocities simultaneously—an impossibility.

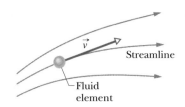

Fig. 14-12   At a certain point, the rising flow of smoke and heated gas changes from steady to turbulent. *(Will McIntyre/Photo Researchers)*

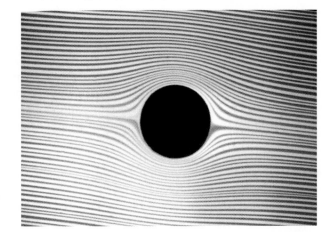

Fig. 14-13   The steady flow of a fluid around a cylinder, as revealed by a dye tracer that was injected into the fluid upstream of the cylinder. *(Courtesy D.H. Peregrine, University of Bristol)*

Fig. 14-14   A fluid element traces out a streamline as it moves. The velocity vector of the element is tangent to the streamline at every point.

Fig. 14-15 Fluid flows from left to right at a steady rate through a tube segment of length L. The fluid's speed is v_1 at the left side and v_2 at the right side. The tube's cross-sectional area is A_1 at the left side and A_2 at the right side. From time t in (a) to time $t + \Delta t$ in (b), the amount of fluid shown in purple enters at the left side and the equal amount of fluid shown in green emerges at the right side.

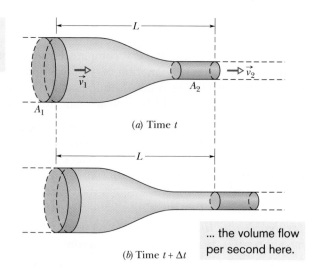

The volume flow per second here must match ...

(a) Time t

... the volume flow per second here.

(b) Time $t + \Delta t$

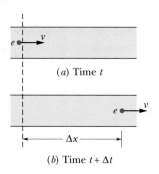

(a) Time t

(b) Time $t + \Delta t$

Fig. 14-16 Fluid flows at a constant speed v through a tube. (a) At time t, fluid element e is about to pass the dashed line. (b) At time $t + \Delta t$, element e is a distance $\Delta x = v\,\Delta t$ from the dashed line.

14-9 The Equation of Continuity

You may have noticed that you can increase the speed of the water emerging from a garden hose by partially closing the hose opening with your thumb. Apparently the speed v of the water depends on the cross-sectional area A through which the water flows.

Here we wish to derive an expression that relates v and A for the steady flow of an ideal fluid through a tube with varying cross section, like that in Fig. 14-15. The flow there is toward the right, and the tube segment shown (part of a longer tube) has length L. The fluid has speeds v_1 at the left end of the segment and v_2 at the right end. The tube has cross-sectional areas A_1 at the left end and A_2 at the right end. Suppose that in a time interval Δt a volume ΔV of fluid enters the tube segment at its left end (that volume is colored purple in Fig. 14-15). Then, because the fluid is incompressible, an identical volume ΔV must emerge from the right end of the segment (it is colored green in Fig. 14-15).

We can use this common volume ΔV to relate the speeds and areas. To do so, we first consider Fig. 14-16, which shows a side view of a tube of *uniform* cross-sectional area A. In Fig. 14-16a, a fluid element e is about to pass through the dashed line drawn across the tube width. The element's speed is v, so during a time interval Δt, the element moves along the tube a distance $\Delta x = v\,\Delta t$. The volume ΔV of fluid that has passed through the dashed line in that time interval Δt is

$$\Delta V = A\,\Delta x = Av\,\Delta t. \tag{14-22}$$

Applying Eq. 14-22 to both the left and right ends of the tube segment in Fig. 14-15, we have

$$\Delta V = A_1 v_1\,\Delta t = A_2 v_2\,\Delta t$$

or

$$A_1 v_1 = A_2 v_2 \quad \text{(equation of continuity)}. \tag{14-23}$$

This relation between speed and cross-sectional area is called the **equation of continuity** for the flow of an ideal fluid. It tells us that the flow speed increases when we decrease the cross-sectional area through which the fluid flows.

Equation 14-23 applies not only to an actual tube but also to any so-called *tube of flow*, or imaginary tube whose boundary consists of streamlines. Such a tube acts like a real tube because no fluid element can cross a streamline; thus, all the fluid within a tube of flow must remain within its boundary. Figure 14-17 shows a tube of flow in which the cross-sectional area increases from area A_1 to area A_2 along the flow direction. From Eq. 14-23 we know that, with the increase in area, the speed must decrease, as is indicated by the greater spacing between streamlines at the right in Fig. 14-17. Similarly, you can see that in Fig. 14-13 the speed of the flow is greatest just above and just below the cylinder.

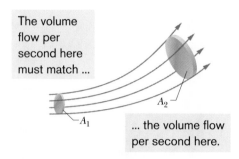

The volume flow per second here must match ...

... the volume flow per second here.

Fig. 14-17 A tube of flow is defined by the streamlines that form the boundary of the tube. The volume flow rate must be the same for all cross sections of the tube of flow.

We can rewrite Eq. 14-23 as

$$R_V = Av = \text{a constant} \qquad \text{(volume flow rate, equation of continuity),} \qquad (14\text{-}24)$$

in which R_V is the **volume flow rate** of the fluid (volume past a given point per unit time). Its SI unit is the cubic meter per second (m³/s). If the density ρ of the fluid is uniform, we can multiply Eq. 14-24 by that density to get the **mass flow rate** R_m (mass per unit time):

$$R_m = \rho R_V = \rho Av = \text{a constant} \qquad \text{(mass flow rate).} \qquad (14\text{-}25)$$

The SI unit of mass flow rate is the kilogram per second (kg/s). Equation 14-25 says that the mass that flows into the tube segment of Fig. 14-15 each second must be equal to the mass that flows out of that segment each second.

✓ CHECKPOINT 3

The figure shows a pipe and gives the volume flow rate (in cm³/s) and the direction of flow for all but one section. What are the volume flow rate and the direction of flow for that section?

Sample Problem

A water stream narrows as it falls

Figure 14-18 shows how the stream of water emerging from a faucet "necks down" as it falls. This change in the horizontal cross-sectional area is characteristic of any laminar (non-turbulent) falling stream because the gravitational force increases the speed of the stream. Here the indicated cross-sectional areas are $A_0 = 1.2$ cm² and $A = 0.35$ cm². The two levels are separated by a vertical distance $h = 45$ mm. What is the volume flow rate from the tap?

The volume flow per second here must match ...

... the volume flow per second here.

Fig. 14-18 As water falls from a tap, its speed increases. Because the volume flow rate must be the same at all horizontal cross sections of the stream, the stream must "neck down" (narrow).

KEY IDEA

The volume flow rate through the higher cross section must be the same as that through the lower cross section.

Calculations: From Eq. 14-24, we have

$$A_0v_0 = Av, \qquad (14\text{-}26)$$

where v_0 and v are the water speeds at the levels corresponding to A_0 and A. From Eq. 2-16 we can also write, because the water is falling freely with acceleration g,

$$v^2 = v_0^2 + 2gh. \qquad (14\text{-}27)$$

Eliminating v between Eqs. 14-26 and 14-27 and solving for v_0, we obtain

$$v_0 = \sqrt{\frac{2ghA^2}{A_0^2 - A^2}}$$

$$= \sqrt{\frac{(2)(9.8 \text{ m/s}^2)(0.045 \text{ m})(0.35 \text{ cm}^2)^2}{(1.2 \text{ cm}^2)^2 - (0.35 \text{ cm}^2)^2}}$$

$$= 0.286 \text{ m/s} = 28.6 \text{ cm/s}.$$

From Eq. 14-24, the volume flow rate R_V is then

$$R_V = A_0v_0 = (1.2 \text{ cm}^2)(28.6 \text{ cm/s})$$

$$= 34 \text{ cm}^3/\text{s}. \qquad \text{(Answer)}$$

 Additional examples, video, and practice available at *WileyPLUS*

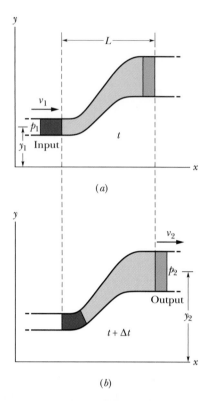

Fig. 14-19 Fluid flows at a steady rate through a length L of a tube, from the input end at the left to the output end at the right. From time t in (a) to time $t + \Delta t$ in (b), the amount of fluid shown in purple enters the input end and the equal amount shown in green emerges from the output end.

14-10 Bernoulli's Equation

Figure 14-19 represents a tube through which an ideal fluid is flowing at a steady rate. In a time interval Δt, suppose that a volume of fluid ΔV, colored purple in Fig. 14-19, enters the tube at the left (or input) end and an identical volume, colored green in Fig. 14-19, emerges at the right (or output) end. The emerging volume must be the same as the entering volume because the fluid is incompressible, with an assumed constant density ρ.

Let y_1, v_1, and p_1 be the elevation, speed, and pressure of the fluid entering at the left, and y_2, v_2, and p_2 be the corresponding quantities for the fluid emerging at the right. By applying the principle of conservation of energy to the fluid, we shall show that these quantities are related by

$$p_1 + \tfrac{1}{2}\rho v_1^2 + \rho g y_1 = p_2 + \tfrac{1}{2}\rho v_2^2 + \rho g y_2. \qquad (14\text{-}28)$$

In general, the term $\tfrac{1}{2}\rho v^2$ is called the fluid's **kinetic energy density** (kinetic energy per unit volume). We can also write Eq. 14-28 as

$$p + \tfrac{1}{2}\rho v^2 + \rho g y = \text{a constant} \qquad \text{(Bernoulli's equation).} \qquad (14\text{-}29)$$

Equations 14-28 and 14-29 are equivalent forms of **Bernoulli's equation,** after Daniel Bernoulli, who studied fluid flow in the 1700s.* Like the equation of continuity (Eq. 14-24), Bernoulli's equation is not a new principle but simply the reformulation of a familiar principle in a form more suitable to fluid mechanics. As a check, let us apply Bernoulli's equation to fluids at rest, by putting $v_1 = v_2 = 0$ in Eq. 14-28. The result is

$$p_2 = p_1 + \rho g(y_1 - y_2),$$

which is Eq. 14-7.

A major prediction of Bernoulli's equation emerges if we take y to be a constant ($y = 0$, say) so that the fluid does not change elevation as it flows. Equation 14-28 then becomes

$$p_1 + \tfrac{1}{2}\rho v_1^2 = p_2 + \tfrac{1}{2}\rho v_2^2, \qquad (14\text{-}30)$$

which tells us that:

> If the speed of a fluid element increases as the element travels along a horizontal streamline, the pressure of the fluid must decrease, and conversely.

Put another way, where the streamlines are relatively close together (where the velocity is relatively great), the pressure is relatively low, and conversely.

The link between a change in speed and a change in pressure makes sense if you consider a fluid element that travels through a tube of various widths. Recall that the element's speed in the narrower regions is fast and its speed in the wider regions is slow. By Newton's second law, forces (or pressures) must cause the changes in speed (the accelerations). When the element nears a narrow region, the higher pressure behind it accelerates it so that it then has a greater speed in the narrow region. When it nears a wide region, the higher pressure ahead of it decelerates it so that it then has a lesser speed in the wide region.

Bernoulli's equation is strictly valid only to the extent that the fluid is ideal. If viscous forces are present, thermal energy will be involved. We take no account of this in the derivation that follows.

---

*For irrotational flow (which we assume), the constant in Eq. 14-29 has the same value for all points within the tube of flow; the points do not have to lie along the same streamline. Similarly, the points 1 and 2 in Eq. 14-28 can lie anywhere within the tube of flow.

Proof of Bernoulli's Equation

Let us take as our system the entire volume of the (ideal) fluid shown in Fig. 14-19. We shall apply the principle of conservation of energy to this system as it moves from its initial state (Fig. 14-19a) to its final state (Fig. 14-19b). The fluid lying between the two vertical planes separated by a distance L in Fig. 14-19 does not change its properties during this process; we need be concerned only with changes that take place at the input and output ends.

First, we apply energy conservation in the form of the work–kinetic energy theorem,

$$W = \Delta K, \tag{14-31}$$

which tells us that the change in the kinetic energy of our system must equal the net work done on the system. The change in kinetic energy results from the change in speed between the ends of the tube and is

$$\Delta K = \tfrac{1}{2}\Delta m\, v_2^2 - \tfrac{1}{2}\Delta m\, v_1^2$$
$$= \tfrac{1}{2}\rho\, \Delta V(v_2^2 - v_1^2), \tag{14-32}$$

in which $\Delta m\ (= \rho\, \Delta V)$ is the mass of the fluid that enters at the input end and leaves at the output end during a small time interval Δt.

The work done on the system arises from two sources. The work W_g done by the gravitational force $(\Delta m\, \vec{g})$ on the fluid of mass Δm during the vertical lift of the mass from the input level to the output level is

$$W_g = -\Delta m\, g(y_2 - y_1)$$
$$= -\rho g\, \Delta V(y_2 - y_1). \tag{14-33}$$

This work is negative because the upward displacement and the downward gravitational force have opposite directions.

Work must also be done *on* the system (at the input end) to push the entering fluid into the tube and *by* the system (at the output end) to push forward the fluid that is located ahead of the emerging fluid. In general, the work done by a force of magnitude F, acting on a fluid sample contained in a tube of area A to move the fluid through a distance Δx, is

$$F\, \Delta x = (pA)(\Delta x) = p(A\, \Delta x) = p\, \Delta V.$$

The work done on the system is then $p_1\, \Delta V$, and the work done by the system is $-p_2\, \Delta V$. Their sum W_p is

$$W_p = -p_2\, \Delta V + p_1\, \Delta V$$
$$= -(p_2 - p_1)\, \Delta V. \tag{14-34}$$

The work–kinetic energy theorem of Eq. 14-31 now becomes

$$W = W_g + W_p = \Delta K.$$

Substituting from Eqs. 14-32, 14-33, and 14-34 yields

$$-\rho g\, \Delta V(y_2 - y_1) - \Delta V(p_2 - p_1) = \tfrac{1}{2}\rho\, \Delta V(v_2^2 - v_1^2).$$

This, after a slight rearrangement, matches Eq. 14-28, which we set out to prove.

✓ CHECKPOINT 4

Water flows smoothly through the pipe shown in the figure, descending in the process. Rank the four numbered sections of pipe according to (a) the volume flow rate R_V through them, (b) the flow speed v through them, and (c) the water pressure p within them, greatest first.

Sample Problem

Bernoulli principle of fluid through a narrowing pipe

Ethanol of density $\rho = 791$ kg/m^3 flows smoothly through a horizontal pipe that tapers (as in Fig. 14-15) in cross-sectional area from $A_1 = 1.20 \times 10^{-3}$ m^2 to $A_2 = A_1/2$. The pressure difference between the wide and narrow sections of pipe is 4120 Pa. What is the volume flow rate R_V of the ethanol?

KEY IDEAS

(1) Because the fluid flowing through the wide section of pipe must entirely pass through the narrow section, the volume flow rate R_V must be the same in the two sections. Thus, from Eq. 14-24,

$$R_V = v_1 A_1 = v_2 A_2. \qquad (14\text{-}35)$$

However, with two unknown speeds, we cannot evaluate this equation for R_V. (2) Because the flow is smooth, we can apply Bernoulli's equation. From Eq. 14-28, we can write

$$p_1 + \tfrac{1}{2}\rho v_1^2 + \rho g y = p_2 + \tfrac{1}{2}\rho v_2^2 + \rho g y, \qquad (14\text{-}36)$$

where subscripts 1 and 2 refer to the wide and narrow sections of pipe, respectively, and y is their common elevation. This equation hardly seems to help because it does not contain the desired R_V and it contains the unknown speeds v_1 and v_2.

Calculations: There is a neat way to make Eq. 14-36 work for us: First, we can use Eq. 14-35 and the fact that $A_2 = A_1/2$

to write

$$v_1 = \frac{R_V}{A_1} \quad \text{and} \quad v_2 = \frac{R_V}{A_2} = \frac{2R_V}{A_1}. \qquad (14\text{-}37)$$

Then we can substitute these expressions into Eq. 14-36 to eliminate the unknown speeds and introduce the desired volume flow rate. Doing this and solving for R_V yield

$$R_V = A_1 \sqrt{\frac{2(p_1 - p_2)}{3\rho}}. \qquad (14\text{-}38)$$

We still have a decision to make: We know that the pressure difference between the two sections is 4120 Pa, but does that mean that $p_1 - p_2$ is 4120 Pa or -4120 Pa? We could guess the former is true, or otherwise the square root in Eq. 14-38 would give us an imaginary number. Instead of guessing, however, let's try some reasoning. From Eq. 14-35 we see that speed v_2 in the narrow section (small A_2) must be greater than speed v_1 in the wider section (larger A_1). Recall that if the speed of a fluid increases as the fluid travels along a horizontal path (as here), the pressure of the fluid must decrease. Thus, p_1 is greater than p_2, and $p_1 - p_2 = 4120$ Pa. Inserting this and known data into Eq. 14-38 gives

$$R_V = 1.20 \times 10^{-3}\,\text{m}^2 \sqrt{\frac{(2)(4120\,\text{Pa})}{(3)(791\,\text{kg/m}^3)}}$$

$$= 2.24 \times 10^{-3}\,\text{m}^3\text{/s}. \qquad \text{(Answer)}$$

Sample Problem

Bernoulli principle for a leaky water tank

In the old West, a desperado fires a bullet into an open water tank (Fig. 14-20), creating a hole a distance h below the water surface. What is the speed v of the water exiting the tank?

KEY IDEAS

(1) This situation is essentially that of water moving (downward) with speed v_0 through a wide pipe (the tank) of cross-sectional area A and then moving (horizontally) with speed v through a narrow pipe (the hole) of cross-sectional area a. (2) Because the water flowing through the wide pipe must entirely pass through the narrow pipe, the volume flow rate R_V must be the same in the two "pipes." (3) We can also relate v to v_0 (and to h) through Bernoulli's equation (Eq. 14-28).

Fig. 14-20 Water pours through a hole in a water tank, at a distance h below the water surface. The pressure at the water surface and at the hole is atmospheric pressure p_0.

Calculations: From Eq. 14-24,

$$R_V = av = Av_0$$

and thus

$$v_0 = \frac{a}{A} v.$$

Because $a \ll A$, we see that $v_0 \ll v$. To apply Bernoulli's equation, we take the level of the hole as our reference level for measuring elevations (and thus gravitational potential energy). Noting that the pressure at the top of the tank and at the bullet hole is the atmospheric pressure p_0 (because both places are exposed to the atmosphere), we write Eq. 14-28 as

$$p_0 + \tfrac{1}{2}\rho v_0^2 + \rho gh = p_0 + \tfrac{1}{2}\rho v^2 + \rho g(0). \qquad (14\text{-}39)$$

(Here the top of the tank is represented by the left side of the equation and the hole by the right side. The zero on the right indicates that the hole is at our reference level.) Before we solve Eq. 14-39 for v, we can use our result that $v_0 \ll v$ to simplify it: We assume that v_0^2, and thus the term $\tfrac{1}{2}\rho v_0^2$ in Eq. 14-39, is negligible relative to the other terms, and we drop it. Solving the remaining equation for v then yields

$$v = \sqrt{2gh}. \qquad \text{(Answer)}$$

This is the same speed that an object would have when falling a height h from rest.

 Additional examples, video, and practice available at *WileyPLUS*

REVIEW & SUMMARY

Density The **density** ρ of any material is defined as the material's mass per unit volume:

$$\rho = \frac{\Delta m}{\Delta V}. \qquad (14\text{-}1)$$

Usually, where a material sample is much larger than atomic dimensions, we can write Eq. 14-1 as

$$\rho = \frac{m}{V}. \qquad (14\text{-}2)$$

Fluid Pressure A **fluid** is a substance that can flow; it conforms to the boundaries of its container because it cannot withstand shearing stress. It can, however, exert a force perpendicular to its surface. That force is described in terms of **pressure** p:

$$p = \frac{\Delta F}{\Delta A}, \qquad (14\text{-}3)$$

in which ΔF is the force acting on a surface element of area ΔA. If the force is uniform over a flat area, Eq. 14-3 can be written as

$$p = \frac{F}{A}. \qquad (14\text{-}4)$$

The force resulting from fluid pressure at a particular point in a fluid has the same magnitude in all directions. **Gauge pressure** is the difference between the actual pressure (or *absolute pressure*) at a point and the atmospheric pressure.

Pressure Variation with Height and Depth Pressure in a fluid at rest varies with vertical position y. For y measured positive upward,

$$p_2 = p_1 + \rho g(y_1 - y_2). \qquad (14\text{-}7)$$

The pressure in a fluid is the same for all points at the same level. If h is the *depth* of a fluid sample below some reference level at which the pressure is p_0, Eq. 14-7 becomes

$$p = p_0 + \rho gh, \qquad (14\text{-}8)$$

where p is the pressure in the sample.

Pascal's Principle A change in the pressure applied to an enclosed fluid is transmitted undiminished to every portion of the fluid and to the walls of the containing vessel.

Archimedes' Principle When a body is fully or partially submerged in a fluid, a buoyant force $\vec{F}_b$ from the surrounding fluid acts on the body. The force is directed upward and has a magnitude given by

$$F_b = m_f g, \qquad (14\text{-}16)$$

where m_f is the mass of the fluid that has been displaced by the body (that is, the fluid that has been pushed out of the way by the body).

When a body floats in a fluid, the magnitude F_b of the (upward) buoyant force on the body is equal to the magnitude F_g of the (downward) gravitational force on the body. The **apparent weight** of a body on which a buoyant force acts is related to its actual weight by

$$\text{weight}_{\text{app}} = \text{weight} - F_b. \qquad (14\text{-}19)$$

Flow of Ideal Fluids An **ideal fluid** is incompressible and lacks viscosity, and its flow is steady and irrotational. A *streamline* is the path followed by an individual fluid particle. A *tube of flow* is a bundle of streamlines. The flow within any tube of flow obeys the **equation of continuity:**

$$R_V = Av = \text{a constant}, \qquad (14\text{-}24)$$

in which R_V is the **volume flow rate**, A is the cross-sectional area of the tube of flow at any point, and v is the speed of the fluid at that point. The **mass flow rate** R_m is

$$R_m = \rho R_V = \rho Av = \text{a constant}. \qquad (14\text{-}25)$$

Bernoulli's Equation Applying the principle of conservation of mechanical energy to the flow of an ideal fluid leads to **Bernoulli's equation:**

$$p + \tfrac{1}{2}\rho v^2 + \rho gy = \text{a constant} \qquad (14\text{-}29)$$

along any tube of flow.

QUESTIONS

1 We fully submerge an irregular 3 kg lump of material in a certain fluid. The fluid that would have been in the space now occupied by the lump has a mass of 2 kg. (a) When we release the lump, does it move upward, move downward, or remain in place? (b) If we next fully submerge the lump in a less dense fluid and again release it, what does it do?

2 Figure 14-21 shows four situations in which a red liquid and a gray liquid are in a U-tube. In one situation the liquids cannot be in static equilibrium. (a) Which situation is that? (b) For the other three situations, assume static equilibrium. For each, is the density of the red liquid greater than, less than, or equal to the density of the gray liquid?

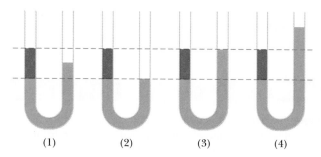

(1)   (2)   (3)   (4)

Fig. 14-21 Question 2.

3 A boat with an anchor on board floats in a swimming pool that is somewhat wider than the boat. Does the pool water level move up, move down, or remain the same if the anchor is (a) dropped into the water or (b) thrown onto the surrounding ground? (c) Does the water level in the pool move upward, move downward, or remain the same if, instead, a cork is dropped from the boat into the water, where it floats?

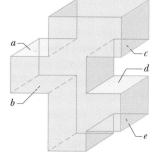

Fig. 14-22 Question 4.

4 Figure 14-22 shows a tank filled with water. Five horizontal floors and ceilings are indicated; all have the same area and are located at distances L, $2L$, or $3L$ below the top of the tank. Rank them according to the force on them due to the water, greatest first.

5 *The teapot effect.* Water poured slowly from a teapot spout can double back under the spout for a considerable distance before detaching and falling. (The water layer is held against the underside of the spout by atmospheric pressure.) In Fig. 14-23, in the water layer inside the spout, point a is at the top of the layer and point b is at the bottom of the layer; in the water layer outside the spout, point c is at the top of the layer and point d is at the bottom of the layer. Rank those four points according to the gauge pressure in the water there, most positive first.

Fig. 14-23 Question 5.

6 Figure 14-24 shows three identical open-top containers filled to the brim with water; toy ducks float in two of them. Rank the containers and contents according to their weight, greatest first.

(a)   (b)   (c)

Fig. 14-24 Question 6.

7 Figure 14-25 shows four arrangements of pipes through which water flows smoothly toward the right. The radii of the pipe sections are indicated. In which arrangements is the net work done on a unit volume of water moving from the leftmost section to the rightmost section (a) zero, (b) positive, and (c) negative?

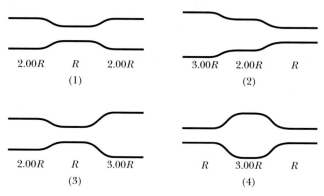

2.00R	R	2.00R		3.00R	2.00R	R
	(1)				(2)	

2.00R	R	3.00R		R	3.00R	R
	(3)				(4)	

Fig. 14-25 Question 7.

8 A rectangular block is pushed facedown into three liquids, in turn. The apparent weight W_{app} of the block versus depth h in the three liquids is plotted in Fig. 14-26. Rank the liquids according to their weight per unit volume, greatest first.

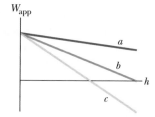

Fig. 14-26 Question 8.

9 Water flows smoothly in a horizontal pipe. Figure 14-27 shows the kinetic energy K of a water element as it moves along an x axis that runs along the pipe. Rank the three lettered sections of the pipe according to the pipe radius, greatest first.

Fig. 14-27 Question 9.

10 We have three containers with different liquids. The gauge pressure p_g versus depth h is plotted in Fig. 14-28 for the liquids. In each container, we will fully submerge a rigid plastic bead. Rank the plots according to the magnitude of the buoyant force on the bead, greatest first.

Fig. 14-28 Question 10.

PROBLEMS

GO   Tutoring problem available (at instructor's discretion) in *WileyPLUS* and WebAssign

SSM   Worked-out solution available in Student Solutions Manual   WWW   Worked-out solution is at

•–•••   Number of dots indicates level of problem difficulty   ILW   Interactive solution is at   http://www.wiley.com/college/halliday

   Additional information available in *The Flying Circus of Physics* and at flyingcircusofphysics.com

sec. 14-3   Density and Pressure

•1 ILW   A fish maintains its depth in fresh water by adjusting the air content of porous bone or air sacs to make its average density the same as that of the water. Suppose that with its air sacs collapsed, a fish has a density of 1.08 g/cm³. To what fraction of its expanded body volume must the fish inflate the air sacs to reduce its density to that of water?

•2   A partially evacuated airtight container has a tight-fitting lid of surface area 77 m² and negligible mass. If the force required to remove the lid is 480 N and the atmospheric pressure is 1.0×10^5 Pa, what is the internal air pressure?

•3 SSM   Find the pressure increase in the fluid in a syringe when a nurse applies a force of 42 N to the syringe's circular piston, which has a radius of 1.1 cm.

•4   Three liquids that will not mix are poured into a cylindrical container. The volumes and densities of the liquids are 0.50 L, 2.6 g/cm³; 0.25 L, 1.0 g/cm³; and 0.40 L, 0.80 g/cm³. What is the force on the bottom of the container due to these liquids? One liter = 1 L = 1000 cm³. (Ignore the contribution due to the atmosphere.)

•5 SSM   An office window has dimensions 3.4 m by 2.1 m. As a result of the passage of a storm, the outside air pressure drops to 0.96 atm, but inside the pressure is held at 1.0 atm. What net force pushes out on the window?

•6   You inflate the front tires on your car to 28 psi. Later, you measure your blood pressure, obtaining a reading of 120/80, the readings being in mm Hg. In metric countries (which is to say, most of the world), these pressures are customarily reported in kilopascals (kPa). In kilopascals, what are (a) your tire pressure and (b) your blood pressure?

••7   In 1654 Otto von Guericke, inventor of the air pump, gave a demonstration before the noblemen of the Holy Roman Empire in which two teams of eight horses could not pull apart two evacuated brass hemispheres. (a) Assuming the hemispheres have (strong) thin walls, so that R in Fig. 14-29 may be considered both the inside and outside radius, show that the force $\vec{F}$ required to pull apart the hemispheres has magnitude $F = \pi R^2 \, \Delta p$, where Δp is the difference between the pressures outside and inside the sphere. (b) Taking R as 30 cm, the inside pressure as 0.10 atm, and the outside pressure as 1.00 atm, find the force magnitude the teams of horses would have had to exert to pull apart the hemispheres. (c) Explain why one team of horses could have proved the point just as well if the hemispheres were attached to a sturdy wall.

Fig. 14-29   Problem 7.

sec. 14-4   Fluids at Rest

•8   *The bends during flight*. Anyone who scuba dives is advised not to fly within the next 24 h because the air mixture for diving can introduce nitrogen to the bloodstream. Without allowing the nitrogen to come out of solution slowly, any sudden air-pressure reduction (such as during airplane ascent) can result in the nitrogen forming bubbles in the blood, creating the *bends*, which can be painful and even fatal. Military special operation forces are especially at risk. What is the change in pressure on such a special-op soldier who must scuba dive at a depth of 20 m in seawater one day and parachute at an altitude of 7.6 km the next day? Assume that the average air density within the altitude range is 0.87 kg/m³.

•9   *Blood pressure in Argentinosaurus*. (a) If this long-necked, gigantic sauropod had a head height of 21 m and a heart height of 9.0 m, what (hydrostatic) gauge pressure in its blood was required at the heart such that the blood pressure at the brain was 80 torr (just enough to perfuse the brain with blood)? Assume the blood had a density of 1.06×10^3 kg/m³. (b) What was the blood pressure (in torr or mm Hg) at the feet?

•10   The plastic tube in Fig. 14-30 has a cross-sectional area of 5.00 cm². The tube is filled with water until the short arm (of length $d = 0.800$ m) is full. Then the short arm is sealed and more water is gradually poured into the long arm. If the seal will pop off when the force on it exceeds 9.80 N, what total height of water in the long arm will put the seal on the verge of popping?

Fig. 14-30 Problems 10 and 81.

•11   *Giraffe bending to drink*. In a giraffe with its head 2.0 m above its heart, and its heart 2.0 m above its feet, the (hydrostatic) gauge pressure in the blood at its heart is 250 torr. Assume that the giraffe stands upright and the blood density is 1.06×10^3 kg/m³. In torr (or mm Hg), find the (gauge) blood pressure (a) at the brain (the pressure is enough to perfuse the brain with blood, to keep the giraffe from fainting) and (b) at the feet (the pressure must be countered by tight-fitting skin acting like a pressure stocking). (c) If the giraffe were to lower its head to drink from a pond without splaying its legs and moving slowly, what would be the increase in the blood pressure in the brain? (Such action would probably be lethal.)

•12   The maximum depth d_{max} that a diver can snorkel is set by the density of the water and the fact that human lungs can function against a maximum pressure difference (between inside and outside the chest cavity) of 0.050 atm. What is the difference in d_{max} for fresh water and the water of the Dead Sea (the saltiest natural water in the world, with a density of 1.5×10^3 kg/m³)?

•13   At a depth of 10.9 km, the Challenger Deep in the Marianas Trench of the Pacific Ocean is the deepest site in any ocean. Yet, in 1960, Donald Walsh and Jacques Piccard reached the Challenger Deep in the bathyscaph *Trieste*. Assuming that seawater has a uniform density of 1024 kg/m³, approximate the hydrostatic pressure (in atmospheres) that the *Trieste* had to withstand. (Even a slight defect in the *Trieste* structure would have been disastrous.)

•14 Calculate the hydrostatic difference in blood pressure between the brain and the foot in a person of height 1.83 m. The density of blood is 1.06×10^3 kg/m³.

•15 What gauge pressure must a machine produce in order to suck mud of density 1800 kg/m³ up a tube by a height of 1.5 m?

•16 *Snorkeling by humans and elephants.* When a person snorkels, the lungs are connected directly to the atmosphere through the snorkel tube and thus are at atmospheric pressure. In atmospheres, what is the difference Δp between this internal air pressure and the water pressure against the body if the length of the snorkel

Fig. 14-31 Problem 16.

tube is (a) 20 cm (standard situation) and (b) 4.0 m (probably lethal situation)? In the latter, the pressure difference causes blood vessels on the walls of the lungs to rupture, releasing blood into the lungs. As depicted in Fig. 14-31, an elephant can safely snorkel through its trunk while swimming with its lungs 4.0 m below the water surface because the membrane around its lungs contains connective tissue that holds and protects the blood vessels, preventing rupturing.

•17 SSM Crew members attempt to escape from a damaged submarine 100 m below the surface. What force must be applied to a pop-out hatch, which is 1.2 m by 0.60 m, to push it out at that depth? Assume that the density of the ocean water is 1024 kg/m³ and the internal air pressure is at 1.00 atm.

•18 In Fig. 14-32, an open tube of length $L = 1.8$ m and cross-sectional area $A = 4.6$ cm² is fixed to the top of a cylindrical barrel of diameter $D = 1.2$ m and height $H = 1.8$ m. The barrel and tube are filled with water (to the top of the tube). Calculate the ratio of the hydrostatic force on the bottom of the barrel to the gravitational force on the water contained in the barrel. Why is that ratio not equal to 1.0? (You need not consider the atmospheric pressure.)

•19 GO A large aquarium of height 5.00 m is filled with fresh water to a depth of 2.00 m. One wall of the aquarium consists of thick plastic 8.00 m wide. By how much does the total force on that wall increase if the aquarium is next filled to a depth of 4.00 m?

Fig. 14-32
Problem 18.

••20 The **L**-shaped tank shown in Fig. 14-33 is filled with water and is open at the top. If $d = 5.0$ m, what is the force due to the water (a) on face A and (b) on face B?

••21 SSM Two identical cylindrical vessels with their bases at the same level each contain a liquid of density 1.30×10^3 kg/m³. The area of each base is 4.00 cm², but in one vessel the liquid height is 0.854 m and in the other it is 1.560 m. Find the work done by the gravitational force in equalizing the levels when the two vessels are connected.

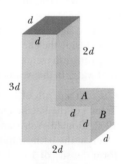

Fig. 14-33
Problem 20.

••22 *g-LOC in dogfights.* When a pilot takes a tight turn at high speed in a modern fighter airplane, the blood pressure at the brain level decreases, blood no longer perfuses the brain, and the blood in the brain drains. If the heart maintains the (hydrostatic) gauge pressure in the aorta at 120 torr (or mm Hg) when the pilot undergoes a horizontal centripetal acceleration of 4g, what is the blood pressure (in torr) at the brain, 30 cm radially inward from the heart? The perfusion in the brain is small enough that the vision switches to black and white and narrows to "tunnel vision" and the pilot can undergo *g*-LOC ("*g*-induced loss of consciousness"). Blood density is 1.06×10^3 kg/m³.

••23 In analyzing certain geological features, it is often appropriate to assume that the pressure at some horizontal *level of compensation,* deep inside Earth, is the same over a large region and is equal to the pressure due to the gravitational force on the overlying material. Thus, the pressure on the level of compensation is given by the fluid pressure formula. This model requires, for one thing, that mountains have *roots* of continental rock extending into the denser

Fig. 14-34 Problem 23.

mantle (Fig. 14-34). Consider a mountain of height $H = 6.0$ km on a continent of thickness $T = 32$ km. The continental rock has a density of 2.9 g/cm³, and beneath this rock the mantle has a density of 3.3 g/cm³. Calculate the depth D of the root. (*Hint:* Set the pressure at points a and b equal; the depth y of the level of compensation will cancel out.)

•••24 GO In Fig. 14-35, water stands at depth $D = 35.0$ m behind the vertical upstream face of a dam of width $W = 314$ m. Find (a) the net horizontal force on the dam from the gauge pressure of the water and (b) the net torque due to that force about a horizontal line through O parallel to the (long) width of the dam. This torque tends to rotate the

Fig. 14-35 Problem 24.

dam around that line, which would cause the dam to fail. (c) Find the moment arm of the torque.

sec. 14-5 Measuring Pressure

•25 In one observation, the column in a mercury barometer (as is shown in Fig. 14-5a) has a measured height h of 740.35 mm. The temperature is $-5.0°C$, at which temperature the density of mercury ρ is 1.3608×10^4 kg/m³. The free-fall acceleration g at the site of the barometer is 9.7835 m/s². What is the atmospheric pressure at that site in pascals and in torr (which is the common unit for barometer readings)?

•26 To suck lemonade of density 1000 kg/m³ up a straw to a maximum height of 4.0 cm, what minimum gauge pressure (in atmospheres) must you produce in your lungs?

••27 SSM What would be the height of the atmosphere if the air density (a) were uniform and (b) decreased linearly to zero with height? Assume that at sea level the air pressure is 1.0 atm and the air density is 1.3 kg/m³.

sec. 14-6 Pascal's Principle

•28 A piston of cross-sectional area a is used in a hydraulic press to exert a small force of magnitude f on the enclosed liquid. A connecting pipe leads to a larger piston of cross-sectional area A (Fig. 14-36). (a) What force magnitude F will the larger piston sustain without moving? (b) If the piston diameters are 3.80 cm and 53.0 cm, what force magnitude on the small piston will balance a 20.0 kN force on the large piston?

Fig. 14-36
Problem 28.

••29 In Fig. 14-37, a spring of spring constant 3.00×10^4 N/m is between a rigid beam and the output piston of a hydraulic lever. An empty container with negligible mass sits on the input piston. The input piston has area A_i, and the output piston has area $18.0A_i$. Initially the spring is at its rest length. How many kilograms of sand must be (slowly) poured into the container to compress the spring by 5.00 cm?

Fig. 14-37 Problem 29.

sec. 14-7 Archimedes' Principle

•30 A 5.00 kg object is released from rest while fully submerged in a liquid. The liquid displaced by the submerged object has a mass of 3.00 kg. How far and in what direction does the object move in 0.200 s, assuming that it moves freely and that the drag force on it from the liquid is negligible?

•31 **SSM** A block of wood floats in fresh water with two-thirds of its volume V submerged and in oil with $0.90V$ submerged. Find the density of (a) the wood and (b) the oil.

•32 In Fig. 14-38, a cube of edge length $L = 0.600$ m and mass 450 kg is suspended by a rope in an open tank of liquid of density 1030 kg/m³. Find (a) the magnitude of the total downward force on the top of the cube from the liquid and the atmosphere, assuming atmospheric pressure is 1.00 atm, (b) the magnitude of the total upward force on the bottom of the cube, and (c) the tension in the rope. (d) Calculate the magnitude of the buoyant force on the cube using Archimedes' principle. What relation exists among all these quantities?

Fig. 14-38 Problem 32.

•33 **SSM** An iron anchor of density 7870 kg/m³ appears 200 N lighter in water than in air. (a) What is the volume of the anchor? (b) How much does it weigh in air?

•34 A boat floating in fresh water displaces water weighing 35.6 kN. (a) What is the weight of the water this boat displaces when floating in salt water of density 1.10×10^3 kg/m³? (b) What is the difference between the volume of fresh water displaced and the volume of salt water displaced?

•35 Three children, each of weight 356 N, make a log raft by lashing together logs of diameter 0.30 m and length 1.80 m. How many logs will be needed to keep them afloat in fresh water? Take the density of the logs to be 800 kg/m³.

••36 In Fig. 14-39a, a rectangular block is gradually pushed facedown into a liquid. The block has height d; on the bottom and top the face area is $A = 5.67$ cm². Figure 14-39b gives the apparent weight W_{app} of the block as a function of the depth h of its lower face. The scale on the vertical axis is set by $W_s = 0.20$ N. What is the density of the liquid?

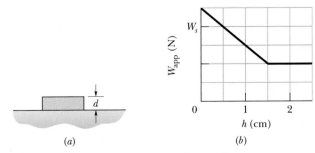

(a)   (b)

Fig. 14-39 Problem 36.

••37 **ILW** A hollow spherical iron shell floats almost completely submerged in water. The outer diameter is 60.0 cm, and the density of iron is 7.87 g/cm³. Find the inner diameter.

••38 **GO** A small solid ball is released from rest while fully submerged in a liquid and then its kinetic energy is measured when it has moved 4.0 cm in the liquid. Figure 14-40 gives the results after many liquids are used: The kinetic energy K is plotted versus the liquid density ρ_{liq}, and $K_s = 1.60$ J sets the scale on the vertical axis. What are (a) the density and (b) the volume of the ball?

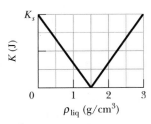

Fig. 14-40 Problem 38.

••39 **SSM** **WWW** A hollow sphere of inner radius 8.0 cm and outer radius 9.0 cm floats half-submerged in a liquid of density 800 kg/m³. (a) What is the mass of the sphere? (b) Calculate the density of the material of which the sphere is made.

••40 *Lurking alligators.* An alligator waits for prey by floating with only the top of its head exposed, so that the prey cannot easily see it. One way it can adjust the extent of sinking is by controlling the size of its lungs. Another way may be by swallowing stones (*gastrolithes*) that then reside in the stomach. Figure 14-41 shows a highly simplified model (a "rhombohedron gater") of mass 130 kg that roams with its head partially exposed. The top head surface has area 0.20 m². If the alligator were to swallow stones with a total mass of 1.0% of its body mass (a typical amount), how far would it sink?

Fig. 14-41 Problem 40.

••41 What fraction of the volume of an iceberg (density 917 kg/m³) would be visible if the iceberg floats (a) in the ocean (salt water, density 1024 kg/m³) and (b) in a river (fresh water, density 1000 kg/m³)? (When salt water freezes to form ice, the salt is excluded. So, an iceberg could provide fresh water to a community.)

••42 A flotation device is in the shape of a right cylinder, with a height of 0.500 m and a face area of 4.00 m² on top and bottom, and its density is 0.400 times that of fresh water. It is initially held fully submerged in fresh water, with its top face at the water surface. Then it is allowed to ascend gradually until it begins to float. How much work does the buoyant force do on the device during the ascent?

••43 When researchers find a reasonably complete fossil of a dinosaur, they can determine the mass and weight of the living dinosaur with a scale model sculpted from plastic and based on the dimensions of the fossil bones. The scale of the model is 1/20; that is, lengths are 1/20 actual length, areas are $(1/20)^2$ ac-

Fig. 14-42   Problem 43.

tual areas, and volumes are $(1/20)^3$ actual volumes. First, the model is suspended from one arm of a balance and weights are added to the other arm until equilibrium is reached. Then the model is fully submerged in water and enough weights are removed from the second arm to reestablish equilibrium (Fig. 14-42). For a model of a particular *T. rex* fossil, 637.76 g had to be removed to reestablish equilibrium. What was the volume of (a) the model and (b) the actual *T. rex*? (c) If the density of *T. rex* was approximately the density of water, what was its mass?

••44 A block of wood has a mass of 3.67 kg and a density of 600 kg/m³. It is to be loaded with lead (1.14×10^4 kg/m³) so that it will float in water with 0.900 of its volume submerged. What mass of lead is needed if the lead is attached to (a) the top of the wood and (b) the bottom of the wood?

••45 An iron casting containing a number of cavities weighs 6000 N in air and 4000 N in water. What is the total volume of all the cavities in the casting? The density of iron (that is, a sample with no cavities) is 7.87 g/cm³.

••46 **GO** Suppose that you release a small ball from rest at a depth of 0.600 m below the surface in a pool of water. If the density of the ball is 0.300 that of water and if the drag force on the ball from the water is negligible, how high above the water surface will the ball shoot as it emerges from the water? (Neglect any transfer of energy to the splashing and waves produced by the emerging ball.)

••47 The volume of air space in the passenger compartment of an 1800 kg car is 5.00 m³. The volume of the motor and front wheels is 0.750 m³, and the volume of the rear wheels, gas tank, and trunk is 0.800 m³; water cannot enter these two regions. The car rolls into a lake. (a) At first, no water enters the passenger compartment. How much of the car, in cubic meters, is below the water surface with the car floating (Fig. 14-43)? (b) As water slowly enters, the car sinks. How many cubic meters of water are in the car as it disappears below the water surface? (The car, with a heavy load in the trunk, remains horizontal.)

Fig. 14-43   Problem 47.

•••48 **GO** Figure 14-44 shows an iron ball suspended by thread of negligible mass from an upright cylinder that floats partially submerged in water. The cylinder has a height of 6.00 cm, a face area of 12.0 cm² on the top and bottom, and a density of 0.30 g/cm³, and 2.00 cm of its height is above the water surface. What is the radius of the iron ball?

Fig. 14-44
Problem 48.

sec. 14-9   The Equation of Continuity

•49 ✈ *Canal effect.* Figure 14-45 shows an anchored barge that extends across a canal by distance $d = 30$ m and into the water by distance $b = 12$ m. The canal has a width $D = 55$ m, a water depth $H = 14$ m, and a uniform water-flow speed $v_i = 1.5$ m/s. Assume that the flow around the barge is uniform. As the water passes the bow, the water level undergoes a dramatic dip known as the canal effect. If the dip has depth $h = 0.80$ m, what is the water speed alongside the boat through the vertical cross

Fig. 14-45   Problem 49.

sections at (a) point *a* and (b) point *b*? The erosion due to the speed increase is a common concern to hydraulic engineers.

•50 Figure 14-46 shows two sections of an old pipe system that runs through a hill, with distances $d_A = d_B = 30$ m and $D = 110$ m. On each side of the hill, the pipe radius is 2.00 cm. However, the radius of the pipe inside the hill is no longer known. To determine it, hydraulic engineers first establish that water flows through the left and right sections at 2.50 m/s. Then they release a dye in the water at point *A* and find that it takes 88.8 s to reach point *B*. What is the average radius of the pipe within the hill?

Fig. 14-46   Problem 50.

•51 **SSM** A garden hose with an internal diameter of 1.9 cm is connected to a (stationary) lawn sprinkler that consists merely of a container with 24 holes, each 0.13 cm in diameter. If the water in the hose has a speed of 0.91 m/s, at what speed does it leave the sprinkler holes?

•52 Two streams merge to form a river. One stream has a width of 8.2 m, depth of 3.4 m, and current speed of 2.3 m/s. The other stream is 6.8 m wide and 3.2 m deep, and flows at 2.6 m/s. If the river has width 10.5 m and speed 2.9 m/s, what is its depth?

••53 **SSM** Water is pumped steadily out of a flooded basement at a speed of 5.0 m/s through a uniform hose of radius 1.0 cm. The hose passes out through a window 3.0 m above the waterline. What is the power of the pump?

••54 The water flowing through a 1.9 cm (inside diameter) pipe flows out through three 1.3 cm pipes. (a) If the flow rates in the three smaller pipes are 26, 19, and 11 L/min, what is the flow rate in the 1.9 cm pipe? (b) What is the ratio of the speed in the 1.9 cm pipe to that in the pipe carrying 26 L/min?

sec. 14-10   Bernoulli's Equation

•55   How much work is done by pressure in forcing 1.4 m³ of water through a pipe having an internal diameter of 13 mm if the difference in pressure at the two ends of the pipe is 1.0 atm?

•56   Suppose that two tanks, 1 and 2, each with a large opening at the top, contain different liquids. A small hole is made in the side of each tank at the same depth h below the liquid surface, but the hole in tank 1 has half the cross-sectional area of the hole in tank 2. (a) What is the ratio ρ_1/ρ_2 of the densities of the liquids if the mass flow rate is the same for the two holes? (b) What is the ratio R_{V1}/R_{V2} of the volume flow rates from the two tanks? (c) At one instant, the liquid in tank 1 is 12.0 cm above the hole. If the tanks are to have *equal* volume flow rates, what height above the hole must the liquid in tank 2 be just then?

•57   SSM   A cylindrical tank with a large diameter is filled with water to a depth $D = 0.30$ m. A hole of cross-sectional area $A = 6.5$ cm² in the bottom of the tank allows water to drain out. (a) What is the rate at which water flows out, in cubic meters per second? (b) At what distance below the bottom of the tank is the cross-sectional area of the stream equal to one-half the area of the hole?

•58   The intake in Fig. 14-47 has cross-sectional area of 0.74 m² and water flow at 0.40 m/s. At the outlet, distance $D = 180$ m below the intake, the cross-sectional area is smaller than at the intake and the water flows out at 9.5 m/s into equipment. What is the pressure difference between inlet and outlet?

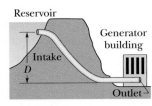

Fig. 14-47   Problem 58.

•59   SSM   Water is moving with a speed of 5.0 m/s through a pipe with a cross-sectional area of 4.0 cm². The water graduall descends 10 m as the pipe cross-sectional area increases to 8.0 cm². (a) What is the speed at the lower level? (b) If the pressure at the upper level is 1.5×10^5 Pa, what is the pressure at the lower level?

•60   Models of torpedoes are sometimes tested in a horizontal pipe of flowing water, much as a wind tunnel is used to test model airplanes. Consider a circular pipe of internal diameter 25.0 cm and a torpedo model aligned along the long axis of the pipe. The model has a 5.00 cm diameter and is to be tested with water flowing past it at 2.50 m/s. (a) With what speed must the water flow in the part of the pipe that is unconstricted by the model? (b) What will the pressure difference be between the constricted and unconstricted parts of the pipe?

•61   ILW   A water pipe having a 2.5 cm inside diameter carries water into the basement of a house at a speed of 0.90 m/s and a pressure of 170 kPa. If the pipe tapers to 1.2 cm and rises to the second floor 7.6 m above the input point, what are the (a) speed and (b) water pressure at the second floor?

••62   A pitot tube (Fig. 14-48) is used to determine the airspeed of an airplane. It consists of an outer tube with a number of small holes B (four are shown) that allow air into the tube; that tube is connected to one arm of a U-tube. The other arm of the U-tube is connected to hole A at the front end of the device, which points in the direction the plane is headed. At A the air becomes stagnant so that $v_A = 0$. At B, however, the speed of the air presumably equals the airspeed v of the plane. (a) Use Bernoulli's equation to show that

$$v = \sqrt{\frac{2\rho gh}{\rho_{air}}},$$

where ρ is the density of the liquid in the U-tube and h is the difference in the liquid levels in that tube. (b) Suppose that the tube contains alcohol and the level difference h is 26.0 cm. What is the plane's speed relative to the air? The density of the air is 1.03 kg/m³ and that of alcohol is 810 kg/m³.

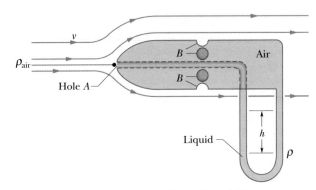

Fig. 14-48   Problems 62 and 63.

••63   A pitot tube (see Problem 62) on a high-altitude aircraft measures a differential pressure of 180 Pa. What is the aircraft's airspeed if the density of the air is 0.031 kg/m³?

••64   GO   In Fig. 14-49, water flows through a horizontal pipe and then out into the atmosphere at a speed $v_1 = 15$ m/s. The diameters of the left and right sections of the pipe are 5.0 cm and 3.0 cm. (a) What volume of water flows into the atmosphere during a 10 min period? In the left section of the pipe, what are (b) the speed v_2 and (c) the gauge pressure?

Fig. 14-49   Problem 64.

••65   SSM   WWW   A *venturi meter* is used to measure the flow speed of a fluid in a pipe. The meter is connected between two sections of the pipe (Fig. 14-50); the cross-sectional area A of the entrance and exit of the meter matches the pipe's cross-sectional area. Between the entrance and exit, the fluid flows from the pipe with speed V and then through a narrow "throat" of cross-

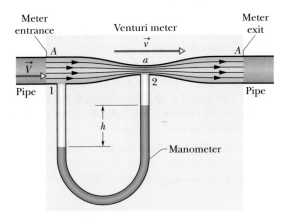

Fig. 14-50   Problems 65 and 66.

sectional area a with speed v. A manometer connects the wider portion of the meter to the narrower portion. The change in the fluid's speed is accompanied by a change Δp in the fluid's pressure, which causes a height difference h of the liquid in the two arms of the manometer. (Here Δp means pressure in the throat minus pressure in the pipe.) (a) By applying Bernoulli's equation and the equation of continuity to points 1 and 2 in Fig. 14-50, show that

$$V = \sqrt{\frac{2a^2\,\Delta p}{\rho(a^2 - A^2)}},$$

where ρ is the density of the fluid. (b) Suppose that the fluid is fresh water, that the cross-sectional areas are 64 cm^2 in the pipe and 32 cm^2 in the throat, and that the pressure is 55 kPa in the pipe and 41 kPa in the throat. What is the rate of water flow in cubic meters per second?

••66 Consider the venturi tube of Problem 65 and Fig. 14-50 without the manometer. Let A equal $5a$. Suppose the pressure p_1 at A is 2.0 atm. Compute the values of (a) the speed V at A and (b) the speed v at a that make the pressure p_2 at a equal to zero. (c) Compute the corresponding volume flow rate if the diameter at A is 5.0 cm. The phenomenon that occurs at a when p_2 falls to nearly zero is known as cavitation. The water vaporizes into small bubbles.

••67 **ILW** In Fig. 14-51, the fresh water behind a reservoir dam has depth $D = 15$ m. A horizontal pipe 4.0 cm in diameter passes through the dam at depth $d = 6.0$ m. A plug secures the pipe opening. (a) Find the magnitude of the frictional force between plug and pipe wall. (b) The plug is removed. What water volume exits the pipe in 3.0 h?

Fig. 14-51 Problem 67.

••68 **GO** Fresh water flows horizontally from pipe section 1 of cross-sectional area A_1 into pipe section 2 of cross-sectional area A_2. Figure 14-52 gives a plot of the pressure difference $p_2 - p_1$ versus the inverse area squared A_1^{-2} that would be expected for a volume flow rate of a certain value if the water flow were laminar under all

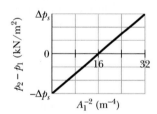

Fig. 14-52 Problem 68.

circumstances. The scale on the vertical axis is set by $\Delta p_s = 300$ kN/m^2. For the conditions of the figure, what are the values of (a) A_2 and (b) the volume flow rate?

••69 A liquid of density 900 kg/m^3 flows through a horizontal pipe that has a cross-sectional area of 1.90×10^{-2} m^2 in region A and a cross-sectional area of 9.50×10^{-2} m^2 in region B. The pressure difference between the two regions is 7.20×10^3 Pa. What are (a) the volume flow rate and (b) the mass flow rate?

••70 **GO** In Fig. 14-53, water flows steadily from the left pipe section

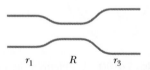

Fig. 14-53 Problem 70.

(radius $r_1 = 2.00R$), through the middle section (radius R), and into the right section (radius $r_3 = 3.00R$). The speed of the water in the middle section is 0.500 m/s. What is the net work done on 0.400 m^3 of the water as it moves from the left section to the right section?

••71 Figure 14-54 shows a stream of water flowing through a hole at depth $h = 10$ cm in a tank holding water to height $H = 40$ cm. (a) At what distance x does the stream strike the floor? (b) At what depth should a second hole be made to give the same value of x? (c) At what depth should a hole be made to maximize x?

Fig. 14-54 Problem 71.

•••72 A very simplified schematic of the rain drainage system for a home is shown in Fig. 14-55. Rain falling on the slanted roof runs off into gutters around the roof edge; it then drains through downspouts (only one is shown) into a main drainage pipe M below the basement, which carries the water to an even larger pipe below the street. In Fig. 14-55, a floor drain in the basement is also connected to drainage pipe M. Suppose the following apply:

1. the downspouts have height $h_1 = 11$ m,
2. the floor drain has height $h_2 = 1.2$ m,
3. pipe M has radius 3.0 cm,
4. the house has side width $w = 30$ m and front length $L = 60$ m,
5. all the water striking the roof goes through pipe M,
6. the initial speed of the water in a downspout is negligible,
7. the wind speed is negligible (the rain falls vertically).

At what rainfall rate, in centimeters per hour, will water from pipe M reach the height of the floor drain and threaten to flood the basement?

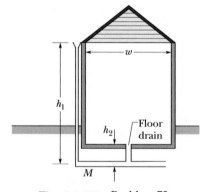

Fig. 14-55 Problem 72.

Additional Problems

73 About one-third of the body of a person floating in the Dead Sea will be above the waterline. Assuming that the human body density is 0.98 g/cm^3, find the density of the water in the Dead Sea. (Why is it so much greater than 1.0 g/cm^3?)

74 A simple open U-tube contains mercury. When 11.2 cm of water is poured into the right arm of the tube, how high above its initial level does the mercury rise in the left arm?

75 If a bubble in sparkling water accelerates upward at the rate of 0.225 m/s^2 and has a radius of 0.500 mm, what is its mass? Assume that the drag force on the bubble is negligible.

76   Suppose that your body has a uniform density of 0.95 times that of water. (a) If you float in a swimming pool, what fraction of your body's volume is above the water surface?

    Quicksand is a fluid produced when water is forced up into sand, moving the sand grains away from one another so they are no longer locked together by friction. Pools of quicksand can form when water drains underground from hills into valleys where there are sand pockets. (b) If you float in a deep pool of quicksand that has a density 1.6 times that of water, what fraction of your body's volume is above the quicksand surface? (c) In particular, are you submerged enough to be unable to breathe?

77   A glass ball of radius 2.00 cm sits at the bottom of a container of milk that has a density of 1.03 g/cm³. The normal force on the ball from the container's lower surface has magnitude 9.48×10^{-2} N. What is the mass of the ball?

78   Caught in an avalanche, a skier is fully submerged in flowing snow of density 96 kg/m³. Assume that the average density of the skier, clothing, and skiing equipment is 1020 kg/m³. What percentage of the gravitational force on the skier is offset by the buoyant force from the snow?

79   An object hangs from a spring balance. The balance registers 30 N in air, 20 N when this object is immersed in water, and 24 N when the object is immersed in another liquid of unknown density. What is the density of that other liquid?

80   In an experiment, a rectangular block with height h is allowed to float in four separate liquids. In the first liquid, which is water, it floats fully submerged. In liquids A, B, and C, it floats with heights $h/2$, $2h/3$, and $h/4$ above the liquid surface, respectively. What are the *relative densities* (the densities relative to that of water) of (a) A, (b) B, and (c) C?

81 SSM Figure 14-30 shows a modified U-tube: the right arm is shorter than the left arm. The open end of the right arm is height $d = 10.0$ cm above the laboratory bench. The radius throughout the tube is 1.50 cm. Water is gradually poured into the open end of the left arm until the water begins to flow out the open end of the right arm. Then a liquid of density 0.80 g/cm³ is gradually added to the left arm until its height in that arm is 8.0 cm (it does not mix with the water). How much water flows out of the right arm?

82   What is the acceleration of a rising hot-air balloon if the ratio of the air density outside the balloon to that inside is 1.39? Neglect the mass of the balloon fabric and the basket.

83   Figure 14-56 shows a *siphon*, which is a device for removing liquid from a container. Tube ABC must initially be filled, but once this has been done, liquid will flow through the tube until the liquid surface in the container is level with the tube opening at A. The liquid has density 1000 kg/m³ and negligible viscosity. The distances shown are $h_1 = 25$ cm, $d = 12$ cm, and $h_2 = 40$ cm. (a) With what speed does the liquid emerge from the tube at C? (b) If the atmospheric pressure is 1.0×10^5 Pa, what is the pressure in the liquid at the topmost point B? (c) Theoretically, what is the greatest possible height h_1 that a siphon can lift water?

Fig. 14-56   Problem 83.

84   When you cough, you expel air at high speed through the trachea and upper bronchi so that the air will remove excess mucus lining the pathway. You produce the high speed by this procedure: You breathe in a large amount of air, trap it by closing the glottis (the narrow opening in the larynx), increase the air pressure by contracting the lungs, partially collapse the trachea and upper bronchi to narrow the pathway, and then expel the air through the pathway by suddenly reopening the glottis. Assume that during the expulsion the volume flow rate is 7.0×10^{-3} m³/s. What multiple of the speed of sound v_s ($= 343$ m/s) is the airspeed through the trachea if the trachea diameter (a) remains its normal value of 14 mm and (b) contracts to 5.2 mm?

85   A tin can has a total volume of 1200 cm³ and a mass of 130 g. How many grams of lead shot of density 11.4 g/cm³ could it carry without sinking in water?

15

OSCILLATIONS

15-1 WHAT IS PHYSICS?

Our world is filled with oscillations in which objects move back and forth repeatedly. Many oscillations are merely amusing or annoying, but many others are dangerous or financially important. Here are a few examples: When a bat hits a baseball, the bat may oscillate enough to sting the batter's hands or even to break apart. When wind blows past a power line, the line may oscillate ("gallop" in electrical engineering terms) so severely that it rips apart, shutting off the power supply to a community. When an airplane is in flight, the turbulence of the air flowing past the wings makes them oscillate, eventually leading to metal fatigue and even failure. When a train travels around a curve, its wheels oscillate horizontally ("hunt" in mechanical engineering terms) as they are forced to turn in new directions (you can hear the oscillations).

When an earthquake occurs near a city, buildings may be set oscillating so severely that they are shaken apart. When an arrow is shot from a bow, the feathers at the end of the arrow manage to snake around the bow staff without hitting it because the arrow oscillates. When a coin drops into a metal collection plate, the coin oscillates with such a familiar ring that the coin's denomination can be determined from the sound. When a rodeo cowboy rides a bull, the cowboy oscillates wildly as the bull jumps and turns (at least the cowboy hopes to be oscillating).

The study and control of oscillations are two of the primary goals of both physics and engineering. In this chapter we discuss a basic type of oscillation called *simple harmonic motion.*

15-2 Simple Harmonic Motion

Figure 15-1a shows a sequence of "snapshots" of a simple oscillating system, a particle moving repeatedly back and forth about the origin of an x axis. In this section we simply describe the motion. Later, we shall discuss how to attain such motion.

One important property of oscillatory motion is its **frequency,** or number of oscillations that are completed each second. The symbol for frequency is f, and its SI unit is the **hertz** (abbreviated Hz), where

$$1 \text{ hertz} = 1 \text{ Hz} = 1 \text{ oscillation per second} = 1 \text{ s}^{-1}. \qquad (15\text{-}1)$$

Related to the frequency is the **period** T of the motion, which is the time for one complete oscillation (or **cycle**); that is,

$$T = \frac{1}{f}. \qquad (15\text{-}2)$$

Any motion that repeats itself at regular intervals is called **periodic motion** or **harmonic motion.** We are interested here in motion that repeats itself in a particular way—namely, like that in Fig. 15-1a. For such motion the displacement

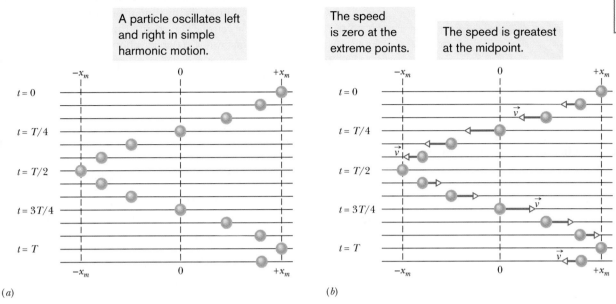

A particle oscillates left and right in simple harmonic motion.

The speed is zero at the extreme points.

The speed is greatest at the midpoint.

(a)

(b)

Rotating the figure reveals that the motion forms a cosine function.

Fig. 15-1 (a) A sequence of "snapshots" (taken at equal time intervals) showing the position of a particle as it oscillates back and forth about the origin of an x axis, between the limits $+x_m$ and $-x_m$. (b) The vector arrows are scaled to indicate the speed of the particle. The speed is maximum when the particle is at the origin and zero when it is at $\pm x_m$. If the time t is chosen to be zero when the particle is at $+x_m$, then the particle returns to $+x_m$ at $t = T$, where T is the period of the motion. The motion is then repeated. (c) Rotating the figure reveals the motion forms a cosine function of time, as shown in (d). (e) The speed (the slope) changes.

(c)

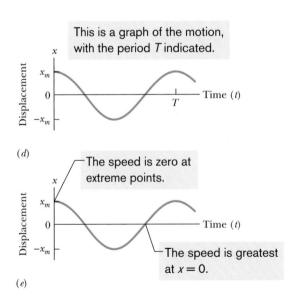

This is a graph of the motion, with the period T indicated.

(d)

The speed is zero at extreme points.

The speed is greatest at $x = 0$.

(e)

x of the particle from the origin is given as a function of time by

$$x(t) = x_m \cos(\omega t + \phi) \quad \text{(displacement)}, \qquad (15\text{-}3)$$

in which x_m, ω, and ϕ are constants. This motion is called **simple harmonic motion** (SHM), a term that means the periodic motion is a sinusoidal function of time. Equation 15-3, in which the sinusoidal function is a cosine function, is graphed in Fig. 15-1d. (We get that graph by rotating Fig. 15-1a counterclockwise by 90°.) The quantities that determine the shape of the graph are displayed in Fig. 15-2 with their names. We now shall define those quantities.

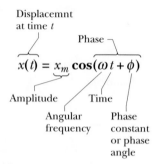

$$x(t) = x_m \cos(\omega t + \phi)$$

Displacement at time t

Phase

Amplitude

Time

Angular frequency

Phase constant or phase angle

Fig. 15-2 A handy reference to the quantities in Eq. 15-3 for simple harmonic motion.

The quantity x_m, called the **amplitude** of the motion, is a positive constant whose value depends on how the motion was started. The subscript m stands for *maximum* because the amplitude is the magnitude of the maximum displacement of the particle in either direction. The cosine function in Eq. 15-3 varies between the limits ± 1; so the displacement $x(t)$ varies between the limits $\pm x_m$.

The time-varying quantity $(\omega t + \phi)$ in Eq. 15-3 is called the **phase** of the motion, and the constant ϕ is called the **phase constant** (or **phase angle**). The value of ϕ depends on the displacement and velocity of the particle at time $t = 0$. For the $x(t)$ plots of Fig. 15-3a, the phase constant ϕ is zero.

To interpret the constant ω, called the **angular frequency** of the motion, we first note that the displacement $x(t)$ must return to its initial value after one period T of the motion; that is, $x(t)$ must equal $x(t + T)$ for all t. To simplify this analysis, let us put $\phi = 0$ in Eq. 15-3. From that equation we then can write

$$x_m \cos \omega t = x_m \cos \omega(t + T). \tag{15-4}$$

The cosine function first repeats itself when its argument (the phase) has increased by 2π rad; so Eq. 15-4 gives us

$$\omega(t + T) = \omega t + 2\pi$$

or

$$\omega T = 2\pi.$$

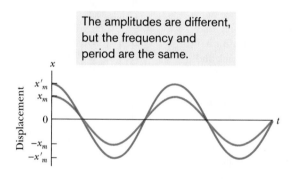

The amplitudes are different, but the frequency and period are the same.

(a)

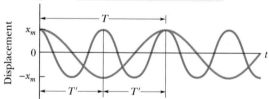

The amplitudes are the same, but the frequencies and periods are different.

(b)

Fig. 15-3 In all three cases, the blue curve is obtained from Eq. 15-3 with $\phi = 0$. (*a*) The red curve differs from the blue curve *only* in that the red-curve amplitude x'_m is greater (the red-curve extremes of displacement are higher and lower). (*b*) The red curve differs from the blue curve *only* in that the red-curve period is $T' = T/2$ (the red curve is compressed horizontally). (*c*) The red curve differs from the blue curve *only* in that for the red curve $\phi = -\pi/4$ rad rather than zero (the negative value of ϕ shifts the red curve to the right).

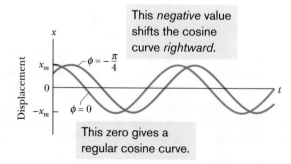

This *negative* value shifts the cosine curve *rightward*.

This zero gives a regular cosine curve.

(c)

Thus, from Eq. 15-2 the angular frequency is

$$\omega = \frac{2\pi}{T} = 2\pi f. \qquad (15\text{-}5)$$

The SI unit of angular frequency is the radian per second. (To be consistent, then, ϕ must be in radians.) Figure 15-3 compares $x(t)$ for two simple harmonic motions that differ either in amplitude, in period (and thus in frequency and angular frequency), or in phase constant.

✔ CHECKPOINT 1

A particle undergoing simple harmonic oscillation of period T (like that in Fig. 15-1) is at $-x_m$ at time $t = 0$. Is it at $-x_m$, at $+x_m$, at 0, between $-x_m$ and 0, or between 0 and $+x_m$ when (a) $t = 2.00T$, (b) $t = 3.50T$, and (c) $t = 5.25T$?

The Velocity of SHM

By differentiating Eq. 15-3, we can find an expression for the velocity of a particle moving with simple harmonic motion; that is,

$$v(t) = \frac{dx(t)}{dt} = \frac{d}{dt}\,[x_m \cos(\omega t + \phi)]$$

or $\qquad\qquad v(t) = -\omega x_m \sin(\omega t + \phi) \quad$ (velocity).$\qquad (15\text{-}6)$

Figure 15-4a is a plot of Eq. 15-3 with $\phi = 0$. Figure 15-4b shows Eq. 15-6, also with $\phi = 0$. Analogous to the amplitude x_m in Eq. 15-3, the positive quantity ωx_m in Eq. 15-6 is called the **velocity amplitude** v_m. As you can see in Fig. 15-4b, the velocity of the oscillating particle varies between the limits $\pm v_m = \pm \omega x_m$. Note also in that figure that the curve of $v(t)$ is *shifted* (to the left) from the curve of $x(t)$ by one-quarter period; when the magnitude of the displacement is greatest (that is, $x(t) = x_m$), the magnitude of the velocity is least (that is, $v(t) = 0$). When the magnitude of the displacement is least (that is, zero), the magnitude of the velocity is greatest (that is, $v_m = \omega x_m$).

The Acceleration of SHM

Knowing the velocity $v(t)$ for simple harmonic motion, we can find an expression for the acceleration of the oscillating particle by differentiating once more. Thus, we have, from Eq. 15-6,

$$a(t) = \frac{dv(t)}{dt} = \frac{d}{dt}\,[-\omega x_m \sin(\omega t + \phi)]$$

or $\qquad\qquad a(t) = -\omega^2 x_m \cos(\omega t + \phi) \quad$ (acceleration).$\qquad (15\text{-}7)$

Figure 15-4c is a plot of Eq. 15-7 for the case $\phi = 0$. The positive quantity $\omega^2 x_m$ in Eq. 15-7 is called the **acceleration amplitude** a_m; that is, the acceleration of the particle varies between the limits $\pm a_m = \pm \omega^2 x_m$, as Fig. 15-4$c$ shows. Note also that the acceleration curve $a(t)$ is shifted (to the left) by $\frac{1}{4}T$ relative to the velocity curve $v(t)$.

We can combine Eqs. 15-3 and 15-7 to yield

$$a(t) = -\omega^2 x(t), \qquad (15\text{-}8)$$

which is the hallmark of simple harmonic motion:

🔔 In SHM, the acceleration is proportional to the displacement but opposite in sign, and the two quantities are related by the square of the angular frequency.

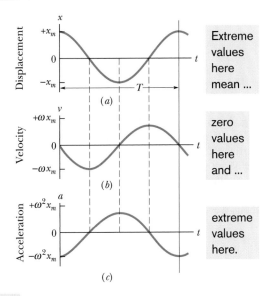

Fig. 15-4 (a) The displacement $x(t)$ of a particle oscillating in SHM with phase angle ϕ equal to zero. The period T marks one complete oscillation. (b) The velocity $v(t)$ of the particle. (c) The acceleration $a(t)$ of the particle.

Thus, as Fig. 15-4 shows, when the displacement has its greatest positive value, the acceleration has its greatest negative value, and conversely. When the displacement is zero, the acceleration is also zero.

15-3 The Force Law for Simple Harmonic Motion

Once we know how the acceleration of a particle varies with time, we can use Newton's second law to learn what force must act on the particle to give it that acceleration. If we combine Newton's second law and Eq. 15-8, we find, for simple harmonic motion,

$$F = ma = -(m\omega^2)x. \tag{15-9}$$

This result—a restoring force that is proportional to the displacement but opposite in sign—is familiar. It is Hooke's law,

$$F = -kx, \tag{15-10}$$

for a spring, the spring constant here being

$$k = m\omega^2. \tag{15-11}$$

We can in fact take Eq. 15-10 as an alternative definition of simple harmonic motion. It says:

 Simple harmonic motion is the motion executed by a particle subject to a force that is proportional to the displacement of the particle but opposite in sign.

The block–spring system of Fig. 15-5 forms a **linear simple harmonic oscillator** (linear oscillator, for short), where "linear" indicates that F is proportional to x rather than to some other power of x. The angular frequency ω of the simple harmonic motion of the block is related to the spring constant k and the mass m of the block by Eq. 15-11, which yields

$$\omega = \sqrt{\frac{k}{m}} \quad \text{(angular frequency)}. \tag{15-12}$$

By combining Eqs. 15-5 and 15-12, we can write, for the **period** of the linear oscillator of Fig. 15-5,

$$T = 2\pi\sqrt{\frac{m}{k}} \quad \text{(period)}. \tag{15-13}$$

Equations 15-12 and 15-13 tell us that a large angular frequency (and thus a small period) goes with a stiff spring (large k) and a light block (small m).

Every oscillating system, be it a diving board or a violin string, has some element of "springiness" and some element of "inertia" or mass, and thus resembles a linear oscillator. In the linear oscillator of Fig. 15-5, these elements are located in separate parts of the system: The springiness is entirely in the spring, which we assume to be massless, and the inertia is entirely in the block, which we assume to be rigid. In a violin string, however, the two elements are both within the string, as you will see in Chapter 16.

Fig. 15-5 A linear simple harmonic oscillator. The surface is frictionless. Like the particle of Fig. 15-1, the block moves in simple harmonic motion once it has been either pulled or pushed away from the $x = 0$ position and released. Its displacement is then given by Eq. 15-3.

☑ CHECKPOINT 2

Which of the following relationships between the force F on a particle and the particle's position x implies simple harmonic oscillation: (a) $F = -5x$, (b) $F = -400x^2$, (c) $F = 10x$, (d) $F = 3x^2$?

Sample Problem

Block–spring SHM, amplitude, acceleration, phase constant

A block whose mass m is 680 g is fastened to a spring whose spring constant k is 65 N/m. The block is pulled a distance $x = 11$ cm from its equilibrium position at $x = 0$ on a frictionless surface and released from rest at $t = 0$.

(a) What are the angular frequency, the frequency, and the period of the resulting motion?

KEY IDEA

The block–spring system forms a linear simple harmonic oscillator, with the block undergoing SHM.

Calculations: The angular frequency is given by Eq. 15-12:

$$\omega = \sqrt{\frac{k}{m}} = \sqrt{\frac{65 \text{ N/m}}{0.68 \text{ kg}}} = 9.78 \text{ rad/s}$$

$$\approx 9.8 \text{ rad/s}. \qquad \text{(Answer)}$$

The frequency follows from Eq. 15-5, which yields

$$f = \frac{\omega}{2\pi} = \frac{9.78 \text{ rad/s}}{2\pi \text{ rad}} = 1.56 \text{ Hz} \approx 1.6 \text{ Hz}. \quad \text{(Answer)}$$

The period follows from Eq. 15-2, which yields

$$T = \frac{1}{f} = \frac{1}{1.56 \text{ Hz}} = 0.64 \text{ s} = 640 \text{ ms}. \quad \text{(Answer)}$$

(b) What is the amplitude of the oscillation?

KEY IDEA

With no friction involved, the mechanical energy of the spring–block system is conserved.

Reasoning: The block is released from rest 11 cm from its equilibrium position, with zero kinetic energy and the elastic potential energy of the system at a maximum. Thus, the block will have zero kinetic energy whenever it is again 11 cm from its equilibrium position, which means it will never be farther than 11 cm from that position. Its maximum displacement is 11 cm:

$$x_m = 11 \text{ cm}. \qquad \text{(Answer)}$$

(c) What is the maximum speed v_m of the oscillating block, and where is the block when it has this speed?

KEY IDEA

The maximum speed v_m is the velocity amplitude ωx_m in Eq. 15-6.

Calculation: Thus, we have

$$v_m = \omega x_m = (9.78 \text{ rad/s})(0.11 \text{ m})$$

$$= 1.1 \text{ m/s}. \qquad \text{(Answer)}$$

This maximum speed occurs when the oscillating block is rushing through the origin; compare Figs. 15-4a and 15-4b, where you can see that the speed is a maximum whenever $x = 0$.

(d) What is the magnitude a_m of the maximum acceleration of the block?

KEY IDEA

The magnitude a_m of the maximum acceleration is the acceleration amplitude $\omega^2 x_m$ in Eq. 15-7.

Calculation: So, we have

$$a_m = \omega^2 x_m = (9.78 \text{ rad/s})^2 (0.11 \text{ m})$$

$$= 11 \text{ m/s}^2. \qquad \text{(Answer)}$$

This maximum acceleration occurs when the block is at the ends of its path. At those points, the force acting on the block has its maximum magnitude; compare Figs. 15-4a and 15-4c, where you can see that the magnitudes of the displacement and acceleration are maximum at the same times.

(e) What is the phase constant ϕ for the motion?

Calculations: Equation 15-3 gives the displacement of the block as a function of time. We know that at time $t = 0$, the block is located at $x = x_m$. Substituting these *initial conditions*, as they are called, into Eq. 15-3 and canceling x_m give us

$$1 = \cos \phi. \qquad (15\text{-}14)$$

Taking the inverse cosine then yields

$$\phi = 0 \text{ rad}. \qquad \text{(Answer)}$$

(Any angle that is an integer multiple of 2π rad also satisfies Eq. 15-14; we chose the smallest angle.)

(f) What is the displacement function $x(t)$ for the spring–block system?

Calculation: The function $x(t)$ is given in general form by Eq. 15-3. Substituting known quantities into that equation gives us

$$x(t) = x_m \cos(\omega t + \phi)$$

$$= (0.11 \text{ m}) \cos[(9.8 \text{ rad/s})t + 0]$$

$$= 0.11 \cos(9.8t), \qquad \text{(Answer)}$$

where x is in meters and t is in seconds.

 Additional examples, video, and practice available at *WileyPLUS*

Finding SHM phase constant from displacement and velocity

At $t = 0$, the displacement $x(0)$ of the block in a linear oscillator like that of Fig. 15-5 is -8.50 cm. (Read $x(0)$ as "x at time zero.") The block's velocity $v(0)$ then is -0.920 m/s, and its acceleration $a(0)$ is $+47.0$ m/s^2.

(a) What is the angular frequency ω of this system?

KEY IDEA

With the block in SHM, Eqs. 15-3, 15-6, and 15-7 give its displacement, velocity, and acceleration, respectively, and each contains ω.

Calculations: Let's substitute $t = 0$ into each to see whether we can solve any one of them for ω. We find

$$x(0) = x_m \cos \phi, \tag{15-15}$$
$$v(0) = -\omega x_m \sin \phi, \tag{15-16}$$

and

$$a(0) = -\omega^2 x_m \cos \phi. \tag{15-17}$$

In Eq. 15-15, ω has disappeared. In Eqs. 15-16 and 15-17, we know values for the left sides, but we do not know x_m and ϕ. However, if we divide Eq. 15-17 by Eq. 15-15, we neatly eliminate both x_m and ϕ and can then solve for ω as

$$\omega = \sqrt{-\frac{a(0)}{x(0)}} = \sqrt{-\frac{47.0 \text{ m/s}^2}{-0.0850 \text{ m}}}$$

$$= 23.5 \text{ rad/s}. \qquad \text{(Answer)}$$

(b) What are the phase constant ϕ and amplitude x_m?

Calculations: We know ω and want ϕ and x_m. If we divide Eq. 15-16 by Eq. 15-15, we eliminate one of those unknowns and reduce the other to a single trig function:

$$\frac{v(0)}{x(0)} = \frac{-\omega x_m \sin \phi}{x_m \cos \phi} = -\omega \tan \phi.$$

Solving for $\tan \phi$, we find

$$\tan \phi = -\frac{v(0)}{\omega x(0)} = -\frac{-0.920 \text{ m/s}}{(23.5 \text{ rad/s})(-0.0850 \text{ m})}$$

$$= -0.461.$$

This equation has two solutions:

$$\phi = -25° \quad \text{and} \quad \phi = 180° + (-25°) = 155°.$$

Normally only the first solution here is displayed by a calculator, but it may not be the physically possible solution. To choose the proper solution, we test them both by using them to compute values for the amplitude x_m. From Eq. 15-15, we find that if $\phi = -25°$, then

$$x_m = \frac{x(0)}{\cos \phi} = \frac{-0.0850 \text{ m}}{\cos(-25°)} = -0.094 \text{ m}.$$

We find similarly that if $\phi = 155°$, then $x_m = 0.094$ m. Because the amplitude of SHM must be a positive constant, the correct phase constant and amplitude here are

$$\phi = 155° \quad \text{and} \quad x_m = 0.094 \text{ m} = 9.4 \text{ cm}. \quad \text{(Answer)}$$

 Additional examples, video, and practice available at *WileyPLUS*

15-4 Energy in Simple Harmonic Motion

In Chapter 8 we saw that the energy of a linear oscillator transfers back and forth between kinetic energy and potential energy, while the sum of the two—the mechanical energy E of the oscillator—remains constant. We now consider this situation quantitatively.

The potential energy of a linear oscillator like that of Fig. 15-5 is associated entirely with the spring. Its value depends on how much the spring is stretched or compressed—that is, on $x(t)$. We can use Eqs. 8-11 and 15-3 to find

$$U(t) = \tfrac{1}{2}kx^2 = \tfrac{1}{2}kx_m^2 \cos^2(\omega t + \phi). \tag{15-18}$$

Caution: A function written in the form $\cos^2 A$ (as here) means $(\cos A)^2$ and is *not* the same as one written $\cos A^2$, which means $\cos(A^2)$.

The kinetic energy of the system of Fig. 15-5 is associated entirely with the block. Its value depends on how fast the block is moving—that is, on $v(t)$. We can use Eq. 15-6 to find

$$K(t) = \tfrac{1}{2}mv^2 = \tfrac{1}{2}m\omega^2 x_m^2 \sin^2(\omega t + \phi). \tag{15-19}$$

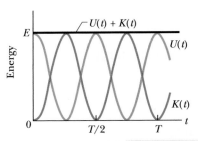

As *time* changes, the energy shifts between the two types, but the total is constant.

Fig. 15-6 (*a*) Potential energy $U(t)$, kinetic energy $K(t)$, and mechanical energy E as functions of time t for a linear harmonic oscillator. Note that all energies are positive and that the potential energy and the kinetic energy peak twice during every period. (*b*) Potential energy $U(x)$, kinetic energy $K(x)$, and mechanical energy E as functions of position x for a linear harmonic oscillator with amplitude x_m. For $x = 0$ the energy is all kinetic, and for $x = \pm x_m$ it is all potential.

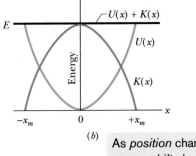

As *position* changes, the energy shifts between the two types, but the total is constant.

If we use Eq. 15-12 to substitute k/m for ω^2, we can write Eq. 15-19 as

$$K(t) = \tfrac{1}{2}mv^2 = \tfrac{1}{2}kx_m^2 \sin^2(\omega t + \phi). \tag{15-20}$$

The mechanical energy follows from Eqs. 15-18 and 15-20 and is

$$E = U + K$$
$$= \tfrac{1}{2}kx_m^2 \cos^2(\omega t + \phi) + \tfrac{1}{2}kx_m^2 \sin^2(\omega t + \phi)$$
$$= \tfrac{1}{2}kx_m^2 [\cos^2(\omega t + \phi) + \sin^2(\omega t + \phi)].$$

For any angle α,

$$\cos^2 \alpha + \sin^2 \alpha = 1.$$

Thus, the quantity in the square brackets above is unity and we have

$$E = U + K = \tfrac{1}{2}kx_m^2. \tag{15-21}$$

The mechanical energy of a linear oscillator is indeed constant and independent of time. The potential energy and kinetic energy of a linear oscillator are shown as functions of time t in Fig. 15-6*a*, and they are shown as functions of displacement x in Fig. 15-6*b*.

You might now understand why an oscillating system normally contains an element of springiness and an element of inertia: The former stores its potential energy and the latter stores its kinetic energy.

 CHECKPOINT 3

In Fig. 15-5, the block has a kinetic energy of 3 J and the spring has an elastic potential energy of 2 J when the block is at $x = +2.0$ cm. (a) What is the kinetic energy when the block is at $x = 0$? What is the elastic potential energy when the block is at (b) $x = -2.0$ cm and (c) $x = -x_m$?

SHM potential energy, kinetic energy, mass dampers

Many tall buildings have *mass dampers*, which are anti-sway devices to prevent them from oscillating in a wind. The device might be a block oscillating at the end of a spring and on a lubricated track. If the building sways, say, eastward, the block also moves eastward but delayed enough so that when it finally moves, the building is then moving back westward. Thus, the motion of the oscillator is out of step with the motion of the building.

Suppose the block has mass $m = 2.72 \times 10^5$ kg and is designed to oscillate at frequency $f = 10.0$ Hz and with amplitude $x_m = 20.0$ cm.

(a) What is the total mechanical energy E of the spring–block system?

The mechanical energy E (the sum of the kinetic energy $K = \frac{1}{2}mv^2$ of the block and the potential energy $U = \frac{1}{2}kx^2$ of the spring) is constant throughout the motion of the oscillator. Thus, we can evaluate E at any point during the motion.

Calculations: Because we are given amplitude x_m of the oscillations, let's evaluate E when the block is at position $x = x_m$, where it has velocity $v = 0$. However, to evaluate U

at that point, we first need to find the spring constant k. From Eq. 15-12 ($\omega = \sqrt{k/m}$) and Eq. 15-5 ($\omega = 2\pi f$), we find

$$k = m\omega^2 = m(2\pi f)^2$$
$$= (2.72 \times 10^5 \text{ kg})(2\pi)^2(10.0 \text{ Hz})^2$$
$$= 1.073 \times 10^9 \text{ N/m}.$$

We can now evaluate E as

$$E = K + U = \tfrac{1}{2}mv^2 + \tfrac{1}{2}kx^2$$
$$= 0 + \tfrac{1}{2}(1.073 \times 10^9 \text{ N/m})(0.20 \text{ m})^2$$
$$= 2.147 \times 10^7 \text{ J} \approx 2.1 \times 10^7 \text{ J}. \qquad \text{(Answer)}$$

(b) What is the block's speed as it passes through the equilibrium point?

Calculations: We want the speed at $x = 0$, where the potential energy is $U = \frac{1}{2}kx^2 = 0$ and the mechanical energy is entirely kinetic energy. So, we can write

$$E = K + U = \tfrac{1}{2}mv^2 + \tfrac{1}{2}kx^2$$
$$2.147 \times 10^7 \text{ J} = \tfrac{1}{2}(2.72 \times 10^5 \text{ kg})v^2 + 0,$$

or
$$v = 12.6 \text{ m/s}. \qquad \text{(Answer)}$$

Because E is entirely kinetic energy, this is the maximum speed v_m.

PLUS Additional examples, video, and practice available at *WileyPLUS*

15-5 An Angular Simple Harmonic Oscillator

Figure 15-7 shows an angular version of a simple harmonic oscillator; the element of springiness or elasticity is associated with the twisting of a suspension wire rather than the extension and compression of a spring as we previously had. The device is called a **torsion pendulum,** with *torsion* referring to the twisting.

If we rotate the disk in Fig. 15-7 by some angular displacement θ from its rest position (where the reference line is at $\theta = 0$) and release it, it will oscillate about that position in **angular simple harmonic motion.** Rotating the disk through an angle θ in either direction introduces a restoring torque given by

$$\tau = -\kappa\theta. \qquad (15\text{-}22)$$

Here κ (Greek *kappa*) is a constant, called the **torsion constant,** that depends on the length, diameter, and material of the suspension wire.

Comparison of Eq. 15-22 with Eq. 15-10 leads us to suspect that Eq. 15-22 is the angular form of Hooke's law, and that we can transform Eq. 15-13, which gives the period of linear SHM, into an equation for the period of angular SHM: We replace the spring constant k in Eq. 15-13 with its equivalent, the constant κ of Eq. 15-22, and we replace the mass m in Eq. 15-13 with *its* equivalent, the rotational inertia I of the oscillating disk. These replacements lead to

$$T = 2\pi\sqrt{\frac{I}{\kappa}} \qquad \text{(torsion pendulum).} \qquad (15\text{-}23)$$

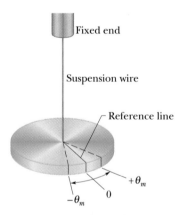

Fig. 15-7 A torsion pendulum is an angular version of a linear simple harmonic oscillator. The disk oscillates in a horizontal plane; the reference line oscillates with angular amplitude θ_m. The twist in the suspension wire stores potential energy as a spring does and provides the restoring torque.

Angular simple harmonic oscillator, rotational inertia, period

Figure 15-8a shows a thin rod whose length L is 12.4 cm and whose mass m is 135 g, suspended at its midpoint from a long wire. Its period T_a of angular SHM is measured to be 2.53 s. An irregularly shaped object, which we call object X, is then hung from the same wire, as in Fig. 15-8b, and its period T_b is found to be 4.76 s. What is the rotational inertia of object X about its suspension axis?

The rotational inertia of either the rod or object X is related to the measured period by Eq. 15-23.

Calculations: In Table 10-2e, the rotational inertia of a thin rod about a perpendicular axis through its midpoint is given as $\frac{1}{12}mL^2$. Thus, we have, for the rod in Fig. 15-8a,

$$I_a = \frac{1}{12}mL^2 = (\tfrac{1}{12})(0.135 \text{ kg})(0.124 \text{ m})^2$$

$$= 1.73 \times 10^{-4} \text{ kg} \cdot \text{m}^2.$$

Now let us write Eq. 15-23 twice, once for the rod and once for object X:

$$T_a = 2\pi\sqrt{\frac{I_a}{\kappa}} \quad \text{and} \quad T_b = 2\pi\sqrt{\frac{I_b}{\kappa}}.$$

The constant κ, which is a property of the wire, is the same for both figures; only the periods and the rotational inertias differ.

Let us square each of these equations, divide the second by the first, and solve the resulting equation for I_b. The result is

$$I_b = I_a\frac{T_b^2}{T_a^2} = (1.73 \times 10^{-4} \text{ kg} \cdot \text{m}^2)\frac{(4.76 \text{ s})^2}{(2.53 \text{ s})^2}$$

$$= 6.12 \times 10^{-4} \text{ kg} \cdot \text{m}^2. \qquad \text{(Answer)}$$

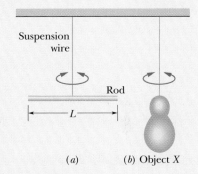

Fig. 15-8 Two torsion pendulums, consisting of (a) a wire and a rod and (b) the same wire and an irregularly shaped object.

PLUS Additional examples, video, and practice available at *WileyPLUS*

15-6 Pendulums

We turn now to a class of simple harmonic oscillators in which the springiness is associated with the gravitational force rather than with the elastic properties of a twisted wire or a compressed or stretched spring.

The Simple Pendulum

If an apple swings on a long thread, does it have simple harmonic motion? If so, what is the period T? To answer, we consider a **simple pendulum,** which consists of a particle of mass m (called the *bob* of the pendulum) suspended from one end of an unstretchable, massless string of length L that is fixed at the other end, as in Fig. 15-9a. The bob is free to swing back and forth in the plane of the page, to the left and right of a vertical line through the pendulum's pivot point.

The forces acting on the bob are the force $\vec{T}$ from the string and the gravitational force $\vec{F}_g$, as shown in Fig. 15-9b, where the string makes an angle θ with the vertical. We resolve $\vec{F}_g$ into a radial component $F_g \cos\theta$ and a component $F_g \sin\theta$ that is tangent to the path taken by the bob. This tangential component produces a restoring torque about the pendulum's pivot point because the component always acts opposite the displacement of the bob so as to bring the bob back toward its central location. That location is called the *equilibrium position* ($\theta = 0$) because the pendulum would be at rest there were it not swinging.

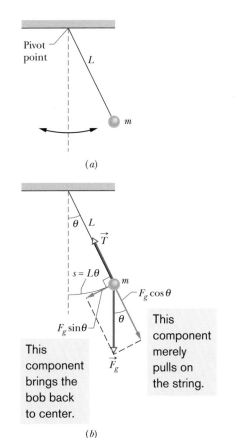

Fig. 15-9 (a) A simple pendulum. (b) The forces acting on the bob are the gravitational force $\vec{F}_g$ and the force $\vec{T}$ from the string. The tangential component $F_g \sin\theta$ of the gravitational force is a restoring force that tends to bring the pendulum back to its central position.

From Eq. 10-41 ($\tau = r_\perp F$), we can write this restoring torque as

$$\tau = -L(F_g \sin \theta), \tag{15-24}$$

where the minus sign indicates that the torque acts to reduce θ and L is the moment arm of the force component $F_g \sin \theta$ about the pivot point. Substituting Eq. 15-24 into Eq. 10-44 ($\tau = I\alpha$) and then substituting mg as the magnitude of F_g, we obtain

$$-L(mg \sin \theta) = I\alpha, \tag{15-25}$$

where I is the pendulum's rotational inertia about the pivot point and α is its angular acceleration about that point.

We can simplify Eq. 15-25 if we assume the angle θ is small, for then we can approximate $\sin \theta$ with θ (expressed in radian measure). (As an example, if $\theta = 5.00° = 0.0873$ rad, then $\sin \theta = 0.0872$, a difference of only about 0.1%.) With that approximation and some rearranging, we then have

$$\alpha = -\frac{mgL}{I}\,\theta. \tag{15-26}$$

This equation is the angular equivalent of Eq. 15-8, the hallmark of SHM. It tells us that the angular acceleration α of the pendulum is proportional to the angular displacement θ but opposite in sign. Thus, as the pendulum bob moves to the right, as in Fig. 15-9a, its acceleration *to the left* increases until the bob stops and begins moving to the left. Then, when it is to the left of the equilibrium position, its acceleration to the right tends to return it to the right, and so on, as it swings back and forth in SHM. More precisely, the motion of a *simple pendulum swinging through only small angles* is approximately SHM. We can state this restriction to small angles another way: The **angular amplitude** θ_m of the motion (the maximum angle of swing) must be small.

Comparing Eqs. 15-26 and 15-8, we see that the angular frequency of the pendulum is $\omega = \sqrt{mgL/I}$. Next, if we substitute this expression for ω into Eq. 15-5 ($\omega = 2\pi/T$), we see that the period of the pendulum may be written as

$$T = 2\pi \sqrt{\frac{I}{mgL}}. \tag{15-27}$$

All the mass of a simple pendulum is concentrated in the mass m of the particle-like bob, which is at radius L from the pivot point. Thus, we can use Eq. 10-33 ($I = mr^2$) to write $I = mL^2$ for the rotational inertia of the pendulum. Substituting this into Eq. 15-27 and simplifying then yield

$$T = 2\pi \sqrt{\frac{L}{g}} \qquad \text{(simple pendulum, small amplitude)}. \tag{15-28}$$

We assume small-angle swinging in this chapter.

The Physical Pendulum

A real pendulum, usually called a **physical pendulum,** can have a complicated distribution of mass. Does it also undergo SHM? If so, what is its period?

Figure 15-10 shows an arbitrary physical pendulum displaced to one side by angle θ. The gravitational force $\vec{F}_g$ acts at its center of mass C, at a distance h from the pivot point O. Comparison of Figs. 15-10 and 15-9b reveals only one important difference between an arbitrary physical pendulum and a simple pendulum. For a physical pendulum the restoring component $F_g \sin \theta$ of the gravitational force has a moment arm of distance h about the pivot point, rather than of string length L. In all other respects, an analysis of the physical pendulum would duplicate our analysis of the simple pendulum up through Eq. 15-27. Again (for small θ_m), we would find that the motion is approximately SHM.

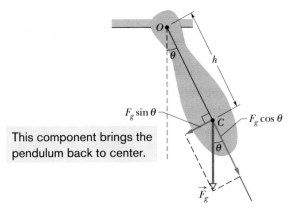

Fig. 15-10 A physical pendulum. The restoring torque is $hF_g \sin \theta$. When $\theta = 0$, center of mass C hangs directly below pivot point O.

$F_g \sin \theta$

This component brings the pendulum back to center.

$F_g \cos \theta$

$\vec{F}_g$

If we replace L with h in Eq. 15-27, we can write the period as

$$T = 2\pi \sqrt{\frac{I}{mgh}} \qquad \text{(physical pendulum, small amplitude).} \qquad (15\text{-}29)$$

As with the simple pendulum, I is the rotational inertia of the pendulum about O. However, now I is not simply mL^2 (it depends on the shape of the physical pendulum), but it is still proportional to m.

A physical pendulum will not swing if it pivots at its center of mass. Formally, this corresponds to putting $h = 0$ in Eq. 15-29. That equation then predicts $T \to \infty$, which implies that such a pendulum will never complete one swing.

Corresponding to any physical pendulum that oscillates about a given pivot point O with period T is a simple pendulum of length L_0 with the same period T. We can find L_0 with Eq. 15-28. The point along the physical pendulum at distance L_0 from point O is called the *center of oscillation* of the physical pendulum for the given suspension point.

Measuring *g*

We can use a physical pendulum to measure the free-fall acceleration g at a particular location on Earth's surface. (Countless thousands of such measurements have been made during geophysical prospecting.)

To analyze a simple case, take the pendulum to be a uniform rod of length L, suspended from one end. For such a pendulum, h in Eq. 15-29, the distance between the pivot point and the center of mass, is $\frac{1}{2}L$. Table 10-2e tells us that the rotational inertia of this pendulum about a perpendicular axis through its center of mass is $\frac{1}{12}mL^2$. From the parallel-axis theorem of Eq. 10-36 ($I = I_{\text{com}} + Mh^2$), we then find that the rotational inertia about a perpendicular axis through one end of the rod is

$$I = I_{\text{com}} + mh^2 = \tfrac{1}{12}mL^2 + m(\tfrac{1}{2}L)^2 = \tfrac{1}{3}mL^2. \qquad (15\text{-}30)$$

If we put $h = \frac{1}{2}L$ and $I = \frac{1}{3}mL^2$ in Eq. 15-29 and solve for g, we find

$$g = \frac{8\pi^2 L}{3T^2}. \qquad (15\text{-}31)$$

Thus, by measuring L and the period T, we can find the value of g at the pendulum's location. (If precise measurements are to be made, a number of refinements are needed, such as swinging the pendulum in an evacuated chamber.)

CHECKPOINT 4

Three physical pendulums, of masses m_0, $2m_0$, and $3m_0$, have the same shape and size and are suspended at the same point. Rank the masses according to the periods of the pendulums, greatest first.

Physical pendulum, period and length

In Fig. 15-11a, a meter stick swings about a pivot point at one end, at distance h from the stick's center of mass.

(a) What is the period of oscillation T?

The stick is not a simple pendulum because its mass is not concentrated in a bob at the end opposite the pivot point—so the stick is a physical pendulum.

Calculations: The period for a physical pendulum is given by Eq. 15-29, for which we need the rotational inertia I of the stick about the pivot point. We can treat the stick as a uniform rod of length L and mass m. Then Eq. 15-30 tells us that $I = \frac{1}{3}mL^2$, and the distance h in Eq. 15-29 is $\frac{1}{2}L$. Substituting these quantities into Eq. 15-29, we find

$$T = 2\pi\sqrt{\frac{I}{mgh}} = 2\pi\sqrt{\frac{\frac{1}{3}mL^2}{mg(\frac{1}{2}L)}} \quad (15\text{-}32)$$

$$= 2\pi\sqrt{\frac{2L}{3g}} \quad (15\text{-}33)$$

$$= 2\pi\sqrt{\frac{(2)(1.00 \text{ m})}{(3)(9.8 \text{ m/s}^2)}} = 1.64 \text{ s}. \quad \text{(Answer)}$$

Note the result is independent of the pendulum's mass m.

(b) What is the distance L_0 between the pivot point O of the stick and the center of oscillation of the stick?

Calculations: We want the length L_0 of the simple pendulum (drawn in Fig. 15-11b) that has the same period as the

Fig. 15-11 (a) A meter stick suspended from one end as a physical pendulum. (b) A simple pendulum whose length L_0 is chosen so that the periods of the two pendulums are equal. Point P on the pendulum of (a) marks the center of oscillation.

physical pendulum (the stick) of Fig. 15-11a. Setting Eqs. 15-28 and 15-33 equal yields

$$T = 2\pi\sqrt{\frac{L_0}{g}} = 2\pi\sqrt{\frac{2L}{3g}}. \quad (15\text{-}34)$$

You can see by inspection that

$$L_0 = \tfrac{2}{3}L \quad (15\text{-}35)$$

$$= (\tfrac{2}{3})(100 \text{ cm}) = 66.7 \text{ cm}. \quad \text{(Answer)}$$

In Fig. 15-11a, point P marks this distance from suspension point O. Thus, point P is the stick's center of oscillation for the given suspension point. Point P would be different for a different suspension choice.

15-7 Simple Harmonic Motion and Uniform Circular Motion

In 1610, Galileo, using his newly constructed telescope, discovered the four principal moons of Jupiter. Over weeks of observation, each moon seemed to him to be moving back and forth relative to the planet in what today we would call simple harmonic motion; the disk of the planet was the midpoint of the motion. The record of Galileo's observations, written in his own hand, is still available. A. P. French of MIT used Galileo's data to work out the position of the moon Callisto relative to Jupiter. In the results shown in Fig. 15-12, the circles are based on Galileo's observations and the curve is a best fit to the data. The curve strongly suggests Eq. 15-3, the displacement function for SHM. A period of about 16.8 days can be measured from the plot.

Actually, Callisto moves with essentially constant speed in an essentially circular orbit around Jupiter. Its true motion—far from being simple har-

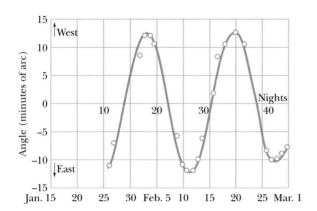

Fig. 15-12   The angle between Jupiter and its moon Callisto as seen from Earth. The circles are based on Galileo's 1610 measurements. The curve is a best fit, strongly suggesting simple harmonic motion. At Jupiter's mean distance from Earth, 10 minutes of arc corresponds to about 2×10^6 km. (Adapted from A. P. French, *Newtonian Mechanics*, W. W. Norton & Company, New York, 1971, p. 288.)

monic—is uniform circular motion. What Galileo saw—and what you can see with a good pair of binoculars and a little patience—is the projection of this uniform circular motion on a line in the plane of the motion. We are led by Galileo's remarkable observations to the conclusion that simple harmonic motion is uniform circular motion viewed edge-on. In more formal language:

> Simple harmonic motion is the projection of uniform circular motion on a diameter of the circle in which the circular motion occurs.

Figure 15-13a gives an example. It shows a *reference particle P′* moving in uniform circular motion with (constant) angular speed ω in a *reference circle*. The radius x_m of the circle is the magnitude of the particle's position vector. At any time t, the angular position of the particle is $\omega t + \phi$, where ϕ is its angular position at $t = 0$.

The projection of particle $P′$ onto the x axis is a point P, which we take to be a second particle. The projection of the position vector of particle $P′$ onto the x axis gives the location $x(t)$ of P. Thus, we find

$$x(t) = x_m \cos(\omega t + \phi), \qquad (15\text{-}36)$$

which is precisely Eq. 15-3. Our conclusion is correct. If reference particle $P′$ moves in uniform circular motion, its projection particle P moves in simple harmonic motion along a diameter of the circle.

Figure 15-13b shows the velocity $\vec{v}$ of the reference particle. From Eq. 10-18 ($v = \omega r$), the magnitude of the velocity vector is ωx_m; its projection on the x axis is

$$v(t) = -\omega x_m \sin(\omega t + \phi), \qquad (15\text{-}37)$$

which is exactly Eq. 15-6. The minus sign appears because the velocity component of P in Fig. 15-13b is directed to the left, in the negative direction of x.

Figure 15-13c shows the radial acceleration $\vec{a}$ of the reference particle. From Eq. 10-23 ($a_r = \omega^2 r$), the magnitude of the radial acceleration vector is $\omega^2 x_m$; its projection on the x axis is

$$a(t) = -\omega^2 x_m \cos(\omega t + \phi), \qquad (15\text{-}38)$$

which is exactly Eq. 15-7. Thus, whether we look at the displacement, the velocity, or the acceleration, the projection of uniform circular motion is indeed simple harmonic motion.

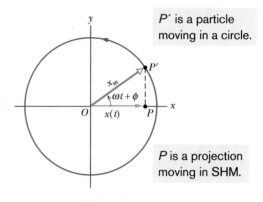

P′ is a particle moving in a circle.

P is a projection moving in SHM.

(a)

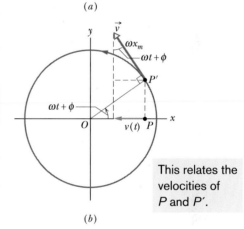

This relates the velocities of P and P′.

(b)

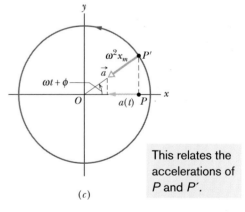

This relates the accelerations of P and P′.

(c)

Fig. 15-13   (a) A reference particle $P′$ moving with uniform circular motion in a reference circle of radius x_m. Its projection P on the x axis executes simple harmonic motion. (b) The projection of the velocity $\vec{v}$ of the reference particle is the velocity of SHM. (c) The projection of the radial acceleration $\vec{a}$ of the reference particle is the acceleration of SHM.

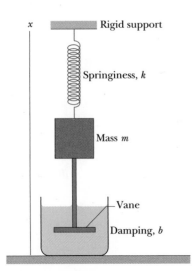

Fig. 15-14 An idealized damped simple harmonic oscillator. A vane immersed in a liquid exerts a damping force on the block as the block oscillates parallel to the x axis.

15-8 Damped Simple Harmonic Motion

A pendulum will swing only briefly underwater, because the water exerts on the pendulum a drag force that quickly eliminates the motion. A pendulum swinging in air does better, but still the motion dies out eventually, because the air exerts a drag force on the pendulum (and friction acts at its support point), transferring energy from the pendulum's motion.

When the motion of an oscillator is reduced by an external force, the oscillator and its motion are said to be **damped.** An idealized example of a damped oscillator is shown in Fig. 15-14, where a block with mass m oscillates vertically on a spring with spring constant k. From the block, a rod extends to a vane (both assumed massless) that is submerged in a liquid. As the vane moves up and down, the liquid exerts an inhibiting drag force on it and thus on the entire oscillating system. With time, the mechanical energy of the block–spring system decreases, as energy is transferred to thermal energy of the liquid and vane.

Let us assume the liquid exerts a **damping force** $\vec{F}_d$ that is proportional to the velocity $\vec{v}$ of the vane and block (an assumption that is accurate if the vane moves slowly). Then, for components along the x axis in Fig. 15-14, we have

$$F_d = -bv, \tag{15-39}$$

where b is a **damping constant** that depends on the characteristics of both the vane and the liquid and has the SI unit of kilogram per second. The minus sign indicates that $\vec{F}_d$ opposes the motion.

The force on the block from the spring is $F_s = -kx$. Let us assume that the gravitational force on the block is negligible relative to F_d and F_s. Then we can write Newton's second law for components along the x axis ($F_{net,x} = ma_x$) as

$$-bv - kx = ma. \tag{15-40}$$

Substituting dx/dt for v and d^2x/dt^2 for a and rearranging give us the differential equation

$$m\frac{d^2x}{dt^2} + b\frac{dx}{dt} + kx = 0. \tag{15-41}$$

The solution of this equation is

$$x(t) = x_m e^{-bt/2m} \cos(\omega' t + \phi), \tag{15-42}$$

where x_m is the amplitude and ω' is the angular frequency of the damped oscillator. This angular frequency is given by

$$\omega' = \sqrt{\frac{k}{m} - \frac{b^2}{4m^2}}. \tag{15-43}$$

If $b = 0$ (there is no damping), then Eq. 15-43 reduces to Eq. 15-12 ($\omega = \sqrt{k/m}$) for the angular frequency of an undamped oscillator, and Eq. 15-42 reduces to Eq. 15-3 for the displacement of an undamped oscillator. If the damping constant is small but not zero (so that $b \ll \sqrt{km}$), then $\omega' \approx \omega$.

We can regard Eq. 15-42 as a cosine function whose amplitude, which is $x_m e^{-bt/2m}$, gradually decreases with time, as Fig. 15-15 suggests. For an undamped oscillator, the mechanical energy is constant and is given by Eq. 15-21 ($E = \frac{1}{2}kx_m^2$). If the oscillator is damped, the mechanical energy is not constant but decreases with time. If the damping is small, we can find $E(t)$ by replacing x_m in Eq. 15-21 with $x_m e^{-bt/2m}$, the amplitude of the damped oscillations. By doing so, we find that

$$E(t) \approx \tfrac{1}{2}kx_m^2 e^{-bt/m}, \tag{15-44}$$

which tells us that, like the amplitude, the mechanical energy decreases exponentially with time.

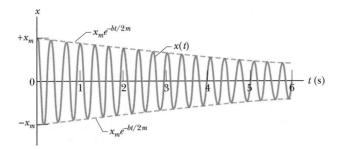

Fig. 15-15 The displacement function $x(t)$ for the damped oscillator of Fig. 15-14. The amplitude, which is $x_m e^{-bt/2m}$, decreases exponentially with time.

✔CHECKPOINT 5

Here are three sets of values for the spring constant, damping constant, and mass for the damped oscillator of Fig. 15-14. Rank the sets according to the time required for the mechanical energy to decrease to one-fourth of its initial value, greatest first.

Set 1	$2k_0$	b_0	m_0
Set 2	k_0	$6b_0$	$4m_0$
Set 3	$3k_0$	$3b_0$	m_0

Sample Problem

Damped harmonic oscillator, time to decay, energy

For the damped oscillator of Fig. 15-14, $m = 250$ g, $k = 85$ N/m, and $b = 70$ g/s.

(a) What is the period of the motion?

KEY IDEA

Because $b \ll \sqrt{km} = 4.6$ kg/s, the period is approximately that of the undamped oscillator.

Calculation: From Eq. 15-13, we then have

$$T = 2\pi \sqrt{\frac{m}{k}} = 2\pi \sqrt{\frac{0.25 \text{ kg}}{85 \text{ N/m}}} = 0.34 \text{ s}. \quad \text{(Answer)}$$

(b) How long does it take for the amplitude of the damped oscillations to drop to half its initial value?

KEY IDEA

The amplitude at time t is displayed in Eq. 15-42 as $x_m e^{-bt/2m}$.

Calculations: The amplitude has the value x_m at $t = 0$. Thus, we must find the value of t for which

$$x_m e^{-bt/2m} = \tfrac{1}{2} x_m.$$

Canceling x_m and taking the natural logarithm of the equation that remains, we have $\ln \tfrac{1}{2}$ on the right side and

$$\ln(e^{-bt/2m}) = -bt/2m$$

on the left side. Thus,

$$t = \frac{-2m \ln \tfrac{1}{2}}{b} = \frac{-(2)(0.25 \text{ kg})(\ln \tfrac{1}{2})}{0.070 \text{ kg/s}}$$

$$= 5.0 \text{ s}. \quad \text{(Answer)}$$

Because $T = 0.34$ s, this is about 15 periods of oscillation.

(c) How long does it take for the mechanical energy to drop to one-half its initial value?

KEY IDEA

From Eq. 15-44, the mechanical energy at time t is $\tfrac{1}{2}kx_m^2 e^{-bt/m}$.

Calculations: The mechanical energy has the value $\tfrac{1}{2}kx_m^2$ at $t = 0$. Thus, we must find the value of t for which

$$\tfrac{1}{2}kx_m^2 e^{-bt/m} = \tfrac{1}{2}(\tfrac{1}{2}kx_m^2).$$

If we divide both sides of this equation by $\tfrac{1}{2}kx_m^2$ and solve for t as we did above, we find

$$t = \frac{-m \ln \tfrac{1}{2}}{b} = \frac{-(0.25 \text{ kg})(\ln \tfrac{1}{2})}{0.070 \text{ kg/s}} = 2.5 \text{ s}. \quad \text{(Answer)}$$

This is exactly half the time we calculated in (b), or about 7.5 periods of oscillation. Figure 15-15 was drawn to illustrate this sample problem.

 Additional examples, video, and practice available at *WileyPLUS*

Fig. 15-16 The displacement amplitude x_m of a forced oscillator varies as the angular frequency ω_d of the driving force is varied. The curves here correspond to three values of the damping constant b.

15-9 Forced Oscillations and Resonance

A person swinging in a swing without anyone pushing it is an example of *free oscillation*. However, if someone pushes the swing periodically, the swing has *forced*, or *driven*, *oscillations*. *Two* angular frequencies are associated with a system undergoing driven oscillations: (1) the *natural* angular frequency ω of the system, which is the angular frequency at which it would oscillate if it were suddenly disturbed and then left to oscillate freely, and (2) the angular frequency ω_d of the external driving force causing the driven oscillations.

We can use Fig. 15-14 to represent an idealized forced simple harmonic oscillator if we allow the structure marked "rigid support" to move up and down at a variable angular frequency ω_d. Such a forced oscillator oscillates at the angular frequency ω_d of the driving force, and its displacement $x(t)$ is given by

$$x(t) = x_m \cos(\omega_d t + \phi),\tag{15-45}$$

where x_m is the amplitude of the oscillations.

How large the displacement amplitude x_m is depends on a complicated function of ω_d and ω. The velocity amplitude v_m of the oscillations is easier to describe: it is greatest when

$$\omega_d = \omega \qquad \text{(resonance)},\tag{15-46}$$

a condition called **resonance**. Equation 15-46 is also *approximately* the condition at which the displacement amplitude x_m of the oscillations is greatest. Thus, if you push a swing at its natural angular frequency, the displacement and velocity amplitudes will increase to large values, a fact that children learn quickly by trial and error. If you push at other angular frequencies, either higher or lower, the displacement and velocity amplitudes will be smaller.

Figure 15-16 shows how the displacement amplitude of an oscillator depends on the angular frequency ω_d of the driving force, for three values of the damping coefficient b. Note that for all three the amplitude is approximately greatest when $\omega_d/\omega = 1$ (the resonance condition of Eq. 15-46). The curves of Fig. 15-16 show that less damping gives a taller and narrower *resonance peak*.

All mechanical structures have one or more natural angular frequencies, and if a structure is subjected to a strong external driving force that matches one of these angular frequencies, the resulting oscillations of the structure may rupture it. Thus, for example, aircraft designers must make sure that none of the natural angular frequencies at which a wing can oscillate matches the angular frequency of the engines in flight. A wing that flaps violently at certain engine speeds would obviously be dangerous.

Resonance appears to be one reason buildings in Mexico City collapsed in September 1985 when a major earthquake (8.1 on the Richter scale) occurred on the western coast of Mexico. The seismic waves from the earthquake should have been too weak to cause extensive damage when they reached Mexico City about 400 km away. However, Mexico City is largely built on an ancient lake bed, where the soil is still soft with water. Although the amplitude of the seismic waves was small in the firmer ground en route to Mexico City, their amplitude substantially increased in the loose soil of the city. Acceleration amplitudes of the waves were as much as $0.20g$, and the angular frequency was (surprisingly) concentrated around 3 rad/s. Not only was the ground severely oscillated, but many intermediate-height buildings had resonant angular frequencies of about 3 rad/s. Most of those buildings collapsed during the shaking (Fig. 15-17), while shorter buildings (with higher resonant angular frequencies) and taller buildings (with lower resonant angular frequencies) remained standing.

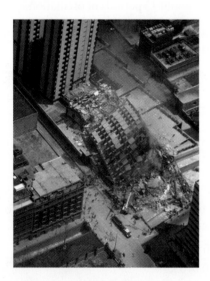

Fig. 15-17 In 1985, buildings of intermediate height collapsed in Mexico City as a result of an earthquake far from the city. Taller and shorter buildings remained standing. *(John T. Barr/Getty Images News and Sport Services)*

REVIEW & SUMMARY

Frequency The *frequency f* of periodic, or oscillatory, motion is the number of oscillations per second. In the SI system, it is measured in hertz:

$$1 \text{ hertz} = 1 \text{ Hz} = 1 \text{ oscillation per second} = 1 \text{ s}^{-1}. \qquad (15\text{-}1)$$

Period The *period T* is the time required for one complete oscillation, or **cycle.** It is related to the frequency by

$$T = \frac{1}{f}. \qquad (15\text{-}2)$$

Simple Harmonic Motion In *simple harmonic motion* (SHM), the displacement $x(t)$ of a particle from its equilibrium position is described by the equation

$$x = x_m \cos(\omega t + \phi) \qquad \text{(displacement)}, \qquad (15\text{-}3)$$

in which x_m is the **amplitude** of the displacement, $\omega t + \phi$ is the **phase** of the motion, and ϕ is the **phase constant.** The **angular frequency** ω is related to the period and frequency of the motion by

$$\omega = \frac{2\pi}{T} = 2\pi f \qquad \text{(angular frequency)}. \qquad (15\text{-}5)$$

Differentiating Eq. 15-3 leads to equations for the particle's SHM velocity and acceleration as functions of time:

$$v = -\omega x_m \sin(\omega t + \phi) \qquad \text{(velocity)} \qquad (15\text{-}6)$$

and

$$a = -\omega^2 x_m \cos(\omega t + \phi) \qquad \text{(acceleration)}. \qquad (15\text{-}7)$$

In Eq. 15-6, the positive quantity ωx_m is the **velocity amplitude** v_m of the motion. In Eq. 15-7, the positive quantity $\omega^2 x_m$ is the **acceleration amplitude** a_m of the motion.

The Linear Oscillator A particle with mass m that moves under the influence of a Hooke's law restoring force given by $F = -kx$ exhibits simple harmonic motion with

$$\omega = \sqrt{\frac{k}{m}} \qquad \text{(angular frequency)} \qquad (15\text{-}12)$$

and

$$T = 2\pi \sqrt{\frac{m}{k}} \qquad \text{(period)}. \qquad (15\text{-}13)$$

Such a system is called a **linear simple harmonic oscillator.**

Energy A particle in simple harmonic motion has, at any time, kinetic energy $K = \frac{1}{2}mv^2$ and potential energy $U = \frac{1}{2}kx^2$. If no friction is present, the mechanical energy $E = K + U$ remains constant even though K and U change.

Pendulums Examples of devices that undergo simple harmonic motion are the **torsion pendulum** of Fig. 15-7, the **simple pendulum** of Fig. 15-9, and the **physical pendulum** of Fig. 15-10. Their periods of oscillation for small oscillations are, respectively,

$$T = 2\pi \sqrt{I/\kappa} \qquad \text{(torsion pendulum)}, \qquad (15\text{-}23)$$

$$T = 2\pi \sqrt{L/g} \qquad \text{(simple pendulum)}, \qquad (15\text{-}28)$$

$$T = 2\pi \sqrt{I/mgh} \qquad \text{(physical pendulum)}. \qquad (15\text{-}29)$$

Simple Harmonic Motion and Uniform Circular Motion Simple harmonic motion is the projection of uniform circular motion onto the diameter of the circle in which the circular motion occurs. Figure 15-13 shows that all parameters of circular motion (position, velocity, and acceleration) project to the corresponding values for simple harmonic motion.

Damped Harmonic Motion The mechanical energy E in a real oscillating system decreases during the oscillations because external forces, such as a drag force, inhibit the oscillations and transfer mechanical energy to thermal energy. The real oscillator and its motion are then said to be **damped.** If the **damping force** is given by $\vec{F}_d = -b\vec{v}$, where $\vec{v}$ is the velocity of the oscillator and b is a **damping constant,** then the displacement of the oscillator is given by

$$x(t) = x_m e^{-bt/2m} \cos(\omega' t + \phi), \qquad (15\text{-}42)$$

where ω', the angular frequency of the damped oscillator, is given by

$$\omega' = \sqrt{\frac{k}{m} - \frac{b^2}{4m^2}}. \qquad (15\text{-}43)$$

If the damping constant is small ($b \ll \sqrt{km}$), then $\omega' \approx \omega$, where ω is the angular frequency of the undamped oscillator. For small b, the mechanical energy E of the oscillator is given by

$$E(t) \approx \tfrac{1}{2}kx_m^2 e^{-bt/m}. \qquad (15\text{-}44)$$

Forced Oscillations and Resonance If an external driving force with angular frequency ω_d acts on an oscillating system with *natural* angular frequency ω, the system oscillates with angular frequency ω_d. The velocity amplitude v_m of the system is greatest when

$$\omega_d = \omega, \qquad (15\text{-}46)$$

a condition called **resonance.** The amplitude x_m of the system is (approximately) greatest under the same condition.

QUESTIONS

1 Which of the following describe ϕ for the SHM of Fig. 15-18a:

(a) $-\pi < \phi < -\pi/2$,

(b) $\pi < \phi < 3\pi/2$,

(c) $-3\pi/2 < \phi < -\pi$?

2 The velocity $v(t)$ of a particle undergoing SHM is graphed in Fig. 15-18b. Is the particle momentarily stationary, headed toward $-x_m$, or headed toward $+x_m$ at (a) point A on the graph and (b) point B? Is the particle at $-x_m$, at $+x_m$, at 0, between $-x_m$ and 0, or between 0 and $+x_m$ when its velocity is represented by (c) point A

and (d) point B? Is the speed of the particle increasing or decreasing at (e) point A and (f) point B?

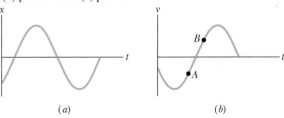

(a)   (b)

Fig. 15-18 Questions 1 and 2.

3 The acceleration $a(t)$ of a particle undergoing SHM is graphed in Fig. 15-19. (a) Which of the labeled points corresponds to the particle at $-x_m$? (b) At point 4, is the velocity of the particle positive, negative, or zero? (c) At point 5, is the particle at $-x_m$, at $+x_m$, at 0, between $-x_m$ and 0, or between 0 and $+x_m$?

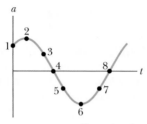

Fig. 15-19 Question 3.

4 Which of the following relationships between the acceleration a and the displacement x of a particle involve SHM: (a) $a = 0.5x$, (b) $a = 400x^2$, (c) $a = -20x$, (d) $a = -3x^2$?

5 You are to complete Fig. 15-20a so that it is a plot of velocity v versus time t for the spring–block oscillator that is shown in Fig. 15-20b for $t = 0$. (a) In Fig. 15-20a, at which lettered point or in what region between the points should the (vertical) v axis intersect the t axis? (For example, should it intersect at point A, or maybe in the region between points A and B?) (b) If the block's velocity is given by $v = -v_m \sin(\omega t + \phi)$, what is the value of ϕ? Make it positive, and if you cannot specify the value (such as $+\pi/2$ rad), then give a range of values (such as between 0 and $\pi/2$).

(a)

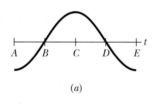

(b)

Fig. 15-20 Question 5.

6 You are to complete Fig. 15-21a so that it is a plot of acceleration a versus time t for the spring–block oscillator that is shown in Fig. 15-21b for $t = 0$. (a) In Fig. 15-21a, at which lettered point or in what region between the points should the (vertical) a axis intersect the t axis? (For example, should it intersect at point A, or maybe in the region between points A and B?) (b) If the block's acceleration is given by $a = -a_m \cos(\omega t + \phi)$, what is the value of ϕ? Make it positive, and if you cannot specify the value (such as $+\pi/2$ rad), then give a range of values (such as between 0 and $\pi/2$).

(a)

(b)

Fig. 15-21 Question 6.

7 Figure 15-22 shows the $x(t)$ curves for three experiments involving a particular spring–box system oscillating in SHM. Rank the curves according to (a) the system's angular frequency, (b) the spring's potential energy at time $t = 0$, (c) the box's kinetic energy at $t = 0$, (d) the box's speed at $t = 0$, and (e) the box's maximum kinetic energy, greatest first.

8 Figure 15-23 shows plots of the kinetic energy K versus position x for three harmonic oscillators that have the same mass. Rank the plots according to

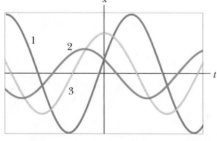

Fig. 15-22 Question 7.

(a) the corresponding spring constant and (b) the corresponding period of the oscillator, greatest first.

9 Figure 15-24 shows three physical pendulums consisting of identical uniform spheres of the same mass that are rigidly connected by identical rods of negligible mass. Each pendulum is vertical and can pivot about suspension point O. Rank the pendulums according to their period of oscillation, greatest first.

10 You are to build the oscillation transfer device shown in Fig. 15-25. It consists of two spring–block systems hanging from a flexible rod. When the spring of system 1 is stretched and then released, the resulting SHM of system 1 at frequency f_1 oscillates the rod. The rod then exerts a driving force on system 2, at the same frequency f_1. You can choose from four springs with spring constants k of 1600, 1500, 1400, and 1200 N/m, and four blocks with masses m of 800, 500, 400, and 200 kg. Mentally determine which spring should go with which block in each of the two systems to maximize the amplitude of oscillations in system 2.

Fig. 15-23 Question 8.

Fig. 15-24 Question 9.

Fig. 15-25 Question 10.

11 In Fig. 15-26, a spring–block system is put into SHM in two experiments. In the first, the block is pulled from the equilibrium position through a displacement d_1 and then released. In the second, it is pulled from the equilibrium position through a greater displacement d_2 and then released. Are the (a) amplitude, (b) period, (c) frequency, (d) maximum kinetic energy, and (e) maximum potential energy in the second experiment greater than, less than, or the same as those in the first experiment?

Fig. 15-26 Question 11.

12 Figure 15-27 gives, for three situations, the displacements $x(t)$ of a pair of simple harmonic oscillators (A and B) that are identical except for phase. For each pair, what phase shift (in radians and in degrees) is needed to shift the curve for A to coincide with the curve for B? Of the many possible answers, choose the shift with the smallest absolute magnitude.

Fig. 15-27 Question 12.

sec. 15-3   The Force Law for Simple Harmonic Motion

•1   An object undergoing simple harmonic motion takes 0.25 s to travel from one point of zero velocity to the next such point. The distance between those points is 36 cm. Calculate the (a) period, (b) frequency, and (c) amplitude of the motion.

•2   A 0.12 kg body undergoes simple harmonic motion of amplitude 8.5 cm and period 0.20 s. (a) What is the magnitude of the maximum force acting on it? (b) If the oscillations are produced by a spring, what is the spring constant?

•3   What is the maximum acceleration of a platform that oscillates at amplitude 2.20 cm and frequency 6.60 Hz?

•4   An automobile can be considered to be mounted on four identical springs as far as vertical oscillations are concerned. The springs of a certain car are adjusted so that the oscillations have a frequency of 3.00 Hz. (a) What is the spring constant of each spring if the mass of the car is 1450 kg and the mass is evenly distributed over the springs? (b) What will be the oscillation frequency if five passengers, averaging 73.0 kg each, ride in the car with an even distribution of mass?

•5   SSM   In an electric shaver, the blade moves back and forth over a distance of 2.0 mm in simple harmonic motion, with frequency 120 Hz. Find (a) the amplitude, (b) the maximum blade speed, and (c) the magnitude of the maximum blade acceleration.

•6   A particle with a mass of 1.00×10^{-20} kg is oscillating with simple harmonic motion with a period of 1.00×10^{-5} s and a maximum speed of 1.00×10^{3} m/s. Calculate (a) the angular frequency and (b) the maximum displacement of the particle.

•7   SSM   A loudspeaker produces a musical sound by means of the oscillation of a diaphragm whose amplitude is limited to 1.00 μm. (a) At what frequency is the magnitude a of the diaphragm's acceleration equal to g? (b) For greater frequencies, is a greater than or less than g?

•8   What is the phase constant for the harmonic oscillator with the position function $x(t)$ given in Fig. 15-28 if the position function has the form $x = x_m \cos(\omega t + \phi)$? The vertical axis scale is set by $x_s = 6.0$ cm.

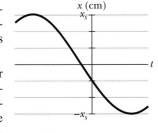

Fig. 15-28   Problem 8.

•9   The function $x = (6.0$ m$) \cos[(3\pi$ rad/s$)t + \pi/3$ rad$]$ gives the simple harmonic motion of a body. At $t = 2.0$ s, what are the (a) displacement, (b) velocity, (c) acceleration, and (d) phase of the motion? Also, what are the (e) frequency and (f) period of the motion?

•10   An oscillating block–spring system takes 0.75 s to begin repeating its motion. Find (a) the period, (b) the frequency in hertz, and (c) the angular frequency in radians per second.

•11   In Fig. 15-29, two identical springs of spring constant 7580 N/m are attached to a block of mass 0.245 kg. What is the frequency of oscillation on the frictionless floor?

Fig. 15-29
Problems 11 and 21.

•12   What is the phase constant for the harmonic oscillator with the velocity function $v(t)$ given in Fig. 15-30 if the position function $x(t)$ has the form $x = x_m \cos(\omega t + \phi)$? The vertical axis scale is set by $v_s = 4.0$ cm/s.

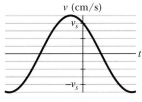

Fig. 15-30   Problem 12.

•13   SSM   An oscillator consists of a block of mass 0.500 kg connected to a spring. When set into oscillation with amplitude 35.0 cm, the oscillator repeats its motion every 0.500 s. Find the (a) period, (b) frequency, (c) angular frequency, (d) spring constant, (e) maximum speed, and (f) magnitude of the maximum force on the block from the spring.

••14   A simple harmonic oscillator consists of a block of mass 2.00 kg attached to a spring of spring constant 100 N/m. When $t = 1.00$ s, the position and velocity of the block are $x = 0.129$ m and $v = 3.415$ m/s. (a) What is the amplitude of the oscillations? What were the (b) position and (c) velocity of the block at $t = 0$ s?

••15   SSM   Two particles oscillate in simple harmonic motion along a common straight-line segment of length A. Each particle has a period of 1.5 s, but they differ in phase by $\pi/6$ rad. (a) How far apart are they (in terms of A) 0.50 s after the lagging particle leaves one end of the path? (b) Are they then moving in the same direction, toward each other, or away from each other?

••16   Two particles execute simple harmonic motion of the same amplitude and frequency along close parallel lines. They pass each other moving in opposite directions each time their displacement is half their amplitude. What is their phase difference?

••17   ILW   An oscillator consists of a block attached to a spring ($k = 400$ N/m). At some time t, the position (measured from the system's equilibrium location), velocity, and acceleration of the block are $x = 0.100$ m, $v = -13.6$ m/s, and $a = -123$ m/s². Calculate (a) the frequency of oscillation, (b) the mass of the block, and (c) the amplitude of the motion.

••18   GO   At a certain harbor, the tides cause the ocean surface to rise and fall a distance d (from highest level to lowest level) in simple harmonic motion, with a period of 12.5 h. How long does it take for the water to fall a distance 0.250d from its highest level?

••19   A block rides on a piston that is moving vertically with simple harmonic motion. (a) If the SHM has period 1.0 s, at what amplitude of motion will the block and piston separate? (b) If the piston has an amplitude of 5.0 cm, what is the maximum frequency for which the block and piston will be in contact continuously?

••20 GO Figure 15-31a is a partial graph of the position function $x(t)$ for a simple harmonic oscillator with an angular frequency of 1.20 rad/s; Fig. 15-31b is a partial graph of the corresponding velocity function $v(t)$. The vertical axis scales are set by $x_s = 5.0$ cm and $v_s = 5.0$ cm/s. What is the phase constant of the SHM if the position function $x(t)$ is in the general form $x = x_m \cos(\omega t + \phi)$?

••21 ILW In Fig. 15-29, two springs are attached to a block that can oscillate over a frictionless floor. If the left spring is removed, the block oscillates at a frequency of 30 Hz. If, instead, the spring on the right is removed, the block oscillates at a frequency of 45 Hz. At what frequency does the block oscillate with both springs attached?

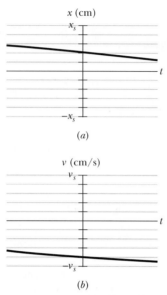

Fig. 15-31 Problem 20.

••22 GO Figure 15-32 shows block 1 of mass 0.200 kg sliding to the right over a frictionless elevated surface at a speed of 8.00 m/s. The block undergoes an elastic collision with stationary block 2, which is attached to a spring of spring constant 1208.5 N/m. (Assume that the spring does not affect the collision.) After the collision, block 2 oscillates in SHM with a period of 0.140 s, and block 1 slides off the opposite end of the elevated surface, landing a distance d from the base of that surface after falling height $h = 4.90$ m. What is the value of d?

Fig. 15-32 Problem 22.

••23 SSM WWW A block is on a horizontal surface (a shake table) that is moving back and forth horizontally with simple harmonic motion of frequency 2.0 Hz. The coefficient of static friction between block and surface is 0.50. How great can the amplitude of the SHM be if the block is not to slip along the surface?

•••24 In Fig. 15-33, two springs are joined and connected to a block of mass 0.245 kg that is set oscillating over a frictionless floor. The springs each have spring constant $k = 6430$ N/m. What is the frequency of the oscillations?

Fig. 15-33 Problem 24.

•••25 GO In Fig. 15-34, a block weighing 14.0 N, which can slide without friction on an incline at angle $\theta = 40.0°$, is connected to the top of the incline by a massless spring of unstretched length 0.450 m and spring constant 120 N/m. (a) How far from the top of

the incline is the block's equilibrium point? (b) If the block is pulled slightly down the incline and released, what is the period of the resulting oscillations?

Fig. 15-34 Problem 25.

•••26 GO In Fig. 15-35, two blocks ($m = 1.8$ kg and $M = 10$ kg) and a spring ($k = 200$ N/m) are arranged on a horizontal, frictionless surface. The coefficient of static friction between the two blocks is 0.40. What amplitude of simple harmonic motion of the spring–blocks system puts the smaller block on the verge of slipping over the larger block?

Fig. 15-35 Problem 26.

sec. 15-4 Energy in Simple Harmonic Motion

•27 SSM When the displacement in SHM is one-half the amplitude x_m, what fraction of the total energy is (a) kinetic energy and (b) potential energy? (c) At what displacement, in terms of the amplitude, is the energy of the system half kinetic energy and half potential energy?

•28 Figure 15-36 gives the one-dimensional potential energy well for a 2.0 kg particle (the function $U(x)$ has the form bx^2 and the vertical axis scale is set by $U_s = 2.0$ J). (a) If the particle passes through the equilibrium position with a velocity of 85 cm/s, will it be turned back before it reaches $x = 15$ cm? (b) If yes, at what position, and if no, what is the speed of the particle at $x = 15$ cm?

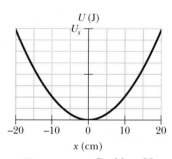

Fig. 15-36 Problem 28.

•29 SSM Find the mechanical energy of a block–spring system having a spring constant of 1.3 N/cm and an oscillation amplitude of 2.4 cm.

•30 An oscillating block–spring system has a mechanical energy of 1.00 J, an amplitude of 10.0 cm, and a maximum speed of 1.20 m/s. Find (a) the spring constant, (b) the mass of the block, and (c) the frequency of oscillation.

•31 ILW A 5.00 kg object on a horizontal frictionless surface is attached to a spring with $k = 1000$ N/m. The object is displaced from equilibrium 50.0 cm horizontally and given an initial velocity of

10.0 m/s back toward the equilibrium position. What are (a) the motion's frequency, (b) the initial potential energy of the block–spring system, (c) the initial kinetic energy, and (d) the motion's amplitude?

•**32** Figure 15-37 shows the kinetic energy K of a simple harmonic oscillator versus its position x. The vertical axis scale is set by $K_s = 4.0$ J. What is the spring constant?

Fig. 15-37 Problem 32.

••**33** GO A block of mass $M = 5.4$ kg, at rest on a horizontal frictionless table, is attached to a rigid support by a spring of constant $k = 6000$ N/m. A bullet of mass $m = 9.5$ g and velocity $\vec{v}$ of magnitude 630 m/s strikes and is embedded in the block (Fig. 15-38). Assuming the compression of the spring is negligible until the bullet is embedded, determine (a) the speed of the block immediately after the collision and (b) the amplitude of the resulting simple harmonic motion.

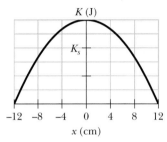

Fig. 15-38 Problem 33.

••**34** GO In Fig. 15-39, block 2 of mass 2.0 kg oscillates on the end of a spring in SHM with a period of 20 ms. The block's position is given by $x = (1.0 \text{ cm}) \cos(\omega t + \pi/2)$. Block 1 of mass 4.0 kg slides toward block 2 with a velocity of magnitude 6.0 m/s, directed along the spring's

Fig. 15-39 Problem 34.

length. The two blocks undergo a completely inelastic collision at time $t = 5.0$ ms. (The duration of the collision is much less than the period of motion.) What is the amplitude of the SHM after the collision?

••**35** A 10 g particle undergoes SHM with an amplitude of 2.0 mm, a maximum acceleration of magnitude 8.0×10^3 m/s², and an unknown phase constant ϕ. What are (a) the period of the motion, (b) the maximum speed of the particle, and (c) the total mechanical energy of the oscillator? What is the magnitude of the force on the particle when the particle is at (d) its maximum displacement and (e) half its maximum displacement?

••**36** If the phase angle for a block–spring system in SHM is $\pi/6$ rad and the block's position is given by $x = x_m \cos(\omega t + \phi)$, what is the ratio of the kinetic energy to the potential energy at time $t = 0$?

•••**37** A massless spring hangs from the ceiling with a small object attached to its lower end. The object is initially held at rest in a position y_i such that the spring is at its rest length. The object is then released from y_i and oscillates up and down, with its lowest position being 10 cm below y_i. (a) What is the frequency of the oscillation? (b) What is the speed of the object when it is 8.0 cm below the initial position? (c) An object of mass 300 g is attached to the

first object, after which the system oscillates with half the original frequency. What is the mass of the first object? (d) How far below y_i is the new equilibrium (rest) position with both objects attached to the spring?

sec. 15-5 An Angular Simple Harmonic Oscillator

•**38** A 95 kg solid sphere with a 15 cm radius is suspended by a vertical wire. A torque of 0.20 N · m is required to rotate the sphere through an angle of 0.85 rad and then maintain that orientation. What is the period of the oscillations that result when the sphere is then released?

••**39** SSM WWW The balance wheel of an old-fashioned watch oscillates with angular amplitude π rad and period 0.500 s. Find (a) the maximum angular speed of the wheel, (b) the angular speed at displacement $\pi/2$ rad, and (c) the magnitude of the angular acceleration at displacement $\pi/4$ rad.

sec. 15-6 Pendulums

•**40** ILW A physical pendulum consists of a meter stick that is pivoted at a small hole drilled through the stick a distance d from the 50 cm mark. The period of oscillation is 2.5 s. Find d.

•**41** SSM In Fig. 15-40, the pendulum consists of a uniform disk with radius $r = 10.0$ cm and mass 500 g attached to a uniform rod with length $L = 500$ mm and mass 270 g. (a) Calculate the rotational inertia of the pendulum about the pivot point. (b) What is the distance between the pivot point and the center of mass of the pendulum? (c) Calculate the period of oscillation.

Fig. 15-40 Problem 41.

•**42** Suppose that a simple pendulum consists of a small 60.0 g bob at the end of a cord of negligible mass. If the angle θ between the cord and the vertical is given by

$$\theta = (0.0800 \text{ rad}) \cos[(4.43 \text{ rad/s})t + \phi],$$

what are (a) the pendulum's length and (b) its maximum kinetic energy?

•**43** (a) If the physical pendulum of Fig. 15-11 and the associated sample problem is inverted and suspended at point P, what is its period of oscillation? (b) Is the period now greater than, less than, or equal to its previous value?

•**44** A physical pendulum consists of two meter-long sticks joined together as shown in Fig. 15-41. What is the pendulum's period of oscillation about a pin inserted through point A at the center of the horizontal stick?

Fig. 15-41 Problem 44.

•45 ⚙ A performer seated on a trapeze is swinging back and forth with a period of 8.85 s. If she stands up, thus raising the center of mass of the *trapeze + performer* system by 35.0 cm, what will be the new period of the system? Treat *trapeze + performer* as a simple pendulum.

•46 A physical pendulum has a center of oscillation at distance $2L/3$ from its point of suspension. Show that the distance between the point of suspension and the center of oscillation for a physical pendulum of any form is I/mh, where I and h have the meanings assigned to them in Eq. 15-29 and m is the mass of the pendulum.

•47 In Fig. 15-42, a physical pendulum consists of a uniform solid disk (of radius $R = 2.35$ cm) supported in a vertical plane by a pivot located a distance $d = 1.75$ cm from the center of the disk. The disk is displaced by a small angle and released. What is the period of the resulting simple harmonic motion?

Fig. 15-42 Problem 47.

••48 A rectangular block, with face lengths $a = 35$ cm and $b = 45$ cm, is to be suspended on a thin horizontal rod running through a narrow hole in the block. The block is then to be set swinging about the rod like a pendulum, through small angles so that it is in SHM. Figure 15-43 shows one possible position of the hole, at distance r from the block's center, along a line connecting the center with a corner. (a) Plot the period of the pendulum versus distance r along that line such that the minimum in the curve is apparent. (b) For what value of r does that minimum occur? There is actually a line of points around the block's center for which the period of swinging has the same minimum value. (c) What shape does that line make?

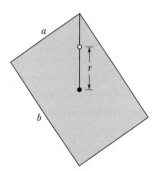

Fig. 15-43 Problem 48.

••49 ⚙ The angle of the pendulum of Fig. 15-9b is given by $\theta = \theta_m \cos[(4.44 \text{ rad/s})t + \phi]$. If at $t = 0$, $\theta = 0.040$ rad and $d\theta/dt = -0.200$ rad/s, what are (a) the phase constant ϕ and (b) the maximum angle θ_m? (*Hint:* Don't confuse the rate $d\theta/dt$ at which θ changes with the ω of the SHM.)

••50 A thin uniform rod (mass = 0.50 kg) swings about an axis that passes through one end of the rod and is perpendicular to the plane of the swing. The rod swings with a period of 1.5 s and an angular amplitude of 10°. (a) What is the length of the rod? (b) What is the maximum kinetic energy of the rod as it swings?

••51 In Fig. 15-44, a stick of length $L = 1.85$ m oscillates as a physical pendulum. (a) What value of distance x between the

stick's center of mass and its pivot point O gives the least period? (b) What is that least period?

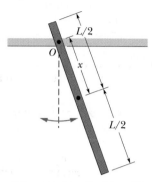

Fig. 15-44 Problem 51.

••52 ⚙ The 3.00 kg cube in Fig. 15-45 has edge lengths $d = 6.00$ cm and is mounted on an axle through its center. A spring ($k = 1200$ N/m) connects the cube's upper corner to a rigid wall. Initially the spring is at its rest length. If the cube is rotated 3° and released, what is the period of the resulting SHM?

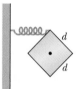

Fig. 15-45 Problem 52.

••53 SSM ILW In the overhead view of Fig. 15-46, a long uniform rod of mass 0.600 kg is free to rotate in a horizontal plane about a vertical axis through its center. A spring with force constant $k = 1850$ N/m is connected horizontally between one end of the rod and a fixed wall. When the rod is in equilibrium, it is parallel to the wall. What is the period of the small oscillations that result when the rod is rotated slightly and released?

Fig. 15-46 Problem 53.

••54 ⚙ In Fig. 15-47a, a metal plate is mounted on an axle through its center of mass. A spring with $k = 2000$ N/m connects a wall with a point on the rim a distance $r = 2.5$ cm from the center of mass. Initially the spring is at its rest length. If the plate is rotated by 7° and released, it rotates about the axle in SHM, with its angular position given by Fig. 15-47b. The horizontal axis scale is set by $t_s = 20$ ms. What is the rotational inertia of the plate about its center of mass?

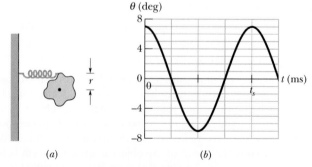

Fig. 15-47 Problem 54.

•••55 A pendulum is formed by pivoting a long thin rod about a point on the rod. In a series of experiments, the period is measured as a function of the distance x between the pivot point and the rod's center. (a) If the rod's length is $L = 2.20$ m and its mass is $m = 22.1$ g, what is the minimum period? (b) If x is chosen to minimize the period and then L is increased, does the period increase, decrease, or remain the same? (c) If, instead, m is increased without L increasing, does the period increase, decrease, or remain the same?

•••56 In Fig. 15-48, a 2.50 kg disk of diameter $D = 42.0$ cm is supported by a rod of length $L = 76.0$ cm and negligible mass that is pivoted at its end. (a) With the massless torsion spring unconnected, what is the period of oscillation? (b) With the torsion spring connected, the rod is vertical at equilibrium. What is the torsion constant of the spring if the period of oscillation has been decreased by 0.500 s?

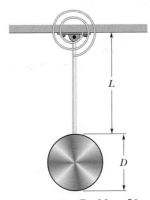

Fig. 15-48 Problem 56.

sec. 15-8   Damped Simple Harmonic Motion

•57 The amplitude of a lightly damped oscillator decreases by 3.0% during each cycle. What percentage of the mechanical energy of the oscillator is lost in each cycle?

•58 In a damped oscillator with $m = 250$ g, $k = 85$ N/m, and $b = 70$ g/s, what is the ratio of the amplitude of the damped oscillations to the initial amplitude at the end of 20 cycles?

•59 **SSM** **WWW** In Fig. 15-14, the block has a mass of 1.50 kg and the spring constant is 8.00 N/m. The damping force is given by $-b(dx/dt)$, where $b = 230$ g/s. The block is pulled down 12.0 cm and released. (a) Calculate the time required for the amplitude of the resulting oscillations to fall to one-third of its initial value. (b) How many oscillations are made by the block in this time?

••60 The suspension system of a 2000 kg automobile "sags" 10 cm when the chassis is placed on it. Also, the oscillation amplitude decreases by 50% each cycle. Estimate the values of (a) the spring constant k and (b) the damping constant b for the spring and shock absorber system of one wheel, assuming each wheel supports 500 kg.

sec. 15-9   Forced Oscillations and Resonance

•61 For Eq. 15-45, suppose the amplitude x_m is given by

$$x_m = \frac{F_m}{[m^2(\omega_d^2 - \omega^2)^2 + b^2\omega_d^2]^{1/2}},$$

where F_m is the (constant) amplitude of the external oscillating force exerted on the spring by the rigid support in Fig. 15-14. At

resonance, what are the (a) amplitude and (b) velocity amplitude of the oscillating object?

•62 Hanging from a horizontal beam are nine simple pendulums of the following lengths: (a) 0.10, (b) 0.30, (c) 0.40, (d) 0.80, (e) 1.2, (f) 2.8, (g) 3.5, (h) 5.0, and (i) 6.2 m. Suppose the beam undergoes horizontal oscillations with angular frequencies in the range from 2.00 rad/s to 4.00 rad/s. Which of the pendulums will be (strongly) set in motion?

••63 A 1000 kg car carrying four 82 kg people travels over a "washboard" dirt road with corrugations 4.0 m apart. The car bounces with maximum amplitude when its speed is 16 km/h. When the car stops, and the people get out, by how much does the car body rise on its suspension?

Additional Problems

64 Although California is known for earthquakes, it has large regions dotted with precariously balanced rocks that would be easily toppled by even a mild earthquake. The rocks have stood this way for thousands of years, suggesting that major earthquakes have not occurred in those regions during that time. If an earthquake were to put such a rock into sinusoidal oscillation (parallel to the ground) with a frequency of 2.2 Hz, an oscillation amplitude of 1.0 cm would cause the rock to topple. What would be the magnitude of the maximum acceleration of the oscillation, in terms of g?

65 A loudspeaker diaphragm is oscillating in simple harmonic motion with a frequency of 440 Hz and a maximum displacement of 0.75 mm. What are the (a) angular frequency, (b) maximum speed, and (c) magnitude of the maximum acceleration?

66 A uniform spring with $k = 8600$ N/m is cut into pieces 1 and 2 of unstretched lengths $L_1 = 7.0$ cm and $L_2 = 10$ cm. What are (a) k_1 and (b) k_2? A block attached to the original spring as in Fig. 15-5 oscillates at 200 Hz. What is the oscillation frequency of the block attached to (c) piece 1 and (d) piece 2?

67 **GO** In Fig. 15-49, three 10 000 kg ore cars are held at rest on a mine railway using a cable that is parallel to the rails, which are inclined at angle $\theta = 30°$. The cable stretches 15 cm just before the coupling between the two lower cars breaks, detaching the lowest car. Assuming that the cable obeys Hooke's law, find the (a) frequency and (b) amplitude of the resulting oscillations of the remaining two cars.

Car that breaks free

Fig. 15-49 Problem 67.

68 A 2.00 kg block hangs from a spring. A 300 g body hung below the block stretches the spring 2.00 cm farther. (a) What is the spring constant? (b) If the 300 g body is removed and the block is set into oscillation, find the period of the motion.

69 **SSM** The piston in the cylinder head of a locomotive has a stroke (twice the amplitude) of 0.76 m. If the piston moves with simple harmonic motion with an angular frequency of 180 rev/min, what is its maximum speed?

70 A wheel is free to rotate about its fixed axle. A spring is attached to one of its spokes a distance r from the axle, as shown in Fig. 15-50. (a) Assuming that the wheel is a hoop of mass m and radius R, what is the angular frequency ω of small oscillations of this system in terms of m, R, r, and the spring constant k? What is ω if (b) $r = R$ and (c) $r = 0$?

Fig. 15-50 Problem 70.

71 A 50.0 g stone is attached to the bottom of a vertical spring and set vibrating. If the maximum speed of the stone is 15.0 cm/s and the period is 0.500 s, find the (a) spring constant of the spring, (b) amplitude of the motion, and (c) frequency of oscillation.

72 A uniform circular disk whose radius R is 12.6 cm is suspended as a physical pendulum from a point on its rim. (a) What is its period? (b) At what radial distance $r < R$ is there a pivot point that gives the same period?

73 **SSM** A vertical spring stretches 9.6 cm when a 1.3 kg block is hung from its end. (a) Calculate the spring constant. This block is then displaced an additional 5.0 cm downward and released from rest. Find the (b) period, (c) frequency, (d) amplitude, and (e) maximum speed of the resulting SHM.

74 A massless spring with spring constant 19 N/m hangs vertically. A body of mass 0.20 kg is attached to its free end and then released. Assume that the spring was unstretched before the body was released. Find (a) how far below the initial position the body descends, and the (b) frequency and (c) amplitude of the resulting SHM.

75 A 4.00 kg block is suspended from a spring with $k = 500$ N/m. A 50.0 g bullet is fired into the block from directly below with a speed of 150 m/s and becomes embedded in the block. (a) Find the amplitude of the resulting SHM. (b) What percentage of the original kinetic energy of the bullet is transferred to mechanical energy of the oscillator?

76 A 55.0 g block oscillates in SHM on the end of a spring with $k = 1500$ N/m according to $x = x_m \cos(\omega t + \phi)$. How long does the block take to move from position $+0.800x_m$ to (a) position $+0.600x_m$ and (b) position $-0.800x_m$?

77 Figure 15-51 gives the position of a 20 g block oscillating in SHM on the end of a spring. The horizontal axis scale is set

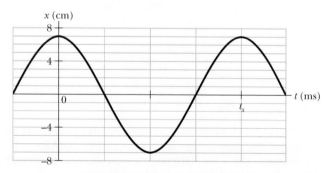

Fig. 15-51 Problems 77 and 78.

by $t_s = 40.0$ ms. What are (a) the maximum kinetic energy of the block and (b) the number of times per second that maximum is reached? (*Hint:* Measuring a slope will probably not be very accurate. Find another approach.)

78 Figure 15-51 gives the position $x(t)$ of a block oscillating in SHM on the end of a spring ($t_s = 40.0$ ms). What are (a) the speed and (b) the magnitude of the radial acceleration of a particle in the corresponding uniform circular motion?

79 Figure 15-52 shows the kinetic energy K of a simple pendulum versus its angle θ from the vertical. The vertical axis scale is set by $K_s = 10.0$ mJ. The pendulum bob has mass 0.200 kg. What is the length of the pendulum?

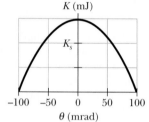

Fig. 15-52 Problem 79.

80 A block is in SHM on the end of a spring, with position given by $x = x_m \cos(\omega t + \phi)$. If $\phi = \pi/5$ rad, then at $t = 0$ what percentage of the total mechanical energy is potential energy?

81 A simple harmonic oscillator consists of a 0.50 kg block attached to a spring. The block slides back and forth along a straight line on a frictionless surface with equilibrium point $x = 0$. At $t = 0$ the block is at $x = 0$ and moving in the positive x direction. A graph of the magnitude of the net force $\vec{F}$ on the block as a function of its position is shown in Fig. 15-53. The vertical scale is set by $F_s = 75.0$ N. What are (a) the amplitude and (b) the period of the motion, (c) the magnitude of the maximum acceleration, and (d) the maximum kinetic energy?

Fig. 15-53 Problem 81.

82 A simple pendulum of length 20 cm and mass 5.0 g is suspended in a race car traveling with constant speed 70 m/s around a circle of radius 50 m. If the pendulum undergoes small oscillations in a radial direction about its equilibrium position, what is the frequency of oscillation?

83 The scale of a spring balance that reads from 0 to 15.0 kg is 12.0 cm long. A package suspended from the balance is found to oscillate vertically with a frequency of 2.00 Hz. (a) What is the spring constant? (b) How much does the package weigh?

84 A 0.10 kg block oscillates back and forth along a straight line on a frictionless horizontal surface. Its displacement from the origin is given by

$$x = (10\ \text{cm}) \cos[(10\ \text{rad/s})t + \pi/2\ \text{rad}].$$

(a) What is the oscillation frequency? (b) What is the maximum speed acquired by the block? (c) At what value of x does this occur? (d) What is the magnitude of the maximum acceleration of the block? (e) At what value of x does this occur? (f) What force, applied to the block by the spring, results in the given oscillation?

85 The end point of a spring oscillates with a period of 2.0 s when a block with mass m is attached to it. When this mass is increased by 2.0 kg, the period is found to be 3.0 s. Find m.

86 The tip of one prong of a tuning fork undergoes SHM of frequency 1000 Hz and amplitude 0.40 mm. For this tip, what is the magnitude of the (a) maximum acceleration, (b) maximum velocity, (c) acceleration at tip displacement 0.20 mm, and (d) velocity at tip displacement 0.20 mm?

87 A flat uniform circular disk has a mass of 3.00 kg and a radius of 70.0 cm. It is suspended in a horizontal plane by a vertical wire attached to its center. If the disk is rotated 2.50 rad about the wire, a torque of 0.0600 N·m is required to maintain that orientation. Calculate (a) the rotational inertia of the disk about the wire, (b) the torsion constant, and (c) the angular frequency of this torsion pendulum when it is set oscillating.

88 A block weighing 20 N oscillates at one end of a vertical spring for which $k = 100$ N/m; the other end of the spring is attached to a ceiling. At a certain instant the spring is stretched 0.30 m beyond its relaxed length (the length when no object is attached) and the block has zero velocity. (a) What is the net force on the block at this instant? What are the (b) amplitude and (c) period of the resulting simple harmonic motion? (d) What is the maximum kinetic energy of the block as it oscillates?

89 A 3.0 kg particle is in simple harmonic motion in one dimension and moves according to the equation

$$x = (5.0 \text{ m}) \cos[(\pi/3 \text{ rad/s})t - \pi/4 \text{ rad}],$$

with t in seconds. (a) At what value of x is the potential energy of the particle equal to half the total energy? (b) How long does the particle take to move to this position x from the equilibrium position?

90 A particle executes linear SHM with frequency 0.25 Hz about the point $x = 0$. At $t = 0$, it has displacement $x = 0.37$ cm and zero velocity. For the motion, determine the (a) period, (b) angular frequency, (c) amplitude, (d) displacement $x(t)$, (e) velocity $v(t)$, (f) maximum speed, (g) magnitude of the maximum acceleration, (h) displacement at $t = 3.0$ s, and (i) speed at $t = 3.0$ s.

91 **SSM** What is the frequency of a simple pendulum 2.0 m long (a) in a room, (b) in an elevator accelerating upward at a rate of 2.0 m/s², and (c) in free fall?

92 A grandfather clock has a pendulum that consists of a thin brass disk of radius $r = 15.00$ cm and mass 1.000 kg that is attached to a long thin rod of negligible mass. The pendulum swings freely about an axis perpendicular to the rod and through the end of the rod opposite the disk, as shown in Fig. 15-54. If the pendulum is to have a period of 2.000 s for small oscillations at a place where $g = 9.800$ m/s², what must be the rod length L to the nearest tenth of a millimeter?

Fig. 15-54 Problem 92.

93 A 4.00 kg block hangs from a spring, extending it 16.0 cm from its unstretched position. (a) What is the spring constant? (b) The block is removed, and a 0.500 kg body is hung from the same spring. If the spring is then stretched and released, what is its period of oscillation?

94 What is the phase constant for SMH with $a(t)$ given in Fig. 15-55 if the position function $x(t)$ has the form $x = x_m \cos(\omega t + \phi)$ and $a_s = 4.0$ m/s²?

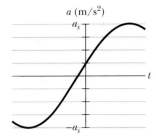

Fig. 15-55 Problem 94.

95 An engineer has an odd-shaped 10 kg object and needs to find its rotational inertia about an axis through its center of mass. The object is supported on a wire stretched along the desired axis. The wire has a torsion constant $\kappa = 0.50$ N·m. If this torsion pendulum oscillates through 20 cycles in 50 s, what is the rotational inertia of the object?

96 A spider can tell when its web has captured, say, a fly because the fly's thrashing causes the web threads to oscillate. A spider can even determine the size of the fly by the frequency of the oscillations. Assume that a fly oscillates on the *capture thread* on which it is caught like a block on a spring. What is the ratio of oscillation frequency for a fly with mass m to a fly with mass 2.5m?

97 A torsion pendulum consists of a metal disk with a wire running through its center and soldered in place. The wire is mounted vertically on clamps and pulled taut. Figure 15-56a gives the magnitude τ of the torque needed to rotate the disk about its center (and thus twist the wire) versus the rotation angle θ. The vertical axis scale is set by $\tau_s = 4.0 \times 10^{-3}$ N·m. The disk is rotated to $\theta = 0.200$ rad and then released. Figure 15-56b shows the resulting oscillation in terms of angular position θ versus time t. The horizontal axis scale is set by $t_s = 0.40$ s. (a) What is the rotational inertia of the disk about its center? (b) What is the maximum angular speed $d\theta/dt$ of the disk? (*Caution:* Do not confuse the (constant) angular frequency of the SHM with the (varying) angular speed of the rotating disk, even though they usually have the same symbol ω. *Hint:* The potential energy U of a torsion pendulum is equal to $\frac{1}{2}\kappa\theta^2$, analogous to $U = \frac{1}{2}kx^2$ for a spring.)

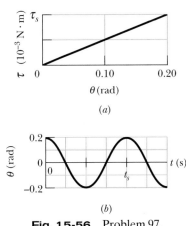

Fig. 15-56 Problem 97.

98   When a 20 N can is hung from the bottom of a vertical spring, it causes the spring to stretch 20 cm. (a) What is the spring constant? (b) This spring is now placed horizontally on a frictionless table. One end of it is held fixed, and the other end is attached to a 5.0 N can. The can is then moved (stretching the spring) and released from rest. What is the period of the resulting oscillation?

99   For a simple pendulum, find the angular amplitude θ_m at which the restoring torque required for simple harmonic motion deviates from the actual restoring torque by 1.0%. (See "Trigonometric Expansions" in Appendix E.)

100   In Fig. 15-57, a solid cylinder attached to a horizontal spring ($k = 3.00$ N/m) rolls without slipping along a horizontal surface. If the system is released from rest when the spring is stretched by 0.250 m, find (a) the translational kinetic en-

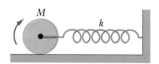

Fig. 15-57   Problem 100.

ergy and (b) the rotational kinetic energy of the cylinder as it passes through the equilibrium position. (c) Show that under these conditions the cylinder's center of mass executes simple harmonic motion with period

$$T = 2\pi \sqrt{\frac{3M}{2k}},$$

where M is the cylinder mass. (*Hint:* Find the time derivative of the total mechanical energy.)

101   SSM   A 1.2 kg block sliding on a horizontal frictionless surface is attached to a horizontal spring with $k = 480$ N/m. Let x be the displacement of the block from the position at which the spring is unstretched. At $t = 0$ the block passes through $x = 0$ with a speed of 5.2 m/s in the positive x direction. What are the (a) frequency and (b) amplitude of the block's motion? (c) Write an expression for x as a function of time.

102   A simple harmonic oscillator consists of an 0.80 kg block attached to a spring ($k = 200$ N/m). The block slides on a horizontal frictionless surface about the equilibrium point $x = 0$ with a total mechanical energy of 4.0 J. (a) What is the amplitude of the oscillation? (b) How many oscillations does the block complete in 10 s? (c) What is the maximum kinetic energy attained by the block? (d) What is the speed of the block at $x = 0.15$ m?

103   A block sliding on a horizontal frictionless surface is

attached to a horizontal spring with a spring constant of 600 N/m. The block executes SHM about its equilibrium position with a period of 0.40 s and an amplitude of 0.20 m. As the block slides through its equilibrium position, a 0.50 kg putty wad is dropped vertically onto the block. If the putty wad sticks to the block, determine (a) the new period of the motion and (b) the new amplitude of the motion.

104   A damped harmonic oscillator consists of a block ($m = 2.00$ kg), a spring ($k = 10.0$ N/m), and a damping force ($F = -bv$). Initially, it oscillates with an amplitude of 25.0 cm; because of the damping, the amplitude falls to three-fourths of this initial value at the completion of four oscillations. (a) What is the value of b? (b) How much energy has been "lost" during these four oscillations?

105   A block weighing 10.0 N is attached to the lower end of a vertical spring ($k = 200.0$ N/m), the other end of which is attached to a ceiling. The block oscillates vertically and has a kinetic energy of 2.00 J as it passes through the point at which the spring is unstretched. (a) What is the period of the oscillation? (b) Use the law of conservation of energy to determine the maximum distance the block moves both above and below the point at which the spring is unstretched. (These are not necessarily the same.) (c) What is the amplitude of the oscillation? (d) What is the maximum kinetic energy of the block as it oscillates?

106   A simple harmonic oscillator consists of a block attached to a spring with $k = 200$ N/m. The block slides on a frictionless surface, with equilibrium point $x = 0$ and amplitude 0.20 m. A graph of the block's velocity v as a function of time t is shown in Fig. 15-58. The horizontal scale is set

Fig. 15-58   Problem 106.

by $t_s = 0.20$ s. What are (a) the period of the SHM, (b) the block's mass, (c) its displacement at $t = 0$, (d) its acceleration at $t = 0.10$ s, and (e) its maximum kinetic energy?

107   The vibration frequencies of atoms in solids at normal temperatures are of the order of 10^{13} Hz. Imagine the atoms to be connected to one another by springs. Suppose that a single silver atom in a solid vibrates with this frequency and that all the other atoms are at rest. Compute the effective spring constant. One mole of silver (6.02×10^{23} atoms) has a mass of 108 g.

WAVES—I

16-1 WHAT IS PHYSICS?

One of the primary subjects of physics is waves. To see how important waves are in the modern world, just consider the music industry. Every piece of music you hear, from some retro-punk band playing in a campus dive to the most eloquent concerto playing on the Web, depends on performers producing waves and your detecting those waves. In between production and detection, the information carried by the waves might need to be transmitted (as in a live performance on the Web) or recorded and then reproduced (as with CDs, DVDs, or the other devices currently being developed in engineering labs worldwide). The financial importance of controlling music waves is staggering, and the rewards to engineers who develop new control techniques can be rich.

This chapter focuses on waves traveling along a stretched string, such as on a guitar. The next chapter focuses on sound waves, such as those produced by a guitar string being played. Before we do all this, though, our first job is to classify the countless waves of the everyday world into basic types.

16-2 Types of Waves

Waves are of three main types:

1. *Mechanical waves.* These waves are most familiar because we encounter them almost constantly; common examples include water waves, sound waves, and seismic waves. All these waves have two central features: They are governed by Newton's laws, and they can exist only within a material medium, such as water, air, and rock.

2. *Electromagnetic waves.* These waves are less familiar, but you use them constantly; common examples include visible and ultraviolet light, radio and television waves, microwaves, x rays, and radar waves. These waves require no material medium to exist. Light waves from stars, for example, travel through the vacuum of space to reach us. All electromagnetic waves travel through a vacuum at the same speed $c = 299\ 792\ 458$ m/s.

3. *Matter waves.* Although these waves are commonly used in modern technology, they are probably very unfamiliar to you. These waves are associated with electrons, protons, and other fundamental particles, and even atoms and molecules. Because we commonly think of these particles as constituting matter, such waves are called matter waves.

Much of what we discuss in this chapter applies to waves of all kinds. However, for specific examples we shall refer to mechanical waves.

16-3 Transverse and Longitudinal Waves

A wave sent along a stretched, taut string is the simplest mechanical wave. If you give one end of a stretched string a single up-and-down jerk, a wave in the form of a single *pulse* travels along the string. This pulse and its motion can occur

Fig. 16-1 (*a*) A single pulse is sent along a stretched string. A typical string element (marked with a dot) moves up once and then down as the pulse passes. The element's motion is perpendicular to the wave's direction of travel, so the pulse is a *transverse wave.* (*b*) A sinusoidal wave is sent along the string. A typical string element moves up and down continuously as the wave passes. This too is a transverse wave.

Fig. 16-2 A sound wave is set up in an air-filled pipe by moving a piston back and forth. Because the oscillations of an element of the air (represented by the dot) are parallel to the direction in which the wave travels, the wave is a *longitudinal wave.*

because the string is under tension. When you pull your end of the string upward, it begins to pull upward on the adjacent section of the string via tension between the two sections. As the adjacent section moves upward, it begins to pull the next section upward, and so on. Meanwhile, you have pulled down on your end of the string. As each section moves upward in turn, it begins to be pulled back downward by neighboring sections that are already on the way down. The net result is that a distortion in the string's shape (a pulse, as in Fig. 16-1*a*) moves along the string at some velocity $\vec{v}$.

If you move your hand up and down in continuous simple harmonic motion, a continuous wave travels along the string at velocity $\vec{v}$. Because the motion of your hand is a sinusoidal function of time, the wave has a sinusoidal shape at any given instant, as in Fig. 16-1*b*; that is, the wave has the shape of a sine curve or a cosine curve.

We consider here only an "ideal" string, in which no friction-like forces within the string cause the wave to die out as it travels along the string. In addition, we assume that the string is so long that we need not consider a wave rebounding from the far end.

One way to study the waves of Fig. 16-1 is to monitor the **wave forms** (shapes of the waves) as they move to the right. Alternatively, we could monitor the motion of an element of the string as the element oscillates up and down while a wave passes through it. We would find that the displacement of every such oscillating string element is *perpendicular* to the direction of travel of the wave, as indicated in Fig. 16-1*b*. This motion is said to be **transverse,** and the wave is said to be a **transverse wave.**

Figure 16-2 shows how a sound wave can be produced by a piston in a long, air-filled pipe. If you suddenly move the piston rightward and then leftward, you can send a pulse of sound along the pipe. The rightward motion of the piston moves the elements of air next to it rightward, changing the air pressure there. The increased air pressure then pushes rightward on the elements of air somewhat farther along the pipe. Moving the piston leftward reduces the air pressure next to it. As a result, first the elements nearest the piston and then farther elements move leftward. Thus, the motion of the air and the change in air pressure travel rightward along the pipe as a pulse.

If you push and pull on the piston in simple harmonic motion, as is being done in Fig. 16-2, a sinusoidal wave travels along the pipe. Because the motion of the elements of air is parallel to the direction of the wave's travel, the motion is said to be **longitudinal,** and the wave is said to be a **longitudinal wave.** In this chapter we focus on transverse waves, and string waves in particular; in Chapter 17 we focus on longitudinal waves, and sound waves in particular.

Both a transverse wave and a longitudinal wave are said to be **traveling waves** because they both travel from one point to another, as from one end of the string to the other end in Fig. 16-1 and from one end of the pipe to the other end in Fig. 16-2. Note that it is the wave that moves from end to end, not the material (string or air) through which the wave moves.

16-4 **Wavelength and Frequency**

To completely describe a wave on a string (and the motion of any element along its length), we need a function that gives the shape of the wave. This means that we need a relation in the form

$$y = h(x, t), \tag{16-1}$$

in which y is the transverse displacement of any string element as a function h of the time t and the position x of the element along the string. In general, a sinusoidal shape like the wave in Fig. 16-1b can be described with h being either a sine or cosine function; both give the same general shape for the wave. In this chapter we use the sine function.

Imagine a sinusoidal wave like that of Fig. 16-1b traveling in the positive direction of an x axis. As the wave sweeps through succeeding elements (that is, very short sections) of the string, the elements oscillate parallel to the y axis. At time t, the displacement y of the element located at position x is given by

$$y(x, t) = y_m \sin(kx - \omega t). \qquad (16\text{-}2)$$

Because this equation is written in terms of position x, it can be used to find the displacements of all the elements of the string as a function of time. Thus, it can tell us the shape of the wave at any given time and how that shape changes as the wave moves along the string.

The names of the quantities in Eq. 16-2 are displayed in Fig. 16-3 and defined next. Before we discuss them, however, let us examine Fig. 16-4, which shows five "snapshots" of a sinusoidal wave traveling in the positive direction of an x axis. The movement of the wave is indicated by the rightward progress of the short arrow pointing to a high point of the wave. From snapshot to snapshot, the short arrow moves to the right with the wave shape, but the string moves *only* parallel to the y axis. To see that, let us follow the motion of the red-dyed string element at $x = 0$. In the first snapshot (Fig. 16-4a), this element is at displacement $y = 0$. In the next snapshot, it is at its extreme downward displacement because a *valley* (or extreme low point) of the wave is passing through it. It then moves back up through $y = 0$. In the fourth snapshot, it is at its extreme upward displacement because a *peak* (or extreme high point) of the wave is passing through it. In the fifth snapshot, it is again at $y = 0$, having completed one full oscillation.

Amplitude and Phase

The **amplitude** y_m of a wave, such as that in Fig. 16-4 , is the magnitude of the maximum displacement of the elements from their equilibrium positions as the wave passes through them. (The subscript m stands for maximum.) Because y_m is a magnitude, it is always a positive quantity, even if it is measured downward instead of upward as drawn in Fig. 16-4a.

The **phase** of the wave is the *argument* $kx - \omega t$ of the sine in Eq. 16-2. As the wave sweeps through a string element at a particular position x, the phase changes linearly with time t. This means that the sine also changes, oscillating between +1 and −1. Its extreme positive value (+1) corresponds to a peak of the wave moving through the element; at that instant the value of y at position x is y_m. Its extreme negative value (−1) corresponds to a valley of the wave moving through the element; at that instant the value of y at position x is $-y_m$. Thus, the sine function and the time-dependent phase of a wave correspond to the oscillation of a string element, and the amplitude of the wave determines the extremes of the element's displacement.

Wavelength and Angular Wave Number

The **wavelength** λ of a wave is the distance (parallel to the direction of the wave's travel) between repetitions of the shape of the wave (or *wave shape*). A typical wavelength is marked in Fig. 16-4a, which is a snapshot of the wave at time $t = 0$. At that time, Eq. 16-2 gives, for the description of the wave shape,

$$y(x, 0) = y_m \sin kx. \qquad (16\text{-}3)$$

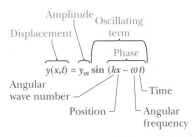

Fig. 16-3 The names of the quantities in Eq. 16-2, for a transverse sinusoidal wave.

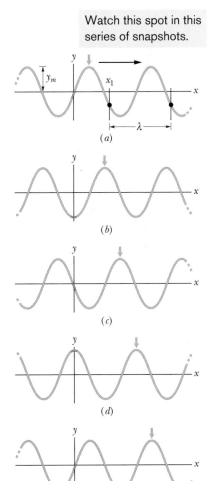

Fig. 16-4 Five "snapshots" of a string wave traveling in the positive direction of an x axis. The amplitude y_m is indicated. A typical wavelength λ, measured from an arbitrary position x_1, is also indicated.

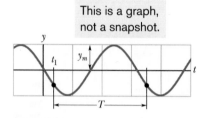

This is a graph, not a snapshot.

Fig. 16-5 A graph of the displacement of the string element at $x = 0$ as a function of time, as the sinusoidal wave of Fig. 16-4 passes through the element. The amplitude y_m is indicated. A typical period T, measured from an arbitrary time t_1, is also indicated.

By definition, the displacement y is the same at both ends of this wavelength—that is, at $x = x_1$ and $x = x_1 + \lambda$. Thus, by Eq. 16-3,

$$y_m \sin kx_1 = y_m \sin k(x_1 + \lambda)$$
$$= y_m \sin(kx_1 + k\lambda). \qquad (16\text{-}4)$$

A sine function begins to repeat itself when its angle (or argument) is increased by 2π rad, so in Eq. 16-4 we must have $k\lambda = 2\pi$, or

$$k = \frac{2\pi}{\lambda} \qquad \text{(angular wave number).} \qquad (16\text{-}5)$$

We call k the **angular wave number** of the wave; its SI unit is the radian per meter, or the inverse meter. (Note that the symbol k here does *not* represent a spring constant as previously.)

Notice that the wave in Fig. 16-4 moves to the right by $\frac{1}{4}\lambda$ from one snapshot to the next. Thus, by the fifth snapshot, it has moved to the right by 1λ.

Period, Angular Frequency, and Frequency

Figure 16-5 shows a graph of the displacement y of Eq. 16-2 versus time t at a certain position along the string, taken to be $x = 0$. If you were to monitor the string, you would see that the single element of the string at that position moves up and down in simple harmonic motion given by Eq. 16-2 with $x = 0$:

$$y(0, t) = y_m \sin(-\omega t)$$
$$= -y_m \sin \omega t \qquad (x = 0). \qquad (16\text{-}6)$$

Here we have made use of the fact that $\sin(-\alpha) = -\sin \alpha$, where α is any angle. Figure 16-5 is a graph of this equation, with displacement plotted versus time; it *does not* show the shape of the wave.

We define the **period** of oscillation T of a wave to be the time any string element takes to move through one full oscillation. A typical period is marked on the graph of Fig. 16-5. Applying Eq. 16-6 to both ends of this time interval and equating the results yield

$$-y_m \sin \omega t_1 = -y_m \sin \omega(t_1 + T)$$
$$= -y_m \sin(\omega t_1 + \omega T). \qquad (16\text{-}7)$$

This can be true only if $\omega T = 2\pi$, or if

$$\omega = \frac{2\pi}{T} \qquad \text{(angular frequency).} \qquad (16\text{-}8)$$

We call ω the **angular frequency** of the wave; its SI unit is the radian per second.

Look back at the five snapshots of a traveling wave in Fig. 16-4. The time between snapshots is $\frac{1}{4}T$. Thus, by the fifth snapshot, every string element has made one full oscillation.

The **frequency** f of a wave is defined as $1/T$ and is related to the angular frequency ω by

$$f = \frac{1}{T} = \frac{\omega}{2\pi} \qquad \text{(frequency).} \qquad (16\text{-}9)$$

Like the frequency of simple harmonic motion in Chapter 15, this frequency f is a number of oscillations per unit time—here, the number made by a string element as the wave moves through it. As in Chapter 15, f is usually measured in hertz or its multiples, such as kilohertz.

CHECKPOINT 1

The figure is a composite of three snapshots, each of a wave traveling along a particular string. The phases for the waves are given by (a) $2x - 4t$, (b) $4x - 8t$, and (c) $8x - 16t$. Which phase corresponds to which wave in the figure?

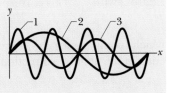

Phase Constant

When a sinusoidal traveling wave is given by the wave function of Eq. 16-2, the wave near $x = 0$ looks like Fig. 16-6a when $t = 0$. Note that at $x = 0$, the displacement is $y = 0$ and the slope is at its maximum positive value. We can generalize Eq. 16-2 by inserting a **phase constant** ϕ in the wave function:

$$y = y_m \sin(kx - \omega t + \phi). \qquad (16\text{-}10)$$

The value of ϕ can be chosen so that the function gives some other displacement and slope at $x = 0$ when $t = 0$. For example, a choice of $\phi = +\pi/5$ rad gives the displacement and slope shown in Fig. 16-6b when $t = 0$. The wave is still sinusoidal with the same values of y_m, k, and ω, but it is now shifted from what you see in Fig. 16-6a (where $\phi = 0$).

16-5 The Speed of a Traveling Wave

Figure 16-7 shows two snapshots of the wave of Eq. 16-2, taken a small time interval Δt apart. The wave is traveling in the positive direction of x (to the right in Fig. 16-7), the entire wave pattern moving a distance Δx in that direction during the interval Δt. The ratio $\Delta x/\Delta t$ (or, in the differential limit, dx/dt) is the **wave speed** v. How can we find its value?

As the wave in Fig. 16-7 moves, each point of the moving wave form, such as point A marked on a peak, retains its displacement y. (Points on the string do not retain their displacement, but points on the wave *form* do.) If point A retains its displacement as it moves, the phase in Eq. 16-2 giving it that displacement must remain a constant:

$$kx - \omega t = \text{a constant.} \qquad (16\text{-}11)$$

Note that although this argument is constant, both x and t are changing. In fact, as t increases, x must also, to keep the argument constant. This confirms that the wave pattern is moving in the positive direction of x.

To find the wave speed v, we take the derivative of Eq. 16-11, getting

$$k \frac{dx}{dt} - \omega = 0$$

or

$$\frac{dx}{dt} = v = \frac{\omega}{k}. \qquad (16\text{-}12)$$

Using Eq. 16-5 ($k = 2\pi/\lambda$) and Eq. 16-8 ($\omega = 2\pi/T$), we can rewrite the wave speed as

$$v = \frac{\omega}{k} = \frac{\lambda}{T} = \lambda f \quad \text{(wave speed).} \qquad (16\text{-}13)$$

The equation $v = \lambda/T$ tells us that the wave speed is one wavelength per period; the wave moves a distance of one wavelength in one period of oscillation.

Equation 16-2 describes a wave moving in the positive direction of x. We can find the equation of a wave traveling in the opposite direction by replacing t in

The effect of the phase constant ϕ is to shift the wave.

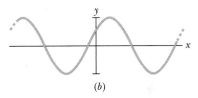

Fig. 16-6 A sinusoidal traveling wave at $t = 0$ with a phase constant ϕ of (a) 0 and (b) $\pi/5$ rad.

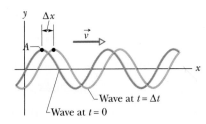

Fig. 16-7 Two snapshots of the wave of Fig. 16-4, at time $t = 0$ and then at time $t = \Delta t$. As the wave moves to the right at velocity $\vec{v}$, the entire curve shifts a distance Δx during Δt. Point A "rides" with the wave form, but the string elements move only up and down.

Eq. 16-2 with $-t$. This corresponds to the condition

$$kx + \omega t = \text{a constant}, \tag{16-14}$$

which (compare Eq. 16-11) requires that x *decrease* with time. Thus, a wave traveling in the negative direction of x is described by the equation

$$y(x, t) = y_m \sin(kx + \omega t). \tag{16-15}$$

If you analyze the wave of Eq. 16-15 as we have just done for the wave of Eq. 16-2, you will find for its velocity

$$\frac{dx}{dt} = -\frac{\omega}{k}. \tag{16-16}$$

The minus sign (compare Eq. 16-12) verifies that the wave is indeed moving in the negative direction of x and justifies our switching the sign of the time variable. Consider now a wave of arbitrary shape, given by

$$y(x, t) = h(kx \pm \omega t), \tag{16-17}$$

where h represents *any* function, the sine function being one possibility. Our previous analysis shows that all waves in which the variables x and t enter into the combination $kx \pm \omega t$ are traveling waves. Furthermore, all traveling waves *must* be of the form of Eq. 16-17. Thus, $y(x, t) = \sqrt{ax + bt}$ represents a possible (though perhaps physically a little bizarre) traveling wave. The function $y(x, t) = \sin(ax^2 - bt)$, on the other hand, does *not* represent a traveling wave.

 CHECKPOINT 2

Here are the equations of three waves:
(1) $y(x, t) = 2\sin(4x - 2t)$, (2) $y(x, t) = \sin(3x - 4t)$, (3) $y(x, t) = 2\sin(3x - 3t)$.
Rank the waves according to their (a) wave speed and (b) maximum speed perpendicular to the wave's direction of travel (the transverse speed), greatest first.

Sample Problem

Transverse wave, amplitude, wavelength, period, velocity

A wave traveling along a string is described by

$$y(x, t) = 0.00327 \sin(72.1x - 2.72t), \tag{16-18}$$

in which the numerical constants are in SI units (0.00327 m, 72.1 rad/m, and 2.72 rad/s).

(a) What is the amplitude of this wave?

KEY IDEA

Equation 16-18 is of the same form as Eq. 16-2,

$$y = y_m \sin(kx - \omega t), \tag{16-19}$$

so we have a sinusoidal wave. By comparing the two equations, we can find the amplitude.

Calculation: We see that

$$y_m = 0.00327 \text{ m} = 3.27 \text{ mm}. \quad \text{(Answer)}$$

(b) What are the wavelength, period, and frequency of this wave?

Calculations: By comparing Eqs. 16-18 and 16-19, we see that the angular wave number and angular frequency are

$$k = 72.1 \text{ rad/m} \quad \text{and} \quad \omega = 2.72 \text{ rad/s}.$$

We then relate wavelength λ to k via Eq. 16-5:

$$\lambda = \frac{2\pi}{k} = \frac{2\pi \text{ rad}}{72.1 \text{ rad/m}}$$
$$= 0.0871 \text{ m} = 8.71 \text{ cm}. \quad \text{(Answer)}$$

Next, we relate T to ω with Eq. 16-8:

$$T = \frac{2\pi}{\omega} = \frac{2\pi \text{ rad}}{2.72 \text{ rad/s}} = 2.31 \text{ s}, \quad \text{(Answer)}$$

and from Eq. 16-9 we have

$$f = \frac{1}{T} = \frac{1}{2.31 \text{ s}} = 0.433 \text{ Hz}. \quad \text{(Answer)}$$

(c) What is the velocity of this wave?

Calculation: The speed of the wave is given by Eq. 16-13:

$$v = \frac{\omega}{k} = \frac{2.72 \text{ rad/s}}{72.1 \text{ rad/m}} = 0.0377 \text{ m/s}$$

$$= 3.77 \text{ cm/s.} \qquad \text{(Answer)}$$

Because the phase in Eq. 16-18 contains the position variable x, the wave is moving along the x axis. Also, because the wave equation is written in the form of Eq. 16-2, the *minus* sign in front of the ωt term indicates that the wave is moving in the *positive* direction of the x axis. (Note that the quantities calculated in (b) and (c) are independent of the amplitude of the wave.)

(d) What is the displacement y of the string at $x = 22.5$ cm and $t = 18.9$ s?

Calculation: Equation 16-18 gives the displacement as a function of position x and time t. Substituting the given values into the equation yields

$$y = 0.00327 \sin(72.1 \times 0.225 - 2.72 \times 18.9)$$

$$= (0.00327 \text{ m}) \sin(-35.1855 \text{ rad})$$

$$= (0.00327 \text{ m})(0.588)$$

$$= 0.00192 \text{ m} = 1.92 \text{ mm.} \qquad \text{(Answer)}$$

Thus, the displacement is positive. (Be sure to change your calculator mode to radians before evaluating the sine. Also, note that we do *not* round off the sine's argument before evaluating the sine. Also note that both terms in the argument are properly in radians, a dimensionless quantity.)

Sample Problem

Transverse wave, transverse velocity, transverse acceleration

In the preceding sample problem, we showed that at $t = 18.9$ s the transverse displacement y of the element of the string at $x = 22.5$ cm due to the wave of Eq. 16-18 is 1.92 mm.

(a) What is u, the transverse velocity of the same element of the string, at that time? (This velocity, which is associated with the transverse oscillation of an element of the string, is in the y direction. Do not confuse it with v, the constant velocity at which the *wave form* travels along the x axis.)

KEY IDEAS

The transverse velocity u is the rate at which the displacement y of the element is changing. In general, that displacement is given by

$$y(x, t) = y_m \sin(kx - \omega t). \qquad (16\text{-}20)$$

For an element at a certain location x, we find the rate of change of y by taking the derivative of Eq. 16-20 with respect to t while treating x as a constant. A derivative taken while one (or more) of the variables is treated as a constant is called a *partial derivative* and is represented by the symbol $\partial/\partial x$ rather than d/dx.

Calculations: Here we have

$$u = \frac{\partial y}{\partial t} = -\omega y_m \cos(kx - \omega t). \qquad (16\text{-}21)$$

Next, substituting numerical values from the preceding sample problem, we obtain

$$u = (-2.72 \text{ rad/s})(3.27 \text{ mm}) \cos(-35.1855 \text{ rad})$$

$$= 7.20 \text{ mm/s.} \qquad \text{(Answer)}$$

Thus, at $t = 18.9$ s, the element of the string at $x = 22.5$ cm is moving in the positive direction of y with a speed of 7.20 mm/s.

(b) What is the transverse acceleration a_y of the same element at that time?

KEY IDEA

The transverse acceleration a_y is the rate at which the transverse velocity of the element is changing.

Calculations: From Eq. 16-21, again treating x as a constant but allowing t to vary, we find

$$a_y = \frac{\partial u}{\partial t} = -\omega^2 y_m \sin(kx - \omega t).$$

Comparison with Eq. 16-20 shows that we can write this as

$$a_y = -\omega^2 y.$$

We see that the transverse acceleration of an oscillating string element is proportional to its transverse displacement but opposite in sign. This is completely consistent with the action of the element itself—namely, that it is moving transversely in simple harmonic motion. Substituting numerical values yields

$$a_y = -(2.72 \text{ rad/s})^2(1.92 \text{ mm})$$

$$= -14.2 \text{ mm/s}^2. \qquad \text{(Answer)}$$

Thus, at $t = 18.9$ s, the element of string at $x = 22.5$ cm is displaced from its equilibrium position by 1.92 mm in the positive y direction and has an acceleration of magnitude 14.2 mm/s^2 in the negative y direction.

 Additional examples, video, and practice available at *WileyPLUS*

16-6 Wave Speed on a Stretched String

The speed of a wave is related to the wave's wavelength and frequency by Eq. 16-13, but *it is set by the properties of the medium*. If a wave is to travel through a medium such as water, air, steel, or a stretched string, it must cause the particles of that medium to oscillate as it passes, which requires both mass (for kinetic energy) and elasticity (for potential energy). Thus, the mass and elasticity determine how fast the wave can travel. Here, we find the wave speed through a medium in terms of these properties in two ways.

Dimensional Analysis

In dimensional analysis we carefully examine the dimensions of all the physical quantities that enter into a given situation to determine the quantities they produce. In this case, we examine mass and elasticity to find a speed v, which has the dimension of length divided by time, or LT^{-1}.

For the mass, we use the mass of a string element, which is the mass m of the string divided by the length l of the string. We call this ratio the *linear density* μ of the string. Thus, $\mu = m/l$, its dimension being mass divided by length, ML^{-1}.

You cannot send a wave along a string unless the string is under tension, which means that it has been stretched and pulled taut by forces at its two ends. The tension τ in the string is equal to the common magnitude of those two forces. As a wave travels along the string, it displaces elements of the string by causing additional stretching, with adjacent sections of string pulling on each other because of the tension. Thus, we can associate the tension in the string with the stretching (elasticity) of the string. The tension and the stretching forces it produces have the dimension of a force—namely, MLT^{-2} (from $F = ma$).

We need to combine μ (dimension ML^{-1}) and τ (dimension MLT^{-2}) to get v (dimension LT^{-1}). A little juggling of various combinations suggests

$$v = C\sqrt{\frac{\tau}{\mu}}, \qquad (16\text{-}22)$$

in which C is a dimensionless constant that cannot be determined with dimensional analysis. In our second approach to determining wave speed, you will see that Eq. 16-22 is indeed correct and that $C = 1$.

Derivation from Newton's Second Law

Instead of the sinusoidal wave of Fig. 16-1b, let us consider a single symmetrical pulse such as that of Fig. 16-8, moving from left to right along a string with speed v. For convenience, we choose a reference frame in which the pulse remains stationary; that is, we run along with the pulse, keeping it constantly in view. In this frame, the string appears to move past us, from right to left in Fig. 16-8, with speed v.

Consider a small string element of length Δl within the pulse, an element that forms an arc of a circle of radius R and subtending an angle 2θ at the center of that circle. A force $\vec{\tau}$ with a magnitude equal to the tension in the string pulls tangentially on this element at each end. The horizontal components of these forces cancel, but the vertical components add to form a radial restoring force $\vec{F}$. In magnitude,

$$F = 2(\tau \sin \theta) \approx \tau(2\theta) = \tau\frac{\Delta l}{R} \qquad \text{(force)}, \qquad (16\text{-}23)$$

where we have approximated $\sin \theta$ as θ for the small angles θ in Fig. 16-8. From that figure, we have also used $2\theta = \Delta l/R$. The mass of the element is given by

$$\Delta m = \mu\,\Delta l \qquad \text{(mass)}, \qquad (16\text{-}24)$$

where μ is the string's linear density.

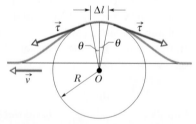

Fig. 16-8 A symmetrical pulse, viewed from a reference frame in which the pulse is stationary and the string appears to move right to left with speed v. We find speed v by applying Newton's second law to a string element of length Δl, located at the top of the pulse.

At the moment shown in Fig. 16-8, the string element Δl is moving in an arc of a circle. Thus, it has a centripetal acceleration toward the center of that circle, given by

$$a = \frac{v^2}{R} \qquad \text{(acceleration)}. \qquad (16\text{-}25)$$

Equations 16-23, 16-24, and 16-25 contain the elements of Newton's second law. Combining them in the form

$$\text{force} = \text{mass} \times \text{acceleration}$$

gives
$$\frac{\tau \, \Delta l}{R} = (\mu \, \Delta l) \, \frac{v^2}{R}.$$

Solving this equation for the speed v yields

$$v = \sqrt{\frac{\tau}{\mu}} \qquad \text{(speed)}, \qquad (16\text{-}26)$$

in exact agreement with Eq. 16-22 if the constant C in that equation is given the value unity. Equation 16-26 gives the speed of the pulse in Fig. 16-8 and the speed of *any* other wave on the same string under the same tension.

Equation 16-26 tells us:

  The speed of a wave along a stretched ideal string depends only on the tension and linear density of the string and not on the frequency of the wave.

The *frequency* of the wave is fixed entirely by whatever generates the wave (for example, the person in Fig. 16-1*b*). The *wavelength* of the wave is then fixed by Eq. 16-13 in the form $\lambda = v/f$.

 CHECKPOINT 3

You send a traveling wave along a particular string by oscillating one end. If you increase the frequency of the oscillations, do (a) the speed of the wave and (b) the wavelength of the wave increase, decrease, or remain the same? If, instead, you increase the tension in the string, do (c) the speed of the wave and (d) the wavelength of the wave increase, decrease, or remain the same?

16-7 Energy and Power of a Wave Traveling Along a String

When we set up a wave on a stretched string, we provide energy for the motion of the string. As the wave moves away from us, it transports that energy as both kinetic energy and elastic potential energy. Let us consider each form in turn.

Kinetic Energy

A string element of mass dm, oscillating transversely in simple harmonic motion as the wave passes through it, has kinetic energy associated with its transverse velocity $\vec{u}$. When the element is rushing through its $y = 0$ position (element b in Fig. 16-9), its transverse velocity—and thus its kinetic energy—is a maximum. When the element is at its extreme position $y = y_m$ (as is element a), its transverse velocity—and thus its kinetic energy—is zero.

Elastic Potential Energy

To send a sinusoidal wave along a previously straight string, the wave must necessarily stretch the string. As a string element of length dx oscillates transversely, its

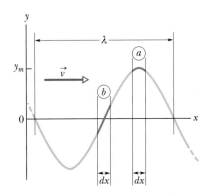

Fig. 16-9   A snapshot of a traveling wave on a string at time $t = 0$. String element a is at displacement $y = y_m$, and string element b is at displacement $y = 0$. The kinetic energy of the string element at each position depends on the transverse velocity of the element. The potential energy depends on the amount by which the string element is stretched as the wave passes through it.

length must increase and decrease in a periodic way if the string element is to fit the sinusoidal wave form. Elastic potential energy is associated with these length changes, just as for a spring.

When the string element is at its $y = y_m$ position (element a in Fig. 16-9), its length has its normal undisturbed value dx, so its elastic potential energy is zero. However, when the element is rushing through its $y = 0$ position, it has maximum stretch and thus maximum elastic potential energy.

Energy Transport

The oscillating string element thus has both its maximum kinetic energy and its maximum elastic potential energy at $y = 0$. In the snapshot of Fig. 16-9, the regions of the string at maximum displacement have no energy, and the regions at zero displacement have maximum energy. As the wave travels along the string, forces due to the tension in the string continuously do work to transfer energy from regions with energy to regions with no energy.

Suppose we set up a wave on a string stretched along a horizontal x axis so that Eq. 16-2 describes the string's displacement. We might send a wave along the string by continuously oscillating one end of the string, as in Fig. 16-1b. In doing so, we continuously provide energy for the motion and stretching of the string—as the string sections oscillate perpendicularly to the x axis, they have kinetic energy and elastic potential energy. As the wave moves into sections that were previously at rest, energy is transferred into those new sections. Thus, we say that the wave *transports* the energy along the string.

The Rate of Energy Transmission

The kinetic energy dK associated with a string element of mass dm is given by

$$dK = \tfrac{1}{2} dm\, u^2, \tag{16-27}$$

where u is the transverse speed of the oscillating string element. To find u, we differentiate Eq. 16-2 with respect to time while holding x constant:

$$u = \frac{\partial y}{\partial t} = -\omega y_m \cos(kx - \omega t). \tag{16-28}$$

Using this relation and putting $dm = \mu\, dx$, we rewrite Eq. 16-27 as

$$dK = \tfrac{1}{2}(\mu\, dx)(-\omega y_m)^2 \cos^2(kx - \omega t). \tag{16-29}$$

Dividing Eq. 16-29 by dt gives the rate at which kinetic energy passes through a string element, and thus the rate at which kinetic energy is carried along by the wave. The ratio dx/dt that then appears on the right of Eq. 16-29 is the wave speed v, so we obtain

$$\frac{dK}{dt} = \tfrac{1}{2}\mu v \omega^2 y_m^2 \cos^2(kx - \omega t). \tag{16-30}$$

The *average* rate at which kinetic energy is transported is

$$\left(\frac{dK}{dt}\right)_{\text{avg}} = \tfrac{1}{2}\mu v \omega^2 y_m^2 \left[\cos^2(kx - \omega t)\right]_{\text{avg}}$$

$$= \tfrac{1}{4}\mu v \omega^2 y_m^2. \tag{16-31}$$

Here we have taken the average over an integer number of wavelengths and have used the fact that the average value of the square of a cosine function over an integer number of periods is $\tfrac{1}{2}$.

Elastic potential energy is also carried along with the wave, and at the same average rate given by Eq. 16-31. Although we shall not examine the proof, you should recall that, in an oscillating system such as a pendulum or a spring–block system, the average kinetic energy and the average potential energy are equal.

The **average power,** which is the average rate at which energy of both kinds is transmitted by the wave, is then

$$P_{avg} = 2 \left(\frac{dK}{dt} \right)_{avg} \tag{16-32}$$

or, from Eq. 16-31,

$$P_{avg} = \tfrac{1}{2} \mu v \omega^2 y_m^2 \quad \text{(average power).} \tag{16-33}$$

The factors μ and v in this equation depend on the material and tension of the string. The factors ω and y_m depend on the process that generates the wave. The dependence of the average power of a wave on the square of its amplitude and also on the square of its angular frequency is a general result, true for waves of all types.

Sample Problem

Average power of a transverse wave

A string has linear density $\mu = 525$ g/m and is under tension $\tau = 45$ N. We send a sinusoidal wave with frequency $f = 120$ Hz and amplitude $y_m = 8.5$ mm along the string. At what average rate does the wave transport energy?

KEY IDEA

The average rate of energy transport is the average power P_{avg} as given by Eq. 16-33.

Calculations: To use Eq. 16-33, we first must calculate an-

gular frequency ω and wave speed v. From Eq. 16-9,

$$\omega = 2\pi f = (2\pi)(120 \text{ Hz}) = 754 \text{ rad/s}.$$

From Eq. 16-26 we have

$$v = \sqrt{\frac{\tau}{\mu}} = \sqrt{\frac{45 \text{ N}}{0.525 \text{ kg/m}}} = 9.26 \text{ m/s}.$$

Equation 16-33 then yields

$$P_{avg} = \tfrac{1}{2} \mu v \omega^2 y_m^2$$
$$= (\tfrac{1}{2})(0.525 \text{ kg/m})(9.26 \text{ m/s})(754 \text{ rad/s})^2(0.0085 \text{ m})^2$$
$$\approx 100 \text{ W.} \qquad \text{(Answer)}$$

 PLUS Additional examples, video, and practice available at *WileyPLUS*

16-8 The Wave Equation

As a wave passes through any element on a stretched string, the element moves perpendicularly to the wave's direction of travel. By applying Newton's second law to the element's motion, we can derive a general differential equation, called the *wave equation*, that governs the travel of waves of any type.

Figure 16-10a shows a snapshot of a string element of mass dm and length ℓ as a wave travels along a string of linear density μ that is stretched along a horizontal x axis. Let us assume that the wave amplitude is small so that the element can be tilted only slightly from the x axis as the wave passes. The force $\vec{F}_2$ on the right end of the element has a magnitude equal to tension τ in the string and is directed slightly upward. The force $\vec{F}_1$ on the left end of the element also has a magnitude equal to the tension τ but is directed slightly downward. Because of the slight curvature of the element, these two forces produce a net force that causes the element to have an upward acceleration a_y. Newton's second law written for y components ($F_{net,y} = ma_y$) gives us

$$F_{2y} - F_{1y} = dm \, a_y. \tag{16-34}$$

Let's analyze this equation in parts.

Mass. The element's mass dm can be written in terms of the string's linear density μ and the element's length ℓ as $dm = \mu\ell$. Because the element can have

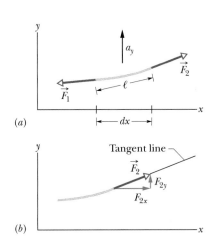

Fig. 16-10 (a) A string element as a sinusoidal transverse wave travels on a stretched string. Forces $\vec{F}_1$ and $\vec{F}_2$ act at the left and right ends, producing acceleration $\vec{a}$ having a vertical component a_y. (b) The force at the element's right end is directed along a tangent to the element's right side.

only a slight tilt, $\ell \approx dx$ (Fig. 16-10a) and we have the approximation

$$dm = \mu\, dx. \tag{16-35}$$

Acceleration. The acceleration a_y in Eq. 16-34 is the second derivative of the displacement y with respect to time:

$$a_y = \frac{d^2 y}{dt^2}. \tag{16-36}$$

Forces. Figure 16-10b shows that $\vec{F}_2$ is tangent to the string at the right end of the string element. Thus we can relate the components of the force to the string slope S_2 at the right end as

$$\frac{F_{2y}}{F_{2x}} = S_2. \tag{16-37}$$

We can also relate the components to the magnitude $F_2 (= \tau)$ with

$$F_2 = \sqrt{F_{2x}^2 + F_{2y}^2}$$

or

$$\tau = \sqrt{F_{2x}^2 + F_{2y}^2}. \tag{16-38}$$

However, because we assume that the element is only slightly tilted, $F_{2y} \ll F_{2x}$ and therefore we can rewrite Eq. 16-38 as

$$\tau = F_{2x}. \tag{16-39}$$

Substituting this into Eq. 16-37 and solving for F_{2y} yield

$$F_{2y} = \tau S_2. \tag{16-40}$$

Similar analysis at the left end of the string element gives us

$$F_{1y} = \tau S_1. \tag{16-41}$$

We can now substitute Eqs. 16-35, 16-36, 16-40, and 16-41 into Eq. 16-34 to write

$$\tau S_2 - \tau S_1 = (\mu\, dx)\frac{d^2 y}{dt^2},$$

or

$$\frac{S_2 - S_1}{dx} = \frac{\mu}{\tau}\frac{d^2 y}{dt^2}. \tag{16-42}$$

Because the string element is short, slopes S_2 and S_1 differ by only a differential amount dS, where S is the slope at any point:

$$S = \frac{dy}{dx}. \tag{16-43}$$

First replacing $S_2 - S_1$ in Eq. 16-42 with dS and then using Eq. 16-43 to substitute dy/dx for S, we find

$$\frac{dS}{dx} = \frac{\mu}{\tau}\frac{d^2 y}{dt^2},$$

$$\frac{d(dy/dx)}{dx} = \frac{\mu}{\tau}\frac{d^2 y}{dt^2},$$

and

$$\frac{\partial^2 y}{\partial x^2} = \frac{\mu}{\tau}\frac{\partial^2 y}{\partial t^2}. \tag{16-44}$$

In the last step, we switched to the notation of partial derivatives because on the left we differentiate only with respect to x and on the right we differentiate only with respect to t. Finally, substituting from Eq. 16-26 ($v = \sqrt{\tau/\mu}$), we find

$$\frac{\partial^2 y}{\partial x^2} = \frac{1}{v^2}\frac{\partial^2 y}{\partial t^2} \qquad \text{(wave equation).} \tag{16-45}$$

This is the general differential equation that governs the travel of waves of all types.

16-9 The Principle of Superposition for Waves

It often happens that two or more waves pass simultaneously through the same region. When we listen to a concert, for example, sound waves from many instruments fall simultaneously on our eardrums. The electrons in the antennas of our radio and television receivers are set in motion by the net effect of many electromagnetic waves from many different broadcasting centers. The water of a lake or harbor may be churned up by waves in the wakes of many boats.

Suppose that two waves travel simultaneously along the same stretched string. Let $y_1(x, t)$ and $y_2(x, t)$ be the displacements that the string would experience if each wave traveled alone. The displacement of the string when the waves overlap is then the algebraic sum

$$y'(x, t) = y_1(x, t) + y_2(x, t). \tag{16-46}$$

This summation of displacements along the string means that

 Overlapping waves algebraically add to produce a **resultant wave** (or **net wave**).

This is another example of the **principle of superposition,** which says that when several effects occur simultaneously, their net effect is the sum of the individual effects.

Figure 16-11 shows a sequence of snapshots of two pulses traveling in opposite directions on the same stretched string. When the pulses overlap, the resultant pulse is their sum. Moreover,

 Overlapping waves do not in any way alter the travel of each other.

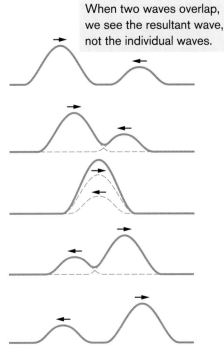

When two waves overlap, we see the resultant wave, not the individual waves.

Fig. 16-11 A series of snapshots that show two pulses traveling in opposite directions along a stretched string. The superposition principle applies as the pulses move through each other.

16-10 Interference of Waves

Suppose we send two sinusoidal waves of the same wavelength and amplitude in the same direction along a stretched string. The superposition principle applies. What resultant wave does it predict for the string?

The resultant wave depends on the extent to which the waves are *in phase* (in step) with respect to each other—that is, how much one wave form is shifted from the other wave form. If the waves are exactly in phase (so that the peaks and valleys of one are exactly aligned with those of the other), they combine to double the displacement of either wave acting alone. If they are exactly out of phase (the peaks of one are exactly aligned with the valleys of the other), they combine to cancel everywhere, and the string remains straight. We call this phenomenon of combining waves **interference,** and the waves are said to **interfere.** (These terms refer only to the wave displacements; the travel of the waves is unaffected.)

Let one wave traveling along a stretched string be given by

$$y_1(x, t) = y_m \sin(kx - \omega t) \tag{16-47}$$

and another, shifted from the first, by

$$y_2(x, t) = y_m \sin(kx - \omega t + \phi). \tag{16-48}$$

These waves have the same angular frequency ω (and thus the same frequency f), the same angular wave number k (and thus the same wavelength λ), and the same amplitude y_m. They both travel in the positive direction of the x axis, with the same speed, given by Eq. 16-26. They differ only by a constant angle ϕ, the phase constant. These waves are said to be *out of phase* by ϕ or to have a *phase difference* of ϕ, or one wave is said to be *phase-shifted* from the other by ϕ.

Displacement
$$y'(x,t) = \underbrace{[2y_m \cos \tfrac{1}{2}\phi]}_{\substack{\text{Magnitude} \\ \text{gives} \\ \text{amplitude}}} \underbrace{\sin(kx - \omega t + \tfrac{1}{2}\phi)}_{\substack{\text{Oscillating} \\ \text{term}}}$$

Fig. 16-12 The resultant wave of Eq. 16-51, due to the interference of two sinusoidal transverse waves, is also a sinusoidal transverse wave, with an amplitude and an oscillating term.

From the principle of superposition (Eq. 16-46), the resultant wave is the algebraic sum of the two interfering waves and has displacement

$$y'(x, t) = y_1(x, t) + y_2(x, t)$$
$$= y_m \sin(kx - \omega t) + y_m \sin(kx - \omega t + \phi). \quad (16\text{-}49)$$

In Appendix E we see that we can write the sum of the sines of two angles α and β as

$$\sin \alpha + \sin \beta = 2 \sin \tfrac{1}{2}(\alpha + \beta) \cos \tfrac{1}{2}(\alpha - \beta). \quad (16\text{-}50)$$

Applying this relation to Eq. 16-49 leads to

$$y'(x, t) = [2y_m \cos \tfrac{1}{2}\phi] \sin(kx - \omega t + \tfrac{1}{2}\phi). \quad (16\text{-}51)$$

As Fig. 16-12 shows, the resultant wave is also a sinusoidal wave traveling in the direction of increasing x. It is the only wave you would actually see on the string (you would *not* see the two interfering waves of Eqs. 16-47 and 16-48).

> If two sinusoidal waves of the same amplitude and wavelength travel in the *same* direction along a stretched string, they interfere to produce a resultant sinusoidal wave traveling in that direction.

The resultant wave differs from the interfering waves in two respects: (1) its phase constant is $\tfrac{1}{2}\phi$, and (2) its amplitude y'_m is the magnitude of the quantity in the brackets in Eq. 16-51:

$$y'_m = |2y_m \cos \tfrac{1}{2}\phi| \quad \text{(amplitude).} \quad (16\text{-}52)$$

If $\phi = 0$ rad (or 0°), the two interfering waves are exactly in phase, as in Fig. 16-13a. Then Eq. 16-51 reduces to

$$y'(x, t) = 2y_m \sin(kx - \omega t) \quad (\phi = 0). \quad (16\text{-}53)$$

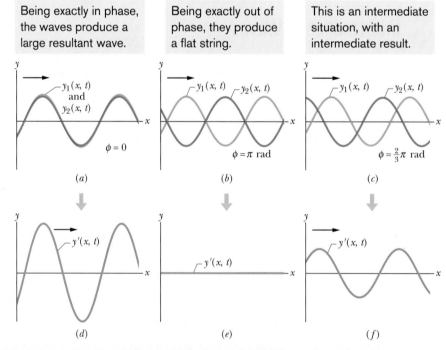

Fig. 16-13 Two identical sinusoidal waves, $y_1(x, t)$ and $y_2(x, t)$, travel along a string in the positive direction of an x axis. They interfere to give a resultant wave $y'(x, t)$. The resultant wave is what is actually seen on the string. The phase difference ϕ between the two interfering waves is (a) 0 rad or 0°, (b) π rad or 180°, and (c) $\tfrac{2}{3}\pi$ rad or 120°. The corresponding resultant waves are shown in (d), (e), and (f).

This resultant wave is plotted in Fig. 16-13d. Note from both that figure and Eq. 16-53 that the amplitude of the resultant wave is twice the amplitude of either interfering wave. That is the greatest amplitude the resultant wave can have, because the cosine term in Eqs. 16-51 and 16-52 has its greatest value (unity) when $\phi = 0$. Interference that produces the greatest possible amplitude is called *fully constructive interference.*

If $\phi = \pi$ rad (or 180°), the interfering waves are exactly out of phase as in Fig. 16-13b. Then $\cos\frac{1}{2}\phi$ becomes $\cos\pi/2 = 0$, and the amplitude of the resultant wave as given by Eq. 16-52 is zero. We then have, for all values of x and t,

$$y'(x, t) = 0 \qquad (\phi = \pi\,\text{rad}). \qquad (16\text{-}54)$$

The resultant wave is plotted in Fig. 16-13e. Although we sent two waves along the string, we see no motion of the string. This type of interference is called *fully destructive interference.*

Because a sinusoidal wave repeats its shape every 2π rad, a phase difference of $\phi = 2\pi$ rad (or 360°) corresponds to a shift of one wave relative to the other wave by a distance equivalent to one wavelength. Thus, phase differences can be described in terms of wavelengths as well as angles. For example, in Fig. 16-13b the waves may be said to be 0.50 wavelength out of phase. Table 16-1 shows some other examples of phase differences and the interference they produce. Note that when interference is neither fully constructive nor fully destructive, it is called *intermediate interference.* The amplitude of the resultant wave is then intermediate between 0 and $2y_m$. For example, from Table 16-1, if the interfering waves have a phase difference of 120° ($\phi = \frac{2}{3}\pi$ rad = 0.33 wavelength), then the resultant wave has an amplitude of y_m, the same as that of the interfering waves (see Figs. 16-13c and f).

Two waves with the same wavelength are in phase if their phase difference is zero or any integer number of wavelengths. Thus, the integer part of any phase difference *expressed in wavelengths* may be discarded. For example, a phase difference of 0.40 wavelength (an intermediate interference, close to fully destructive interference) is equivalent in every way to one of 2.40 wavelengths, and so the simpler of the two numbers can be used in computations.

Table 16-1

Phase Difference and Resulting Interference Types[a]

Phase Difference, in			Amplitude of Resultant Wave	Type of Interference
Degrees	Radians	Wavelengths		
0	0	0	$2y_m$	Fully constructive
120	$\frac{2}{3}\pi$	0.33	y_m	Intermediate
180	π	0.50	0	Fully destructive
240	$\frac{4}{3}\pi$	0.67	y_m	Intermediate
360	2π	1.00	$2y_m$	Fully constructive
865	15.1	2.40	$0.60y_m$	Intermediate

[a]The phase difference is between two otherwise identical waves, with amplitude y_m, moving in the same direction.

CHECKPOINT 4

Here are four possible phase differences between two identical waves, expressed in wavelengths: 0.20, 0.45, 0.60, and 0.80. Rank them according to the amplitude of the resultant wave, greatest first.

Interference of two waves, same direction, same amplitude

Two identical sinusoidal waves, moving in the same direction along a stretched string, interfere with each other. The amplitude y_m of each wave is 9.8 mm, and the phase difference ϕ between them is 100°.

(a) What is the amplitude y'_m of the resultant wave due to the interference, and what is the type of this interference?

These are identical sinusoidal waves traveling in the *same direction* along a string, so they interfere to produce a sinusoidal traveling wave.

Calculations: Because they are identical, the waves have the *same amplitude*. Thus, the amplitude y'_m of the resultant wave is given by Eq. 16-52:

$$y'_m = |2y_m \cos \tfrac{1}{2}\phi| = |(2)(9.8 \text{ mm}) \cos(100°/2)|$$
$$= 13 \text{ mm}. \qquad \text{(Answer)}$$

We can tell that the interference is *intermediate* in two ways. The phase difference is between 0 and 180°, and, correspondingly, the amplitude y'_m is between 0 and $2y_m$ (= 19.6 mm).

(b) What phase difference, in radians and wavelengths, will give the resultant wave an amplitude of 4.9 mm?

Calculations: Now we are given y'_m and seek ϕ. From Eq. 16-52,

$$y'_m = |2y_m \cos \tfrac{1}{2}\phi|,$$

we now have

$$4.9 \text{ mm} = (2)(9.8 \text{ mm}) \cos \tfrac{1}{2}\phi,$$

which gives us (with a calculator in the radian mode)

$$\phi = 2 \cos^{-1} \frac{4.9 \text{ mm}}{(2)(9.8 \text{ mm})}$$
$$= \pm 2.636 \text{ rad} \approx \pm 2.6 \text{ rad}. \qquad \text{(Answer)}$$

There are two solutions because we can obtain the same resultant wave by letting the first wave *lead* (travel ahead of) or *lag* (travel behind) the second wave by 2.6 rad. In wavelengths, the phase difference is

$$\frac{\phi}{2\pi \text{ rad/wavelength}} = \frac{\pm 2.636 \text{ rad}}{2\pi \text{ rad/wavelength}}$$
$$= \pm 0.42 \text{ wavelength}. \qquad \text{(Answer)}$$

 Additional examples, video, and practice available at *WileyPLUS*

16-11 **Phasors**

We can represent a string wave (or any other type of wave) vectorially with a **phasor.** In essence, a phasor is a vector that has a magnitude equal to the amplitude of the wave and that rotates around an origin; the angular speed of the phasor is equal to the angular frequency ω of the wave. For example, the wave

$$y_1(x, t) = y_{m1} \sin(kx - \omega t) \qquad (16\text{-}55)$$

is represented by the phasor shown in Figs. 16-14a to d. The magnitude of the phasor is the amplitude y_{m1} of the wave. As the phasor rotates around the origin at angular speed ω, its projection y_1 on the vertical axis varies sinusoidally, from a maximum of y_{m1} through zero to a minimum of $-y_{m1}$ and then back to y_{m1}. This variation corresponds to the sinusoidal variation in the displacement y_1 of any point along the string as the wave passes through that point.

When two waves travel along the same string in the same direction, we can represent them and their resultant wave in a *phasor diagram*. The phasors in Fig. 16-14e represent the wave of Eq. 16-55 and a second wave given by

$$y_2(x, t) = y_{m2} \sin(kx - \omega t + \phi). \qquad (16\text{-}56)$$

This second wave is phase-shifted from the first wave by phase constant ϕ. Because the phasors rotate at the same angular speed ω, the angle between the two phasors is always ϕ. If ϕ is a *positive* quantity, then the phasor for wave 2 *lags* the phasor for wave 1 as they rotate, as drawn in Fig. 16-14e. If ϕ is a negative quantity, then the phasor for wave 2 *leads* the phasor for wave 1.

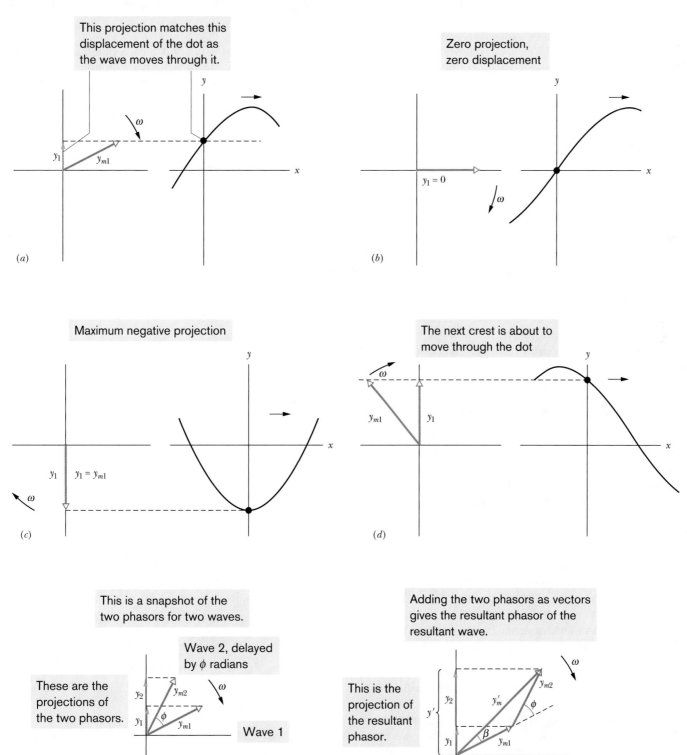

Fig. 16-14 (a)–(d) A phasor of magnitude y_{m1} rotating about an origin at angular speed ω represents a sinusoidal wave. The phasor's projection y_1 on the vertical axis represents the displacement of a point through which the wave passes. (e) A second phasor, also of angular speed ω but of magnitude y_{m2} and rotating at a constant angle ϕ from the first phasor, represents a second wave, with a phase constant ϕ. (f) The resultant wave is represented by the vector sum y'_m of the two phasors.

Because waves y_1 and y_2 have the same angular wave number k and angular frequency ω, we know from Eqs. 16-51 and 16-52 that their resultant is of the form

$$y'(x, t) = y'_m \sin(kx - \omega t + \beta), \qquad (16\text{-}57)$$

where y'_m is the amplitude of the resultant wave and β is its phase constant. To find the values of y'_m and β, we would have to sum the two combining waves, as we did to obtain Eq. 16-51. To do this on a phasor diagram, we vectorially add the two phasors at any instant during their rotation, as in Fig. 16-14f where phasor y_{m2} has been shifted to the head of phasor y_{m1}. The magnitude of the vector sum equals the amplitude y'_m in Eq. 16-57. The angle between the vector sum and the phasor for y_1 equals the phase constant β in Eq. 16-57.

Note that, in contrast to the method of Section 16-10:

 We can use phasors to combine waves *even if their amplitudes are different.*

Sample Problem

Interference of two waves, same direction, phasors, any amplitudes

Two sinusoidal waves $y_1(x, t)$ and $y_2(x, t)$ have the same wavelength and travel together in the same direction along a string. Their amplitudes are $y_{m1} = 4.0$ mm and $y_{m2} = 3.0$ mm, and their phase constants are 0 and $\pi/3$ rad, respectively. What are the amplitude y'_m and phase constant β of the resultant wave? Write the resultant wave in the form of Eq. 16-57.

KEY IDEAS

(1) The two waves have a number of properties in common: Because they travel along the same string, they must have the same speed v, as set by the tension and linear density of the string according to Eq. 16-26. With the same wavelength λ, they have the same angular wave number k $(= 2\pi/\lambda)$. Also, with the same wave number k and speed v, they must have the same angular frequency ω $(= kv)$.

(2) The waves (call them waves 1 and 2) can be represented by phasors rotating at the same angular speed ω about an origin. Because the phase constant for wave 2 is *greater* than that for wave 1 by $\pi/3$, phasor 2 must *lag* phasor 1 by $\pi/3$ rad in their clockwise rotation, as shown in Fig. 16-15a. The resultant wave due to the interference of waves 1 and 2 can then be represented by a phasor that is the vector sum of phasors 1 and 2.

Calculations: To simplify the vector summation, we drew phasors 1 and 2 in Fig. 16-15a at the instant when phasor 1 lies along the horizontal axis. We then drew lagging phasor 2 at positive angle $\pi/3$ rad. In Fig. 16-15b we shifted phasor 2 so its tail is at the head of phasor 1. Then we can draw the phasor y'_m of the resultant wave from the tail of phasor 1 to the head of phasor 2. The phase constant β is the angle phasor y'_m makes with phasor 1.

To find values for y'_m and β, we can sum phasors 1 and 2 as vectors on a calculator. Here we shall sum them by components. For the horizontal components we have

Add the phasors as vectors.

Fig. 16-15 (a) Two phasors of magnitudes y_{m1} and y_{m2} and with phase difference $\pi/3$. (b) Vector addition of these phasors at any instant during their rotation gives the magnitude y'_m of the phasor for the resultant wave.

$$y'_{mh} = y_{m1} \cos 0 + y_{m2} \cos \pi/3$$
$$= 4.0 \text{ mm} + (3.0 \text{ mm}) \cos \pi/3 = 5.50 \text{ mm}.$$

For the vertical components we have

$$y'_{mv} = y_{m1} \sin 0 + y_{m2} \sin \pi/3$$
$$= 0 + (3.0 \text{ mm}) \sin \pi/3 = 2.60 \text{ mm}.$$

Thus, the resultant wave has an amplitude of

$$y'_m = \sqrt{(5.50 \text{ mm})^2 + (2.60 \text{ mm})^2}$$
$$= 6.1 \text{ mm} \qquad \text{(Answer)}$$

and a phase constant of

$$\beta = \tan^{-1} \frac{2.60 \text{ mm}}{5.50 \text{ mm}} = 0.44 \text{ rad}. \qquad \text{(Answer)}$$

From Fig. 16-15b, phase constant β is a *positive* angle relative to phasor 1. Thus, the resultant wave *lags* wave 1 in their travel by phase constant $\beta = +0.44$ rad. From Eq. 16-57, we can write the resultant wave as

$$y'(x, t) = (6.1 \text{ mm}) \sin(kx - \omega t + 0.44 \text{ rad}). \qquad \text{(Answer)}$$

 Additional examples, video, and practice available at *WileyPLUS*

16-12 Standing Waves

In Section 16-10, we discussed two sinusoidal waves of the same wavelength and amplitude traveling *in the same direction* along a stretched string. What if they travel in opposite directions? We can again find the resultant wave by applying the superposition principle.

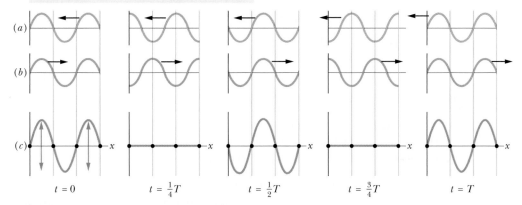

As the waves move through each other, some points never move and some move the most.

Fig. 16-16   (*a*) Five snapshots of a wave traveling to the left, at the times *t* indicated below part (*c*) (*T* is the period of oscillation). (*b*) Five snapshots of a wave identical to that in (*a*) but traveling to the right, at the same times *t*. (*c*) Corresponding snapshots for the superposition of the two waves on the same string. At $t = 0, \frac{1}{2}T$, and T, fully constructive interference occurs because of the alignment of peaks with peaks and valleys with valleys. At $t = \frac{1}{4}T$ and $\frac{3}{4}T$, fully destructive interference occurs because of the alignment of peaks with valleys. Some points (the nodes, marked with dots) never oscillate; some points (the antinodes) oscillate the most.

Figure 16-16 suggests the situation graphically. It shows the two combining waves, one traveling to the left in Fig. 16-16*a*, the other to the right in Fig. 16-16*b*. Figure 16-16*c* shows their sum, obtained by applying the superposition principle graphically. The outstanding feature of the resultant wave is that there are places along the string, called **nodes,** where the string never moves. Four such nodes are marked by dots in Fig. 16-16*c*. Halfway between adjacent nodes are **antinodes,** where the amplitude of the resultant wave is a maximum. Wave patterns such as that of Fig. 16-16*c* are called **standing waves** because the wave patterns do not move left or right; the locations of the maxima and minima do not change.

> If two sinusoidal waves of the same amplitude and wavelength travel in *opposite* directions along a stretched string, their interference with each other produces a standing wave.

To analyze a standing wave, we represent the two combining waves with the equations

$$y_1(x, t) = y_m \sin(kx - \omega t) \tag{16-58}$$

and

$$y_2(x, t) = y_m \sin(kx + \omega t). \tag{16-59}$$

The principle of superposition gives, for the combined wave,

$$y'(x, t) = y_1(x, t) + y_2(x, t) = y_m \sin(kx - \omega t) + y_m \sin(kx + \omega t).$$

Applying the trigonometric relation of Eq. 16-50 leads to Fig. 16-17 and

$$y'(x, t) = [2y_m \sin kx] \cos \omega t. \tag{16-60}$$

Displacement

$$y'(x,t) = \underbrace{[2y_m \sin kx]}_{\substack{\text{Magnitude} \\ \text{gives} \\ \text{amplitude} \\ \text{at position } x}} \underbrace{\cos \omega t}_{\substack{\text{Oscillating} \\ \text{term}}}$$

Fig. 16-17   The resultant wave of Eq. 16-60 is a standing wave and is due to the interference of two sinusoidal waves of the same amplitude and wavelength that travel in opposite directions.

There are two ways a pulse can reflect from the end of a string.

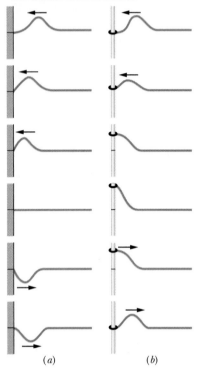

(a) (b)

Fig. 16-18 (*a*) A pulse incident from the right is reflected at the left end of the string, which is tied to a wall. Note that the reflected pulse is inverted from the incident pulse. (*b*) Here the left end of the string is tied to a ring that can slide without friction up and down the rod. Now the pulse is not inverted by the reflection.

This equation does not describe a traveling wave because it is not of the form of Eq. 16-17. Instead, it describes a standing wave.

The quantity $2y_m \sin kx$ in the brackets of Eq. 16-60 can be viewed as the amplitude of oscillation of the string element that is located at position x. However, since an amplitude is always positive and $\sin kx$ can be negative, we take the absolute value of the quantity $2y_m \sin kx$ to be the amplitude at x.

In a traveling sinusoidal wave, the amplitude of the wave is the same for all string elements. That is not true for a standing wave, in which the amplitude *varies with position*. In the standing wave of Eq. 16-60, for example, the amplitude is zero for values of kx that give $\sin kx = 0$. Those values are

$$kx = n\pi, \quad \text{for } n = 0, 1, 2, \ldots . \quad (16\text{-}61)$$

Substituting $k = 2\pi/\lambda$ in this equation and rearranging, we get

$$x = n\frac{\lambda}{2}, \quad \text{for } n = 0, 1, 2, \ldots \quad \text{(nodes)}, \quad (16\text{-}62)$$

as the positions of zero amplitude—the nodes—for the standing wave of Eq. 16-60. Note that adjacent nodes are separated by $\lambda/2$, half a wavelength.

The amplitude of the standing wave of Eq. 16-60 has a maximum value of $2y_m$, which occurs for values of kx that give $|\sin kx| = 1$. Those values are

$$kx = \tfrac{1}{2}\pi, \tfrac{3}{2}\pi, \tfrac{5}{2}\pi, \ldots$$
$$= (n + \tfrac{1}{2})\pi, \quad \text{for } n = 0, 1, 2, \ldots . \quad (16\text{-}63)$$

Substituting $k = 2\pi/\lambda$ in Eq. 16-63 and rearranging, we get

$$x = \left(n + \frac{1}{2}\right)\frac{\lambda}{2}, \quad \text{for } n = 0, 1, 2, \ldots \quad \text{(antinodes)}, \quad (16\text{-}64)$$

as the positions of maximum amplitude—the antinodes—of the standing wave of Eq. 16-60. The antinodes are separated by $\lambda/2$ and are located halfway between pairs of nodes.

Reflections at a Boundary

We can set up a standing wave in a stretched string by allowing a traveling wave to be reflected from the far end of the string so that the wave travels back through itself. The incident (original) wave and the reflected wave can then be described by Eqs. 16-58 and 16-59, respectively, and they can combine to form a pattern of standing waves.

In Fig. 16-18, we use a single pulse to show how such reflections take place. In Fig. 16-18*a*, the string is fixed at its left end. When the pulse arrives at that end, it exerts an upward force on the support (the wall). By Newton's third law, the support exerts an opposite force of equal magnitude on the string. This second force generates a pulse at the support, which travels back along the string in the direction opposite that of the incident pulse. In a "hard" reflection of this kind, there must be a node at the support because the string is fixed there. The reflected and incident pulses must have opposite signs, so as to cancel each other at that point.

In Fig. 16-18*b*, the left end of the string is fastened to a light ring that is free to slide without friction along a rod. When the incident pulse arrives, the ring moves up the rod. As the ring moves, it pulls on the string, stretching the string and producing a reflected pulse with the same sign and amplitude as the incident pulse. Thus, in such a "soft" reflection, the incident and reflected pulses reinforce each other, creating an antinode at the end of the string; the maximum displacement of the ring is twice the amplitude of either of these two pulses.

16-13 Standing Waves and Resonance

Consider a string, such as a guitar string, that is stretched between two clamps. Suppose we send a continuous sinusoidal wave of a certain frequency along the string, say, toward the right. When the wave reaches the right end, it reflects and begins to travel back to the left. That left-going wave then overlaps the wave that is still traveling to the right. When the left-going wave reaches the left end, it reflects again and the newly reflected wave begins to travel to the right, overlapping the left-going and right-going waves. In short, we very soon have many overlapping traveling waves, which interfere with one another.

For certain frequencies, the interference produces a standing wave pattern (or **oscillation mode**) with nodes and large antinodes like those in Fig. 16-19. Such a standing wave is said to be produced at **resonance,** and the string is said to *resonate* at these certain frequencies, called **resonant frequencies.** If the string is oscillated at some frequency other than a resonant frequency, a standing wave is not set up. Then the interference of the right-going and left-going traveling waves results in only small (perhaps imperceptible) oscillations of the string.

Fig. 16-19 Stroboscopic photographs reveal (imperfect) standing wave patterns on a string being made to oscillate by an oscillator at the left end. The patterns occur at certain frequencies of oscillation. *(Richard Megna/Fundamental Photographs)*

Let a string be stretched between two clamps separated by a fixed distance L. To find expressions for the resonant frequencies of the string, we note that a node must exist at each of its ends, because each end is fixed and cannot oscillate. The simplest pattern that meets this key requirement is that in Fig. 16-20a, which shows the string at both its extreme displacements (one solid and one dashed, together forming a single "loop"). There is only one antinode, which is at the center of the string. Note that half a wavelength spans the length L, which we take to be the string's length. Thus, for this pattern, $\lambda/2 = L$. This condition tells us that if the left-going and right-going traveling waves are to set up this pattern by their interference, they must have the wavelength $\lambda = 2L$.

A second simple pattern meeting the requirement of nodes at the fixed ends is shown in Fig. 16-20b. This pattern has three nodes and two antinodes and is said

(a) $L = \dfrac{\lambda}{2}$

First harmonic

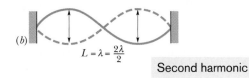

(b) $L = \lambda = \dfrac{2\lambda}{2}$

Second harmonic

(c) $L = \dfrac{3\lambda}{2}$

Third harmonic

Fig. 16-20 A string, stretched between two clamps, is made to oscillate in standing wave patterns. (a) The simplest possible pattern consists of one *loop*, which refers to the composite shape formed by the string in its extreme displacements (the solid and dashed lines). (b) The next simplest pattern has two loops. (c) The next has three loops.

Fig. 16-21 One of many possible standing wave patterns for a kettledrum head, made visible by dark powder sprinkled on the drumhead. As the head is set into oscillation at a single frequency by a mechanical oscillator at the upper left of the photograph, the powder collects at the nodes, which are circles and straight lines in this two-dimensional example. *(Courtesy Thomas D. Rossing, Northern Illinois University)*

to be a two-loop pattern. For the left-going and right-going waves to set it up, they must have a wavelength $\lambda = L$. A third pattern is shown in Fig. 16-20c. It has four nodes, three antinodes, and three loops, and the wavelength is $\lambda = \frac{2}{3}L$. We could continue this progression by drawing increasingly more complicated patterns. In each step of the progression, the pattern would have one more node and one more antinode than the preceding step, and an additional $\lambda/2$ would be fitted into the distance L.

Thus, a standing wave can be set up on a string of length L by a wave with a wavelength equal to one of the values

$$\lambda = \frac{2L}{n}, \qquad \text{for } n = 1, 2, 3, \ldots . \tag{16-65}$$

The resonant frequencies that correspond to these wavelengths follow from Eq. 16-13:

$$f = \frac{v}{\lambda} = n\frac{v}{2L}, \qquad \text{for } n = 1, 2, 3, \ldots . \tag{16-66}$$

Here v is the speed of traveling waves on the string.

Equation 16-66 tells us that the resonant frequencies are integer multiples of the lowest resonant frequency, $f = v/2L$, which corresponds to $n = 1$. The oscillation mode with that lowest frequency is called the *fundamental mode* or the *first harmonic*. The *second harmonic* is the oscillation mode with $n = 2$, the *third harmonic* is that with $n = 3$, and so on. The frequencies associated with these modes are often labeled f_1, f_2, f_3, and so on. The collection of all possible oscillation modes is called the **harmonic series,** and n is called the **harmonic number** of the nth harmonic.

For a given string under a given tension, each resonant frequency corresponds to a particular oscillation pattern. Thus, if the frequency is in the audible range, you can hear the shape of the string. Resonance can also occur in two dimensions (such as on the surface of the kettledrum in Fig. 16-21) and in three dimensions (such as in the wind-induced swaying and twisting of a tall building).

✔CHECKPOINT 6

In the following series of resonant frequencies, one frequency (lower than 400 Hz) is missing: 150, 225, 300, 375 Hz. (a) What is the missing frequency? (b) What is the frequency of the seventh harmonic?

Sample Problem

Resonance of transverse waves, standing waves, harmonics

Figure 16-22 shows a pattern of resonant oscillation of a string of mass $m = 2.500$ g and length $L = 0.800$ m and that is under tension $\tau = 325.0$ N. What is the wavelength λ of the transverse waves producing the standing-wave pattern, and what is the harmonic number n? What is the frequency f of the transverse waves and of the oscillations of the moving string elements? What is the maximum magnitude of the transverse velocity u_m of the element oscillating at coordinate $x = 0.180$ m (note the x axis in the figure)? At what point

Fig. 16-22 Resonant oscillation of a string under tension.

during the element's oscillation is the transverse velocity maximum?

(1) The traverse waves that produce a standing-wave pattern must have a wavelength such that an integer number n of half-wavelengths fit into the length L of the string. (2) The frequency of those waves and of the oscillations of the string elements is given by Eq. 16-66 ($f = nv/2L$). (3) The displacement of a string element as a function of position x and time t is given by Eq. 16-60:

$$y'(x,t) = [2y_m \sin kx] \cos \omega t. \tag{16-67}$$

Wavelength and harmonic number: In Fig. 16-22, the solid line, which is effectively a snapshot (or freeze frame) of the oscillations, reveals that 2 full wavelengths fit into the length $L = 0.800$ m of the string. Thus, we have

$$2\lambda = L,$$

or

$$\lambda = \frac{L}{2}. \tag{16-68}$$

$$= \frac{0.800 \text{ m}}{2} = 0.400 \text{ m}. \quad \text{(Answer)}$$

By counting the number of loops (or half-wavelengths) in Fig. 16-22, we see that the harmonic number is

$$n = 4. \quad \text{(Answer)}$$

We reach the same conclusion by comparing Eqs. 16-68 and 16-65 ($\lambda = 2L/n$). Thus, the string is oscillating in its fourth harmonic.

Frequency: We can get the frequency f of the transverse waves from Eq. 16-13 ($v = \lambda f$) if we first find the speed v of the waves. That speed is given by Eq. 16-26, but we must substitute m/L for the unknown linear density μ. We obtain

$$v = \sqrt{\frac{\tau}{\mu}} = \sqrt{\frac{\tau}{m/L}} = \sqrt{\frac{\tau L}{m}}$$

$$= \sqrt{\frac{(325 \text{ N})(0.800 \text{ m})}{2.50 \times 10^{-3} \text{ kg}}} = 322.49 \text{ m/s}.$$

After rearranging Eq. 16-13, we write

$$f = \frac{v}{\lambda} = \frac{322.49 \text{ m/s}}{0.400 \text{ m}}$$

$$= 806.2 \text{ Hz} \approx 806 \text{ Hz}. \quad \text{(Answer)}$$

Note that we get the same answer by substituting into Eq. 16-66:

$$f = n\frac{v}{2L} = 4\frac{322.49 \text{ m/s}}{2(0.800 \text{ m})}$$

$$= 806 \text{ Hz}. \quad \text{(Answer)}$$

Now note that this 806 Hz is not only the frequency of the waves producing the fourth harmonic but also it is said to *be* the fourth harmonic, as in the statement, "The fourth harmonic of this oscillating string is 806 Hz." It is also the frequency of the string elements as they oscillate vertically in the figure in simple harmonic motion, just as a block on a vertical spring would oscillate in simple harmonic motion. Finally, it is also the frequency of the sound you would hear from the string as the oscillating string elements periodically push against the air, sending out sound waves.

Transverse velocity: The displacement y' of the string element located at coordinate x is given by Eq. 16-67 as a function of time t. The term $\cos \omega t$ contains the dependence on time and thus provides the "motion" of the standing wave. The term $2y_m \sin kx$ sets the extent of the motion—that is, the amplitude. The greatest amplitude occurs at an antinode, where $\sin kx$ is $+1$ or -1 and thus the greatest amplitude is $2y_m$. From Fig. 16-22, we see that $2y_m = 4.00$ mm, which tells us that $y_m = 2.00$ mm.

We want the transverse velocity—the velocity of a string element parallel to the y axis. To find it, we take the time derivative of Eq. 16-67:

$$u(x,t) = \frac{\partial y'}{\partial t} = \frac{\partial}{\partial t}[(2y_m \sin kx)\cos \omega t]$$

$$= [-2y_m\omega \sin kx]\sin \omega t. \tag{16-69}$$

Here the term $\sin \omega t$ provides the variation with time and the term $-2y_m\omega \sin kx$ provides the extent of that variation. We want the absolute magnitude of that extent:

$$u_m = |-2y_m\omega \sin kx|.$$

To evaluate this for the element at $x = 0.180$ m, we first note that $y_m = 2.00$ mm, $k = 2\pi/\lambda = 2\pi/(0.400$ m$)$, and $\omega = 2\pi f = 2\pi(806.2$ Hz$)$. Then the maximum speed of the element at $x = 0.180$ m is

$$u_m = \left| -2(2.00 \times 10^{-3} \text{ m})(2\pi)(806.2 \text{ Hz}) \right.$$

$$\left. \times \sin\left(\frac{2\pi}{0.400 \text{ m}}(0.180 \text{ m})\right) \right|$$

$$= 6.26 \text{ m/s}. \quad \text{(Answer)}$$

To determine when the string element has this maximum speed, we could investigate Eq. 16-69. However, a little thought can save a lot of work. The element is undergoing simple harmonic motion and must come to a momentary stop at its extreme upward position and extreme downward position. It has the greatest speed as it zips through the midpoint of its oscillation, just as a block does in a block–spring oscillator.

PLUS Additional examples, video, and practice available at *WileyPLUS*

Transverse and Longitudinal Waves Mechanical waves can exist only in material media and are governed by Newton's laws. **Transverse** mechanical waves, like those on a stretched string, are waves in which the particles of the medium oscillate perpendicular to the wave's direction of travel. Waves in which the particles of the medium oscillate parallel to the wave's direction of travel are **longitudinal** waves.

Sinusoidal Waves A sinusoidal wave moving in the positive direction of an x axis has the mathematical form

$$y(x, t) = y_m \sin(kx - \omega t), \qquad (16\text{-}2)$$

where y_m is the **amplitude** of the wave, k is the **angular wave number,** ω is the **angular frequency,** and $kx - \omega t$ is the **phase.** The **wavelength** λ is related to k by

$$k = \frac{2\pi}{\lambda}. \qquad (16\text{-}5)$$

The **period** T and **frequency** f of the wave are related to ω by

$$\frac{\omega}{2\pi} = f = \frac{1}{T}. \qquad (16\text{-}9)$$

Finally, the **wave speed** v is related to these other parameters by

$$v = \frac{\omega}{k} = \frac{\lambda}{T} = \lambda f. \qquad (16\text{-}13)$$

Equation of a Traveling Wave Any function of the form

$$y(x, t) = h(kx \pm \omega t) \qquad (16\text{-}17)$$

can represent a **traveling wave** with a wave speed given by Eq. 16-13 and a wave shape given by the mathematical form of h. The plus sign denotes a wave traveling in the negative direction of the x axis, and the minus sign a wave traveling in the positive direction.

Wave Speed on Stretched String The speed of a wave on a stretched string is set by properties of the string. The speed on a string with tension τ and linear density μ is

$$v = \sqrt{\frac{\tau}{\mu}}. \qquad (16\text{-}26)$$

Power The **average power** of, or average rate at which energy is transmitted by, a sinusoidal wave on a stretched string is given by

$$P_{\text{avg}} = \tfrac{1}{2}\mu v \omega^2 y_m^2. \qquad (16\text{-}33)$$

Superposition of Waves When two or more waves traverse the same medium, the displacement of any particle of the medium is the sum of the displacements that the individual waves would give it.

Interference of Waves Two sinusoidal waves on the same string exhibit **interference,** adding or canceling according to the principle of superposition. If the two are traveling in the same direction and have the same amplitude y_m and frequency (hence the same wavelength) but differ in phase by a **phase constant** ϕ, the result is a single wave with this same frequency:

$$y'(x, t) = [2y_m \cos \tfrac{1}{2}\phi] \sin(kx - \omega t + \tfrac{1}{2}\phi). \qquad (16\text{-}51)$$

If $\phi = 0$, the waves are exactly in phase and their interference is fully constructive; if $\phi = \pi$ rad, they are exactly out of phase and their interference is fully destructive.

Phasors A wave $y(x, t)$ can be represented with a **phasor.** This is a vector that has a magnitude equal to the amplitude y_m of the wave and that rotates about an origin with an angular speed equal to the angular frequency ω of the wave. The projection of the rotating phasor on a vertical axis gives the displacement y of a point along the wave's travel.

Standing Waves The interference of two identical sinusoidal waves moving in opposite directions produces **standing waves.** For a string with fixed ends, the standing wave is given by

$$y'(x, t) = [2y_m \sin kx] \cos \omega t. \qquad (16\text{-}60)$$

Standing waves are characterized by fixed locations of zero displacement called **nodes** and fixed locations of maximum displacement called **antinodes.**

Resonance Standing waves on a string can be set up by reflection of traveling waves from the ends of the string. If an end is fixed, it must be the position of a node. This limits the frequencies at which standing waves will occur on a given string. Each possible frequency is a **resonant frequency,** and the corresponding standing wave pattern is an **oscillation mode.** For a stretched string of length L with fixed ends, the resonant frequencies are

$$f = \frac{v}{\lambda} = n\frac{v}{2L}, \qquad \text{for } n = 1, 2, 3, \ldots. \qquad (16\text{-}66)$$

The oscillation mode corresponding to $n = 1$ is called the *fundamental mode* or the *first harmonic*; the mode corresponding to $n = 2$ is the *second harmonic*; and so on.

1 The following four waves are sent along strings with the same linear densities (x is in meters and t is in seconds). Rank the waves according to (a) their wave speed and (b) the tension in the strings along which they travel, greatest first:

(1) $y_1 = (3 \text{ mm}) \sin(x - 3t)$,   (3) $y_3 = (1 \text{ mm}) \sin(4x - t)$,

(2) $y_2 = (6 \text{ mm}) \sin(2x - t)$,   (4) $y_4 = (2 \text{ mm}) \sin(x - 2t)$.

2 In Fig. 16-23, wave 1 consists of a rectangular peak of height 4 units and width d, and a rectangular valley of depth 2 units and width d. The wave travels rightward along an x axis. Choices 2, 3, and 4 are similar waves, with the same heights, depths, and widths, that will travel leftward along that axis and through wave 1. Right-going wave 1 and one of the left-going waves will interfere as they pass through each other. With which left-going wave will the interference give, for an instant, (a) the deepest valley, (b) a flat line, and (c) a flat peak $2d$ wide?

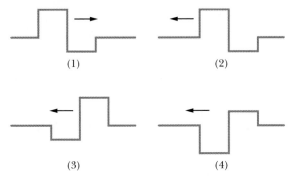

Fig. 16-23   Question 2.

3   Figure 16-24*a* gives a snapshot of a wave traveling in the direction of positive x along a string under tension. Four string elements are indicated by the lettered points. For each of those elements, determine whether, at the instant of the snapshot, the element is moving upward or downward or is momentarily at rest. (*Hint:* Imagine the wave as it moves through the four string elements, as if you were watching a video of the wave as it traveled rightward.)

Figure 16-24*b* gives the displacement of a string element located at, say, $x = 0$ as a function of time. At the lettered times, is the element moving upward or downward or is it momentarily at rest?

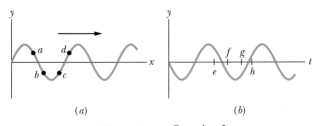

Fig. 16-24   Question 3.

4   Figure 16-25 shows three waves that are *separately* sent along a string that is stretched under a certain tension along an x axis. Rank the waves according to their (a) wavelengths, (b) speeds, and (c) angular frequencies, greatest first.

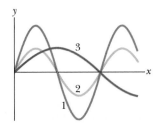

Fig. 16-25   Question 4.

5   If you start with two sinusoidal waves of the same amplitude traveling in phase on a string and then somehow phase-shift one of them by 5.4 wavelengths, what type of interference will occur on the string?

6   The amplitudes and phase differences for four pairs of waves of equal wavelengths are (a) 2 mm, 6 mm, and π rad; (b) 3 mm, 5 mm, and π rad; (c) 7 mm, 9 mm, and π rad; (d) 2 mm, 2 mm, and 0 rad. Each pair travels in the same direction along the same string. Without written calculation, rank the four pairs according to the amplitude of their resultant wave, greatest first. (*Hint:* Construct phasor diagrams.)

7   A sinusoidal wave is sent along a cord under tension, transporting energy at the average rate of $P_{\mathrm{avg},1}$. Two waves, identical to that first one, are then to be sent along the cord with a phase difference ϕ of either 0, 0.2 wavelength, or 0.5 wavelength. (a) With only mental calculation, rank those choices of ϕ according to the average rate at which the waves will transport energy, greatest first. (b) For the first choice of ϕ, what is the average rate in terms of $P_{\mathrm{avg},1}$?

8   (a) If a standing wave on a string is given by

$$y'(t) = (3 \text{ mm}) \sin(5x) \cos(4t),$$

is there a node or an antinode of the oscillations of the string at $x = 0$? (b) If the standing wave is given by

$$y'(t) = (3 \text{ mm}) \sin(5x + \pi/2) \cos(4t),$$

is there a node or an antinode at $x = 0$?

9   Strings A and B have identical lengths and linear densities, but string B is under greater tension than string A. Figure 16-26 shows four situations, (*a*) through (*d*), in which standing wave patterns exist on the two strings. In which situations is there the possibility that strings A and B are oscillating at the same resonant frequency?

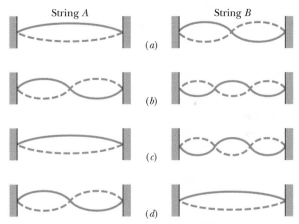

Fig. 16-26   Question 9.

10   If you set up the seventh harmonic on a string, (a) how many nodes are present, and (b) is there a node, antinode, or some intermediate state at the midpoint? If you next set up the sixth harmonic, (c) is its resonant wavelength longer or shorter than that for the seventh harmonic, and (d) is the resonant frequency higher or lower?

11   Figure 16-27 shows phasor diagrams for three situations in which two waves travel along the same string. All six waves have the same amplitude. Rank the situations according to the amplitude of the net wave on the string, greatest first.

Fig. 16-27   Question 11.

GO	Tutoring problem available (at instructor's discretion) in *WileyPLUS* and WebAssign
SSM	Worked-out solution available in Student Solutions Manual	WWW	Worked-out solution is at
• – •••	Number of dots indicates level of problem difficulty	ILW	Interactive solution is at	http://www.wiley.com/college/halliday
	Additional information available in *The Flying Circus of Physics* and at flyingcircusofphysics.com		

sec. 16-5 The Speed of a Traveling Wave

•1 If a wave $y(x, t) = (6.0 \text{ mm}) \sin(kx + (600 \text{ rad/s})t + \phi)$ travels along a string, how much time does any given point on the string take to move between displacements $y = +2.0$ mm and $y = -2.0$ mm?

•2 *A human wave.* During sporting events within large, densely packed stadiums, spectators will send a wave (or pulse) around the stadium (Fig. 16-28). As the wave reaches a group of spectators, they stand with a cheer and then sit. At any instant, the width w of the wave

Fig. 16-28 Problem 2.

is the distance from the leading edge (people are just about to stand) to the trailing edge (people have just sat down). Suppose a human wave travels a distance of 853 seats around a stadium in 39 s, with spectators requiring about 1.8 s to respond to the wave's passage by standing and then sitting. What are (a) the wave speed v (in seats per second) and (b) width w (in number of seats)?

•3 A wave has an angular frequency of 110 rad/s and a wavelength of 1.80 m. Calculate (a) the angular wave number and (b) the speed of the wave.

•4 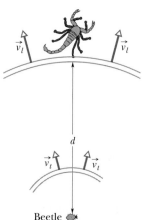 A sand scorpion can detect the motion of a nearby beetle (its prey) by the waves the motion sends along the sand surface (Fig. 16-29). The waves are of two types: transverse waves traveling at $v_t = 50$ m/s and longitudinal waves traveling at $v_l = 150$ m/s. If a sudden motion sends out such waves, a scorpion can tell the distance of the beetle from the difference Δt in the arrival times of the waves at its leg nearest the beetle. If $\Delta t = 4.0$ ms, what is the beetle's distance?

Fig. 16-29 Problem 4.

•5 A sinusoidal wave travels along a string. The time for a particular point to move from maximum displacement to zero is 0.170 s. What are the (a) period and (b) frequency? (c) The wavelength is 1.40 m; what is the wave speed?

••6 GO A sinusoidal wave travels along a string under tension. Figure 16-30 gives the slopes along the string at time $t = 0$. The scale of the x axis is set by $x_s = 0.80$ m. What is the amplitude of the wave?

Fig. 16-30 Problem 6.

••7 A transverse sinusoidal wave is moving along a string in the positive direction of an x axis with a speed of 80 m/s. At $t = 0$, the

string particle at $x = 0$ has a transverse displacement of 4.0 cm from its equilibrium position and is not moving. The maximum transverse speed of the string particle at $x = 0$ is 16 m/s. (a) What is the frequency of the wave? (b) What is the wavelength of the wave? If $y(x, t) = y_m \sin(kx \pm \omega t + \phi)$ is the form of the wave equation, what are (c) y_m, (d) k, (e) ω, (f) ϕ, and (g) the correct choice of sign in front of ω?

••8 GO Figure 16-31 shows the transverse velocity u versus time t of the point on a string at $x = 0$, as a wave passes through it. The scale on the vertical axis is set by $u_s = 4.0$ m/s. The wave has the form $y(x, t) = y_m \sin(kx - \omega t + \phi)$. What is ϕ? (*Caution:* A calculator does not always give the proper inverse trig function, so check your answer by substituting it and an assumed value of ω into $y(x, t)$ and then plotting the function.)

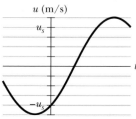
Fig. 16-31 Problem 8.

••9 A sinusoidal wave moving along a string is shown twice in Fig. 16-32, as crest A travels in the positive direction of an x axis by distance $d = 6.0$ cm in 4.0 ms. The tick marks along the axis are separated by 10 cm; height $H = 6.00$ mm. If the wave equation is of the form $y(x, t) = y_m \sin(kx \pm \omega t)$, what are (a) y_m, (b) k, (c) ω, and (d) the correct choice of sign in front of ω?

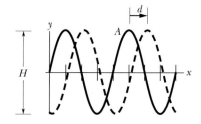
Fig. 16-32 Problem 9.

••10 The equation of a transverse wave traveling along a very long string is $y = 6.0 \sin(0.020\pi x + 4.0\pi t)$, where x and y are expressed in centimeters and t is in seconds. Determine (a) the amplitude, (b) the wavelength, (c) the frequency, (d) the speed, (e) the direction of propagation of the wave, and (f) the maximum transverse speed of a particle in the string. (g) What is the transverse displacement at $x = 3.5$ cm when $t = 0.26$ s?

••11 GO A sinusoidal transverse wave of wavelength 20 cm travels along a string in the positive direction of an x axis. The displacement y of the string particle at $x = 0$ is given in Fig. 16-33 as a function of time t. The scale of the vertical axis is set by $y_s = 4.0$ cm. The wave equation is to be in the form $y(x, t) =$

Fig. 16-33 Problem 11.

$y_m \sin(kx \pm \omega t + \phi)$. (a) At $t = 0$, is a plot of y versus x in the shape of a positive sine function or a negative sine function? What are (b) y_m, (c) k, (d) ω, (e) ϕ, (f) the sign in front of ω, and (g) the speed of the wave? (h) What is the transverse velocity of the particle at $x = 0$ when $t = 5.0$ s?

•12   The function $y(x, t) = (15.0 \text{ cm}) \cos(\pi x - 15\pi t)$, with x in meters and t in seconds, describes a wave on a taut string. What is the transverse speed for a point on the string at an instant when that point has the displacement $y = +12.0$ cm?

••13 ILW A sinusoidal wave of frequency 500 Hz has a speed of 350 m/s. (a) How far apart are two points that differ in phase by $\pi/3$ rad? (b) What is the phase difference between two displacements at a certain point at times 1.00 ms apart?

sec. 16-6   Wave Speed on a Stretched String

•14   The equation of a transverse wave on a string is

$$y = (2.0 \text{ mm}) \sin[(20 \text{ m}^{-1})x - (600 \text{ s}^{-1})t].$$

The tension in the string is 15 N. (a) What is the wave speed? (b) Find the linear density of this string in grams per meter.

•15 SSM WWW A stretched string has a mass per unit length of 5.00 g/cm and a tension of 10.0 N. A sinusoidal wave on this string has an amplitude of 0.12 mm and a frequency of 100 Hz and is traveling in the negative direction of an x axis. If the wave equation is of the form $y(x, t) = y_m \sin(kx \pm \omega t)$, what are (a) y_m, (b) k, (c) ω, and (d) the correct choice of sign in front of ω?

•16   The speed of a transverse wave on a string is 170 m/s when the string tension is 120 N. To what value must the tension be changed to raise the wave speed to 180 m/s?

•17   The linear density of a string is 1.6×10^{-4} kg/m. A transverse wave on the string is described by the equation

$$y = (0.021 \text{ m}) \sin[(2.0 \text{ m}^{-1})x + (30 \text{ s}^{-1})t].$$

What are (a) the wave speed and (b) the tension in the string?

•18   The heaviest and lightest strings on a certain violin have linear densities of 3.0 and 0.29 g/m. What is the ratio of the diameter of the heaviest string to that of the lightest string, assuming that the strings are of the same material?

•19 SSM What is the speed of a transverse wave in a rope of length 2.00 m and mass 60.0 g under a tension of 500 N?

•20   The tension in a wire clamped at both ends is doubled without appreciably changing the wire's length between the clamps. What is the ratio of the new to the old wave speed for transverse waves traveling along this wire?

••21 ILW A 100 g wire is held under a tension of 250 N with one end at $x = 0$ and the other at $x = 10.0$ m. At time $t = 0$, pulse 1 is sent along the wire from the end at $x = 10.0$ m. At time $t = 30.0$ ms, pulse 2 is sent along the wire from the end at $x = 0$. At what position x do the pulses begin to meet?

••22   A sinusoidal wave is traveling on a string with speed 40 cm/s. The displacement of the particles of the string at $x = 10$ cm varies with time according to $y = (5.0 \text{ cm}) \sin[1.0 - (4.0 \text{ s}^{-1})t]$. The linear density of the string is 4.0 g/cm. What are (a) the frequency and (b) the wavelength of the wave? If the wave equation is of the form $y(x, t) = y_m \sin(kx \pm \omega t)$, what are (c) y_m, (d) k, (e) ω, and (f) the correct choice of sign in front of ω? (g) What is the tension in the string?

••23 SSM ILW A sinusoidal transverse wave is traveling along a string in the negative direction of an x axis. Figure 16-34 shows a plot of the displacement as a function of position at time $t = 0$; the scale of the y axis is set by $y_s = 4.0$ cm. The string tension is 3.6 N, and its linear density is 25 g/m. Find the (a) amplitude, (b) wavelength, (c) wave speed, and (d) period of the wave. (e) Find the maximum transverse speed of a particle in the string. If the wave is of the form $y(x, t) = y_m \sin(kx \pm \omega t + \phi)$, what are (f) k, (g) ω, (h) ϕ, and (i) the correct choice of sign in front of ω?

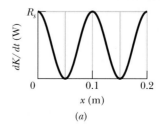

Fig. 16-34   Problem 23.

•••24   In Fig. 16-35a, string 1 has a linear density of 3.00 g/m, and string 2 has a linear density of 5.00 g/m. They are under tension due to the hanging block of mass $M = 500$ g. Calculate the wave speed on (a) string 1 and (b) string 2. (*Hint:* When a string loops halfway around a pulley, it pulls on the pulley with a net force that is twice the tension in the string.) Next the block is divided into two blocks (with $M_1 + M_2 = M$) and the apparatus is rearranged as shown in Fig. 16-35b. Find (c) M_1 and (d) M_2 such that the wave speeds in the two strings are equal.

•••25   A uniform rope of mass m and length L hangs from a ceiling.

Fig. 16-35   Problem 24.

(a) Show that the speed of a transverse wave on the rope is a function of y, the distance from the lower end, and is given by $v = \sqrt{gy}$. (b) Show that the time a transverse wave takes to travel the length of the rope is given by $t = 2\sqrt{L/g}$.

sec. 16-7   Energy and Power of a Wave Traveling Along a String

•26   A string along which waves can travel is 2.70 m long and has a mass of 260 g. The tension in the string is 36.0 N. What must be the frequency of traveling waves of amplitude 7.70 mm for the average power to be 85.0 W?

••27 GO A sinusoidal wave is sent along a string with a linear density of 2.0 g/m. As it travels, the kinetic energies of the mass elements along the string vary. Figure 16-36a gives the

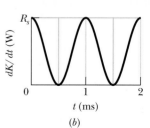

Fig. 16-36   Problem 27.

rate dK/dt at which kinetic energy passes through the string elements at a particular instant, plotted as a function of distance x along the string. Figure 16-36b is similar except that it gives the rate at which kinetic energy passes through a particular mass element (at a particular location), plotted as a function of time t. For both figures, the scale on the vertical (rate) axis is set by $R_s = 10$ W. What is the amplitude of the wave?

sec. 16-8 The Wave Equation

•28 Use the wave equation to find the speed of a wave given by

$$y(x, t) = (3.00 \text{ mm}) \sin[(4.00 \text{ m}^{-1})x - (7.00 \text{ s}^{-1})t].$$

••29 Use the wave equation to find the speed of a wave given by

$$y(x, t) = (2.00 \text{ mm})[(20 \text{ m}^{-1})x - (4.0 \text{ s}^{-1})t]^{0.5}.$$

•••30 Use the wave equation to find the speed of a wave given in terms of the general function $h(x, t)$:

$$y(x, t) = (4.00 \text{ mm}) h[(30 \text{ m}^{-1})x + (6.0 \text{ s}^{-1})t].$$

sec. 16-10 Interference of Waves

•31 SSM Two identical traveling waves, moving in the same direction, are out of phase by $\pi/2$ rad. What is the amplitude of the resultant wave in terms of the common amplitude y_m of the two combining waves?

•32 What phase difference between two identical traveling waves, moving in the same direction along a stretched string, results in the combined wave having an amplitude 1.50 times that of the common amplitude of the two combining waves? Express your answer in (a) degrees, (b) radians, and (c) wavelengths.

••33 Two sinusoidal waves with the same amplitude of 9.00 mm and the same wavelength travel together along a string that is stretched along an x axis. Their resultant wave is shown twice in Fig. 16-37, as valley A travels in the negative direction of the x axis by distance $d = 56.0$ cm in 8.0 ms. The tick marks along the axis are separated by 10 cm, and height H is 8.0 mm. Let the equation for one wave be of the form $y(x, t) = y_m \sin(kx \pm \omega t + \phi_1)$, where $\phi_1 = 0$ and you must choose the correct sign in front of ω. For the equation for the other wave, what are (a) y_m, (b) k, (c) ω, (d) ϕ_2, and (e) the sign in front of ω?

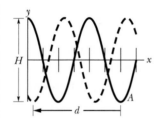

Fig. 16-37 Problem 33.

•••34 A sinusoidal wave of angular frequency 1200 rad/s and amplitude 3.00 mm is sent along a cord with linear density 2.00 g/m and tension 1200 N. (a) What is the average rate at which energy is transported by the wave to the opposite end of the cord? (b) If, simultaneously, an identical wave travels along an adjacent, identical cord, what is the total average rate at which energy is transported to the opposite ends of the two cords by the waves? If, instead, those two waves are sent along the *same* cord simultaneously, what is the total average rate at which they transport energy when their phase difference is (c) 0, (d) 0.4π rad, and (e) π rad?

sec. 16-11 Phasors

•35 SSM Two sinusoidal waves of the same frequency travel in the same direction along a string. If $y_{m1} = 3.0$ cm, $y_{m2} = 4.0$ cm, $\phi_1 = 0$, and $\phi_2 = \pi/2$ rad, what is the amplitude of the resultant wave?

••36 Four waves are to be sent along the same string, in the same direction:

$$y_1(x, t) = (4.00 \text{ mm}) \sin(2\pi x - 400\pi t)$$
$$y_2(x, t) = (4.00 \text{ mm}) \sin(2\pi x - 400\pi t + 0.7\pi)$$
$$y_3(x, t) = (4.00 \text{ mm}) \sin(2\pi x - 400\pi t + \pi)$$
$$y_4(x, t) = (4.00 \text{ mm}) \sin(2\pi x - 400\pi t + 1.7\pi).$$

What is the amplitude of the resultant wave?

••37 GO These two waves travel along the same string:

$$y_1(x, t) = (4.60 \text{ mm}) \sin(2\pi x - 400\pi t)$$
$$y_2(x, t) = (5.60 \text{ mm}) \sin(2\pi x - 400\pi t + 0.80\pi \text{ rad}).$$

What are (a) the amplitude and (b) the phase angle (relative to wave 1) of the resultant wave? (c) If a third wave of amplitude 5.00 mm is also to be sent along the string in the same direction as the first two waves, what should be its phase angle in order to maximize the amplitude of the new resultant wave?

••38 Two sinusoidal waves of the same frequency are to be sent in the same direction along a taut string. One wave has an amplitude of 5.0 mm, the other 8.0 mm. (a) What phase difference ϕ_1 between the two waves results in the smallest amplitude of the resultant wave? (b) What is that smallest amplitude? (c) What phase difference ϕ_2 results in the largest amplitude of the resultant wave? (d) What is that largest amplitude? (e) What is the resultant amplitude if the phase angle is $(\phi_1 - \phi_2)/2$?

••39 Two sinusoidal waves of the same period, with amplitudes of 5.0 and 7.0 mm, travel in the same direction along a stretched string; they produce a resultant wave with an amplitude of 9.0 mm. The phase constant of the 5.0 mm wave is 0. What is the phase constant of the 7.0 mm wave?

sec. 16-13 Standing Waves and Resonance

•40 Two sinusoidal waves with identical wavelengths and amplitudes travel in opposite directions along a string with a speed of 10 cm/s. If the time interval between instants when the string is flat is 0.50 s, what is the wavelength of the waves?

•41 SSM A string fixed at both ends is 8.40 m long and has a mass of 0.120 kg. It is subjected to a tension of 96.0 N and set oscillating. (a) What is the speed of the waves on the string? (b) What is the longest possible wavelength for a standing wave? (c) Give the frequency of that wave.

•42 A string under tension τ_i oscillates in the third harmonic at frequency f_3, and the waves on the string have wavelength λ_3. If the tension is increased to $\tau_f = 4\tau_i$ and the string is again made to oscillate in the third harmonic, what then are (a) the frequency of oscillation in terms of f_3 and (b) the wavelength of the waves in terms of λ_3?

•43 SSM WWW What are (a) the lowest frequency, (b) the second lowest frequency, and (c) the third lowest frequency for standing waves on a wire that is 10.0 m long, has a mass of 100 g, and is stretched under a tension of 250 N?

•44 A 125 cm length of string has mass 2.00 g and tension 7.00 N. (a) What is the wave speed for this string? (b) What is the lowest resonant frequency of this string?

•45 SSM ILW A string that is stretched between fixed supports separated by 75.0 cm has resonant frequencies of 420 and 315 Hz, with no intermediate resonant frequencies. What are (a) the lowest resonant frequency and (b) the wave speed?

•46 String A is stretched between two clamps separated by distance L. String B, with the same linear density and under the same tension as string A, is stretched between two clamps separated by distance $4L$. Consider the first eight harmonics of string B. For which of these eight harmonics of B (if any) does the frequency match the frequency of (a) A's first harmonic, (b) A's second harmonic, and (c) A's third harmonic?

•47 One of the harmonic frequencies for a particular string under tension is 325 Hz. The next higher harmonic frequency is 390 Hz. What harmonic frequency is next higher after the harmonic frequency 195 Hz?

•48 🛩 If a transmission line in a cold climate collects ice, the increased diameter tends to cause vortex formation in a passing wind. The air pressure variations in the vortexes tend to cause the line to oscillate (*gallop*), especially if the frequency of the variations matches a resonant frequency of the line. In long lines, the resonant frequencies are so close that almost any wind speed can set up a resonant mode vigorous enough to pull down support towers or cause the line to *short out* with an adjacent line. If a transmission line has a length of 347 m, a linear density of 3.35 kg/m, and a tension of 65.2 MN, what are (a) the frequency of the fundamental mode and (b) the frequency difference between successive modes?

•49 ILW A nylon guitar string has a linear density of 7.20 g/m and is under a tension of 150 N. The fixed supports are distance $D = 90.0$ cm apart. The string is oscillating in the standing wave pattern shown in Fig. 16-38. Calculate the (a) speed, (b) wavelength, and (c) frequency of the traveling waves whose superposition gives this standing wave.

Fig. 16-38 Problem 49.

••50 For a certain transverse standing wave on a long string, an antinode is at $x = 0$ and an adjacent node is at $x = 0.10$ m. The displacement $y(t)$ of the string particle at $x = 0$ is shown in Fig. 16-39, where the scale of the y axis is set by $y_s = 4.0$ cm. When $t = 0.50$ s, what is the displacement of the string particle at (a) $x = 0.20$ m and (b) $x = 0.30$ m? What is the transverse velocity of the string particle at $x = 0.20$ m at (c) $t = 0.50$ s and (d) $t = 1.0$ s? (e) Sketch the standing wave at $t = 0.50$ s for the range $x = 0$ to $x = 0.40$ m.

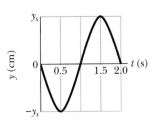

Fig. 16-39 Problem 50.

••51 SSM WWW Two waves are generated on a string of length 3.0 m to produce a three-loop standing wave with an amplitude of 1.0 cm. The wave speed is 100 m/s. Let the equation for one of the waves be of the form $y(x, t) = y_m \sin(kx + \omega t)$. In the equation for the other wave, what are (a) y_m, (b) k, (c) ω, and (d) the sign in front of ω?

••52 A rope, under a tension of 200 N and fixed at both ends, oscillates in a second-harmonic standing wave pattern. The displacement of the rope is given by

$$y = (0.10 \text{ m})(\sin \pi x/2) \sin 12\pi t,$$

where $x = 0$ at one end of the rope, x is in meters, and t is in seconds. What are (a) the length of the rope, (b) the speed of the waves on the rope, and (c) the mass of the rope? (d) If the rope oscillates in a third-harmonic standing wave pattern, what will be the period of oscillation?

••53 A string oscillates according to the equation

$$y' = (0.50 \text{ cm}) \sin\left[\left(\frac{\pi}{3} \text{ cm}^{-1}\right)x\right] \cos[(40\pi \text{ s}^{-1})t].$$

What are the (a) amplitude and (b) speed of the two waves (identical except for direction of travel) whose superposition gives this oscillation? (c) What is the distance between nodes? (d) What is the transverse speed of a particle of the string at the position $x = 1.5$ cm when $t = \frac{9}{8}$ s?

••54 GO Two sinusoidal waves with the same amplitude and wavelength travel through each other along a string that is stretched along an x axis. Their resultant wave is shown twice in Fig. 16-40, as the antinode A travels from an extreme upward displacement to an extreme downward displacement in 6.0 ms. The tick marks along the axis are separated by 10 cm; height H is 1.80 cm. Let the equation for one of the two waves be of the form $y(x, t) = y_m \sin(kx + \omega t)$. In the equation for the other wave, what are (a) y_m, (b) k, (c) ω, and (d) the sign in front of ω?

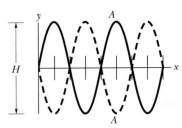

Fig. 16-40 Problem 54.

••55 The following two waves are sent in opposite directions on a horizontal string so as to create a standing wave in a vertical plane:

$$y_1(x, t) = (6.00 \text{ mm}) \sin(4.00\pi x - 400\pi t)$$
$$y_2(x, t) = (6.00 \text{ mm}) \sin(4.00\pi x + 400\pi t),$$

with x in meters and t in seconds. An antinode is located at point A. In the time interval that point takes to move from maximum upward displacement to maximum downward displacement, how far does each wave move along the string?

••56 A standing wave pattern on a string is described by

$$y(x, t) = 0.040 (\sin 5\pi x)(\cos 40\pi t),$$

where x and y are in meters and t is in seconds. For $x \geq 0$, what is the location of the node with the (a) smallest, (b) second smallest, and (c) third smallest value of x? (d) What is the period of the oscillatory motion of any (nonnode) point? What are the (e) speed and (f) amplitude of the two traveling waves that interfere to produce this wave? For $t \geq 0$, what are the (g) first, (h) second, and (i) third time that all points on the string have zero transverse velocity?

••57 A generator at one end of a very long string creates a wave given by

$$y = (6.0 \text{ cm}) \cos \frac{\pi}{2} [(2.00 \text{ m}^{-1})x + (8.00 \text{ s}^{-1})t],$$

and a generator at the other end creates the wave

$$y = (6.0 \text{ cm}) \cos \frac{\pi}{2} [(2.00 \text{ m}^{-1})x - (8.00 \text{ s}^{-1})t].$$

Calculate the (a) frequency, (b) wavelength, and (c) speed of each wave. For $x \geq 0$, what is the location of the node having the (d) smallest, (e) second smallest, and (f) third smallest value of x? For $x \geq 0$, what is the location of the antinode having the (g) smallest, (h) second smallest, and (i) third smallest value of x?

••58 **GO** In Fig. 16-41, a string, tied to a sinusoidal oscillator at P and running over a support at Q, is stretched by a block of mass m. Separation $L = 1.20$ m, linear density $\mu = 1.6$ g/m, and the oscillator frequency $f = 120$ Hz. The amplitude of the motion at P is small enough for that point to be considered a node. A node also exists at Q. (a) What mass m allows the oscillator to set up the fourth harmonic on the string? (b) What standing wave mode, if any, can be set up if $m = 1.00$ kg?

Fig. 16-41   Problems 58 and 60.

•••59 In Fig. 16-42, an aluminum wire, of length $L_1 = 60.0$ cm, cross-sectional area 1.00×10^{-2} cm^2, and density 2.60 g/cm^3, is joined to a steel wire, of density 7.80 g/cm^3 and the same cross-sectional area. The compound wire, loaded with a block of mass $m = 10.0$ kg, is arranged so that the distance L_2 from the joint to the supporting pulley is 86.6 cm. Transverse waves are set up on the wire by an external source of variable frequency; a node is located at the pulley. (a) Find the lowest frequency that generates a standing wave having the joint as one of the nodes. (b) How many nodes are observed at this frequency?

Fig. 16-42   Problem 59.

•••60 **GO** In Fig. 16-41, a string, tied to a sinusoidal oscillator at P and running over a support at Q, is stretched by a block of mass m. The separation L between P and Q is 1.20 m, and the frequency f of the oscillator is fixed at 120 Hz. The amplitude of the motion at P is small enough for that point to be considered a node. A node also exists at Q. A standing wave appears when the mass of the hanging block is 286.1 g or 447.0 g, but not for any intermediate mass. What is the linear density of the string?

Additional Problems

61 In an experiment on standing waves, a string 90 cm long is attached to the prong of an electrically driven tuning fork that oscillates perpendicular to the length of the string at a frequency of 60 Hz. The mass of the string is 0.044 kg. What tension must the string be under (weights are attached to the other end) if it is to oscillate in four loops?

62 A sinusoidal transverse wave traveling in the positive direction of an x axis has an amplitude of 2.0 cm, a wavelength of 10 cm, and a frequency of 400 Hz. If the wave equation is of the form $y(x, t) = y_m \sin(kx \pm \omega t)$, what are (a) y_m, (b) k, (c) ω, and (d) the correct choice of sign in front of ω? What are (e) the maximum transverse speed of a point on the cord and (f) the speed of the wave?

63 A wave has a speed of 240 m/s and a wavelength of 3.2 m. What are the (a) frequency and (b) period of the wave?

64 The equation of a transverse wave traveling along a string is

$$y = 0.15 \sin(0.79x - 13t),$$

in which x and y are in meters and t is in seconds. (a) What is the displacement y at $x = 2.3$ m, $t = 0.16$ s? A second wave is to be added to the first wave to produce standing waves on the string. If the wave equation for the second wave is of the form $y(x, t) = y_m \sin(kx \pm \omega t)$, what are (b) y_m, (c) k, (d) ω, and (e) the correct choice of sign in front of ω for this second wave? (f) What is the displacement of the resultant standing wave at $x = 2.3$ m, $t = 0.16$ s?

65 The equation of a transverse wave traveling along a string is

$$y = (2.0 \text{ mm}) \sin[(20 \text{ m}^{-1})x - (600 \text{ s}^{-1})t].$$

Find the (a) amplitude, (b) frequency, (c) velocity (including sign), and (d) wavelength of the wave. (e) Find the maximum transverse speed of a particle in the string.

66 Figure 16-43 shows the displacement y versus time t of the point on a string at $x = 0$, as a wave passes through that point. The scale of the y axis is set by $y_s = 6.0$ mm. The wave is given by $y(x, t) = y_m \sin(kx - \omega t + \phi)$. What is ϕ? (Caution: A calculator does not always give the proper inverse trig function, so check your answer by substituting it and an assumed value of ω into $y(x, t)$ and then plotting the function.)

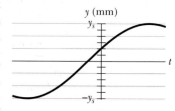

Fig. 16-43   Problem 66.

67 Two sinusoidal waves, identical except for phase, travel in the same direction along a string, producing the net wave $y'(x, t) = (3.0 \text{ mm}) \sin(20x - 4.0t + 0.820 \text{ rad})$, with x in meters and t in seconds. What are (a) the wavelength λ of the two waves, (b) the phase difference between them, and (c) their amplitude y_m?

68 A single pulse, given by $h(x - 5.0t)$, is shown in Fig. 16-44 for $t = 0$. The scale of the vertical axis is set by $h_s = 2$. Here x is in centimeters and t is in seconds. What are the (a) speed and (b) direction of travel of the pulse? (c) Plot $h(x - 5t)$ as a function of x for $t = 2$ s. (d) Plot $h(x - 5t)$ as a function of t for $x = 10$ cm.

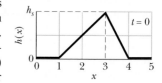

Fig. 16-44   Problem 68.

69 **SSM** Three sinusoidal waves of the same frequency travel along a string in the positive direction of an x axis. Their amplitudes are y_1, $y_1/2$, and $y_1/3$, and their phase constants are 0, $\pi/2$, and π, respectively. What are the (a) amplitude and (b) phase constant of the resultant wave? (c) Plot the wave form of the resultant wave at $t = 0$, and discuss its behavior as t increases.

70 **GO** Figure 16-45 shows transverse acceleration a_y versus time t of the point on a string at $x = 0$, as a wave in the form of $y(x, t) = y_m \sin(kx - \omega t + \phi)$ passes through that point. The scale of the vertical axis is set by $a_s = 400$ m/s^2. What is ϕ? (Caution: A calculator does not

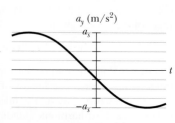

Fig. 16-45   Problem 70.

always give the proper inverse trig function, so check your answer by substituting it and an assumed value of ω into $y(x, t)$ and then plotting the function.)

71 A transverse sinusoidal wave is generated at one end of a long, horizontal string by a bar that moves up and down through a distance of 1.00 cm. The motion is continuous and is repeated regularly 120 times per second. The string has linear density 120 g/m and is kept under a tension of 90.0 N. Find the maximum value of (a) the transverse speed u and (b) the transverse component of the tension τ.

(c) Show that the two maximum values calculated above occur at the same phase values for the wave. What is the transverse displacement y of the string at these phases? (d) What is the maximum rate of energy transfer along the string? (e) What is the transverse displacement y when this maximum transfer occurs? (f) What is the minimum rate of energy transfer along the string? (g) What is the transverse displacement y when this minimum transfer occurs?

72 Two sinusoidal 120 Hz waves, of the same frequency and amplitude, are to be sent in the positive direction of an x axis that is directed along a cord under tension. The waves can be sent in phase, or they can be phase-shifted. Figure 16-46 shows the amplitude y' of the resulting wave versus the distance of the shift (how far one wave is shifted from the other wave). The scale of the vertical axis is set by $y'_s = 6.0$ mm. If the equations for the two waves are of the form $y(x, t) = y_m \sin(kx \pm \omega t)$, what are (a) y_m, (b) k, (c) ω, and (d) the correct choice of sign in front of ω?

Fig. 16-46 Problem 72.

73 At time $t = 0$ and at position $x = 0$ m along a string, a traveling sinusoidal wave with an angular frequency of 440 rad/s has displacement $y = +4.5$ mm and transverse velocity $u = -0.75$ m/s. If the wave has the general form $y(x, t) = y_m \sin(kx - \omega t + \phi)$, what is phase constant ϕ?

74 Energy is transmitted at rate P_1 by a wave of frequency f_1 on a string under tension τ_1. What is the new energy transmission rate P_2 in terms of P_1 (a) if the tension is increased to $\tau_2 = 4\tau_1$ and (b) if, instead, the frequency is decreased to $f_2 = f_1/2$?

75 (a) What is the fastest transverse wave that can be sent along a steel wire? For safety reasons, the maximum tensile stress to which steel wires should be subjected is 7.00×10^8 N/m². The density of steel is 7800 kg/m³. (b) Does your answer depend on the diameter of the wire?

76 A standing wave results from the sum of two transverse traveling waves given by

$$y_1 = 0.050 \cos(\pi x - 4\pi t)$$

and $$y_2 = 0.050 \cos(\pi x + 4\pi t),$$

where x, y_1, and y_2 are in meters and t is in seconds. (a) What is the smallest positive value of x that corresponds to a node? Beginning at $t = 0$, what is the value of the (b) first, (c) second, and (d) third time the particle at $x = 0$ has zero velocity?

77 SSM The type of rubber band used inside some baseballs and golf balls obeys Hooke's law over a wide range of elongation of the band. A segment of this material has an unstretched length ℓ and a mass m. When a force F is applied, the band stretches an additional

length $\Delta\ell$. (a) What is the speed (in terms of m, $\Delta\ell$, and the spring constant k) of transverse waves on this stretched rubber band? (b) Using your answer to (a), show that the time required for a transverse pulse to travel the length of the rubber band is proportional to $1/\sqrt{\Delta\ell}$ if $\Delta\ell \ll \ell$ and is constant if $\Delta\ell \gg \ell$.

78 The speed of electromagnetic waves (which include visible light, radio, and x rays) in vacuum is 3.0×10^8 m/s. (a) Wavelengths of visible light waves range from about 400 nm in the violet to about 700 nm in the red. What is the range of frequencies of these waves? (b) The range of frequencies for shortwave radio (for example, FM radio and VHF television) is 1.5 to 300 MHz. What is the corresponding wavelength range? (c) X-ray wavelengths range from about 5.0 nm to about 1.0×10^{-2} nm. What is the frequency range for x rays?

79 SSM A 1.50 m wire has a mass of 8.70 g and is under a tension of 120 N. The wire is held rigidly at both ends and set into oscillation. (a) What is the speed of waves on the wire? What is the wavelength of the waves that produce (b) one-loop and (c) two-loop standing waves? What is the frequency of the waves that produce (d) one-loop and (e) two-loop standing waves?

80 When played in a certain manner, the lowest resonant frequency of a certain violin string is concert A (440 Hz). What is the frequency of the (a) second and (b) third harmonic of the string?

81 A sinusoidal transverse wave traveling in the negative direction of an x axis has an amplitude of 1.00 cm, a frequency of 550 Hz, and a speed of 330 m/s. If the wave equation is of the form $y(x, t) = y_m \sin(kx \pm \omega t)$, what are (a) y_m, (b) ω, (c) k, and (d) the correct choice of sign in front of ω?

82 Two sinusoidal waves of the same wavelength travel in the same direction along a stretched string. For wave 1, $y_m = 3.0$ mm and $\phi = 0$; for wave 2, $y_m = 5.0$ mm and $\phi = 70°$. What are the (a) amplitude and (b) phase constant of the resultant wave?

83 SSM A sinusoidal transverse wave of amplitude y_m and wavelength λ travels on a stretched cord. (a) Find the ratio of the maximum particle speed (the speed with which a single particle in the cord moves transverse to the wave) to the wave speed. (b) Does this ratio depend on the material of which the cord is made?

84 Oscillation of a 600 Hz tuning fork sets up standing waves in a string clamped at both ends. The wave speed for the string is 400 m/s. The standing wave has four loops and an amplitude of 2.0 mm. (a) What is the length of the string? (b) Write an equation for the displacement of the string as a function of position and time.

85 A 120 cm length of string is stretched between fixed supports. What are the (a) longest, (b) second longest, and (c) third longest wavelength for waves traveling on the string if standing waves are to be set up? (d) Sketch those standing waves.

86 (a) Write an equation describing a sinusoidal transverse wave traveling on a cord in the positive direction of a y axis with an angular wave number of 60 cm⁻¹, a period of 0.20 s, and an amplitude of 3.0 mm. Take the transverse direction to be the z direction. (b) What is the maximum transverse speed of a point on the cord?

87 A wave on a string is described by

$$y(x, t) = 15.0 \sin(\pi x/8 - 4\pi t),$$

where x and y are in centimeters and t is in seconds. (a) What is the transverse speed for a point on the string at $x = 6.00$ cm when $t = 0.250$ s? (b) What is the maximum transverse speed of any point on the string? (c) What is the magnitude of the transverse acceleration for a point on the string at $x = 6.00$ cm when $t = 0.250$ s? (d) What is the magnitude of the maximum transverse acceleration for any point on the string?

88 ~~~~ *Body armor.* When a high-speed projectile such as a bullet or bomb fragment strikes modern body armor, the fabric of the armor stops the projectile and prevents penetration by quickly spreading the projectile's energy over a large area. This spreading is done by longitudinal and transverse pulses that move *radially* from the impact point, where the projectile pushes a cone-shaped dent into the fabric. The longitudinal pulse, racing along the fibers of the fabric at speed v_l ahead of the denting, causes the fibers to thin and stretch, with material flowing radially inward into the dent. One such radial fiber is shown in Fig. 16-47a. Part of the projectile's energy goes into this motion and stretching. The transverse pulse, moving at a slower speed v_t, is due to the denting. As the projectile increases the dent's depth, the dent increases in radius, causing the material in the fibers to move in the same direction as the projectile (perpendicular to the transverse pulse's direction of travel). The rest of the projectile's energy goes into this motion. All the energy that does not eventually go into permanently deforming the fibers ends up as thermal energy.

Figure 16-47b is a graph of speed v versus time t for a bullet of mass 10.2 g fired from a .38 Special revolver directly into body armor. The scales of the vertical and horizontal axes are set by $v_s = 300$ m/s and $t_s = 40.0$ μs. Take $v_l = 2000$ m/s, and assume that the half-angle θ of the conical dent is 60°. At the end of the collision, what are the radii of (a) the thinned region and (b) the dent (assuming that the person wearing the armor remains stationary)?

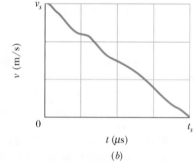

Fig. 16-47 Problem 88.

89 Two waves are described by

$$y_1 = 0.30 \sin[\pi(5x - 200)t]$$

and $$y_2 = 0.30 \sin[\pi(5x - 200t) + \pi/3],$$

where $y_1, y_2,$ and x are in meters and t is in seconds. When these two waves are combined, a traveling wave is produced. What are the (a) amplitude, (b) wave speed, and (c) wavelength of that traveling wave?

90 A certain transverse sinusoidal wave of wavelength 20 cm is moving in the positive direction of an x axis. The transverse velocity of the particle at $x = 0$ as a function of time is shown in Fig. 16-48, where the scale of the vertical axis is set by $u_s = 5.0$ cm/s. What are the (a) wave speed, (b) amplitude, and (c) frequency? (d) Sketch the wave between $x = 0$ and $x = 20$ cm at $t = 2.0$ s.

Fig. 16-48 Problem 90.

91 **SSM** In a demonstration, a 1.2 kg horizontal rope is fixed in place at its two ends ($x = 0$ and $x = 2.0$ m) and made to oscillate up and down in the fundamental mode, at frequency 5.0 Hz. At $t = 0$, the point at $x = 1.0$ m has zero displacement and is moving upward in the positive direction of a y axis with a transverse velocity of 5.0 m/s. What are (a) the amplitude of the motion of that point and (b) the tension in the rope? (c) Write the standing wave equation for the fundamental mode.

92 Two waves,

$$y_1 = (2.50 \text{ mm}) \sin[(25.1 \text{ rad/m})x - (440 \text{ rad/s})t]$$

and $$y_2 = (1.50 \text{ mm}) \sin[(25.1 \text{ rad/m})x + (440 \text{ rad/s})t],$$

travel along a stretched string. (a) Plot the resultant wave as a function of t for $x = 0$, $\lambda/8$, $\lambda/4$, $3\lambda/8$, and $\lambda/2$, where λ is the wavelength. The graphs should extend from $t = 0$ to a little over one period. (b) The resultant wave is the superposition of a standing wave and a traveling wave. In which direction does the traveling wave move? (c) How can you change the original waves so the resultant wave is the superposition of standing and traveling waves with the same amplitudes as before but with the traveling wave moving in the opposite direction? Next, use your graphs to find the place at which the oscillation amplitude is (d) maximum and (e) minimum. (f) How is the maximum amplitude related to the amplitudes of the original two waves? (g) How is the minimum amplitude related to the amplitudes of the original two waves?

93 A traveling wave on a string is described by

$$y = 2.0 \sin\left[2\pi\left(\frac{t}{0.40} + \frac{x}{80}\right)\right],$$

where x and y are in centimeters and t is in seconds. (a) For $t = 0$, plot y as a function of x for $0 \leq x \leq 160$ cm. (b) Repeat (a) for $t = 0.05$ s and $t = 0.10$ s. From your graphs, determine (c) the wave speed and (d) the direction in which the wave is traveling.

WAVES—II

WHAT IS PHYSICS?

The physics of sound waves is the basis of countless studies in the research journals of many fields. Here are just a few examples. Some physiologists are concerned with how speech is produced, how speech impairment might be corrected, how hearing loss can be alleviated, and even how snoring is produced. Some acoustic engineers are concerned with improving the acoustics of cathedrals and concert halls, with reducing noise near freeways and road construction, and with reproducing music by speaker systems. Some aviation engineers are concerned with the shock waves produced by supersonic aircraft and the aircraft noise produced in communities near an airport. Some medical researchers are concerned with how noises produced by the heart and lungs can signal a medical problem in a patient. Some paleontologists are concerned with how a dinosaur's fossil might reveal the dinosaur's vocalizations. Some military engineers are concerned with how the sounds of sniper fire might allow a soldier to pinpoint the sniper's location, and, on the gentler side, some biologists are concerned with how a cat purrs.

To begin our discussion of the physics of sound, we must first answer the question "What *are* sound waves?"

17-2 Sound Waves

As we saw in Chapter 16, mechanical waves are waves that require a material medium to exist. There are two types of mechanical waves: *Transverse waves* involve oscillations perpendicular to the direction in which the wave travels; *longitudinal waves* involve oscillations parallel to the direction of wave travel.

In this book, a **sound wave** is defined roughly as any longitudinal wave. Seismic prospecting teams use such waves to probe Earth's crust for oil. Ships carry sound-ranging gear (sonar) to detect underwater obstacles. Submarines use sound waves to stalk other submarines, largely by listening for the characteristic noises produced by the propulsion system. Figure 17-1 suggests how

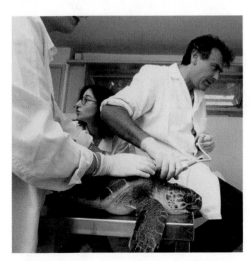

Fig. 17-1 A loggerhead turtle is being checked with ultrasound (which has a frequency above your hearing range); an image of its interior is being produced on a monitor off to the right. (*Mauro Fermariello/SPL/Photo Researchers*)

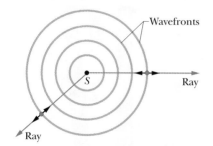

Fig. 17-2 A sound wave travels from a point source S through a three-dimensional medium. The wavefronts form spheres centered on S; the rays are radial to S. The short, double-headed arrows indicate that elements of the medium oscillate parallel to the rays.

sound waves can be used to explore the soft tissues of an animal or human body. In this chapter we shall focus on sound waves that travel through the air and that are audible to people.

Figure 17-2 illustrates several ideas that we shall use in our discussions. Point S represents a tiny sound source, called a *point source*, that emits sound waves in all directions. The *wavefronts* and *rays* indicate the direction of travel and the spread of the sound waves. **Wavefronts** are surfaces over which the oscillations due to the sound wave have the same value; such surfaces are represented by whole or partial circles in a two-dimensional drawing for a point source. **Rays** are directed lines perpendicular to the wavefronts that indicate the direction of travel of the wavefronts. The short double arrows superimposed on the rays of Fig. 17-2 indicate that the longitudinal oscillations of the air are parallel to the rays.

Near a point source like that of Fig. 17-2, the wavefronts are spherical and spread out in three dimensions, and there the waves are said to be *spherical*. As the wavefronts move outward and their radii become larger, their curvature decreases. Far from the source, we approximate the wavefronts as planes (or lines on two-dimensional drawings), and the waves are said to be *planar*.

17-3 The Speed of Sound

The speed of any mechanical wave, transverse or longitudinal, depends on both an inertial property of the medium (to store kinetic energy) and an elastic property of the medium (to store potential energy). Thus, we can generalize Eq. 16-26, which gives the speed of a transverse wave along a stretched string, by writing

$$v = \sqrt{\frac{\tau}{\mu}} = \sqrt{\frac{\text{elastic property}}{\text{inertial property}}}, \tag{17-1}$$

where (for transverse waves) τ is the tension in the string and μ is the string's linear density. If the medium is air and the wave is longitudinal, we can guess that the inertial property, corresponding to μ, is the volume density ρ of air. What shall we put for the elastic property?

In a stretched string, potential energy is associated with the periodic stretching of the string elements as the wave passes through them. As a sound wave passes through air, potential energy is associated with periodic compressions and expansions of small volume elements of the air. The property that determines the extent to which an element of a medium changes in volume when the pressure (force per unit area) on it changes is the **bulk modulus** B, defined (from Eq. 12-25) as

$$B = -\frac{\Delta p}{\Delta V/V} \qquad \text{(definition of bulk modulus).} \tag{17-2}$$

Here $\Delta V/V$ is the fractional change in volume produced by a change in pressure Δp. As explained in Section 14-3, the SI unit for pressure is the newton per square meter, which is given a special name, the *pascal* (Pa). From Eq. 17-2 we see that the unit for B is also the pascal. The signs of Δp and ΔV are always opposite: When we increase the pressure on an element (Δp is positive), its volume decreases (ΔV is negative). We include a minus sign in Eq. 17-2 so that B is always a positive quantity. Now substituting B for τ and ρ for μ in Eq. 17-1 yields

$$v = \sqrt{\frac{B}{\rho}} \qquad \text{(speed of sound)} \tag{17-3}$$

as the speed of sound in a medium with bulk modulus B and density ρ. Table 17-1 lists the speed of sound in various media.

The density of water is almost 1000 times greater than the density of air. If this were the only relevant factor, we would expect from Eq. 17-3 that the speed of sound in water would be considerably less than the speed of sound in air. However, Table 17-1 shows us that the reverse is true. We conclude (again from Eq. 17-3) that the bulk modulus of water must be more than 1000 times greater than that of air. This is indeed the case. Water is much more incompressible than air, which (see Eq. 17-2) is another way of saying that its bulk modulus is much greater.

Formal Derivation of Eq. 17-3

We now derive Eq. 17-3 by direct application of Newton's laws. Let a single pulse in which air is compressed travel (from right to left) with speed v through the air in a long tube, like that in Fig. 16-2. Let us run along with the pulse at that speed, so that the pulse appears to stand still in our reference frame. Figure 17-3a shows the situation as it is viewed from that frame. The pulse is standing still, and air is moving at speed v through it from left to right.

Let the pressure of the undisturbed air be p and the pressure inside the pulse be $p + \Delta p$, where Δp is positive due to the compression. Consider an element of air of thickness Δx and face area A, moving toward the pulse at speed v. As this element enters the pulse, the leading face of the element encounters a region of higher pressure, which slows the element to speed $v + \Delta v$, in which Δv is negative. This slowing is complete when the rear face of the element reaches the pulse, which requires time interval

$$\Delta t = \frac{\Delta x}{v}. \tag{17-4}$$

Let us apply Newton's second law to the element. During Δt, the average force on the element's trailing face is pA toward the right, and the average force on the leading face is $(p + \Delta p)A$ toward the left (Fig. 17-3b). Therefore, the average net force on the element during Δt is

$$F = pA - (p + \Delta p)A$$
$$= -\Delta p\, A \quad \text{(net force)}. \tag{17-5}$$

The minus sign indicates that the net force on the air element is directed to the left in Fig. 17-3b. The volume of the element is $A\, \Delta x$, so with the aid of Eq. 17-4, we can write its mass as

$$\Delta m = \rho\, \Delta V = \rho A\, \Delta x = \rho A v\, \Delta t \quad \text{(mass)}. \tag{17-6}$$

The average acceleration of the element during Δt is

$$a = \frac{\Delta v}{\Delta t} \quad \text{(acceleration)}. \tag{17-7}$$

Table 17-1

The Speed of Sound[a]

Medium	Speed (m/s)
Gases	
Air (0°C)	331
Air (20°C)	343
Helium	965
Hydrogen	1284
Liquids	
Water (0°C)	1402
Water (20°C)	1482
Seawater[b]	1522
Solids	
Aluminum	6420
Steel	5941
Granite	6000

[a]At 0°C and 1 atm pressure, except where noted.

[b]At 20°C and 3.5% salinity.

Fig. 17-3 A compression pulse is sent from right to left down a long air-filled tube. The reference frame of the figure is chosen so that the pulse is at rest and the air moves from left to right. (a) An element of air of width Δx moves toward the pulse with speed v. (b) The leading face of the element enters the pulse. The forces acting on the leading and trailing faces (due to air pressure) are shown.

Thus, from Newton's second law ($F = ma$), we have, from Eqs. 17-5, 17-6, and 17-7,

$$-\Delta p\, A = (\rho A v\, \Delta t)\,\frac{\Delta v}{\Delta t}, \tag{17-8}$$

which we can write as

$$\rho v^2 = -\frac{\Delta p}{\Delta v/v}. \tag{17-9}$$

The air that occupies a volume $V\,(= Av\,\Delta t)$ outside the pulse is compressed by an amount $\Delta V\,(= A\,\Delta v\,\Delta t)$ as it enters the pulse. Thus,

$$\frac{\Delta V}{V} = \frac{A\,\Delta v\,\Delta t}{Av\,\Delta t} = \frac{\Delta v}{v}. \tag{17-10}$$

Substituting Eq. 17-10 and then Eq. 17-2 into Eq. 17-9 leads to

$$\rho v^2 = -\frac{\Delta p}{\Delta v/v} = -\frac{\Delta p}{\Delta V/V} = B. \tag{17-11}$$

Solving for v yields Eq. 17-3 for the speed of the air toward the right in Fig. 17-3, and thus for the actual speed of the pulse toward the left.

17-4 Traveling Sound Waves

Here we examine the displacements and pressure variations associated with a sinusoidal sound wave traveling through air. Figure 17-4a displays such a wave traveling rightward through a long air-filled tube. Recall from Chapter 16 that we can produce such a wave by sinusoidally moving a piston at the left end of the tube (as in Fig. 16-2). The piston's rightward motion moves the element of air next to the piston face and compresses that air; the piston's leftward motion allows the element of air to move back to the left and the pressure to decrease. As each element of air pushes on the next element in turn, the right–left motion of the air and the change in its pressure travel along the tube as a sound wave.

Consider the thin element of air of thickness Δx shown in Fig. 17-4b. As the wave travels through this portion of the tube, the element of air oscillates left

Fig. 17-4 (a) A sound wave, traveling through a long air-filled tube with speed v, consists of a moving, periodic pattern of expansions and compressions of the air. The wave is shown at an arbitrary instant. (b) A horizontally expanded view of a short piece of the tube. As the wave passes, an air element of thickness Δx oscillates left and right in simple harmonic motion about its equilibrium position. At the instant shown in (b), the element happens to be displaced a distance s to the right of its equilibrium position. Its maximum displacement, either right or left, is s_m.

and right in simple harmonic motion about its equilibrium position. Thus, the oscillations of each air element due to the traveling sound wave are like those of a string element due to a transverse wave, except that the air element oscillates *longitudinally* rather than *transversely*. Because string elements oscillate parallel to the y axis, we write their displacements in the form $y(x, t)$. Similarly, because air elements oscillate parallel to the x axis, we could write their displacements in the confusing form $x(x, t)$, but we shall use $s(x, t)$ instead.

To show that the displacements $s(x, t)$ are sinusoidal functions of x and t, we can use either a sine function or a cosine function. In this chapter we use a cosine function, writing

$$s(x, t) = s_m \cos(kx - \omega t). \qquad (17\text{-}12)$$

Figure 17-5a labels the various parts of this equation. In it, s_m is the **displacement amplitude**—that is, the maximum displacement of the air element to either side of its equilibrium position (see Fig. 17-4b). The angular wave number k, angular frequency ω, frequency f, wavelength λ, speed v, and period T for a sound (longitudinal) wave are defined and interrelated exactly as for a transverse wave, except that λ is now the distance (again along the direction of travel) in which the pattern of compression and expansion due to the wave begins to repeat itself (see Fig. 17-4a). (We assume s_m is much less than λ.)

As the wave moves, the air pressure at any position x in Fig. 17-4a varies sinusoidally, as we prove next. To describe this variation we write

$$\Delta p(x, t) = \Delta p_m \sin(kx - \omega t). \qquad (17\text{-}13)$$

Figure 17-5b labels the various parts of this equation. A negative value of Δp in Eq. 17-13 corresponds to an expansion of the air, and a positive value to a compression. Here Δp_m is the **pressure amplitude,** which is the maximum increase or decrease in pressure due to the wave; Δp_m is normally very much less than the pressure p present when there is no wave. As we shall prove, the pressure amplitude Δp_m is related to the displacement amplitude s_m in Eq. 17-12 by

$$\Delta p_m = (v \rho \omega) s_m. \qquad (17\text{-}14)$$

Figure 17-6 shows plots of Eqs. 17-12 and 17-13 at $t = 0$; with time, the two curves would move rightward along the horizontal axes. Note that the displacement and pressure variation are $\pi/2$ rad (or 90°) out of phase. Thus, for example, the pressure variation Δp at any point along the wave is zero when the displacement there is a maximum.

✔CHECKPOINT 1

When the oscillating air element in Fig. 17-4b is moving rightward through the point of zero displacement, is the pressure in the element at its equilibrium value, just beginning to increase, or just beginning to decrease?

Derivation of Eqs. 17-13 and 17-14

Figure 17-4b shows an oscillating element of air of cross-sectional area A and thickness Δx, with its center displaced from its equilibrium position by distance s. From Eq. 17-2 we can write, for the pressure variation in the displaced element,

$$\Delta p = -B \frac{\Delta V}{V}. \qquad (17\text{-}15)$$

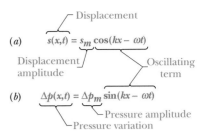

Fig. 17-5 (*a*) The displacement function and (*b*) the pressure-variation function of a traveling sound wave consist of an amplitude and an oscillating term.

Fig. 17-6 (*a*) A plot of the displacement function (Eq. 17-12) for $t = 0$. (*b*) A similar plot of the pressure-variation function (Eq. 17-13). Both plots are for a 1000 Hz sound wave whose pressure amplitude is at the threshold of pain.

The quantity V in Eq. 17-15 is the volume of the element, given by

$$V = A\,\Delta x. \qquad (17\text{-}16)$$

The quantity ΔV in Eq. 17-15 is the change in volume that occurs when the element is displaced. This volume change comes about because the displacements of the two faces of the element are not quite the same, differing by some amount Δs. Thus, we can write the change in volume as

$$\Delta V = A\,\Delta s. \qquad (17\text{-}17)$$

Substituting Eqs. 17-16 and 17-17 into Eq. 17-15 and passing to the differential limit yield

$$\Delta p = -B\,\frac{\Delta s}{\Delta x} = -B\,\frac{\partial s}{\partial x}. \qquad (17\text{-}18)$$

The symbols ∂ indicate that the derivative in Eq. 17-18 is a *partial derivative*, which tells us how s changes with x when the time t is fixed. From Eq. 17-12 we then have, treating t as a constant,

$$\frac{\partial s}{\partial x} = \frac{\partial}{\partial x}\,[s_m\cos(kx - \omega t)] = -ks_m\sin(kx - \omega t). \qquad (17\text{-}19)$$

Substituting this quantity for the partial derivative in Eq. 17-18 yields

$$\Delta p = Bks_m\sin(kx - \omega t).$$

This tells us that the pressure varies as a sinusoidal function of time and that the amplitude of the variation is equal to the terms in front of the sine function. Setting $\Delta p_m = Bks_m$, this yields Eq. 17-13, which we set out to prove.

Using Eq. 17-3, we can now write

$$\Delta p_m = (Bk)s_m = (v^2\rho k)s_m.$$

Equation 17-14, which we also wanted to prove, follows at once if we substitute ω/v for k from Eq. 16-12.

Sample Problem

Pressure amplitude, displacement amplitude

The maximum pressure amplitude Δp_m that the human ear can tolerate in loud sounds is about 28 Pa (which is very much less than the normal air pressure of about 10^5 Pa). What is the displacement amplitude s_m for such a sound in air of density $\rho = 1.21$ kg/m^3, at a frequency of 1000 Hz and a speed of 343 m/s?

KEY IDEA

The displacement amplitude s_m of a sound wave is related to the pressure amplitude Δp_m of the wave according to Eq. 17-14.

Calculations: Solving that equation for s_m yields

$$s_m = \frac{\Delta p_m}{v\rho\omega} = \frac{\Delta p_m}{v\rho(2\pi f)}.$$

Substituting known data then gives us

$$s_m = \frac{28\text{ Pa}}{(343\text{ m/s})(1.21\text{ kg/m}^3)(2\pi)(1000\text{ Hz})}$$
$$= 1.1 \times 10^{-5}\text{ m} = 11\ \mu\text{m}. \qquad \text{(Answer)}$$

That is only about one-seventh the thickness of a book page. Obviously, the displacement amplitude of even the loudest sound that the ear can tolerate is very small. Temporary exposure to such loud sound produces temporary hearing loss, probably due to a decrease in blood supply to the inner ear. Prolonged exposure produces permanent damage.

The pressure amplitude Δp_m for the *faintest* detectable sound at 1000 Hz is 2.8×10^{-5} Pa. Proceeding as above leads to $s_m = 1.1 \times 10^{-11}$ m or 11 pm, which is about one-tenth the radius of a typical atom. The ear is indeed a sensitive detector of sound waves.

 Additional examples, video, and practice available at *WileyPLUS*

17-5 Interference

Like transverse waves, sound waves can undergo interference. Let us consider, in particular, the interference between two identical sound waves traveling in the same direction. Figure 17-7a shows how we can set up such a situation: Two point sources S_1 and S_2 emit sound waves that are in phase and of identical wavelength λ. Thus, the sources themselves are said to be in phase; that is, as the waves emerge from the sources, their displacements are always identical. We are interested in the waves that then travel through point P in Fig. 17-7a. We assume that the distance to P is much greater than the distance between the sources so that we can approximate the waves as traveling in the same direction at P.

If the waves traveled along paths with identical lengths to reach point P, they would be in phase there. As with transverse waves, this means that they would undergo fully constructive interference there. However, in Fig. 17-7a, path L_2 traveled by the wave from S_2 is longer than path L_1 traveled by the wave from S_1. The difference in path lengths means that the waves may not be in phase at point P. In other words, their phase difference ϕ at P depends on their **path length difference** $\Delta L = |L_2 - L_1|$.

To relate phase difference ϕ to path length difference ΔL, we recall (from Section 16-4) that a phase difference of 2π rad corresponds to one wavelength. Thus, we can write the proportion

$$\frac{\phi}{2\pi} = \frac{\Delta L}{\lambda}, \tag{17-20}$$

from which

$$\phi = \frac{\Delta L}{\lambda} 2\pi. \tag{17-21}$$

Fully constructive interference occurs when ϕ is zero, 2π, or any integer multiple of 2π. We can write this condition as

$$\phi = m(2\pi), \quad \text{for } m = 0, 1, 2, \ldots \quad \text{(fully constructive interference)}. \tag{17-22}$$

From Eq. 17-21, this occurs when the ratio $\Delta L/\lambda$ is

$$\frac{\Delta L}{\lambda} = 0, 1, 2, \ldots \quad \text{(fully constructive interference)}. \tag{17-23}$$

For example, if the path length difference $\Delta L = |L_2 - L_1|$ in Fig. 17-7a is equal to 2λ, then $\Delta L/\lambda = 2$ and the waves undergo fully constructive interference at point P (Fig. 17-7b). The interference is fully constructive because the wave from S_2 is phase-shifted relative to the wave from S_1 by 2λ, putting the two waves *exactly in phase* at P.

Fully destructive interference occurs when ϕ is an odd multiple of π:

$$\phi = (2m + 1)\pi, \quad \text{for } m = 0, 1, 2, \ldots \quad \text{(fully destructive interference)}. \tag{17-24}$$

From Eq. 17-21, this occurs when the ratio $\Delta L/\lambda$ is

$$\frac{\Delta L}{\lambda} = 0.5, 1.5, 2.5, \ldots \quad \text{(fully destructive interference)}. \tag{17-25}$$

For example, if the path length difference $\Delta L = |L_2 - L_1|$ in Fig. 17-7a is equal to 2.5λ, then $\Delta L/\lambda = 2.5$ and the waves undergo fully destructive interference at point P (Fig. 17-7c). The interference is fully destructive because the wave from S_2 is phase-shifted relative to the wave from S_1 by 2.5 wavelengths, which puts the two waves *exactly out of phase* at P.

Of course, two waves could produce intermediate interference as, say, when $\Delta L/\lambda = 1.2$. This would be closer to fully constructive interference ($\Delta L/\lambda = 1.0$) than to fully destructive interference ($\Delta L/\lambda = 1.5$).

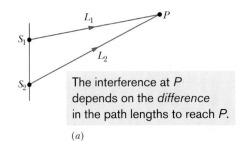

The interference at P depends on the *difference* in the path lengths to reach P.

(a)

If the difference is equal to, say, 2.0λ, then the waves arrive exactly in phase. This is how transverse waves would look.

(b)

If the difference is equal to, say, 2.5λ, then the waves arrive exactly out of phase. This is how transverse waves would look.

(c)

Fig. 17-7 (a) Two point sources S_1 and S_2 emit spherical sound waves in phase. The rays indicate that the waves pass through a common point P. The waves (represented with *transverse* waves) arrive at P (b) exactly in phase and (c) exactly out of phase.

Sample Problem

Interference points along a big circle

In Fig. 17-8a, two point sources S_1 and S_2, which are in phase and separated by distance $D = 1.5\lambda$, emit identical sound waves of wavelength λ.

(a) What is the path length difference of the waves from S_1 and S_2 at point P_1, which lies on the perpendicular bisector of distance D, at a distance greater than D from the sources (Fig.17-8b)? (That is, what is the difference in the distance from source S_1 to point P_1 and the distance from source S_2 to P_1?) What type of interference occurs at P_1?

Reasoning: Because the waves travel identical distances to reach P_1, their path length difference is

$$\Delta L = 0. \qquad \text{(Answer)}$$

From Eq. 17-23, this means that the waves undergo fully constructive interference at P_1 because they start in phase at the sources and reach P_1 in phase.

(b) What are the path length difference and type of interference at point P_2 in Fig. 17-8c?

Reasoning: The wave from S_1 travels the extra distance D $(= 1.5\lambda)$ to reach P_2. Thus, the path length difference is

$$\Delta L = 1.5\lambda. \qquad \text{(Answer)}$$

From Eq. 17-25, this means that the waves are exactly out of phase at P_2 and undergo fully destructive interference there.

(c) Figure 17-8d shows a circle with a radius much greater than D, centered on the midpoint between sources S_1 and S_2. What is the number of points N around this circle at which the interference is fully constructive? (That is, at how many points do the waves arrive exactly in phase?)

Reasoning: Imagine that, starting at point a, we move clockwise along the circle to point d. As we move to point d, the path length difference ΔL increases and so the type of interference changes. From (a), we know that the path length difference is $\Delta L = 0\lambda$ at point a. From (b), we know that $\Delta L = 1.5\lambda$ at point d. Thus, there must be one point along the circle between a and d at which $\Delta L = \lambda$, as indicated in Fig. 17-8e. From Eq. 17-23, fully constructive interference occurs at that point. Also, there can be no other point along the way from point a to point d at which fully constructive interference occurs, because there is no other integer than 1 between 0 at point a and 1.5 at point d.

We can now use symmetry to locate the other points of interference along the rest of the circle (Fig. 17-8f). Symmetry about line cd gives us point b, at which $\Delta L = 0\lambda$. (That point is on the perpendicular bisector of distance D, just like point a, and thus the path length difference from the sources to point b must be zero.) Also, there are three more points at which $\Delta L = \lambda$. In all (Fig. 17-8g) we have

$$N = 6. \qquad \text{(Answer)}$$

 PLUS Additional examples, video, and practice available at *WileyPLUS*

17-6 Intensity and Sound Level

If you have ever tried to sleep while someone played loud music nearby, you are well aware that there is more to sound than frequency, wavelength, and speed. There is also intensity. The **intensity** I of a sound wave at a surface is the average rate per unit area at which energy is transferred by the wave through or onto the surface. We can write this as

$$I = \frac{P}{A}, \qquad (17\text{-}26)$$

where P is the time rate of energy transfer (the power) of the sound wave and A is the area of the surface intercepting the sound. As we shall derive shortly, the intensity I is related to the displacement amplitude s_m of the sound wave by

$$I = \tfrac{1}{2}\rho v \omega^2 s_m^2. \qquad (17\text{-}27)$$

Variation of Intensity with Distance

How intensity varies with distance from a real sound source is often complex. Some real sources (like loudspeakers) may transmit sound only in particular

Fig. 17-8 (a) Two point sources S_1 and S_2, separated by distance D, emit spherical sound waves in phase. (b) The waves travel equal distances to reach point P_1. (c) Point P_2 is on the line extending through S_1 and S_2. (d) We move around a large circle. (e) Another point of fully constructive interference. (f) Using symmetry to determine other points. (g) The six points of fully constructive interference.

directions, and the environment usually produces echoes (reflected sound waves) that overlap the direct sound waves. In some situations, however, we can ignore echoes and assume that the sound source is a point source that emits the sound isotropically—that is, with equal intensity in all directions. The wavefronts spreading from such an isotropic point source S at a particular instant are shown in Fig. 17-9.

Let us assume that the mechanical energy of the sound waves is conserved as they spread from this source. Let us also center an imaginary sphere of radius r on the source, as shown in Fig. 17-9. All the energy emitted by the source must pass through the surface of the sphere. Thus, the time rate at which energy is transferred through the surface by the sound waves must equal the time rate at which energy is emitted by the source (that is, the power P_s of the source). From Eq. 17-26, the intensity I at the sphere must then be

$$I = \frac{P_s}{4\pi r^2},\qquad(17\text{-}28)$$

where $4\pi r^2$ is the area of the sphere. Equation 17-28 tells us that the intensity of sound from an isotropic point source decreases with the square of the distance r from the source.

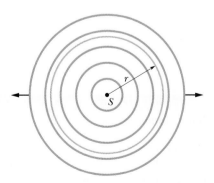

Fig. 17-9 A point source S emits sound waves uniformly in all directions. The waves pass through an imaginary sphere of radius r that is centered on S.

The figure indicates three small patches 1, 2, and 3 that lie on the surfaces of two imaginary spheres; the spheres are centered on an isotropic point source S of sound. The rates at which energy is transmitted through the three patches by the sound waves are equal. Rank the patches according to (a) the intensity of the sound on them and (b) their area, greatest first.

Sound can cause the wall of a drinking glass to oscillate. If the sound produces a standing wave of oscillations and if the intensity of the sound is large enough, the glass will shatter. *(Ben Rose/The Image Bank/Getty Images)*

The Decibel Scale

The displacement amplitude at the human ear ranges from about 10^{-5} m for the loudest tolerable sound to about 10^{-11} m for the faintest detectable sound, a ratio of 10^6. From Eq. 17-27 we see that the intensity of a sound varies as the *square* of its amplitude, so the ratio of intensities at these two limits of the human auditory system is 10^{12}. Humans can hear over an enormous range of intensities.

We deal with such an enormous range of values by using logarithms. Consider the relation

$$y = \log x,$$

in which x and y are variables. It is a property of this equation that if we *multiply* x by 10, then y increases by 1. To see this, we write

$$y' = \log(10x) = \log 10 + \log x = 1 + y.$$

Similarly, if we multiply x by 10^{12}, y increases by only 12.

Thus, instead of speaking of the intensity I of a sound wave, it is much more convenient to speak of its **sound level** β, defined as

$$\beta = (10 \text{ dB}) \log \frac{I}{I_0}. \tag{17-29}$$

Here dB is the abbreviation for **decibel,** the unit of sound level, a name that was chosen to recognize the work of Alexander Graham Bell. I_0 in Eq. 17-29 is a standard reference intensity ($= 10^{-12}$ W/m²), chosen because it is near the lower limit of the human range of hearing. For $I = I_0$, Eq. 17-29 gives $\beta = 10 \log 1 = 0$, so our standard reference level corresponds to zero decibels. Then β increases by 10 dB every time the sound intensity increases by an order of magnitude (a factor of 10). Thus, $\beta = 40$ corresponds to an intensity that is 10^4 times the standard reference level. Table 17-2 lists the sound levels for a variety of environments.

Derivation of Eq. 17-27

Consider, in Fig. 17-4a, a thin slice of air of thickness dx, area A, and mass dm, oscillating back and forth as the sound wave of Eq. 17-12 passes through it. The kinetic energy dK of the slice of air is

$$dK = \tfrac{1}{2} dm \, v_s^2. \tag{17-30}$$

Here v_s is not the speed of the wave but the speed of the oscillating element of air, obtained from Eq. 17-12 as

$$v_s = \frac{\partial s}{\partial t} = -\omega s_m \sin(kx - \omega t).$$

Using this relation and putting $dm = \rho A \, dx$ allow us to rewrite Eq. 17-30 as

$$dK = \tfrac{1}{2}(\rho A \, dx)(-\omega s_m)^2 \sin^2(kx - \omega t). \tag{17-31}$$

Table 17-2	
Some Sound Levels (dB)	
Hearing threshold	0
Rustle of leaves	10
Conversation	60
Rock concert	110
Pain threshold	120
Jet engine	130

Dividing Eq. 17-31 by dt gives the rate at which kinetic energy moves along with the wave. As we saw in Chapter 16 for transverse waves, dx/dt is the wave speed v, so we have

$$\frac{dK}{dt} = \tfrac{1}{2}\rho A v \omega^2 s_m^2 \sin^2(kx - \omega t). \tag{17-32}$$

The *average* rate at which kinetic energy is transported is

$$\left(\frac{dK}{dt}\right)_{avg} = \tfrac{1}{2}\rho A v \omega^2 s_m^2 [\sin^2(kx - \omega t)]_{avg}$$

$$= \tfrac{1}{4}\rho A v \omega^2 s_m^2. \tag{17-33}$$

To obtain this equation, we have used the fact that the average value of the square of a sine (or a cosine) function over one full oscillation is $\tfrac{1}{2}$.

We assume that *potential* energy is carried along with the wave at this same average rate. The wave intensity I, which is the average rate per unit area at which energy of both kinds is transmitted by the wave, is then, from Eq. 17-33,

$$I = \frac{2(dK/dt)_{avg}}{A} = \tfrac{1}{2}\rho v \omega^2 s_m^2,$$

which is Eq. 17-27, the equation we set out to derive.

Sample Problem

Intensity change with distance, cylindrical sound wave

An electric spark jumps along a straight line of length $L = 10$ m, emitting a pulse of sound that travels radially outward from the spark. (The spark is said to be a *line source* of sound.) The power of this acoustic emission is $P_s = 1.6 \times 10^4$ W.

(a) What is the intensity I of the sound when it reaches a distance $r = 12$ m from the spark?

KEY IDEAS

(1) Let us center an imaginary cylinder of radius $r = 12$ m and length $L = 10$ m (open at both ends) on the spark, as shown in Fig. 17-10. Then the intensity I at the cylindrical surface is the ratio P/A, where P is the time rate at which sound energy passes through the surface and A is the surface area. (2) We assume that the principle of conservation of energy applies to the sound energy. This means that the rate P at which energy is transferred through the cylinder must equal the rate P_s at which energy is emitted by the source.

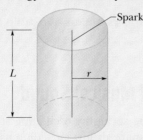

Fig. 17-10 A spark along a straight line of length L emits sound waves radially outward. The waves pass through an imaginary cylinder of radius r and length L that is centered on the spark.

Calculations: Putting these ideas together and noting that the area of the cylindrical surface is $A = 2\pi r L$, we have

$$I = \frac{P}{A} = \frac{P_s}{2\pi r L}. \tag{17-34}$$

This tells us that the intensity of the sound from a line source decreases with distance r (and not with the square of distance r as for a point source). Substituting the given data, we find

$$I = \frac{1.6 \times 10^4\ \text{W}}{2\pi(12\ \text{m})(10\ \text{m})}$$

$$= 21.2\ \text{W/m}^2 \approx 21\ \text{W/m}^2. \tag{Answer}$$

(b) At what time rate P_d is sound energy intercepted by an acoustic detector of area $A_d = 2.0$ cm^2, aimed at the spark and located a distance $r = 12$ m from the spark?

Calculations: We know that the intensity of sound at the detector is the ratio of the energy transfer rate P_d there to the detector's area A_d:

$$I = \frac{P_d}{A_d}. \tag{17-35}$$

We can imagine that the detector lies on the cylindrical surface of (a). Then the sound intensity at the detector is the intensity I ($= 21.2$ W/m^2) at the cylindrical surface. Solving Eq. 17-35 for P_d gives us

$$P_d = (21.2\ \text{W/m}^2)(2.0 \times 10^{-4}\ \text{m}^2) = 4.2\ \text{mW}. \tag{Answer}$$

 Additional examples, video, and practice available at *WileyPLUS*

Decibels, sound level, change in intensity

Many veteran rockers suffer from acute hearing damage because of the high sound levels they endured for years while playing music near loudspeakers or listening to music on headphones. Some, like Ted Nugent, can no longer hear in a damaged ear. Others, like Peter Townshend of the Who, have a continuous ringing sensation (tinnitus). Recently, many rockers, such as Lars Ulrich of Metallica (Fig. 17-11), began wearing special earplugs to protect their hearing during performances. If an earplug decreases the sound level of the sound waves by 20 dB, what is the ratio of the final intensity I_f of the waves to their initial intensity I_i?

Fig. 17-11 Lars Ulrich of Metallica is an advocate for the organization HEAR (Hearing Education and Awareness for Rockers), which warns about the damage high sound levels can have on hearing. (*Tim Mosenfelder/Getty Images News and Sport Services*)

KEY IDEA

For both the final and initial waves, the sound level β is related to the intensity by the definition of sound level in Eq. 17-29.

Calculations: For the final waves we have

$$\beta_f = (10 \text{ dB}) \log \frac{I_f}{I_0},$$

and for the initial waves we have

$$\beta_i = (10 \text{ dB}) \log \frac{I_i}{I_0}.$$

The difference in the sound levels is

$$\beta_f - \beta_i = (10 \text{ dB}) \left(\log \frac{I_f}{I_0} - \log \frac{I_i}{I_0} \right). \qquad (17\text{-}36)$$

Using the identity

$$\log \frac{a}{b} - \log \frac{c}{d} = \log \frac{ad}{bc},$$

we can rewrite Eq. 17-36 as

$$\beta_f - \beta_i = (10 \text{ dB}) \log \frac{I_f}{I_i}. \qquad (17\text{-}37)$$

Rearranging and then substituting the given decrease in sound level as $\beta_f - \beta_i = -20$ dB, we find

$$\log \frac{I_f}{I_i} = \frac{\beta_f - \beta_i}{10 \text{ dB}} = \frac{-20 \text{ dB}}{10 \text{ dB}} = -2.0.$$

We next take the antilog of the far left and far right sides of this equation. (Although the antilog $10^{-2.0}$ can be evaluated mentally, you could use a calculator by keying in 10^-2.0 or using the 10^x key.) We find

$$\frac{I_f}{I_i} = \log^{-1}(-2.0) = 0.010. \qquad \text{(Answer)}$$

Thus, the earplug reduces the intensity of the sound waves to 0.010 of their initial intensity, which is a decrease of two orders of magnitude.

 Additional examples, video, and practice available at *WileyPLUS*

17-7 Sources of Musical Sound

Musical sounds can be set up by oscillating strings (guitar, piano, violin), membranes (kettledrum, snare drum), air columns (flute, oboe, pipe organ, and the digeridoo of Fig. 17-12), wooden blocks or steel bars (marimba, xylophone), and many other oscillating bodies. Most common instruments involve more than a single oscillating part.

Recall from Chapter 16 that standing waves can be set up on a stretched string that is fixed at both ends. They arise because waves traveling along the string are reflected back onto the string at each end. If the wavelength of the waves is suitably matched to the length of the string, the superposition of waves traveling in opposite directions produces a standing wave pattern (or oscillation mode). The wavelength required of the waves for such a match is one that corresponds to a *resonant frequency* of the string. The advantage of setting up standing waves is that the string then oscillates with a large, sustained amplitude, pushing back and forth against the surrounding air and thus generating a noticeable sound wave with the same frequency as the oscillations of the string. This production of sound is of obvious importance to, say, a guitarist.

We can set up standing waves of sound in an air-filled pipe in a similar way. As sound waves travel through the air in the pipe, they are reflected at each end and travel back through the pipe. (The reflection occurs even if an end is open, but the reflection is not as complete as when the end is closed.) If the wavelength of the sound waves is suitably matched to the length of the pipe, the superposition of waves traveling in opposite directions through the pipe sets up a standing wave pattern. The wavelength required of the sound waves for such a match is one that corresponds to a resonant frequency of the pipe. The advantage of such a standing wave is that the air in the pipe oscillates with a large, sustained amplitude, emitting at any open end a sound wave that has the same frequency as the oscillations in the pipe. This emission of sound is of obvious importance to, say, an organist.

Many other aspects of standing sound wave patterns are similar to those of string waves: The closed end of a pipe is like the fixed end of a string in that there must be a node (zero displacement) there, and the open end of a pipe is like the end of a string attached to a freely moving ring, as in Fig. 16-18b, in that there must be an antinode there. (Actually, the antinode for the open end of a pipe is located slightly beyond the end, but we shall not dwell on that detail.)

Fig. 17-12 The air column within a digeridoo ("a pipe") oscillates when the instrument is played. (*Alamy Images*)

Antinodes (maximum oscillation) occur at the open ends.

$\lambda = 2L$

First harmonic

Fig. 17-13 (a) The simplest standing wave pattern of displacement for (longitudinal) sound waves in a pipe with both ends open has an antinode (A) across each end and a node (N) across the middle. (The longitudinal displacements represented by the double arrows are greatly exaggerated.) (b) The corresponding standing wave pattern for (transverse) string waves.

The simplest standing wave pattern that can be set up in a pipe with two open ends is shown in Fig. 17-13a. There is an antinode across each open end, as required. There is also a node across the middle of the pipe. An easier way of representing this standing longitudinal sound wave is shown in Fig. 17-13b—by drawing it as a standing transverse string wave.

The standing wave pattern of Fig. 17-13a is called the *fundamental mode* or *first harmonic*. For it to be set up, the sound waves in a pipe of length L must have a wavelength given by $L = \lambda/2$, so that $\lambda = 2L$. Several more standing sound wave patterns for a pipe with two open ends are shown in Fig. 17-14a using string wave representations. The *second harmonic* requires sound waves of wavelength $\lambda = L$, the *third harmonic* requires wavelength $\lambda = 2L/3$, and so on.

$n = 2$ Second $\lambda = 2L/2 = L$

$n = 3$ Third $\lambda = 2L/3$

$n = 4$ Fourth $\lambda = 2L/4 = L/2$

(a)

Two open ends— any harmonic

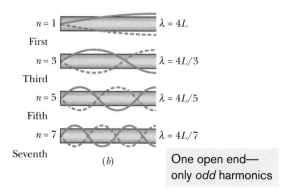

$n = 1$ First $\lambda = 4L$

$n = 3$ Third $\lambda = 4L/3$

$n = 5$ Fifth $\lambda = 4L/5$

$n = 7$ Seventh $\lambda = 4L/7$

(b)

One open end— only *odd* harmonics

Fig. 17-14 Standing wave patterns for string waves superimposed on pipes to represent standing sound wave patterns in the pipes. (a) With *both* ends of the pipe open, any harmonic can be set up in the pipe. (b) With only *one* end open, only odd harmonics can be set up.

Fig. 17-15 The saxophone and violin families, showing the relations between instrument length and frequency range. The frequency range of each instrument is indicated by a horizontal bar along a frequency scale suggested by the keyboard at the bottom; the frequency increases toward the right.

More generally, the resonant frequencies for a pipe of length L with two open ends correspond to the wavelengths

$$\lambda = \frac{2L}{n}, \qquad \text{for } n = 1, 2, 3, \ldots, \tag{17-38}$$

where n is called the *harmonic number*. Letting v be the speed of sound, we write the resonant frequencies for a pipe with two open ends as

$$f = \frac{v}{\lambda} = \frac{nv}{2L}, \qquad \text{for } n = 1, 2, 3, \ldots \quad \text{(pipe, two open ends).} \tag{17-39}$$

Figure 17-14*b* shows (using string wave representations) some of the standing sound wave patterns that can be set up in a pipe with only one open end. As required, across the open end there is an antinode and across the closed end there is a node. The simplest pattern requires sound waves having a wavelength given by $L = \lambda/4$, so that $\lambda = 4L$. The next simplest pattern requires a wavelength given by $L = 3\lambda/4$, so that $\lambda = 4L/3$, and so on.

More generally, the resonant frequencies for a pipe of length L with only one open end correspond to the wavelengths

$$\lambda = \frac{4L}{n}, \qquad \text{for } n = 1, 3, 5, \ldots, \tag{17-40}$$

in which the harmonic number n *must be an odd number*. The resonant frequencies are then given by

$$f = \frac{v}{\lambda} = \frac{nv}{4L}, \qquad \text{for } n = 1, 3, 5, \ldots \quad \text{(pipe, one open end).} \tag{17-41}$$

Note again that only odd harmonics can exist in a pipe with one open end. For example, the second harmonic, with $n = 2$, cannot be set up in such a pipe. Note also that for such a pipe the adjective in a phrase such as "the third harmonic" still refers to the harmonic number n (and not to, say, the third possible harmonic). Finally note that Eqs. 17-38 and 17-39 for two open ends contain the number 2 and any integer value of n, but Eqs. 17-40 and 17-41 for one open end contain the number 4 and only odd values of n.

The length of a musical instrument reflects the range of frequencies over which the instrument is designed to function, and smaller length implies higher frequencies. Figure 17-15, for example, shows the saxophone and violin families, with their frequency ranges suggested by the piano keyboard. Note that, for every instrument, there is overlap with its higher- and lower-frequency neighbors.

In any oscillating system that gives rise to a musical sound, whether it is a violin string or the air in an organ pipe, the fundamental and one or more of the higher harmonics are usually generated simultaneously. Thus, you hear them together—that is, superimposed as a net wave. When different instruments are played at the same note, they produce the same fundamental frequency but different intensities for the higher harmonics. For example, the fourth harmonic of middle C might be relatively loud on one instrument and relatively quiet or even missing on another. Thus, because different instruments produce different net waves, they sound different to you even when they are played at the same note. That would be the case for the two net waves shown in Fig. 17-16, which were produced at the same note by different instruments.

(*a*)

(*b*)

Time

Fig. 17-16 The wave forms produced by (*a*) a flute and (*b*) an oboe when played at the same note, with the same first harmonic frequency.

✔ CHECKPOINT 3

Pipe A, with length L, and pipe B, with length $2L$, both have two open ends. Which harmonic of pipe B has the same frequency as the fundamental of pipe A?

Sound resonance in double-open pipe and single-open pipe

Weak background noises from a room set up the fundamental standing wave in a cardboard tube of length $L = 67.0$ cm with two open ends. Assume that the speed of sound in the air within the tube is 343 m/s.

(a) What frequency do you hear from the tube?

KEY IDEA

With both pipe ends open, we have a symmetric situation in which the standing wave has an antinode at each end of the tube. The standing wave pattern (in string wave style) is that of Fig. 17-13b.

Calculation: The frequency is given by Eq. 17-39 with $n = 1$ for the fundamental mode:

$$f = \frac{nv}{2L} = \frac{(1)(343 \text{ m/s})}{2(0.670 \text{ m})} = 256 \text{ Hz.} \qquad \text{(Answer)}$$

If the background noises set up any higher harmonics, such as the second harmonic, you also hear frequencies that are

integer multiples of 256 Hz. (Thus, the lowest frequency is this fundamental frequency of 256 Hz.)

(b) If you jam your ear against one end of the tube, what fundamental frequency do you hear from the tube?

KEY IDEA

With your ear effectively closing one end of the tube, we have an asymmetric situation—an antinode still exists at the open end, but a node is now at the other (closed) end. The standing wave pattern is the top one in Fig. 17-14b.

Calculation: The frequency is given by Eq. 17-41 with $n = 1$ for the fundamental mode:

$$f = \frac{nv}{4L} = \frac{(1)(343 \text{ m/s})}{4(0.670 \text{ m})} = 128 \text{ Hz.} \qquad \text{(Answer)}$$

If the background noises set up any higher harmonics, they will be *odd* multiples of 128 Hz. That means that the frequency of 256 Hz (which is an even multiple) cannot now occur.

 Additional examples, video, and practice available at *WileyPLUS*

17-8 Beats

If we listen, a few minutes apart, to two sounds whose frequencies are, say, 552 and 564 Hz, most of us cannot tell one from the other. However, if the sounds reach our ears simultaneously, what we hear is a sound whose frequency is 558 Hz, the *average* of the two combining frequencies. We also hear a striking variation in the intensity of this sound—it increases and decreases in slow, wavering **beats** that repeat at a frequency of 12 Hz, the *difference* between the two combining frequencies. Figure 17-17 shows this beat phenomenon.

Let the time-dependent variations of the displacements due to two sound waves of equal amplitude s_m be

$$s_1 = s_m \cos \omega_1 t \qquad \text{and} \qquad s_2 = s_m \cos \omega_2 t, \qquad (17\text{-}42)$$

where $\omega_1 > \omega_2$. From the superposition principle, the resultant displacement is

$$s = s_1 + s_2 = s_m(\cos \omega_1 t + \cos \omega_2 t).$$

Using the trigonometric identity (see Appendix E)

$$\cos \alpha + \cos \beta = 2 \cos[\tfrac{1}{2}(\alpha - \beta)] \cos[\tfrac{1}{2}(\alpha + \beta)]$$

allows us to write the resultant displacement as

$$s = 2s_m \cos[\tfrac{1}{2}(\omega_1 - \omega_2)t] \cos[\tfrac{1}{2}(\omega_1 + \omega_2)t]. \qquad (17\text{-}43)$$

If we write

$$\omega' = \tfrac{1}{2}(\omega_1 - \omega_2) \qquad \text{and} \qquad \omega = \tfrac{1}{2}(\omega_1 + \omega_2), \qquad (17\text{-}44)$$

we can then write Eq. 17-43 as

$$s(t) = [2s_m \cos \omega' t] \cos \omega t. \qquad (17\text{-}45)$$

Fig. 17-17 (a, b) The pressure variations Δp of two sound waves as they would be detected separately. The frequencies of the waves are nearly equal. (c) The resultant pressure variation if the two waves are detected simultaneously.

We now assume that the angular frequencies ω_1 and ω_2 of the combining waves are almost equal, which means that $\omega \gg \omega'$ in Eq. 17-44. We can then regard Eq. 17-45 as a cosine function whose angular frequency is ω and whose amplitude (which is not constant but varies with angular frequency ω') is the absolute value of the quantity in the brackets.

A maximum amplitude will occur whenever $\cos \omega' t$ in Eq. 17-45 has the value $+1$ or -1, which happens twice in each repetition of the cosine function. Because $\cos \omega' t$ has angular frequency ω', the angular frequency ω_{beat} at which beats occur is $\omega_{\text{beat}} = 2\omega'$. Then, with the aid of Eq. 17-44, we can write

$$\omega_{\text{beat}} = 2\omega' = (2)(\tfrac{1}{2})(\omega_1 - \omega_2) = \omega_1 - \omega_2.$$

Because $\omega = 2\pi f$, we can recast this as

$$f_{\text{beat}} = f_1 - f_2 \quad \text{(beat frequency)}. \tag{17-46}$$

Musicians use the beat phenomenon in tuning instruments. If an instrument is sounded against a standard frequency (for example, the note called "concert A" played on an orchestra's first oboe) and tuned until the beat disappears, the instrument is in tune with that standard. In musical Vienna, concert A (440 Hz) is available as a telephone service for the city's many musicians.

Sample Problem

Beat frequencies and penguins finding one another

When an emperor penguin returns from a search for food, how can it find its mate among the thousands of penguins huddled together for warmth in the harsh Antarctic weather? It is not by sight, because penguins all look alike, even to a penguin.

The answer lies in the way penguins vocalize. Most birds vocalize by using only one side of their two-sided vocal organ, called the *syrinx*. Emperor penguins, however, vocalize by using both sides simultaneously. Each side sets up acoustic standing waves in the bird's throat and mouth, much like in a pipe with two open ends. Suppose that the frequency of the first harmonic produced by side A is $f_{A1} = 432$ Hz and the frequency of the first harmonic produced by side B is $f_{B1} = 371$ Hz. What is the beat frequency between those two first-harmonic frequencies and between the two second-harmonic frequencies?

KEY IDEA

The beat frequency between two frequencies is their difference, as given by Eq. 17-46 ($f_{\text{beat}} = f_1 - f_2$).

Calculations: For the two first-harmonic frequencies f_{A1} and f_{B1}, the beat frequency is

$$f_{\text{beat,1}} = f_{A1} - f_{B1} = 432 \text{ Hz} - 371 \text{ Hz}$$
$$= 61 \text{ Hz}. \qquad \text{(Answer)}$$

Because the standing waves in the penguin are effectively in a pipe with two open ends, the resonant frequencies are given by Eq. 17-39 ($f = nv/2L$), in which L is the (unknown) length of the effective pipe. The first-harmonic frequency is $f_1 = v/2L$, and the second-harmonic frequency is $f_2 = 2v/2L$. Comparing these two frequencies, we see that, in general,

$$f_2 = 2f_1.$$

For the penguin, the second harmonic of side A has frequency $f_{A2} = 2f_{A1}$ and the second harmonic of side B has frequency $f_{B2} = 2f_{B1}$. Using Eq. 17-46 with frequencies f_{A2} and f_{B2}, we find that the corresponding beat frequency associated with the second harmonics is

$$f_{\text{beat,2}} = f_{A2} - f_{B2} = 2f_{A1} - 2f_{B1}$$
$$= 2(432 \text{ Hz}) - 2(371 \text{ Hz})$$
$$= 122 \text{ Hz}. \qquad \text{(Answer)}$$

Experiments indicate that penguins can perceive such large beat frequencies (humans cannot hear a beat frequency any higher than about 12 Hz). Thus, a penguin's cry can be rich with different harmonics and different beat frequencies, allowing the voice to be recognized even among the voices of thousands of other, closely huddled penguins.

PLUS Additional examples, video, and practice available at *WileyPLUS*

17-9 **The Doppler Effect**

A police car is parked by the side of the highway, sounding its 1000 Hz siren. If you are also parked by the highway, you will hear that same frequency. However, if there is relative motion between you and the police car, either toward or away from each other, you will hear a different frequency. For example, if you are driving *toward* the police car at 120 km/h (about 75 mi/h), you will hear a *higher* frequency (1096 Hz, an *increase* of 96 Hz). If you are driving *away from* the police car at that same speed, you will hear a *lower* frequency (904 Hz, a *decrease* of 96 Hz).

These motion-related frequency changes are examples of the **Doppler effect.** The effect was proposed (although not fully worked out) in 1842 by Austrian physicist Johann Christian Doppler. It was tested experimentally in 1845 by Buys Ballot in Holland, "using a locomotive drawing an open car with several trumpeters."

The Doppler effect holds not only for sound waves but also for electromagnetic waves, including microwaves, radio waves, and visible light. Here, however, we shall consider only sound waves, and we shall take as a reference frame the body of air through which these waves travel. This means that we shall measure the speeds of a source S of sound waves and a detector D of those waves *relative to that body of air*. (Unless otherwise stated, the body of air is stationary relative to the ground, so the speeds can also be measured relative to the ground.) We shall assume that S and D move either directly toward or directly away from each other, at speeds less than the speed of sound.

If either the detector or the source is moving, or both are moving, the emitted frequency f and the detected frequency f' are related by

$$f' = f\frac{v \pm v_D}{v \pm v_S} \quad \text{(general Doppler effect)}, \qquad (17\text{-}47)$$

where v is the speed of sound through the air, v_D is the detector's speed relative to the air, and v_S is the source's speed relative to the air. The choice of plus or minus signs is set by this rule:

> When the motion of detector or source is toward the other, the sign on its speed must give an upward shift in frequency. When the motion of detector or source is away from the other, the sign on its speed must give a downward shift in frequency.

In short, *toward* means *shift up,* and *away* means *shift down.*

Here are some examples of the rule. If the detector moves toward the source, use the plus sign in the numerator of Eq. 17-47 to get a shift up in the frequency. If it moves away, use the minus sign in the numerator to get a shift down. If it is stationary, substitute 0 for v_D. If the source moves toward the detector, use the minus sign in the denominator of Eq. 17-47 to get a shift up in the frequency. If it moves away, use the plus sign in the denominator to get a shift down. If the source is stationary, substitute 0 for v_S.

Next, we derive equations for the Doppler effect for the following two specific situations and then derive Eq. 17-47 for the general situation.

1. When the detector moves relative to the air and the source is stationary relative to the air, the motion changes the frequency at which the detector intercepts wavefronts and thus changes the detected frequency of the sound wave.

2. When the source moves relative to the air and the detector is stationary relative to the air, the motion changes the wavelength of the sound wave and thus changes the detected frequency (recall that frequency is related to wavelength).

Shift up: The detector moves *toward* the source.

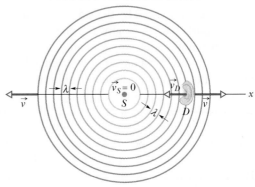

Fig. 17-18 A stationary source of sound S emits spherical wavefronts, shown one wavelength apart, that expand outward at speed v. A sound detector D, represented by an ear, moves with velocity $\vec{v}_D$ toward the source. The detector senses a higher frequency because of its motion.

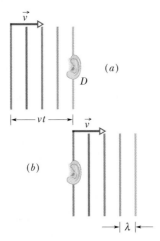

Fig. 17-19 The wavefronts of Fig. 17-18, assumed planar, (*a*) reach and (*b*) pass a stationary detector D; they move a distance vt to the right in time t.

Fig. 17-20 Wavefronts traveling to the right (*a*) reach and (*b*) pass detector D, which moves in the opposite direction. In time t, the wavefronts move a distance vt to the right and D moves a distance $v_D t$ to the left.

Detector Moving, Source Stationary

In Fig. 17-18, a detector D (represented by an ear) is moving at speed v_D toward a stationary source S that emits spherical wavefronts, of wavelength λ and frequency f, moving at the speed v of sound in air. The wavefronts are drawn one wavelength apart. The frequency detected by detector D is the rate at which D intercepts wavefronts (or individual wavelengths). If D were stationary, that rate would be f, but since D is moving into the wavefronts, the rate of interception is greater, and thus the detected frequency f' is greater than f.

Let us for the moment consider the situation in which D is stationary (Fig. 17-19). In time t, the wavefronts move to the right a distance vt. The number of wavelengths in that distance vt is the number of wavelengths intercepted by D in time t, and that number is vt/λ. The rate at which D intercepts wavelengths, which is the frequency f detected by D, is

$$f = \frac{vt/\lambda}{t} = \frac{v}{\lambda}. \tag{17-48}$$

In this situation, with D stationary, there is no Doppler effect—the frequency detected by D is the frequency emitted by S.

Now let us again consider the situation in which D moves in the direction opposite the wavefront velocity (Fig. 17-20). In time t, the wavefronts move to the right a distance vt as previously, but now D moves to the left a distance $v_D t$. Thus, in this time t, the distance moved by the wavefronts relative to D is $vt + v_D t$. The number of wavelengths in this relative distance $vt + v_D t$ is the number of wavelengths intercepted by D in time t and is $(vt + v_D t)/\lambda$. The *rate* at which D intercepts wavelengths in this situation is the frequency f', given by

$$f' = \frac{(vt + v_D t)/\lambda}{t} = \frac{v + v_D}{\lambda}. \tag{17-49}$$

From Eq. 17-48, we have $\lambda = v/f$. Then Eq. 17-49 becomes

$$f' = \frac{v + v_D}{v/f} = f \frac{v + v_D}{v}. \tag{17-50}$$

Note that in Eq. 17-50, $f' > f$ unless $v_D = 0$ (the detector is stationary).

Similarly, we can find the frequency detected by D if D moves away from the source. In this situation, the wavefronts move a distance $vt - v_D t$ relative to D in time t, and f' is given by

$$f' = f \frac{v - v_D}{v}. \tag{17-51}$$

In Eq. 17-51, $f' < f$ unless $v_D = 0$. We can summarize Eqs. 17-50 and 17-51 with

$$f' = f \frac{v \pm v_D}{v} \qquad \text{(detector moving, source stationary).} \tag{17-52}$$

Source Moving, Detector Stationary

Let detector D be stationary with respect to the body of air, and let source S move toward D at speed v_S (Fig. 17-21). The motion of S changes the wavelength of the sound waves it emits and thus the frequency detected by D.

To see this change, let $T\,(= 1/f)$ be the time between the emission of any pair of successive wavefronts W_1 and W_2. During T, wavefront W_1 moves a distance vT and the source moves a distance $v_S T$. At the end of T, wavefront W_2 is emitted. In the direction in which S moves, the distance between W_1 and W_2, which is the wavelength λ' of the waves moving in that direction, is $vT - v_S T$. If D detects those waves, it detects frequency f' given by

$$f' = \frac{v}{\lambda'} = \frac{v}{vT - v_S T} = \frac{v}{v/f - v_S/f}$$

$$= f\frac{v}{v - v_S}.$$ (17-53)

Note that f' must be greater than f unless $v_S = 0$.

In the direction opposite that taken by S, the wavelength λ' of the waves is again the distance between successive waves but now that distance is $vT + v_S T$. If D detects those waves, it detects frequency f' given by

$$f' = f\frac{v}{v + v_S}.$$ (17-54)

Now f' must be less than f unless $v_S = 0$.

We can summarize Eqs. 17-53 and 17-54 with

$$f' = f\frac{v}{v \pm v_S} \qquad \text{(source moving, detector stationary).}$$ (17-55)

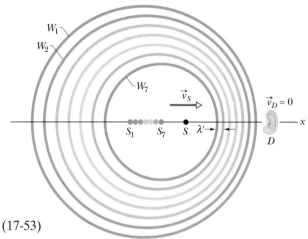

Shift up: The source moves *toward* the detector.

Fig. 17-21 A detector D is stationary, and a source S is moving toward it at speed v_S. Wavefront W_1 was emitted when the source was at S_1, wavefront W_7 when it was at S_7. At the moment depicted, the source is at S. The detector senses a higher frequency because the moving source, chasing its own wavefronts, emits a reduced wavelength λ' in the direction of its motion.

General Doppler Effect Equation

We can now derive the general Doppler effect equation by replacing f in Eq. 17-55 (the source frequency) with f' of Eq. 17-52 (the frequency associated with motion of the detector). That simple replacement gives us Eq. 17-47 for the general Doppler effect.

That general equation holds not only when both detector and source are moving but also in the two specific situations we just discussed. For the situation in which the detector is moving and the source is stationary, substitution of $v_S = 0$ into Eq. 17-47 gives us Eq. 17-52, which we previously found. For the situation in which the source is moving and the detector is stationary, substitution of $v_D = 0$ into Eq. 17-47 gives us Eq. 17-55, which we previously found. Thus, Eq. 17-47 is the equation to remember.

 CHECKPOINT 4

The figure indicates the directions of motion of a sound source and a detector for six situations in stationary air. For each situation, is the detected frequency greater than or less than the emitted frequency, or can't we tell without more information about the actual speeds?

Sample Problem

Double Doppler shift in the echoes used by bats

Bats navigate and search out prey by emitting, and then detecting reflections of, ultrasonic waves, which are sound waves with frequencies greater than can be heard by a human. Suppose a bat emits ultrasound at frequency $f_{be} = 82.52$ kHz while flying with velocity $\vec{v}_b = (9.00 \text{ m/s})\hat{i}$ as it chases a moth that flies with velocity $\vec{v}_m = (8.00 \text{ m/s})\hat{i}$. What frequency f_{md} does the moth detect? What frequency f_{bd} does the bat detect in the returning echo from the moth?

KEY IDEAS

The frequency is shifted by the relative motion of the bat and moth. Because they move along a single axis, the shifted frequency is given by Eq. 17-47 for the general Doppler effect. Motion *toward* tends to shift the frequency *up*, and motion *away* tends to shift the frequency *down*.

Detection by moth: The general Doppler equation is

$$f' = f \frac{v \pm v_D}{v \pm v_S}. \tag{17-56}$$

Here, the detected frequency f' that we want to find is the frequency f_{md} detected by the moth. On the right side of the equation, the emitted frequency f is the bat's emission frequency $f_{be} = 82.52$ kHz, the speed of sound is $v = 343$ m/s, the speed v_D of the detector is the moth's speed $v_m = 8.00$ m/s, and the speed v_S of the source is the bat's speed $v_b = 9.00$ m/s.

These substitutions into Eq. 17-56 are easy to make. However, the decisions about the plus and minus signs can be tricky. Think in terms of *toward* and *away*. We have the speed of the moth (the detector) in the numerator of Eq.

17-56. The moth moves *away* from the bat, which tends to lower the detected frequency. Because the speed is in the numerator, we choose the minus sign to meet that tendency (the numerator becomes smaller). These reasoning steps are shown in Table 17-3.

We have the speed of the bat in the denominator of Eq. 17-56. The bat moves *toward* the moth, which tends to increase the detected frequency. Because the speed is in the denominator, we choose the minus sign to meet that tendency (the denominator becomes smaller).

With these substitutions and decisions, we have

$$f_{md} = f_{be} \frac{v - v_m}{v - v_b}$$

$$= (82.52 \text{ kHz}) \frac{343 \text{ m/s} - 8.00 \text{ m/s}}{343 \text{ m/s} - 9.00 \text{ m/s}}$$

$$= 82.767 \text{ kHz} \approx 82.8 \text{ kHz}. \tag{Answer}$$

Detection of echo by bat: In the echo back to the bat, the moth acts as a source of sound, emitting at the frequency f_{md} we just calculated. So now the moth is the source (moving *away*) and the bat is the detector (moving *toward*). The reasoning steps are shown in Table 17-3. To find the frequency f_{bd} detected by the bat, we write Eq. 17-56 as

$$f_{bd} = f_{md} \frac{v + v_b}{v + v_m}$$

$$= (82.767 \text{ kHz}) \frac{343 \text{ m/s} + 9.00 \text{ m/s}}{343 \text{ m/s} + 8.00 \text{ m/s}}$$

$$= 83.00 \text{ kHz} \approx 83.0 \text{ kHz}. \tag{Answer}$$

Some moths evade bats by "jamming" the detection system with ultrasonic clicks.

Table 17-3

Bat to Moth			Echo Back to Bat	
Detector	Source		Detector	Source
moth	bat		bat	moth
speed $v_D = v_m$	speed $v_S = v_b$		speed $v_D = v_b$	speed $v_S = v_m$
away	toward		toward	away
shift down	shift up		shift up	shift down
numerator	denominator		numerator	denominator
minus	minus		plus	plus

 Additional examples, video, and practice available at *WileyPLUS*

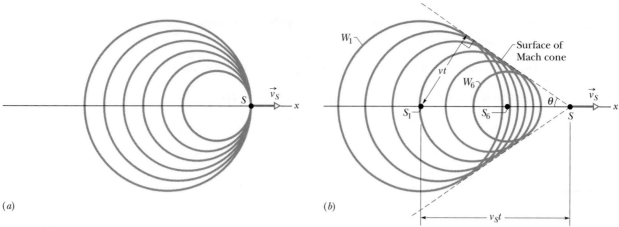

Fig. 17-22 (*a*) A source of sound *S* moves at speed v_S equal to the speed of sound and thus as fast as the wavefronts it generates. (*b*) A source *S* moves at speed v_S faster than the speed of sound and thus faster than the wavefronts. When the source was at position S_1 it generated wavefront W_1, and at position S_6 it generated W_6. All the spherical wavefronts expand at the speed of sound *v* and bunch along the surface of a cone called the Mach cone, forming a shock wave. The surface of the cone has half-angle θ and is tangent to all the wavefronts.

17-10 Supersonic Speeds, Shock Waves

If a source is moving toward a stationary detector at a speed equal to the speed of sound—that is, if $v_S = v$—Eqs. 17-47 and 17-55 predict that the detected frequency f' will be infinitely great. This means that the source is moving so fast that it keeps pace with its own spherical wavefronts, as Fig. 17-22*a* suggests. What happens when the speed of the source *exceeds* the speed of sound?

For such *supersonic* speeds, Eqs. 17-47 and 17-55 no longer apply. Figure 17-22*b* depicts the spherical wavefronts that originated at various positions of the source. The radius of any wavefront in this figure is *vt*, where *v* is the speed of sound and *t* is the time that has elapsed since the source emitted that wavefront. Note that all the wavefronts bunch along a V-shaped envelope in the two-dimensional drawing of Fig. 17-22*b*. The wavefronts actually extend in three dimensions, and the bunching actually forms a cone called the *Mach cone*. A *shock wave* is said to exist along the surface of this cone, because the bunching of wavefronts causes an abrupt rise and fall of air pressure as the surface passes through any point. From Fig. 17-22*b*, we see that the half-angle θ of the cone, called the *Mach cone angle*, is given by

$$\sin \theta = \frac{vt}{v_S t} = \frac{v}{v_S} \quad \text{(Mach cone angle)}. \quad (17\text{-}57)$$

The ratio v_S/v is called the *Mach number*. When you hear that a particular plane has flown at Mach 2.3, it means that its speed was 2.3 times the speed of sound in the air through which the plane was flying. The shock wave generated by a supersonic aircraft (Fig. 17-23) or projectile produces a burst of sound, called a *sonic boom,* in which the air pressure first suddenly increases and then suddenly decreases below normal before returning to normal. Part of the sound that is heard when a rifle is fired is the sonic boom produced by the bullet. A sonic boom can also be heard from a long bullwhip when it is snapped quickly: Near the end of the whip's motion, its tip is moving faster than sound and produces a small sonic boom—the *crack* of the whip.

Fig. 17-23 Shock waves produced by the wings of a Navy FA 18 jet. The shock waves are visible because the sudden decrease in air pressure in them caused water molecules in the air to condense, forming a fog. (*U.S. Navy photo by Ensign John Gay*)

Sound Waves Sound waves are longitudinal mechanical waves that can travel through solids, liquids, or gases. The speed v of a sound wave in a medium having **bulk modulus** B and density ρ is

$$v = \sqrt{\frac{B}{\rho}} \qquad \text{(speed of sound)}. \tag{17-3}$$

In air at 20°C, the speed of sound is 343 m/s.

A sound wave causes a longitudinal displacement s of a mass element in a medium as given by

$$s = s_m \cos(kx - \omega t), \tag{17-12}$$

where s_m is the **displacement amplitude** (maximum displacement) from equilibrium, $k = 2\pi/\lambda$, and $\omega = 2\pi f$, λ and f being the wavelength and frequency, respectively, of the sound wave. The sound wave also causes a pressure change Δp of the medium from the equilibrium pressure:

$$\Delta p = \Delta p_m \sin(kx - \omega t), \tag{17-13}$$

where the **pressure amplitude** is

$$\Delta p_m = (v\rho\omega)s_m. \tag{17-14}$$

Interference The interference of two sound waves with identical wavelengths passing through a common point depends on their phase difference ϕ there. If the sound waves were emitted in phase and are traveling in approximately the same direction, ϕ is given by

$$\phi = \frac{\Delta L}{\lambda} 2\pi, \tag{17-21}$$

where ΔL is their **path length difference** (the difference in the distances traveled by the waves to reach the common point). Fully constructive interference occurs when ϕ is an integer multiple of 2π,

$$\phi = m(2\pi), \qquad \text{for } m = 0, 1, 2, \ldots, \tag{17-22}$$

and, equivalently, when ΔL is related to wavelength λ by

$$\frac{\Delta L}{\lambda} = 0, 1, 2, \ldots. \tag{17-23}$$

Fully destructive interference occurs when ϕ is an odd multiple of π,

$$\phi = (2m + 1)\pi, \qquad \text{for } m = 0, 1, 2, \ldots, \tag{17-24}$$

and, equivalently, when ΔL is related to λ by

$$\frac{\Delta L}{\lambda} = 0.5, 1.5, 2.5, \ldots. \tag{17-25}$$

Sound Intensity The **intensity** I of a sound wave at a surface is the average rate per unit area at which energy is transferred by the wave through or onto the surface:

$$I = \frac{P}{A}, \tag{17-26}$$

where P is the time rate of energy transfer (power) of the sound wave and A is the area of the surface intercepting the sound. The intensity I is related to the displacement amplitude s_m of the sound wave by

$$I = \tfrac{1}{2}\rho v\omega^2 s_m^2. \tag{17-27}$$

The intensity at a distance r from a point source that emits sound waves of power P_s is

$$I = \frac{P_s}{4\pi r^2}. \tag{17-28}$$

Sound Level in Decibels The *sound level* β in *decibels* (dB) is defined as

$$\beta = (10 \text{ dB}) \log \frac{I}{I_0}, \tag{17-29}$$

where I_0 ($= 10^{-12}$ W/m²) is a reference intensity level to which all intensities are compared. For every factor-of-10 increase in intensity, 10 dB is added to the sound level.

Standing Wave Patterns in Pipes Standing sound wave patterns can be set up in pipes. A pipe open at both ends will resonate at frequencies

$$f = \frac{v}{\lambda} = \frac{nv}{2L}, \qquad n = 1, 2, 3, \ldots, \tag{17-39}$$

where v is the speed of sound in the air in the pipe. For a pipe closed at one end and open at the other, the resonant frequencies are

$$f = \frac{v}{\lambda} = \frac{nv}{4L}, \qquad n = 1, 3, 5, \ldots. \tag{17-41}$$

Beats *Beats* arise when two waves having slightly different frequencies, f_1 and f_2, are detected together. The beat frequency is

$$f_{\text{beat}} = f_1 - f_2. \tag{17-46}$$

The Doppler Effect The *Doppler effect* is a change in the observed frequency of a wave when the source or the detector moves relative to the transmitting medium (such as air). For sound the observed frequency f' is given in terms of the source frequency f by

$$f' = f\frac{v \pm v_D}{v \pm v_S} \qquad \text{(general Doppler effect)}, \tag{17-47}$$

where v_D is the speed of the detector relative to the medium, v_S is that of the source, and v is the speed of sound in the medium. The signs are chosen such that f' tends to be *greater* for motion toward and *less* for motion away.

Shock Wave If the speed of a source relative to the medium exceeds the speed of sound in the medium, the Doppler equation no longer applies. In such a case, shock waves result. The half-angle θ of the Mach cone is given by

$$\sin \theta = \frac{v}{v_S} \qquad \text{(Mach cone angle)}. \tag{17-57}$$

QUESTIONS

1 In a first experiment, a sinusoidal sound wave is sent through a long tube of air, transporting energy at the average rate of $P_{avg,1}$. In a second experiment, two other sound waves, identical to the first one, are to be sent simultaneously through the tube with a phase difference ϕ of either $0, 0.2$ wavelength, or 0.5 wavelength between the waves. (a) With only mental calculation, rank those choices of ϕ according to the average rate at which the waves will transport energy, greatest first. (b) For the first choice of ϕ, what is the average rate in terms of $P_{avg,1}$?

2 In Fig. 17-24, two point sources S_1 and S_2, which are in phase, emit identical sound waves of wavelength 2.0 m. In terms of wavelengths, what is the phase difference between the waves arriving at point P if (a) $L_1 = 38$ m and $L_2 = 34$ m, and (b) $L_1 = 39$ m and $L_2 = 36$ m? (c) Assuming that the source separation is much smaller than L_1 and L_2, what type of interference occurs at P in situations (a) and (b)?

Fig. 17-24   Question 2.

3 In Fig. 17-25, three long tubes (A, B, and C) are filled with different gases under different pressures. The ratio of the bulk modulus to the density is indicated for each gas in terms of a basic value B_0/ρ_0. Each tube has a piston at its left end that can send a sound pulse through the tube (as in Fig. 16-2). The three pulses are sent simultaneously. Rank the tubes according to the time of arrival of the pulses at the open right ends of the tubes, earliest first.

Fig. 17-25   Question 3.

4 The sixth harmonic is set up in a pipe. (a) How many open ends does the pipe have (it has at least one)? (b) Is there a node, antinode, or some intermediate state at the midpoint?

5 In Fig. 17-26, pipe A is made to oscillate in its third harmonic by a small internal sound source. Sound emitted at the right end happens to resonate four nearby pipes, each with only one open end (they are *not* drawn to scale). Pipe B oscillates in its lowest harmonic, pipe C in its second lowest harmonic, pipe D in its third lowest harmonic, and pipe E in its fourth lowest harmonic. Without

Fig. 17-26   Question 5.

computation, rank all five pipes according to their length, greatest first. (*Hint:* Draw the standing waves to scale and then draw the pipes to scale.)

6 Pipe A has length L and one open end. Pipe B has length $2L$ and two open ends. Which harmonics of pipe B have a frequency that matches a resonant frequency of pipe A?

7 Figure 17-27 shows a moving sound source S that emits at a certain frequency, and four stationary sound detectors. Rank the detectors according to the frequency of the sound they detect from the source, greatest first.

Fig. 17-27   Question 7.

8 A friend rides, in turn, the rims of three fast merry-go-rounds while holding a sound source that emits isotropically at a certain frequency. You stand far from each merry-go-round. The frequency you hear for each of your friend's three rides varies as the merry-go-round rotates. The variations in frequency for the three rides are given by the three curves in Fig. 17-28. Rank the curves according to (a) the linear speed v of the sound source, (b) the angular speeds ω of the merry-go-rounds, and (c) the radii r of the merry-go-rounds, greatest first.

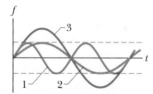

Fig. 17-28   Question 8.

9 For a particular tube, here are four of the six harmonic frequencies below 1000 Hz: 300, 600, 750, and 900 Hz. What two frequencies are missing from the list?

10 Figure 17-29 shows a stretched string of length L and pipes a, b, c, and d of lengths $L, 2L, L/2$, and $L/2$, respectively. The string's tension is adjusted until the speed of waves on the string equals the speed of sound waves in the air. The fundamental mode of oscillation is then set up on the string. In which pipe will the sound produced by the string cause resonance, and what oscillation mode will that sound set up?

Fig. 17-29   Question 10.

Where needed in the problems, use

$$\text{speed of sound in air} = 343 \text{ m/s}$$

and $$\text{density of air} = 1.21 \text{ kg/m}^3$$

unless otherwise specified.

sec. 17-3   The Speed of Sound

•1   Two spectators at a soccer game in Montjuic Stadium see, and a moment later hear, the ball being kicked on the playing field. The time delay for spectator A is 0.23 s, and for spectator B it is 0.12 s. Sight lines from the two spectators to the player kicking the ball meet at an angle of 90°. How far are (a) spectator A and (b) spectator B from the player? (c) How far are the spectators from each other?

•2   What is the bulk modulus of oxygen if 32.0 g of oxygen occupies 22.4 L and the speed of sound in the oxygen is 317 m/s?

•3   ✈ When the door of the Chapel of the Mausoleum in Hamilton, Scotland, is slammed shut, the last echo heard by someone standing just inside the door reportedly comes 15 s later. (a) If that echo were due to a single reflection off a wall opposite the door, how far from the door would that wall be? (b) If, instead, the wall is 25.7 m away, how many reflections (back and forth) correspond to the last echo?

•4   A column of soldiers, marching at 120 paces per minute, keep in step with the beat of a drummer at the head of the column. The soldiers in the rear end of the column are striding forward with the left foot when the drummer is advancing with the right foot. What is the approximate length of the column?

••5   **SSM** **ILW** Earthquakes generate sound waves inside Earth. Unlike a gas, Earth can experience both transverse (S) and longitudinal (P) sound waves. Typically, the speed of S waves is about 4.5 km/s, and that of P waves 8.0 km/s. A seismograph records P and S waves from an earthquake. The first P waves arrive 3.0 min before the first S waves. If the waves travel in a straight line, how far away does the earthquake occur?

••6   A man strikes one end of a thin rod with a hammer. The speed of sound in the rod is 15 times the speed of sound in air. A woman, at the other end with her ear close to the rod, hears the sound of the blow twice with a 0.12 s interval between; one sound comes through the rod and the other comes through the air alongside the rod. If the speed of sound in air is 343 m/s, what is the length of the rod?

••7   **SSM** **WWW** A stone is dropped into a well. The splash is heard 3.00 s later. What is the depth of the well?

••8   ✈ *Hot chocolate effect.* Tap a metal spoon inside a mug of water and note the frequency f_i you hear. Then add a spoonful of powder (say, chocolate mix or instant coffee) and tap again as you stir the powder. The frequency you hear has a lower value f_s because the tiny air bubbles released by the powder change the water's bulk modulus. As the bubbles reach the water surface and dis-appear, the frequency gradually shifts back to its initial value. During the effect, the bubbles don't appreciably change the water's density or volume or the sound's wavelength. Rather, they change the value of dV/dp—that is, the differential change in volume due to the differential change in the pressure caused by the sound wave in the water. If $f_s/f_i = 0.333$, what is the ratio $(dV/dp)_s/(dV/dp)_i$?

sec. 17-4   Traveling Sound Waves

•9   If the form of a sound wave traveling through air is

$$s(x, t) = (6.0 \text{ nm}) \cos(kx + (3000 \text{ rad/s})t + \phi),$$

how much time does any given air molecule along the path take to move between displacements $s = +2.0$ nm and $s = -2.0$ nm?

•10   ✈ *Underwater illusion.* One clue used by your brain to determine the direction of a source of sound is the time delay Δt between the arrival of the sound at the ear closer to the source and the arrival at the farther ear. Assume that the source is distant so that a wavefront from it is approximately planar when it reaches you, and let D represent the separation between your ears. (a) If the source is located at angle θ in front of you (Fig. 17-30), what is Δt in terms of D and the speed of sound v in air? (b) If you are submerged in water and the sound source is directly to your right, what is Δt in terms of D and the speed of sound v_w in water? (c) Based on the time-delay clue, your brain interprets the submerged sound to arrive at an angle θ from the forward direction. Evaluate θ for fresh water at 20°C.

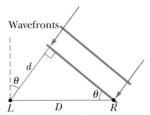
Fig. 17-30   Problem 10.

•11   **SSM** Diagnostic ultrasound of frequency 4.50 MHz is used to examine tumors in soft tissue. (a) What is the wavelength in air of such a sound wave? (b) If the speed of sound in tissue is 1500 m/s, what is the wavelength of this wave in tissue?

•12   The pressure in a traveling sound wave is given by the equation

$$\Delta p = (1.50 \text{ Pa}) \sin \pi[(0.900 \text{ m}^{-1})x - (315 \text{ s}^{-1})t].$$

Find the (a) pressure amplitude, (b) frequency, (c) wavelength, and (d) speed of the wave.

••13   A sound wave of the form $s = s_m \cos(kx - \omega t + \phi)$ travels at 343 m/s through air in a long horizontal tube. At one instant, air molecule A at $x = 2.000$ m is at its maximum positive displacement of 6.00 nm and air molecule B at $x = 2.070$ m is at a positive displacement of 2.00 nm. All the molecules between A and B are at intermediate displacements. What is the frequency of the wave?

••14   Figure 17-31 shows the output from a pressure monitor mounted at a point along the path taken by a sound wave of a single frequency traveling at 343 m/s through air with a uniform density of 1.21 kg/m³. The vertical axis scale is set by

$\Delta p_s = 4.0$ mPa. If the *displacement* function of the wave is $s(x, t) = s_m \cos(kx - \omega t)$, what are (a) s_m, (b) k, and (c) ω? The air is then cooled so that its density is 1.35 kg/m^3 and the speed of a sound wave through it is 320 m/s. The sound source again emits the sound wave at the same frequency and same pressure amplitude. What now are (d) s_m, (e) k, and (f) ω?

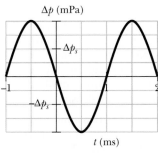

Fig. 17-31   Problem 14.

••15 [airplane icon]   A handclap on stage in an amphitheater sends out sound waves that scatter from terraces of width $w = 0.75$ m (Fig. 17-32). The sound returns to the stage as a periodic series of pulses, one from each terrace; the parade of pulses sounds like a played note. (a) Assuming that all the rays in Fig. 17-32 are horizontal, find the frequency at which the pulses return (that is, the frequency of the perceived note). (b) If the width w of the terraces were smaller, would the frequency be higher or lower?

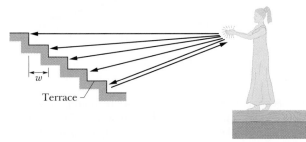

Fig. 17-32   Problem 15.

sec. 17-5   Interference

•16   Two sound waves, from two different sources with the same frequency, 540 Hz, travel in the same direction at 330 m/s. The sources are in phase. What is the phase difference of the waves at a point that is 4.40 m from one source and 4.00 m from the other?

••17 ILW [airplane icon]   Two loud speakers are located 3.35 m apart on an outdoor stage. A listener is 18.3 m from one and 19.5 m from the other. During the sound check, a signal generator drives the two speakers in phase with the same amplitude and frequency. The transmitted frequency is swept through the audible range (20 Hz to 20 kHz). (a) What is the lowest frequency $f_{min,1}$ that gives minimum signal (destructive interference) at the listener's location? By what number must $f_{min,1}$ be multiplied to get (b) the second lowest frequency $f_{min,2}$ that gives minimum signal and (c) the third lowest frequency $f_{min,3}$ that gives minimum signal? (d) What is the lowest frequency $f_{max,1}$ that gives maximum signal (constructive interference) at the listener's location? By what number must $f_{max,1}$ be multiplied to get (e) the second lowest frequency $f_{max,2}$ that gives maximum signal and (f) the third lowest frequency $f_{max,3}$ that gives maximum signal?

••18 [GO]   In Fig. 17-33, sound waves A and B, both of wavelength λ, are initially in phase and travel-

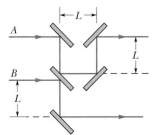

Fig. 17-33   Problem 18.

ing rightward, as indicated by the two rays. Wave A is reflected from four surfaces but ends up traveling in its original direction. Wave B ends in that direction after reflecting from two surfaces. Let distance L in the figure be expressed as a multiple q of λ: $L = q\lambda$. What are the (a) smallest and (b) second smallest values of q that put A and B exactly out of phase with each other after the reflections?

••19   Figure 17-34 shows two isotropic point sources of sound, S_1 and S_2. The sources emit waves in phase at wavelength 0.50 m; they are separated by $D = 1.75$ m. If we move a sound detector along a large circle centered at the midpoint between the sources, at how many points do waves arrive at the detector (a) exactly in phase and (b) exactly out of phase?

Fig. 17-34
Problems 19 and 105.

••20   Figure 17-35 shows four isotropic point sources of sound that are uniformly spaced on an x axis. The sources emit sound at the same wavelength λ and same amplitude s_m, and they emit in phase. A point P is shown on the x axis. Assume that as the sound waves travel to P, the decrease in their amplitude is negligible. What multiple of s_m is the amplitude of the net wave at P if distance d in the figure is (a) $\lambda/4$, (b) $\lambda/2$, and (c) λ?

Fig. 17-35   Problem 20.

••21 SSM   In Fig. 17-36, two speakers separated by distance $d_1 = 2.00$ m are in phase. Assume the amplitudes of the sound waves from the speakers are approximately the same at the listener's ear at distance $d_2 = 3.75$ m directly in front of one speaker. Consider the full audible range for normal hearing, 20 Hz to 20 kHz. (a) What is the lowest frequency $f_{min,1}$ that gives minimum signal (destructive interference) at the listener's ear? By what number must $f_{min,1}$ be multiplied to get (b) the second lowest frequency $f_{min,2}$ that gives minimum signal and (c) the third lowest frequency $f_{min,3}$ that gives minimum signal? (d) What is the lowest frequency $f_{max,1}$ that gives maximum signal (constructive interference) at the listener's ear? By what number must $f_{max,1}$ be multiplied to get (e) the second lowest frequency $f_{max,2}$ that gives maximum signal and (f) the third lowest frequency $f_{max,3}$ that gives maximum signal?

Fig. 17-36   Problem 21.

••22   In Fig. 17-37, sound with a 40.0 cm wavelength travels rightward from a source and through a tube that consists of a straight portion and a half-circle. Part of the sound wave travels through the half-circle and then rejoins the rest of the wave, which goes directly through the straight portion. This rejoining results in interference. What is the smallest radius r that results in an intensity minimum at the detector?

Fig. 17-37   Problem 22.

•••**23** GO Figure 17-38 shows two point sources S_1 and S_2 that emit sound of wavelength $\lambda = 2.00$ m. The emissions are isotropic and in phase, and the separation between the sources is $d = 16.0$ m. At any point P on the x axis, the wave from S_1 and the wave from S_2 interfere. When P is very far away ($x \approx \infty$), what are (a) the phase difference between the arriving waves from S_1 and S_2 and (b) the type of interference they produce? Now move point P along the x axis toward S_1. (c) Does the phase difference between the waves increase or decrease? At what distance x do the waves have a phase difference of (d) 0.50λ, (e) 1.00λ, and (f) 1.50λ?

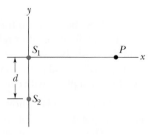

Fig. 17-38 Problem 23.

sec. 17-6 Intensity and Sound Level

•**24** Suppose that the sound level of a conversation is initially at an angry 70 dB and then drops to a soothing 50 dB. Assuming that the frequency of the sound is 500 Hz, determine the (a) initial and (b) final sound intensities and the (c) initial and (d) final sound wave amplitudes.

•**25** A sound wave of frequency 300 Hz has an intensity of 1.00 μW/m². What is the amplitude of the air oscillations caused by this wave?

•**26** A 1.0 W point source emits sound waves isotropically. Assuming that the energy of the waves is conserved, find the intensity (a) 1.0 m from the source and (b) 2.5 m from the source.

•**27** SSM WWW A certain sound source is increased in sound level by 30.0 dB. By what multiple is (a) its intensity increased and (b) its pressure amplitude increased?

•**28** Two sounds differ in sound level by 1.00 dB. What is the ratio of the greater intensity to the smaller intensity?

•**29** SSM A source emits sound waves isotropically. The intensity of the waves 2.50 m from the source is 1.91×10^{-4} W/m². Assuming that the energy of the waves is conserved, find the power of the source.

•**30** The source of a sound wave has a power of 1.00 μW. If it is a point source, (a) what is the intensity 3.00 m away and (b) what is the sound level in decibels at that distance?

•**31** GO When you "crack" a knuckle, you suddenly widen the knuckle cavity, allowing more volume for the synovial fluid inside it and causing a gas bubble suddenly to appear in the fluid. The sudden production of the bubble, called "cavitation," produces a sound pulse—the cracking sound. Assume that the sound is transmitted uniformly in all directions and that it fully passes from the knuckle interior to the outside. If the pulse has a sound level of 62 dB at your ear, estimate the rate at which energy is produced by the cavitation.

•**32** Approximately a third of people with normal hearing have ears that continuously emit a low-intensity sound outward through the ear canal. A person with such *spontaneous otoacoustic emission* is rarely aware of the sound, except perhaps in a noise-free environment, but occasionally the emission is loud enough to be heard by someone else nearby. In one observation, the sound wave had a frequency of 1665 Hz and a pressure amplitude of 1.13×10^{-3} Pa. What were (a) the displacement amplitude and (b) the intensity of the wave emitted by the ear?

•**33** Male *Rana catesbeiana* bullfrogs are known for their loud mating call. The call is emitted not by the frog's mouth but by its eardrums, which lie on the surface of the head. And, surprisingly, the sound has nothing to do with the frog's inflated throat. If the emitted sound has a frequency of 260 Hz and a sound level of 85 dB (near the eardrum), what is the amplitude of the eardrum's oscillation? The air density is 1.21 kg/m³.

••**34** GO Two atmospheric sound sources A and B emit isotropically at constant power. The sound levels β of their emissions are plotted in Fig. 17-39 versus the radial distance r from the sources. The vertical axis scale is set by $\beta_1 = 85.0$ dB and $\beta_2 = 65.0$ dB. What are (a) the ratio of the larger power to the smaller power and (b) the sound level difference at $r = 10$ m?

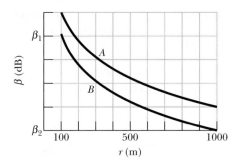

Fig. 17-39 Problem 34.

••**35** A point source emits 30.0 W of sound isotropically. A small microphone intercepts the sound in an area of 0.750 cm², 200 m from the source. Calculate (a) the sound intensity there and (b) the power intercepted by the microphone.

••**36** *Party hearing.* As the number of people at a party increases, you must raise your voice for a listener to hear you against the *background noise* of the other partygoers. However, once you reach the level of yelling, the only way you can be heard is if you move closer to your listener, into the listener's "personal space." Model the situation by replacing you with an isotropic point source of fixed power P and replacing your listener with a point that absorbs part of your sound waves. These points are initially separated by $r_i = 1.20$ m. If the background noise increases by $\Delta\beta = 5$ dB, the sound level at your listener must also increase. What separation r_f is then required?

•••**37** A sound source sends a sinusoidal sound wave of angular frequency 3000 rad/s and amplitude 12.0 nm through a tube of air. The internal radius of the tube is 2.00 cm. (a) What is the average rate at which energy (the sum of the kinetic and potential energies) is transported to the opposite end of the tube? (b) If, simultaneously, an identical wave travels along an adjacent, identical tube, what is the total average rate at which energy is transported to the opposite ends of the two tubes by the waves? If, instead, those two waves are sent along the *same* tube simultaneously, what is the total average rate at which they transport energy when their phase difference is (c) 0, (d) 0.40π rad, and (e) π rad?

sec. 17-7 Sources of Musical Sound

•**38** The water level in a vertical glass tube 1.00 m long can be adjusted to any position in the tube. A tuning fork vibrating at 686 Hz is held just over the open top end of the tube, to set up a standing wave of sound in the air-filled top portion of the tube. (That air-filled top portion acts as a tube with one end closed and the other end open.) (a) For how many different positions of the water level will sound from the fork set up resonance in the tube's air-filled portion, which acts as a pipe with one end closed (by the water) and the other end open? What are the (b) least and (c) second least water heights in the tube for resonance to occur?

•39 SSM ILW (a) Find the speed of waves on a violin string of mass 800 mg and length 22.0 cm if the fundamental frequency is 920 Hz. (b) What is the tension in the string? For the fundamental, what is the wavelength of (c) the waves on the string and (d) the sound waves emitted by the string?

•40 Organ pipe *A*, with both ends open, has a fundamental frequency of 300 Hz. The third harmonic of organ pipe *B*, with one end open, has the same frequency as the second harmonic of pipe *A*. How long are (a) pipe *A* and (b) pipe *B*?

•41 A violin string 15.0 cm long and fixed at both ends oscillates in its $n = 1$ mode. The speed of waves on the string is 250 m/s, and the speed of sound in air is 348 m/s. What are the (a) frequency and (b) wavelength of the emitted sound wave?

•42 A sound wave in a fluid medium is reflected at a barrier so that a standing wave is formed. The distance between nodes is 3.8 cm, and the speed of propagation is 1500 m/s. Find the frequency of the sound wave.

•43 SSM In Fig. 17-40, *S* is a small loudspeaker driven by an audio oscillator with a frequency that is varied from 1000 Hz to 2000 Hz, and *D* is a cylindrical pipe with two open ends and a length of 45.7 cm. The speed of sound in the air-filled pipe is 344 m/s. (a) At how many frequencies does the sound from the loudspeaker set up resonance in the pipe? What are the (b) lowest and (c) second lowest frequencies at which resonance occurs?

Fig. 17-40
Problem 43.

•44 The crest of a *Parasaurolophus* dinosaur skull contains a nasal passage in the shape of a long, bent tube open at both ends. The dinosaur may have used the passage to produce sound by setting up the fundamental mode in it. (a) If the nasal passage in a certain *Parasaurolophus* fossil is 2.0 m long, what frequency would have been produced? (b) If that dinosaur could be recreated (as in *Jurassic Park*), would a person with a hearing range of 60 Hz to 20 kHz be able to hear that fundamental mode and, if so, would the sound be high or low frequency? Fossil skulls that contain shorter nasal passages are thought to be those of the female *Parasaurolophus*. (c) Would that make the female's fundamental frequency higher or lower than the male's?

•45 In pipe *A*, the ratio of a particular harmonic frequency to the next lower harmonic frequency is 1.2. In pipe *B*, the ratio of a particular harmonic frequency to the next lower harmonic frequency is 1.4. How many open ends are in (a) pipe *A* and (b) pipe *B*?

••46 GO Pipe *A*, which is 1.20 m long and open at both ends, oscillates at its third lowest harmonic frequency. It is filled with air for which the speed of sound is 343 m/s. Pipe *B*, which is closed at one end, oscillates at its second lowest harmonic frequency. This frequency of *B* happens to match the frequency of *A*. An *x* axis extends along the interior of *B*, with $x = 0$ at the closed end. (a) How many nodes are along that axis? What are the (b) smallest and (c) second smallest value of *x* locating those nodes? (d) What is the fundamental frequency of *B*?

••47 A well with vertical sides and water at the bottom resonates at 7.00 Hz and at no lower frequency. (The air-filled portion of the well acts as a tube with one closed end and one open end.) The air in the well has a density of 1.10 kg/m³ and a bulk modulus of 1.33×10^5 Pa. How far down in the well is the water surface?

••48 One of the harmonic frequencies of tube *A* with two open ends is 325 Hz. The next-highest harmonic frequency is 390 Hz. (a) What harmonic frequency is next highest after the harmonic frequency 195 Hz? (b) What is the number of this next-highest harmonic?

One of the harmonic frequencies of tube *B* with only one open end is 1080 Hz. The next-highest harmonic frequency is 1320 Hz. (c) What harmonic frequency is next highest after the harmonic frequency 600 Hz? (d) What is the number of this next-highest harmonic?

••49 SSM A violin string 30.0 cm long with linear density 0.650 g/m is placed near a loudspeaker that is fed by an audio oscillator of variable frequency. It is found that the string is set into oscillation only at the frequencies 880 and 1320 Hz as the frequency of the oscillator is varied over the range 500–1500 Hz. What is the tension in the string?

••50 GO A tube 1.20 m long is closed at one end. A stretched wire is placed near the open end. The wire is 0.330 m long and has a mass of 9.60 g. It is fixed at both ends and oscillates in its fundamental mode. By resonance, it sets the air column in the tube into oscillation at that column's fundamental frequency. Find (a) that frequency and (b) the tension in the wire.

sec. 17-8 Beats

•51 The A string of a violin is a little too tightly stretched. Beats at 4.00 per second are heard when the string is sounded together with a tuning fork that is oscillating accurately at concert A (440 Hz). What is the period of the violin string oscillation?

•52 A tuning fork of unknown frequency makes 3.00 beats per second with a standard fork of frequency 384 Hz. The beat frequency decreases when a small piece of wax is put on a prong of the first fork. What is the frequency of this fork?

••53 SSM Two identical piano wires have a fundamental frequency of 600 Hz when kept under the same tension. What fractional increase in the tension of one wire will lead to the occurrence of 6.0 beats/s when both wires oscillate simultaneously?

••54 You have five tuning forks that oscillate at close but different frequencies. What are the (a) maximum and (b) minimum number of different beat frequencies you can produce by sounding the forks two at a time, depending on how the frequencies differ?

sec. 17-9 The Doppler Effect

•55 ILW A whistle of frequency 540 Hz moves in a circle of radius 60.0 cm at an angular speed of 15.0 rad/s. What are the (a) lowest and (b) highest frequencies heard by a listener a long distance away, at rest with respect to the center of the circle?

•56 An ambulance with a siren emitting a whine at 1600 Hz overtakes and passes a cyclist pedaling a bike at 2.44 m/s. After being passed, the cyclist hears a frequency of 1590 Hz. How fast is the ambulance moving?

•57 A state trooper chases a speeder along a straight road; both vehicles move at 160 km/h. The siren on the trooper's vehicle produces sound at a frequency of 500 Hz. What is the Doppler shift in the frequency heard by the speeder?

••58 A sound source *A* and a reflecting surface *B* move directly toward each other. Relative to the air, the speed of source *A* is 29.9 m/s, the speed of surface *B* is 65.8 m/s, and the speed of sound is 329 m/s. The source emits waves at frequency 1200 Hz as measured in the source frame. In the reflector frame, what are the (a) frequency and (b) wavelength of the arriving sound waves? In the source frame, what are the (c) frequency and (d) wavelength of the sound waves reflected back to the source?

••59 **GO** In Fig. 17-41, a French submarine and a U.S. submarine move toward each other during maneuvers in motionless water in the North Atlantic. The French sub moves at speed $v_F = 50.00$ km/h, and the U.S. sub at $v_{US} = 70.00$ km/h. The French sub sends out a sonar signal (sound wave in water) at 1.000×10^3 Hz. Sonar waves travel at 5470 km/h. (a) What is the signal's frequency as detected by the U.S. sub? (b) What frequency is detected by the French sub in the signal reflected back to it by the U.S. sub?

Fig. 17-41 Problem 59.

••60 A stationary motion detector sends sound waves of frequency 0.150 MHz toward a truck approaching at a speed of 45.0 m/s. What is the frequency of the waves reflected back to the detector?

••61 A bat is flitting about in a cave, navigating via ultrasonic bleeps. Assume that the sound emission frequency of the bat is 39 000 Hz. During one fast swoop directly toward a flat wall surface, the bat is moving at 0.025 times the speed of sound in air. What frequency does the bat hear reflected off the wall?

••62 Figure 17-42 shows four tubes with lengths 1.0 m or 2.0 m, with one or two open ends as drawn. The third harmonic is set up in each tube, and some of the sound that escapes from them is detected by detector *D*, which moves directly away from the tubes. In terms of the speed of sound *v*, what speed must the detector have such that the detected frequency of the sound from (a) tube 1, (b) tube 2, (c) tube 3, and (d) tube 4 is equal to the tube's fundamental frequency?

Fig. 17-42 Problem 62.

••63 **ILW** An acoustic burglar alarm consists of a source emitting waves of frequency 28.0 kHz. What is the beat frequency between the source waves and the waves reflected from an intruder walking at an average speed of 0.950 m/s directly away from the alarm?

••64 A stationary detector measures the frequency of a sound source that first moves at constant velocity directly toward the detector and then (after passing the detector) directly away from it. The emitted frequency is *f*. During the approach the detected frequency is f'_{app} and during the recession it is f'_{rec}. If $(f'_{app} - f'_{rec})/f = 0.500$, what is the ratio v_s/v of the speed of the source to the speed of sound?

•••65 **GO** A 2000 Hz siren and a civil defense official are both at rest with respect to the ground. What frequency does the official hear if the wind is blowing at 12 m/s (a) from source to official and (b) from official to source?

•••66 Two trains are traveling toward each other at 30.5 m/s relative to the ground. One train is blowing a whistle at 500 Hz. (a) What frequency is heard on the other train in still air? (b) What frequency is heard on the other train if the wind is blowing at 30.5 m/s toward the whistle and away from the listener? (c) What frequency is heard if the wind direction is reversed?

•••67 **SSM** **WWW** A girl is sitting near the open window of a train that is moving at a velocity of 10.00 m/s to the east. The girl's uncle stands near the tracks and watches the train move away. The locomotive whistle emits sound at frequency 500.0 Hz. The air is still. (a) What frequency does the uncle hear? (b) What frequency does the girl hear? A wind begins to blow from the east at 10.00 m/s. (c) What frequency does the uncle now hear? (d) What frequency does the girl now hear?

sec. 17-10 Supersonic Speeds, Shock Waves

•68 The shock wave off the cockpit of the FA 18 in Fig. 17-23 has an angle of about 60°. The airplane was traveling at about 1350 km/h when the photograph was taken. Approximately what was the speed of sound at the airplane's altitude?

•69 **SSM** A jet plane passes over you at a height of 5000 m and a speed of Mach 1.5. (a) Find the Mach cone angle (the sound speed is 331 m/s). (b) How long after the jet passes directly overhead does the shock wave reach you?

••70 A plane flies at 1.25 times the speed of sound. Its sonic boom reaches a man on the ground 1.00 min after the plane passes directly overhead. What is the altitude of the plane? Assume the speed of sound to be 330 m/s.

Additional Problems

71 At a distance of 10 km, a 100 Hz horn, assumed to be an isotropic point source, is barely audible. At what distance would it begin to cause pain?

72 A bullet is fired with a speed of 685 m/s. Find the angle made by the shock cone with the line of motion of the bullet.

73 A sperm whale (Fig. 17-43*a*) vocalizes by producing a series of clicks. Actually, the whale makes only a single sound near the front of its head to start the series. Part of that sound then emerges from the head into the water to become the first click of the series. The rest of the sound travels backward through the sper-

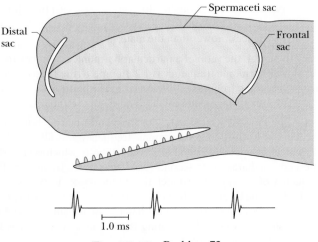

Fig. 17-43 Problem 73.

maceti sac (a body of fat), reflects from the frontal sac (an air layer), and then travels forward through the spermaceti sac. When it reaches the distal sac (another air layer) at the front of the head, some of the sound escapes into the water to form the second click, and the rest is sent back through the spermaceti sac (and ends up forming later clicks).

Figure 17-43b shows a strip-chart recording of a series of clicks. A unit time interval of 1.0 ms is indicated on the chart. Assuming that the speed of sound in the spermaceti sac is 1372 m/s, find the length of the spermaceti sac. From such a calculation, marine scientists estimate the length of a whale from its click series.

74 The average density of Earth's crust 10 km beneath the continents is 2.7 g/cm³. The speed of longitudinal seismic waves at that depth, found by timing their arrival from distant earthquakes, is 5.4 km/s. Use this information to find the bulk modulus of Earth's crust at that depth. For comparison, the bulk modulus of steel is about 16×10^{10} Pa.

75 A certain loudspeaker system emits sound isotropically with a frequency of 2000 Hz and an intensity of 0.960 mW/m² at a distance of 6.10 m. Assume that there are no reflections. (a) What is the intensity at 30.0 m? At 6.10 m, what are (b) the displacement amplitude and (c) the pressure amplitude?

76 Find the ratios (greater to smaller) of the (a) intensities, (b) pressure amplitudes, and (c) particle displacement amplitudes for two sounds whose sound levels differ by 37 dB.

77 In Fig. 17-44, sound waves A and B, both of wavelength λ, are initially in phase and traveling rightward, as indicated by the two rays. Wave A is reflected from four surfaces but ends up traveling in its original direction. What multiple of wavelength λ is the smallest value of distance L in the figure that puts A and B exactly out of phase with each other after the reflections?

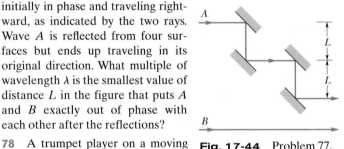

Fig. 17-44   Problem 77.

78 A trumpet player on a moving railroad flatcar moves toward a second trumpet player standing alongside the track while both play a 440 Hz note. The sound waves heard by a stationary observer between the two players have a beat frequency of 4.0 beats/s. What is the flatcar's speed?

79 In Fig. 17-45, sound of wavelength 0.850 m is emitted isotropically by point source S. Sound ray 1 extends directly to detector D, at distance L = 10.0 m. Sound ray 2 extends to D via a reflection (effectively, a "bouncing") of the sound at a flat surface. That reflection occurs on a perpendicular bisector to the SD line, at distance d from the line. Assume that the reflection shifts the sound wave by 0.500λ. For what least value of d (other than zero) do the direct sound and the reflected sound arrive at D (a) exactly out of phase and (b) exactly in phase?

Fig. 17-45   Problem 79.

80 [GO] A detector initially moves at constant velocity directly toward a stationary sound source and then (after passing it) directly from it. The emitted frequency is f. During the approach the detected frequency is f'_{app} and during the recession it is f'_{rec}. If the frequencies are related by $(f'_{app} - f'_{rec})/f = 0.500$, what is the ratio v_D/v of the speed of the detector to the speed of sound?

81 [SSM] (a) If two sound waves, one in air and one in (fresh) water, are equal in intensity and angular frequency, what is the ratio of the pressure amplitude of the wave in water to that of the wave in air? Assume the water and the air are at 20°C. (See Table 14-1.) (b) If the pressure amplitudes are equal instead, what is the ratio of the intensities of the waves?

82 A continuous sinusoidal longitudinal wave is sent along a very long coiled spring from an attached oscillating source. The wave travels in the negative direction of an x axis; the source frequency is 25 Hz; at any instant the distance between successive points of maximum expansion in the spring is 24 cm; the maximum longitudinal displacement of a spring particle is 0.30 cm; and the particle at x = 0 has zero displacement at time t = 0. If the wave is written in the form $s(x, t) = s_m \cos(kx \pm \omega t)$, what are (a) s_m, (b) k, (c) ω, (d) the wave speed, and (e) the correct choice of sign in front of ω?

83 [SSM] Ultrasound, which consists of sound waves with frequencies above the human audible range, can be used to produce an image of the interior of a human body. Moreover, ultrasound can be used to measure the speed of the blood in the body; it does so by comparing the frequency of the ultrasound sent into the body with the frequency of the ultrasound reflected back to the body's surface by the blood. As the blood pulses, this detected frequency varies.

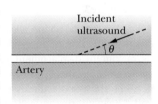

Fig. 17-46   Problem 83.

Suppose that an ultrasound image of the arm of a patient shows an artery that is angled at θ = 20° to the ultrasound's line of travel (Fig. 17-46). Suppose also that the frequency of the ultrasound reflected by the blood in the artery is increased by a maximum of 5495 Hz from the original ultrasound frequency of 5.000 000 MHz. (a) In Fig. 17-46, is the direction of the blood flow rightward or leftward? (b) The speed of sound in the human arm is 1540 m/s. What is the maximum speed of the blood? (Hint: The Doppler effect is caused by the component of the blood's velocity along the ultrasound's direction of travel.) (c) If angle θ were greater, would the reflected frequency be greater or less?

84 The speed of sound in a certain metal is v_m. One end of a long pipe of that metal of length L is struck a hard blow. A listener at the other end hears two sounds, one from the wave that travels along the pipe's metal wall and the other from the wave that travels through the air inside the pipe. (a) If v is the speed of sound in air, what is the time interval Δt between the arrivals of the two sounds at the listener's ear? (b) If Δt = 1.00 s and the metal is steel, what is the length L?

85 [image] An avalanche of sand along some rare desert sand dunes can produce a booming that is loud enough to be heard 10 km away. The booming apparently results from a periodic oscillation of the sliding layer of sand—the layer's thickness expands and contracts. If the emitted frequency is 90 Hz, what are (a) the period of the thickness oscillation and (b) the wavelength of the sound?

86 A sound source moves along an *x* axis, between detectors *A* and *B*. The wavelength of the sound detected at *A* is 0.500 that of the sound detected at *B*. What is the ratio v_s/v of the speed of the source to the speed of sound?

87 **SSM** A siren emitting a sound of frequency 1000 Hz moves away from you toward the face of a cliff at a speed of 10 m/s. Take the speed of sound in air as 330 m/s. (a) What is the frequency of the sound you hear coming directly from the siren? (b) What is the frequency of the sound you hear reflected off the cliff? (c) What is the beat frequency between the two sounds? Is it perceptible (less than 20 Hz)?

88 At a certain point, two waves produce pressure variations given by $\Delta p_1 = \Delta p_m \sin \omega t$ and $\Delta p_2 = \Delta p_m \sin(\omega t - \phi)$. At this point, what is the ratio $\Delta p_r/\Delta p_m$, where Δp_r is the pressure amplitude of the resultant wave, if ϕ is (a) 0, (b) $\pi/2$, (c) $\pi/3$, and (d) $\pi/4$?

89 Two sound waves with an amplitude of 12 nm and a wavelength of 35 cm travel in the same direction through a long tube, with a phase difference of $\pi/3$ rad. What are the (a) amplitude and (b) wavelength of the net sound wave produced by their interference? If, instead, the sound waves travel through the tube in opposite directions, what are the (c) amplitude and (d) wavelength of the net wave?

90 A sinusoidal sound wave moves at 343 m/s through air in the positive direction of an *x* axis. At one instant, air molecule *A* is at its maximum displacement in the negative direction of the axis while air molecule *B* is at its equilibrium position. The separation between those molecules is 15.0 cm, and the molecules between *A* and *B* have intermediate displacements in the negative direction of the axis. (a) What is the frequency of the sound wave?

In a similar arrangement, for a different sinusoidal sound wave, air molecule *C* is at its maximum displacement in the positive direction while molecule *D* is at its maximum displacement in the negative direction. The separation between the molecules is again 15.0 cm, and the molecules between *C* and *D* have intermediate displacements. (b) What is the frequency of the sound wave?

91 Two identical tuning forks can oscillate at 440 Hz. A person is located somewhere on the line between them. Calculate the beat frequency as measured by this individual if (a) she is standing still and the tuning forks move in the same direction along the line at 3.00 m/s, and (b) the tuning forks are stationary and the listener moves along the line at 3.00 m/s.

92 You can estimate your distance from a lightning stroke by counting the seconds between the flash you see and the thunder you later hear. By what integer should you divide the number of seconds to get the distance in kilometers?

93 **SSM** Figure 17-47 shows an air-filled, acoustic interferometer, used to demonstrate the interference of sound waves. Sound source *S* is an oscillating diaphragm; *D* is a sound detector, such as the ear or a microphone. Path *SBD* can be varied in length, but path *SAD* is fixed. At *D*, the sound wave coming along path *SBD* interferes with that coming along path *SAD*. In one demonstration, the sound intensity at *D* has a minimum value of 100 units at one position of the movable arm and continuously climbs to a maximum value of 900 units when that arm is shifted by 1.65 cm. Find (a) the frequency of the sound emit-

Fig. 17-47 Problem 93.

ted by the source and (b) the ratio of the amplitude at *D* of the *SAD* wave to that of the *SBD* wave. (c) How can it happen that these waves have different amplitudes, considering that they originate at the same source?

94 On July 10, 1996, a granite block broke away from a wall in Yosemite Valley and, as it began to slide down the wall, was launched into projectile motion. Seismic waves produced by its impact with the ground triggered seismographs as far away as 200 km. Later measurements indicated that the block had a mass between 7.3×10^7 kg and 1.7×10^8 kg and that it landed 500 m vertically below the launch point and 30 m horizontally from it. (The launch angle is not known.) (a) Estimate the block's kinetic energy just before it landed.

Consider two types of seismic waves that spread from the impact point—a hemispherical *body wave* traveled through the ground in an expanding hemisphere and a cylindrical *surface wave* traveled along the ground in an expanding shallow vertical cylinder (Fig. 17-48). Assume that the impact lasted 0.50 s, the vertical cylinder had a depth *d* of 5.0 m, and each wave type received 20% of the energy the block had just before impact. Neglecting any mechanical energy loss the waves experienced as they traveled, determine the intensities of (b) the body wave and (c) the surface wave when they reached a seismograph 200 km away. (d) On the basis of these results, which wave is more easily detected on a distant seismograph?

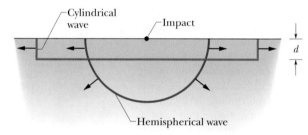
Fig. 17-48 Problem 94.

95 **SSM** The sound intensity is 0.0080 W/m² at a distance of 10 m from an isotropic point source of sound. (a) What is the power of the source? (b) What is the sound intensity 5.0 m from the source? (c) What is the sound level 10 m from the source?

96 Four sound waves are to be sent through the same tube of air, in the same direction:

$$s_1(x, t) = (9.00 \text{ nm}) \cos(2\pi x - 700\pi t)$$
$$s_2(x, t) = (9.00 \text{ nm}) \cos(2\pi x - 700\pi t + 0.7\pi)$$
$$s_3(x, t) = (9.00 \text{ nm}) \cos(2\pi x - 700\pi t + \pi)$$
$$s_4(x, t) = (9.00 \text{ nm}) \cos(2\pi x - 700\pi t + 1.7\pi).$$

What is the amplitude of the resultant wave? (*Hint:* Use a phasor diagram to simplify the problem.)

97 Straight line *AB* connects two point sources that are 5.00 m apart, emit 300 Hz sound waves of the same amplitude, and emit exactly out of phase. (a) What is the shortest distance between the midpoint of *AB* and a point on *AB* where the interfering waves cause maximum oscillation of the air molecules? What are the (b) second and (c) third shortest distances?

98 A point source that is stationary on an *x* axis emits a sinusoidal sound wave at a frequency of 686 Hz and speed 343 m/s. The wave travels radially outward from the source, causing air molecules to oscillate radially inward and outward. Let us define a wavefront as a line that connects points where the air molecules have the maximum, radially outward displacement. At any given instant, the wavefronts are concentric circles that are centered on the source. (a) Along x, what is the adjacent wavefront separation? Next, the source moves along x at a speed of 110 m/s. Along x, what are the wavefront separations (b) in front of and (c) behind the source?

99 You are standing at a distance D from an isotropic point source of sound. You walk 50.0 m toward the source and observe that the intensity of the sound has doubled. Calculate the distance D.

100 Pipe A has only one open end; pipe B is four times as long and has two open ends. Of the lowest 10 harmonic numbers n_B of pipe B, what are the (a) smallest, (b) second smallest, and (c) third smallest values at which a harmonic frequency of B matches one of the harmonic frequencies of A?

101 A pipe 0.60 m long and closed at one end is filled with an unknown gas. The third lowest harmonic frequency for the pipe is 750 Hz. (a) What is the speed of sound in the unknown gas? (b) What is the fundamental frequency for this pipe when it is filled with the unknown gas?

102 A sound wave travels out uniformly in all directions from a point source. (a) Justify the following expression for the displacement s of the transmitting medium at any distance r from the source:

$$s = \frac{b}{r} \sin k(r - vt),$$

where b is a constant. Consider the speed, direction of propagation, periodicity, and intensity of the wave. (b) What is the dimension of the constant b?

103 A police car is chasing a speeding Porsche 911. Assume that the Porsche's maximum speed is 80.0 m/s and the police car's is 54.0 m/s. At the moment both cars reach their maximum speed, what frequency will the Porsche driver hear if the frequency of the police car's siren is 440 Hz? Take the speed of sound in air to be 340 m/s.

104 Suppose a spherical loudspeaker emits sound isotropically at 10 W into a room with completely absorbent walls, floor, and ceiling (an *anechoic chamber*). (a) What is the intensity of the sound at distance $d = 3.0$ m from the center of the source? (b) What is the ratio of the wave amplitude at $d = 4.0$ m to that at $d = 3.0$ m?

105 In Fig. 17-34, S_1 and S_2 are two isotropic point sources of sound. They emit waves in phase at wavelength 0.50 m; they are separated by $D = 1.60$ m. If we move a sound detector along a large circle centered at the midpoint between the sources, at how many points do waves arrive at the detector (a) exactly in phase and (b) exactly out of phase?

18

TEMPERATURE, HEAT, AND THE FIRST LAW OF THERMODYNAMICS

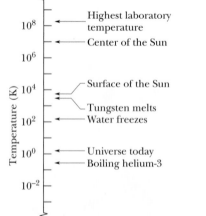

Fig. 18-1 Some temperatures on the Kelvin scale. Temperature $T = 0$ corresponds to $10^{-\infty}$ and cannot be plotted on this logarithmic scale.

18-1 WHAT IS PHYSICS?

One of the principal branches of physics and engineering is **thermodynamics**, which is the study and application of the *thermal energy* (often called the *internal energy*) of systems. One of the central concepts of thermodynamics is temperature, which we begin to explore in the next section. Since childhood, you have been developing a working knowledge of thermal energy and temperature. For example, you know to be cautious with hot foods and hot stoves and to store perishable foods in cool or cold compartments. You also know how to control the temperature inside home and car, and how to protect yourself from wind chill and heat stroke.

Examples of how thermodynamics figures into everyday engineering and science are countless. Automobile engineers are concerned with the heating of a car engine, such as during a NASCAR race. Food engineers are concerned both with the proper heating of foods, such as pizzas being microwaved, and with the proper cooling of foods, such as TV dinners being quickly frozen at a processing plant. Geologists are concerned with the transfer of thermal energy in an El Niño event and in the gradual warming of ice expanses in the Arctic and Antarctic. Agricultural engineers are concerned with the weather conditions that determine whether the agriculture of a country thrives or vanishes. Medical engineers are concerned with how a patient's temperature might distinguish between a benign viral infection and a cancerous growth.

The starting point in our discussion of thermodynamics is the concept of temperature and how it is measured.

18-2 Temperature

Temperature is one of the seven SI base quantities. Physicists measure temperature on the **Kelvin scale,** which is marked in units called *kelvins*. Although the temperature of a body apparently has no upper limit, it does have a lower limit; this limiting low temperature is taken as the zero of the Kelvin temperature scale. Room temperature is about 290 kelvins, or 290 K as we write it, above this *absolute zero*. Figure 18-1 shows a wide range of temperatures.

When the universe began 13.7 billion years ago, its temperature was about 10^{39} K. As the universe expanded it cooled, and it has now reached an average temperature of about 3 K. We on Earth are a little warmer than that because we happen to live near a star. Without our Sun, we too would be at 3 K (or, rather, we could not exist).

18-3 The Zeroth Law of Thermodynamics

The properties of many bodies change as we alter their temperature, perhaps by moving them from a refrigerator to a warm oven. To give a few examples: As their temperature increases, the volume of a liquid increases, a metal rod grows a little longer, and the electrical resistance of a wire increases, as does the pressure exerted by a confined gas. We can use any one of these properties as the basis of an instrument that will help us pin down the concept of temperature.

Figure 18-2 shows such an instrument. Any resourceful engineer could design and construct it, using any one of the properties listed above. The instrument is fitted with a digital readout display and has the following properties: If you heat it (say, with a Bunsen burner), the displayed number starts to increase; if you then put it into a refrigerator, the displayed number starts to decrease. The instrument is not calibrated in any way, and the numbers have (as yet) no physical meaning. The device is a *thermoscope* but not (as yet) a *thermometer*.

Suppose that, as in Fig. 18-3a, we put the thermoscope (which we shall call body *T*) into intimate contact with another body (body *A*). The entire system is confined within a thick-walled insulating box. The numbers displayed by the thermoscope roll by until, eventually, they come to rest (let us say the reading is "137.04") and no further change takes place. In fact, we suppose that every measurable property of body *T* and of body *A* has assumed a stable, unchanging value. Then we say that the two bodies are in *thermal equilibrium* with each other. Even though the displayed readings for body *T* have not been calibrated, we conclude that bodies *T* and *A* must be at the same (unknown) temperature.

Suppose that we next put body *T* into intimate contact with body *B* (Fig. 18-3b) and find that the two bodies come to thermal equilibrium *at the same reading of the thermoscope*. Then bodies *T* and *B* must be at the same (still unknown) temperature. If we now put bodies *A* and *B* into intimate contact (Fig. 18-3c), are they immediately in thermal equilibrium with each other? Experimentally, we find that they are.

The experimental fact shown in Fig. 18-3 is summed up in the **zeroth law of thermodynamics:**

> If bodies *A* and *B* are each in thermal equilibrium with a third body *T*, then *A* and *B* are in thermal equilibrium with each other.

In less formal language, the message of the zeroth law is: "Every body has a property called **temperature.** When two bodies are in thermal equilibrium, their temperatures are equal. And vice versa." We can now make our thermoscope (the third body *T*) into a thermometer, confident that its readings will have physical meaning. All we have to do is calibrate it.

We use the zeroth law constantly in the laboratory. If we want to know whether the liquids in two beakers are at the same temperature, we measure the temperature of each with a thermometer. We do not need to bring the two liquids into intimate contact and observe whether they are or are not in thermal equilibrium.

The zeroth law, which has been called a logical afterthought, came to light only in the 1930s, long after the first and second laws of thermodynamics had been discovered and numbered. Because the concept of temperature is fundamental to those two laws, the law that establishes temperature as a valid concept should have the lowest number—hence the zero.

Thermally sensitive element

Fig. 18-2 A thermoscope. The numbers increase when the device is heated and decrease when it is cooled. The thermally sensitive element could be—among many possibilities—a coil of wire whose electrical resistance is measured and displayed.

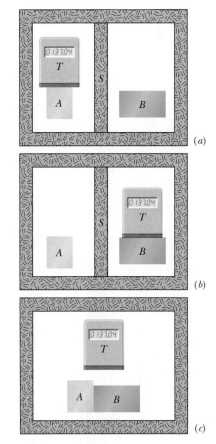

Fig. 18-3 (a) Body *T* (a thermoscope) and body *A* are in thermal equilibrium. (Body *S* is a thermally insulating screen.) (b) Body *T* and body *B* are also in thermal equilibrium, at the same reading of the thermoscope. (c) If (a) and (b) are true, the zeroth law of thermodynamics states that body *A* and body *B* are also in thermal equilibrium.

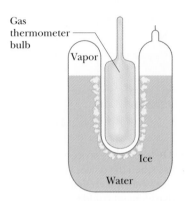

Fig. 18-4 A triple-point cell, in which solid ice, liquid water, and water vapor coexist in thermal equilibrium. By international agreement, the temperature of this mixture has been defined to be 273.16 K. The bulb of a constant-volume gas thermometer is shown inserted into the well of the cell.

18-4 Measuring Temperature

Here we first define and measure temperatures on the Kelvin scale. Then we calibrate a thermoscope so as to make it a thermometer.

The Triple Point of Water

To set up a temperature scale, we pick some reproducible thermal phenomenon and, quite arbitrarily, assign a certain Kelvin temperature to its environment; that is, we select a *standard fixed point* and give it a standard fixed-point *temperature*. We could, for example, select the freezing point or the boiling point of water but, for technical reasons, we select instead the **triple point of water.**

Liquid water, solid ice, and water vapor (gaseous water) can coexist, in thermal equilibrium, at only one set of values of pressure and temperature. Figure 18-4 shows a triple-point cell, in which this so-called triple point of water can be achieved in the laboratory. By international agreement, the triple point of water has been assigned a value of 273.16 K as the standard fixed-point temperature for the calibration of thermometers; that is,

$$T_3 = 273.16 \text{ K} \qquad \text{(triple-point temperature)}, \qquad (18\text{-}1)$$

in which the subscript 3 means "triple point." This agreement also sets the size of the kelvin as 1/273.16 of the difference between the triple-point temperature of water and absolute zero.

Note that we do not use a degree mark in reporting Kelvin temperatures. It is 300 K (not 300°K), and it is read "300 kelvins" (not "300 degrees Kelvin"). The usual SI prefixes apply. Thus, 0.0035 K is 3.5 mK. No distinction in nomenclature is made between Kelvin temperatures and temperature differences, so we can write, "the boiling point of sulfur is 717.8 K" and "the temperature of this water bath was raised by 8.5 K."

The Constant-Volume Gas Thermometer

The standard thermometer, against which all other thermometers are calibrated, is based on the pressure of a gas in a fixed volume. Figure 18-5 shows such a **constant-volume gas thermometer;** it consists of a gas-filled bulb connected by a tube to a mercury manometer. By raising and lowering reservoir R, the mercury level in the left arm of the U-tube can always be brought to the zero of the scale to keep the gas volume constant (variations in the gas volume can affect temperature measurements).

The temperature of any body in thermal contact with the bulb (such as the liquid surrounding the bulb in Fig. 18-5) is then defined to be

$$T = Cp, \qquad (18\text{-}2)$$

in which p is the pressure exerted by the gas and C is a constant. From Eq. 14-10, the pressure p is

$$p = p_0 - \rho g h, \qquad (18\text{-}3)$$

in which p_0 is the atmospheric pressure, ρ is the density of the mercury in the manometer, and h is the measured difference between the mercury levels in the two arms of the tube.* (The minus sign is used in Eq. 18-3 because pressure p is measured *above* the level at which the pressure is p_0.)

Fig. 18-5 A constant-volume gas thermometer, its bulb immersed in a liquid whose temperature T is to be measured.

---

*For pressure units, we shall use units introduced in Section 14-3. The SI unit for pressure is the newton per square meter, which is called the pascal (Pa). The pascal is related to other common pressure units by

$$1 \text{ atm} = 1.01 \times 10^5 \text{ Pa} = 760 \text{ torr} = 14.7 \text{ lb/in.}^2.$$

Fig. 18-6 Temperatures measured by a constant-volume gas thermometer, with its bulb immersed in boiling water. For temperature calculations using Eq. 18-5, pressure p_3 was measured at the triple point of water. Three different gases in the thermometer bulb gave generally different results at different gas pressures, but as the amount of gas was decreased (decreasing p_3), all three curves converged to 373.125 K.

If we next put the bulb in a triple-point cell (Fig. 18-4), the temperature now being measured is

$$T_3 = Cp_3, \tag{18-4}$$

in which p_3 is the gas pressure now. Eliminating C between Eqs. 18-2 and 18-4 gives us the temperature as

$$T = T_3\left(\frac{p}{p_3}\right) = (273.16 \text{ K})\left(\frac{p}{p_3}\right) \quad \text{(provisional)}. \tag{18-5}$$

We still have a problem with this thermometer. If we use it to measure, say, the boiling point of water, we find that different gases in the bulb give slightly different results. However, as we use smaller and smaller amounts of gas to fill the bulb, the readings converge nicely to a single temperature, no matter what gas we use. Figure 18-6 shows this convergence for three gases.

Thus the recipe for measuring a temperature with a gas thermometer is

$$T = (273.16 \text{ K})\left(\lim_{\text{gas} \to 0} \frac{p}{p_3}\right). \tag{18-6}$$

The recipe instructs us to measure an unknown temperature T as follows: Fill the thermometer bulb with an arbitrary amount of *any* gas (for example, nitrogen) and measure p_3 (using a triple-point cell) and p, the gas pressure at the temperature being measured. (Keep the gas volume the same.) Calculate the ratio p/p_3. Then repeat both measurements with a smaller amount of gas in the bulb, and again calculate this ratio. Continue this way, using smaller and smaller amounts of gas, until you can extrapolate to the ratio p/p_3 that you would find if there were approximately no gas in the bulb. Calculate the temperature T by substituting that extrapolated ratio into Eq. 18-6. (The temperature is called the *ideal gas temperature*.)

18-5 The Celsius and Fahrenheit Scales

So far, we have discussed only the Kelvin scale, used in basic scientific work. In nearly all countries of the world, the Celsius scale (formerly called the centigrade scale) is the scale of choice for popular and commercial use and much scientific use. Celsius temperatures are measured in degrees, and the Celsius degree has the same size as the kelvin. However, the zero of the Celsius scale is shifted to a more convenient value than absolute zero. If T_C represents a Celsius temperature

Table 18-1		
Some Corresponding Temperatures		
Temperature	°C	°F
Boiling point of water[a]	100	212
Normal body temperature	37.0	98.6
Accepted comfort level	20	68
Freezing point of water[a]	0	32
Zero of Fahrenheit scale	≈ -18	0
Scales coincide	-40	-40

[a]Strictly, the boiling point of water on the Celsius scale is 99.975°C, and the freezing point is 0.00°C. Thus, there is slightly less than 100 C° between those two points.

Fig. 18-7 The Kelvin, Celsius, and Fahrenheit temperature scales compared.

and T a Kelvin temperature, then

$$T_C = T - 273.15°. \tag{18-7}$$

In expressing temperatures on the Celsius scale, the degree symbol is commonly used. Thus, we write 20.00°C for a Celsius reading but 293.15 K for a Kelvin reading.

The Fahrenheit scale, used in the United States, employs a smaller degree than the Celsius scale and a different zero of temperature. You can easily verify both these differences by examining an ordinary room thermometer on which both scales are marked. The relation between the Celsius and Fahrenheit scales is

$$T_F = \tfrac{9}{5}T_C + 32°, \tag{18-8}$$

where T_F is Fahrenheit temperature. Converting between these two scales can be done easily by remembering a few corresponding points, such as the freezing and boiling points of water (Table 18-1). Figure 18-7 compares the Kelvin, Celsius, and Fahrenheit scales.

We use the letters C and F to distinguish measurements and degrees on the two scales. Thus,

$$0°C = 32°F$$

means that 0° on the Celsius scale measures the same temperature as 32° on the Fahrenheit scale, whereas

$$5 \, C° = 9 \, F°$$

means that a temperature difference of 5 Celsius degrees (note the degree symbol appears *after* C) is equivalent to a temperature difference of 9 Fahrenheit degrees.

✔**CHECKPOINT 1**

The figure here shows three linear temperature scales with the freezing and boiling points of water indicated. (a) Rank the degrees on these scales by size, greatest first. (b) Rank the following temperatures, highest first: 50°X, 50°W, and 50°Y.

Sample Problem

Conversion between two temperature scales

Suppose you come across old scientific notes that describe a temperature scale called Z on which the boiling point of water is 65.0°Z and the freezing point is −14.0°Z. To what temperature on the Fahrenheit scale would a temperature of $T = -98.0°Z$ correspond? Assume that the Z scale is linear; that is, the size of a Z degree is the same everywhere on the Z scale.

Fig. 18-8 An unknown temperature scale compared with the Fahrenheit temperature scale.

KEY IDEA

A conversion factor between two (linear) temperature scales can be calculated by using two known (benchmark) temperatures, such as the boiling and freezing points of water. The number of degrees between the known temperatures on one scale is equivalent to the number of degrees between them on the other scale.

Calculations: We begin by relating the given temperature T to *either* known temperature on the Z scale. Since $T = -98.0°Z$ is closer to the freezing point (−14.0°Z) than to the boiling point (65.0°Z), we use the freezing point. Then we note that the T we seek is *below this point* by $-14.0°Z - (-98.0°Z) = 84.0 \text{ Z}°$ the (Fig. 18-8). (Read this difference as "84.0 Z degrees.")

Next, we set up a conversion factor between the Z and Fahrenheit scales to convert this difference. To do so, we use *both* known temperatures on the Z scale and the corre-

sponding temperatures on the Fahrenheit scale. On the Z scale, the difference between the boiling and freezing points is $65.0°Z - (-14.0°Z) = 79.0 \text{ Z}°$. On the Fahrenheit scale, it is $212°F - 32.0°F = 180 \text{ F}°$. Thus, a temperature difference of $79.0 \text{ Z}°$ is equivalent to a temperature difference of $180 \text{ F}°$ (Fig. 18-8), and we can use the ratio $(180 \text{ F}°)/(79.0 \text{ Z}°)$ as our conversion factor.

Now, since T is below the freezing point by $84.0 \text{ Z}°$, it must also be below the freezing point by

$$(84.0 \text{ Z}°)\frac{180 \text{ F}°}{79.0 \text{ Z}°} = 191 \text{ F}°.$$

Because the freezing point is at 32.0°F, this means that

$$T = 32.0°F - 191 \text{ F}° = -159°F. \qquad \text{(Answer)}$$

 PLUS Additional examples, video, and practice available at *WileyPLUS*

18-6 Thermal Expansion

You can often loosen a tight metal jar lid by holding it under a stream of hot water. Both the metal of the lid and the glass of the jar expand as the hot water adds energy to their atoms. (With the added energy, the atoms can move a bit farther from one another than usual, against the spring-like interatomic forces that hold every solid together.) However, because the atoms in the metal move farther apart than those in the glass, the lid expands more than the jar and thus is loosened.

Such **thermal expansion** of materials with an increase in temperature must be anticipated in many common situations. When a bridge is subject to large seasonal changes in temperature, for example, sections of the bridge are separated by *expansion slots* so that the sections have room to expand on hot days without the bridge buckling. When a dental cavity is filled, the filling material must have the same thermal expansion properties as the surrounding tooth; otherwise, consuming cold ice cream and then hot coffee would be very painful. When the Concorde aircraft (Fig. 18-9) was built, the design had to allow for the thermal expansion of the fuselage during supersonic flight because of frictional heating by the passing air.

The thermal expansion properties of some materials can be put to common use. Thermometers and thermostats may be based on the differences in expansion

Fig. 18-9 When a Concorde flew faster than the speed of sound, thermal expansion due to the rubbing by passing air increased the aircraft's length by about 12.5 cm. (The temperature increased to about 128°C at the aircraft nose and about 90°C at the tail, and cabin windows were noticeably warm to the touch.) (*Hugh Thomas/BWP Media/Getty Images News and Sport Services*)

Table 18-2			
Some Coefficients of Linear Expansion[a]			
Substance	$\alpha\,(10^{-6}/C°)$	Substance	$\alpha\,(10^{-6}/C°)$
Ice (at 0°C)	51	Steel	11
Lead	29	Glass (ordinary)	9
Aluminum	23	Glass (Pyrex)	3.2
Brass	19	Diamond	1.2
Copper	17	Invar[b]	0.7
Concrete	12	Fused quartz	0.5

[a]Room temperature values except for the listing for ice.

[b]This alloy was designed to have a low coefficient of expansion. The word is a shortened form of "invariable."

Fig. 18-10 (a) A bimetal strip, consisting of a strip of brass and a strip of steel welded together, at temperature T_0. (b) The strip bends as shown at temperatures above this reference temperature. Below the reference temperature the strip bends the other way. Many thermostats operate on this principle, making and breaking an electrical contact as the temperature rises and falls.

between the components of a *bimetal strip* (Fig. 18-10). Also, the familiar liquid-in-glass thermometers are based on the fact that liquids such as mercury and alcohol expand to a different (greater) extent than their glass containers.

Linear Expansion

If the temperature of a metal rod of length L is raised by an amount ΔT, its length is found to increase by an amount

$$\Delta L = L\alpha\,\Delta T, \tag{18-9}$$

in which α is a constant called the **coefficient of linear expansion.** The coefficient α has the unit "per degree" or "per kelvin" and depends on the material. Although α varies somewhat with temperature, for most practical purposes it can be taken as constant for a particular material. Table 18-2 shows some coefficients of linear expansion. Note that the unit C° there could be replaced with the unit K.

The thermal expansion of a solid is like photographic enlargement except it is in three dimensions. Figure 18-11b shows the (exaggerated) thermal expansion of a steel ruler. Equation 18-9 applies to every linear dimension of the ruler, including its edge, thickness, diagonals, and the diameters of the circle etched on it and the circular hole cut in it. If the disk cut from that hole originally fits snugly in the hole, it will continue to fit snugly if it undergoes the same temperature increase as the ruler.

Volume Expansion

If all dimensions of a solid expand with temperature, the volume of that solid must also expand. For liquids, volume expansion is the only meaningful expansion parameter. If the temperature of a solid or liquid whose volume is V is increased by an amount ΔT, the increase in volume is found to be

$$\Delta V = V\beta\,\Delta T, \tag{18-10}$$

where β is the **coefficient of volume expansion** of the solid or liquid. The coefficients of volume expansion and linear expansion for a solid are related by

$$\beta = 3\alpha. \tag{18-11}$$

The most common liquid, water, does not behave like other liquids. Above about 4°C, water expands as the temperature rises, as we would expect. Between 0 and about 4°C, however, water *contracts* with increasing temperature. Thus, at about 4°C, the density of water passes through a maximum. At all other temperatures, the density of water is less than this maximum value.

Fig. 18-11 The same steel ruler at two different temperatures. When it expands, the scale, the numbers, the thickness, and the diameters of the circle and circular hole are all increased by the same factor. (The expansion has been exaggerated for clarity.)

This behavior of water is the reason lakes freeze from the top down rather than from the bottom up. As water on the surface is cooled from, say, 10°C toward the freezing point, it becomes denser ("heavier") than lower water and sinks to the bottom. Below 4°C, however, further cooling makes the water then on the surface *less* dense ("lighter") than the lower water, so it stays on the surface until it freezes. Thus the surface freezes while the lower water is still liquid. If lakes froze from the bottom up, the ice so formed would tend not to melt completely during the summer, because it would be insulated by the water above. After a few years, many bodies of open water in the temperate zones of Earth would be frozen solid all year round—and aquatic life could not exist.

✔**CHECKPOINT 2**

The figure here shows four rectangular metal plates, with sides of L, $2L$, or $3L$. They are all made of the same material, and their temperature is to be increased by the same amount. Rank the plates according to the expected increase in (a) their vertical heights and (b) their areas, greatest first.

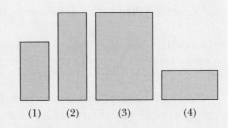

(1)   (2)   (3)   (4)

Sample Problem

Thermal expansion of a volume

On a hot day in Las Vegas, an oil trucker loaded 37 000 L of diesel fuel. He encountered cold weather on the way to Payson, Utah, where the temperature was 23.0 K lower than in Las Vegas, and where he delivered his entire load. How many liters did he deliver? The coefficient of volume expansion for diesel fuel is $9.50 \times 10^{-4}/\text{C}°$, and the coefficient of linear expansion for his steel truck tank is $11 \times 10^{-6}/\text{C}°$.

KEY IDEA

The volume of the diesel fuel depends directly on the temperature. Thus, because the temperature decreased, the volume of the fuel did also, as given by Eq. 18-10 ($\Delta V = V\beta \Delta T$).

Calculations: We find

$$\Delta V = (37\,000\ \text{L})(9.50 \times 10^{-4}/\text{C}°)(-23.0\ \text{K}) = -808\ \text{L}.$$

Thus, the amount delivered was

$$V_{\text{del}} = V + \Delta V = 37\,000\ \text{L} - 808\ \text{L}$$
$$= 36\,190\ \text{L}. \qquad \text{(Answer)}$$

Note that the thermal expansion of the steel tank has nothing to do with the problem. Question: Who paid for the "missing" diesel fuel?

 Additional examples, video, and practice available at *WileyPLUS*

18-7 Temperature and Heat

If you take a can of cola from the refrigerator and leave it on the kitchen table, its temperature will rise—rapidly at first but then more slowly—until the temperature of the cola equals that of the room (the two are then in thermal equilibrium). In the same way, the temperature of a cup of hot coffee, left sitting on the table, will fall until it also reaches room temperature.

In generalizing this situation, we describe the cola or the coffee as a *system* (with temperature T_S) and the relevant parts of the kitchen as the *environment* (with temperature T_E) of that system. Our observation is that if T_S is not equal to T_E, then T_S will change (T_E can also change some) until the two temperatures are equal and thus thermal equilibrium is reached.

Such a change in temperature is due to a change in the thermal energy of the system because of a transfer of energy between the system and the system's environment. (Recall that *thermal energy* is an internal energy that consists of the kinetic and potential energies associated with the random motions of the atoms, molecules, and other microscopic bodies within an object.) The transferred energy is called **heat** and is symbolized Q. Heat is *positive* when energy is transferred to a system's thermal energy from its environment (we say that heat is absorbed by the system). Heat is *negative* when energy is transferred from a system's thermal energy to its environment (we say that heat is released or lost by the system).

This transfer of energy is shown in Fig. 18-12. In the situation of Fig. 18-12a, in which $T_S > T_E$, energy is transferred from the system to the environment, so Q is negative. In Fig. 18-12b, in which $T_S = T_E$, there is no such transfer, Q is zero, and heat is neither released nor absorbed. In Fig. 18-12c, in which $T_S < T_E$, the transfer is to the system from the environment; so Q is positive.

We are led then to this definition of heat:

> Heat is the energy transferred between a system and its environment because of a temperature difference that exists between them.

Recall that energy can also be transferred between a system and its environment as *work W* via a force acting on a system. Heat and work, unlike temperature, pressure, and volume, are not intrinsic properties of a system. They have meaning only as they describe the transfer of energy into or out of a system. Similarly, the phrase "a $600 transfer" has meaning if it describes the transfer to or from an account, not what is in the account, because the account holds money, not a transfer. Here, it is proper to say: "During the last 3 min, 15 J of heat was transferred to the system from its environment" or "During the last minute, 12 J of work was done on the system by its environment." It is meaningless to say: "This system contains 450 J of heat" or "This system contains 385 J of work."

Before scientists realized that heat is transferred energy, heat was measured in terms of its ability to raise the temperature of water. Thus, the **calorie** (cal) was defined as the amount of heat that would raise the temperature of 1 g of water from 14.5°C to 15.5°C. In the British system, the corresponding unit of heat was

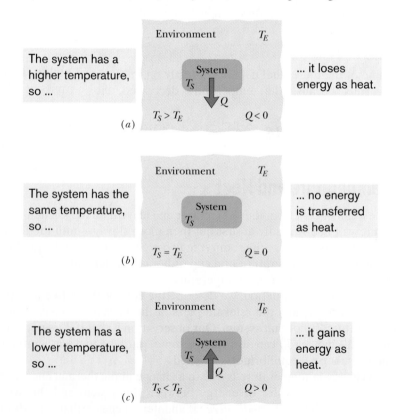

Fig. 18-12 If the temperature of a system exceeds that of its environment as in (a), heat Q is lost by the system to the environment until thermal equilibrium (b) is established. (c) If the temperature of the system is below that of the environment, heat is absorbed by the system until thermal equilibrium is established.

the **British thermal unit** (Btu), defined as the amount of heat that would raise the temperature of 1 lb of water from 63°F to 64°F.

In 1948, the scientific community decided that since heat (like work) is transferred energy, the SI unit for heat should be the one we use for energy—namely, the **joule.** The calorie is now defined to be 4.1868 J (exactly), with no reference to the heating of water. (The "calorie" used in nutrition, sometimes called the Calorie (Cal), is really a kilocalorie.) The relations among the various heat units are

$$1 \text{ cal} = 3.968 \times 10^{-3} \text{ Btu} = 4.1868 \text{ J.} \qquad (18\text{-}12)$$

18-8 The Absorption of Heat by Solids and Liquids

Heat Capacity

The **heat capacity** C of an object is the proportionality constant between the heat Q that the object absorbs or loses and the resulting temperature change ΔT of the object; that is,

$$Q = C \Delta T = C(T_f - T_i), \qquad (18\text{-}13)$$

in which T_i and T_f are the initial and final temperatures of the object. Heat capacity C has the unit of energy per degree or energy per kelvin. The heat capacity C of, say, a marble slab used in a bun warmer might be 179 cal/C°, which we can also write as 179 cal/K or as 749 J/K.

The word "capacity" in this context is really misleading in that it suggests analogy with the capacity of a bucket to hold water. *That analogy is false,* and you should not think of the object as "containing" heat or being limited in its ability to absorb heat. Heat transfer can proceed without limit as long as the necessary temperature difference is maintained. The object may, of course, melt or vaporize during the process.

Specific Heat

Two objects made of the same material—say, marble—will have heat capacities proportional to their masses. It is therefore convenient to define a "heat capacity per unit mass" or **specific heat** c that refers not to an object but to a unit mass of the material of which the object is made. Equation 18-13 then becomes

$$Q = cm \Delta T = cm(T_f - T_i). \qquad (18\text{-}14)$$

Through experiment we would find that although the heat capacity of a particular marble slab might be 179 cal/C° (or 749 J/K), the specific heat of marble itself (in that slab or in any other marble object) is 0.21 cal/g · C° (or 880 J/kg · K).

From the way the calorie and the British thermal unit were initially defined, the specific heat of water is

$$c = 1 \text{ cal/g} \cdot \text{C}° = 1 \text{ Btu/lb} \cdot \text{F}° = 4186.8 \text{ J/kg} \cdot \text{K.} \qquad (18\text{-}15)$$

Table 18-3 shows the specific heats of some substances at room temperature. Note that the value for water is relatively high. The specific heat of any substance actually depends somewhat on temperature, but the values in Table 18-3 apply reasonably well in a range of temperatures near room temperature.

Molar Specific Heat

In many instances the most convenient unit for specifying the amount of a substance is the mole (mol), where

$$1 \text{ mol} = 6.02 \times 10^{23} \text{ elementary units}$$

Table 18-3

Some Specific Heats and Molar Specific Heats at Room Temperature

Substance	Specific Heat $\dfrac{\text{cal}}{\text{g} \cdot \text{K}}$	Specific Heat $\dfrac{\text{J}}{\text{kg} \cdot \text{K}}$	Molar Specific Heat $\dfrac{\text{J}}{\text{mol} \cdot \text{K}}$
Elemental Solids			
Lead	0.0305	128	26.5
Tungsten	0.0321	134	24.8
Silver	0.0564	236	25.5
Copper	0.0923	386	24.5
Aluminum	0.215	900	24.4
Other Solids			
Brass	0.092	380	
Granite	0.19	790	
Glass	0.20	840	
Ice (−10°C)	0.530	2220	
Liquids			
Mercury	0.033	140	
Ethyl alcohol	0.58	2430	
Seawater	0.93	3900	
Water	1.00	4187	

✓ CHECKPOINT 3

A certain amount of heat Q will warm 1 g of material A by 3 C° and 1 g of material B by 4 C°. Which material has the greater specific heat?

of *any* substance. Thus 1 mol of aluminum means 6.02×10^{23} atoms (the atom is the elementary unit), and 1 mol of aluminum oxide means 6.02×10^{23} molecules (the molecule is the elementary unit of the compound).

When quantities are expressed in moles, specific heats must also involve moles (rather than a mass unit); they are then called **molar specific heats.** Table 18-3 shows the values for some elemental solids (each consisting of a single element) at room temperature.

An Important Point

In determining and then using the specific heat of any substance, we need to know the conditions under which energy is transferred as heat. For solids and liquids, we usually assume that the sample is under constant pressure (usually atmospheric) during the transfer. It is also conceivable that the sample is held at constant volume while the heat is absorbed. This means that thermal expansion of the sample is prevented by applying external pressure. For solids and liquids, this is very hard to arrange experimentally, but the effect can be calculated, and it turns out that the specific heats under constant pressure and constant volume for any solid or liquid differ usually by no more than a few percent. Gases, as you will see, have quite different values for their specific heats under constant-pressure conditions and under constant-volume conditions.

Heats of Transformation

When energy is absorbed as heat by a solid or liquid, the temperature of the sample does not necessarily rise. Instead, the sample may change from one *phase*, or *state*, to another. Matter can exist in three common states: In the *solid state*, the molecules of a sample are locked into a fairly rigid structure by their mutual attraction. In the *liquid state*, the molecules have more energy and move about more. They may form brief clusters, but the sample does not have a rigid structure and can flow or settle into a container. In the *gas*, or *vapor*, *state*, the molecules have even more energy, are free of one another, and can fill up the full volume of a container.

To *melt* a solid means to change it from the solid state to the liquid state. The process requires energy because the molecules of the solid must be freed from their rigid structure. Melting an ice cube to form liquid water is a common example. To *freeze* a liquid to form a solid is the reverse of melting and requires that energy be removed from the liquid, so that the molecules can settle into a rigid structure.

To *vaporize* a liquid means to change it from the liquid state to the vapor (gas) state. This process, like melting, requires energy because the molecules must be freed from their clusters. Boiling liquid water to transfer it to water vapor (or steam—a gas of individual water molecules) is a common example. *Condensing* a gas to form a liquid is the reverse of vaporizing; it requires that energy be removed from the gas, so that the molecules can cluster instead of flying away from one another.

The amount of energy per unit mass that must be transferred as heat when a sample completely undergoes a phase change is called the **heat of transformation** L. Thus, when a sample of mass m completely undergoes a phase change, the total energy transferred is

$$Q = Lm. \qquad (18\text{-}16)$$

When the phase change is from liquid to gas (then the sample must absorb heat) or from gas to liquid (then the sample must release heat), the heat of transformation is called the **heat of vaporization** L_V. For water at its normal boiling or condensation temperature,

$$L_V = 539 \text{ cal/g} = 40.7 \text{ kJ/mol} = 2256 \text{ kJ/kg}. \qquad (18\text{-}17)$$

When the phase change is from solid to liquid (then the sample must absorb heat) or from liquid to solid (then the sample must release heat), the heat of transformation is called the **heat of fusion** L_F. For water at its normal freezing or melting temperature,

$$L_F = 79.5 \text{ cal/g} = 6.01 \text{ kJ/mol} = 333 \text{ kJ/kg}. \qquad (18\text{-}18)$$

Table 18-4 shows the heats of transformation for some substances.

Table 18-4

Some Heats of Transformation

Substance	Melting		Boiling	
	Melting Point (K)	Heat of Fusion L_F (kJ/kg)	Boiling Point (K)	Heat of Vaporization L_V (kJ/kg)
Hydrogen	14.0	58.0	20.3	455
Oxygen	54.8	13.9	90.2	213
Mercury	234	11.4	630	296
Water	273	333	373	2256
Lead	601	23.2	2017	858
Silver	1235	105	2323	2336
Copper	1356	207	2868	4730

Sample Problem

Hot slug in water, coming to equilibrium

A copper slug whose mass m_c is 75 g is heated in a laboratory oven to a temperature T of 312°C. The slug is then dropped into a glass beaker containing a mass $m_w = 220$ g of water. The heat capacity C_b of the beaker is 45 cal/K. The initial temperature T_i of the water and the beaker is 12°C. Assuming that the slug, beaker, and water are an isolated system and the water does not vaporize, find the final temperature T_f of the system at thermal equilibrium.

KEY IDEAS

(1) Because the system is isolated, the system's total energy cannot change and only internal transfers of thermal energy can occur. (2) Because nothing in the system undergoes a phase change, the thermal energy transfers can only change the temperatures.

Calculations: To relate the transfers to the temperature changes, we can use Eqs. 18-13 and 18-14 to write

for the water: $Q_w = c_w m_w(T_f - T_i)$;   (18-19)

for the beaker: $Q_b = C_b(T_f - T_i)$;   (18-20)

for the copper: $Q_c = c_c m_c(T_f - T)$.   (18-21)

Because the total energy of the system cannot change, the sum of these three energy transfers is zero:

$$Q_w + Q_b + Q_c = 0. \qquad (18\text{-}22)$$

Substituting Eqs. 18-19 through 18-21 into Eq. 18-22 yields

$$c_w m_w(T_f - T_i) + C_b(T_f - T_i) + c_c m_c(T_f - T) = 0. \qquad (18\text{-}23)$$

Temperatures are contained in Eq. 18-23 only as differences. Thus, because the differences on the Celsius and Kelvin scales are identical, we can use either of those scales in this equation. Solving it for T_f, we obtain

$$T_f = \frac{c_c m_c T + C_b T_i + c_w m_w T_i}{c_w m_w + C_b + c_c m_c}.$$

Using Celsius temperatures and taking values for c_c and c_w from Table 18-3, we find the numerator to be

$(0.0923 \text{ cal/g·K})(75 \text{ g})(312°C) + (45 \text{ cal/K})(12°C)$
$\qquad + (1.00 \text{ cal/g·K})(220 \text{ g})(12°C) = 5339.8 \text{ cal},$

and the denominator to be

$(1.00 \text{ cal/g·K})(220 \text{ g}) + 45 \text{ cal/K}$
$\qquad + (0.0923 \text{ cal/g·K})(75 \text{ g}) = 271.9 \text{ cal/C°}.$

We then have

$$T_f = \frac{5339.8 \text{ cal}}{271.9 \text{ cal/C°}} = 19.6°C \approx 20°C. \qquad \text{(Answer)}$$

From the given data you can show that

$$Q_w \approx 1670 \text{ cal}, \qquad Q_b \approx 342 \text{ cal}, \qquad Q_c \approx -2020 \text{ cal}.$$

Apart from rounding errors, the algebraic sum of these three heat transfers is indeed zero, as Eq. 18-22 requires.

 Additional examples, video, and practice available at *WileyPLUS*

Heat to change temperature and state

(a) How much heat must be absorbed by ice of mass $m = 720$ g at $-10°C$ to take it to the liquid state at $15°C$?

The heating process is accomplished in three steps: (1) The ice cannot melt at a temperature below the freezing point—so initially, any energy transferred to the ice as heat can only increase the temperature of the ice, until $0°C$ is reached. (2) The temperature then cannot increase until all the ice melts—so any energy transferred to the ice as heat now can only change ice to liquid water, until all the ice melts. (3) Now the energy transferred to the liquid water as heat can only increase the temperature of the liquid water.

Warming the ice: The heat Q_1 needed to increase the temperature of the ice from the initial value $T_i = -10°C$ to a final value $T_f = 0°C$ (so that the ice can then melt) is given by Eq. 18-14 ($Q = cm\,\Delta T$). Using the specific heat of ice c_{ice} in Table 18-3 gives us

$$Q_1 = c_{ice}m(T_f - T_i)$$
$$= (2220 \text{ J/kg} \cdot \text{K})(0.720 \text{ kg})[0°C - (-10°C)]$$
$$= 15\,984 \text{ J} \approx 15.98 \text{ kJ}.$$

Melting the ice: The heat Q_2 needed to melt all the ice is given by Eq. 18-16 ($Q = Lm$). Here L is the heat of fusion L_F, with the value given in Eq. 18-18 and Table 18-4. We find

$$Q_2 = L_Fm = (333 \text{ kJ/kg})(0.720 \text{ kg}) \approx 239.8 \text{ kJ}.$$

Warming the liquid: The heat Q_3 needed to increase the temperature of the water from the initial value $T_i = 0°C$ to the final value $T_f = 15°C$ is given by Eq. 18-14 (with the specific heat of liquid water c_{liq}):

$$Q_3 = c_{liq}m(T_f - T_i)$$
$$= (4186.8 \text{ J/kg} \cdot \text{K})(0.720 \text{ kg})(15°C - 0°C)$$
$$= 45\,217 \text{ J} \approx 45.22 \text{ kJ}.$$

Total: The total required heat Q_{tot} is the sum of the amounts required in the three steps:

$$Q_{tot} = Q_1 + Q_2 + Q_3$$
$$= 15.98 \text{ kJ} + 239.8 \text{ kJ} + 45.22 \text{ kJ}$$
$$\approx 300 \text{ kJ}. \qquad \text{(Answer)}$$

Note that the heat required to melt the ice is much greater than the heat required to raise the temperature of either the ice or the liquid water.

(b) If we supply the ice with a total energy of only 210 kJ (as heat), what are the final state and temperature of the water?

From step 1, we know that 15.98 kJ is needed to raise the temperature of the ice to the melting point. The remaining heat Q_{rem} is then 210 kJ − 15.98 kJ, or about 194 kJ. From step 2, we can see that this amount of heat is insufficient to melt all the ice. Because the melting of the ice is incomplete, we must end up with a mixture of ice and liquid; the temperature of the mixture must be the freezing point, $0°C$.

Calculations: We can find the mass m of ice that is melted by the available energy Q_{rem} by using Eq. 18-16 with L_F:

$$m = \frac{Q_{rem}}{L_F} = \frac{194 \text{ kJ}}{333 \text{ kJ/kg}} = 0.583 \text{ kg} \approx 580 \text{ g}.$$

Thus, the mass of the ice that remains is 720 g − 580 g, or 140 g, and we have

$$580 \text{ g water} \quad \text{and} \quad 140 \text{ g ice,} \quad \text{at } 0°C. \qquad \text{(Answer)}$$

 Additional examples, video, and practice available at *WileyPLUS*

18-9 A Closer Look at Heat and Work

Here we look in some detail at how energy can be transferred as heat and work between a system and its environment. Let us take as our system a gas confined to a cylinder with a movable piston, as in Fig. 18-13. The upward force on the piston due to the pressure of the confined gas is equal to the weight of lead shot loaded onto the top of the piston. The walls of the cylinder are made of insulating material that does not allow any transfer of energy as heat. The bottom of the cylinder rests on a reservoir for thermal energy, a *thermal reservoir* (perhaps a hot plate) whose temperature T you can control by turning a knob.

The system (the gas) starts from an *initial state i*, described by a pressure p_i, a volume V_i, and a temperature T_i. You want to change the system to a *final state f*, described by a pressure p_f, a volume V_f, and a temperature T_f. The procedure by which you change the system from its initial state to its final state is called a *thermodynamic process*. During such a process, energy may be transferred into the system from the thermal reservoir (positive heat) or vice versa (negative heat). Also, work can be done by the system to raise the loaded piston (positive work) or lower it (negative work). We assume that all such changes occur slowly, with the result that the system is always in (approximate) thermal equilibrium (that is, every part of the system is always in thermal equilibrium with every other part).

Suppose that you remove a few lead shot from the piston of Fig. 18-13, allowing the gas to push the piston and remaining shot upward through a differential displacement $d\vec{s}$ with an upward force $\vec{F}$. Since the displacement is tiny, we can assume that $\vec{F}$ is constant during the displacement. Then $\vec{F}$ has a magnitude that is equal to pA, where p is the pressure of the gas and A is the face area of the piston. The differential work dW done by the gas during the displacement is

$$dW = \vec{F} \cdot d\vec{s} = (pA)(ds) = p(A\,ds)$$
$$= p\,dV, \tag{18-24}$$

in which dV is the differential change in the volume of the gas due to the movement of the piston. When you have removed enough shot to allow the gas to change its volume from V_i to V_f, the total work done by the gas is

$$W = \int dW = \int_{V_i}^{V_f} p\,dV. \tag{18-25}$$

During the volume change, the pressure and temperature may also change. To evaluate Eq. 18-25 directly, we would need to know how pressure varies with volume for the actual process by which the system changes from state i to state f.

There are actually many ways to take the gas from state i to state f. One way is shown in Fig. 18-14a, which is a plot of the pressure of the gas versus its volume and

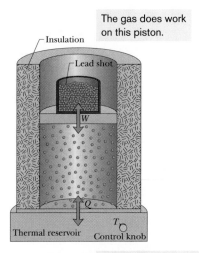

The gas does work on this piston.

Fig. 18-13 A gas is confined to a cylinder with a movable piston. Heat Q can be added to or withdrawn from the gas by regulating the temperature T of the adjustable thermal reservoir. Work W can be done by the gas by raising or lowering the piston.

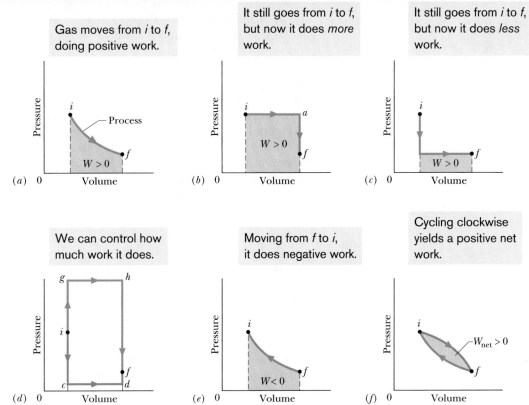

Fig. 18-14 (a) The shaded area represents the work W done by a system as it goes from an initial state i to a final state f. Work W is positive because the system's volume increases. (b) W is still positive, but now greater. (c) W is still positive, but now smaller. (d) W can be even smaller (path *icdf*) or larger (path *ighf*). (e) Here the system goes from state f to state i as the gas is compressed to less volume by an external force. The work W done *by* the system is now negative. (f) The net work W_{net} done by the system during a complete cycle is represented by the shaded area.

which is called a *p-V* diagram. In Fig. 18-14*a*, the curve indicates that the pressure decreases as the volume increases. The integral in Eq. 18-25 (and thus the work *W* done by the gas) is represented by the shaded area under the curve between points *i* and *f*. Regardless of what exactly we do to take the gas along the curve, that work is positive, due to the fact that the gas increases its volume by forcing the piston upward.

Another way to get from state *i* to state *f* is shown in Fig. 18-14*b*. There the change takes place in two steps—the first from state *i* to state *a*, and the second from state *a* to state *f*.

Step *ia* of this process is carried out at constant pressure, which means that you leave undisturbed the lead shot that ride on top of the piston in Fig. 18-13. You cause the volume to increase (from V_i to V_f) by slowly turning up the temperature control knob, raising the temperature of the gas to some higher value T_a. (Increasing the temperature increases the force from the gas on the piston, moving it upward.) During this step, positive work is done by the expanding gas (to lift the loaded piston) and heat is absorbed by the system from the thermal reservoir (in response to the arbitrarily small temperature differences that you create as you turn up the temperature). This heat is positive because it is added to the system.

Step *af* of the process of Fig. 18-14*b* is carried out at constant volume, so you must wedge the piston, preventing it from moving. Then as you use the control knob to decrease the temperature, you find that the pressure drops from p_a to its final value p_f. During this step, heat is lost by the system to the thermal reservoir.

For the overall process *iaf*, the work *W*, which is positive and is carried out only during step *ia*, is represented by the shaded area under the curve. Energy is transferred as heat during both steps *ia* and *af*, with a net energy transfer *Q*.

Figure 18-14*c* shows a process in which the previous two steps are carried out in reverse order. The work *W* in this case is smaller than for Fig. 18-14*b*, as is the net heat absorbed. Figure 18-14*d* suggests that you can make the work done by the gas as small as you want (by following a path like *icdf*) or as large as you want (by following a path like *ighf*).

To sum up: A system can be taken from a given initial state to a given final state by an infinite number of processes. Heat may or may not be involved, and in general, the work *W* and the heat *Q* will have different values for different processes. We say that heat and work are *path-dependent* quantities.

Figure 18-14*e* shows an example in which negative work is done by a system as some external force compresses the system, reducing its volume. The absolute value of the work done is still equal to the area beneath the curve, but because the gas is *compressed,* the work done by the gas is negative.

Figure 18-14*f* shows a *thermodynamic cycle* in which the system is taken from some initial state *i* to some other state *f* and then back to *i*. The net work done by the system during the cycle is the sum of the *positive* work done during the expansion and the *negative* work done during the compression. In Fig. 18-14*f*, the net work is positive because the area under the expansion curve (*i* to *f*) is greater than the area under the compression curve (*f* to *i*).

✔ **CHECKPOINT 4**

The *p-V* diagram here shows six curved paths (connected by vertical paths) that can be followed by a gas. Which two of the curved paths should be part of a closed cycle (those curved paths plus connecting vertical paths) if the net work done by the gas during the cycle is to be at its maximum positive value?

18-10 The First Law of Thermodynamics

You have just seen that when a system changes from a given initial state to a given final state, both the work W and the heat Q depend on the nature of the process. Experimentally, however, we find a surprising thing. *The quantity $Q - W$ is the same for all processes.* It depends only on the initial and final states and does not depend at all on how the system gets from one to the other. All other combinations of Q and W, including Q alone, W alone, $Q + W$, and $Q - 2W$, are *path dependent;* only the quantity $Q - W$ is not.

The quantity $Q - W$ must represent a change in some intrinsic property of the system. We call this property the *internal energy E_{int}* and we write

$$\Delta E_{int} = E_{int,f} - E_{int,i} = Q - W \qquad \text{(first law)}. \qquad (18\text{-}26)$$

Equation 18-26 is the **first law of thermodynamics.** If the thermodynamic system undergoes only a differential change, we can write the first law as*

$$dE_{int} = dQ - dW \qquad \text{(first law)}. \qquad (18\text{-}27)$$

⟶ The internal energy E_{int} of a system tends to increase if energy is added as heat Q and tends to decrease if energy is lost as work W done by the system.

In Chapter 8, we discussed the principle of energy conservation as it applies to isolated systems—that is, to systems in which no energy enters or leaves the system. The first law of thermodynamics is an extension of that principle to systems that are *not* isolated. In such cases, energy may be transferred into or out of the system as either work W or heat Q. In our statement of the first law of thermodynamics above, we assume that there are no changes in the kinetic energy or the potential energy of the system as a whole; that is, $\Delta K = \Delta U = 0$.

Before this chapter, the term *work* and the symbol W always meant the work done *on* a system. However, starting with Eq. 18-24 and continuing through the next two chapters about thermodynamics, we focus on the work done *by* a system, such as the gas in Fig. 18-13.

The work done *on* a system is always the negative of the work done *by* the system, so if we rewrite Eq. 18-26 in terms of the work W_{on} done *on* the system, we have $\Delta E_{int} = Q + W_{on}$. This tells us the following: The internal energy of a system tends to increase if heat is absorbed by the system or if positive work is done *on* the system. Conversely, the internal energy tends to decrease if heat is lost by the system or if negative work is done *on* the system.

✓ CHECKPOINT 5

The figure here shows four paths on a p-V diagram along which a gas can be taken from state i to state f. Rank the paths according to (a) the change ΔE_{int} in the internal energy of the gas, (b) the work W done by the gas, and (c) the magnitude of the energy transferred as heat Q between the gas and its environment, greatest first.

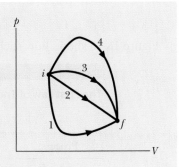

*Here dQ and dW, unlike dE_{int}, are not true differentials; that is, there are no such functions as $Q(p,V)$ and $W(p,V)$ that depend only on the state of the system. The quantities dQ and dW are called *inexact differentials* and are usually represented by the symbols $đQ$ and $đW$. For our purposes, we can treat them simply as infinitesimally small energy transfers.

We slowly remove lead shot, allowing an expansion without any heat transfer.

Fig. 18-15 An adiabatic expansion can be carried out by slowly removing lead shot from the top of the piston. Adding lead shot reverses the process at any stage.

Fig. 18-16 The initial stage of a free-expansion process. After the stopcock is opened, the gas fills both chambers and eventually reaches an equilibrium state.

18-11 Some Special Cases of the First Law of Thermodynamics

Here are four thermodynamic processes as summarized in Table 18-5.

1. *Adiabatic processes.* An adiabatic process is one that occurs so rapidly or occurs in a system that is so well insulated that *no transfer of energy as heat* occurs between the system and its environment. Putting $Q = 0$ in the first law (Eq. 18-26) yields

$$\Delta E_{int} = -W \qquad \text{(adiabatic process).} \qquad (18\text{-}28)$$

This tells us that if work is done *by* the system (that is, if W is positive), the internal energy of the system decreases by the amount of work. Conversely, if work is done *on* the system (that is, if W is negative), the internal energy of the system increases by that amount.

Figure 18-15 shows an idealized adiabatic process. Heat cannot enter or leave the system because of the insulation. Thus, the only way energy can be transferred between the system and its environment is by work. If we remove shot from the piston and allow the gas to expand, the work done by the system (the gas) is positive and the internal energy of the gas decreases. If, instead, we add shot and compress the gas, the work done by the system is negative and the internal energy of the gas increases.

2. *Constant-volume processes.* If the volume of a system (such as a gas) is held constant, that system can do no work. Putting $W = 0$ in the first law (Eq. 18-26) yields

$$\Delta E_{int} = Q \qquad \text{(constant-volume process).} \qquad (18\text{-}29)$$

Thus, if heat is absorbed by a system (that is, if Q is positive), the internal energy of the system increases. Conversely, if heat is lost during the process (that is, if Q is negative), the internal energy of the system must decrease.

3. *Cyclical processes.* There are processes in which, after certain interchanges of heat and work, the system is restored to its initial state. In that case, no intrinsic property of the system—including its internal energy—can possibly change. Putting $\Delta E_{int} = 0$ in the first law (Eq. 18-26) yields

$$Q = W \qquad \text{(cyclical process).} \qquad (18\text{-}30)$$

Thus, the net work done during the process must exactly equal the net amount of energy transferred as heat; the store of internal energy of the system remains unchanged. Cyclical processes form a closed loop on a *p-V* plot, as shown in Fig. 18-14*f*. We discuss such processes in detail in Chapter 20.

4. *Free expansions.* These are adiabatic processes in which no transfer of heat occurs between the system and its environment and no work is done on or by the system. Thus, $Q = W = 0$, and the first law requires that

$$\Delta E_{int} = 0 \qquad \text{(free expansion).} \qquad (18\text{-}31)$$

Figure 18-16 shows how such an expansion can be carried out. A gas, which is

Table 18-5

The First Law of Thermodynamics: Four Special Cases

The Law: $\Delta E_{int} = Q - W$ (Eq. 18-26)

Process	Restriction	Consequence
Adiabatic	$Q = 0$	$\Delta E_{int} = -W$
Constant volume	$W = 0$	$\Delta E_{int} = Q$
Closed cycle	$\Delta E_{int} = 0$	$Q = W$
Free expansion	$Q = W = 0$	$\Delta E_{int} = 0$

in thermal equilibrium within itself, is initially confined by a closed stopcock to one half of an insulated double chamber; the other half is evacuated. The stopcock is opened, and the gas expands freely to fill both halves of the chamber. No heat is transferred to or from the gas because of the insulation. No work is done by the gas because it rushes into a vacuum and thus does not meet any pressure.

A free expansion differs from all other processes we have considered because it cannot be done slowly and in a controlled way. As a result, at any given instant during the sudden expansion, the gas is not in thermal equilibrium and its pressure is not uniform. Thus, although we can plot the initial and final states on a p-V diagram, we cannot plot the expansion itself.

CHECKPOINT 6

For one complete cycle as shown in the p-V diagram here, are (a) ΔE_{int} for the gas and (b) the net energy transferred as heat Q positive, negative, or zero?

Sample Problem

First law of thermodynamics: work, heat, internal energy change

Let 1.00 kg of liquid water at 100°C be converted to steam at 100°C by boiling at standard atmospheric pressure (which is 1.00 atm or 1.01×10^5 Pa) in the arrangement of Fig. 18-17. The volume of that water changes from an initial value of 1.00×10^{-3} m³ as a liquid to 1.671 m³ as steam.

(a) How much work is done by the system during this process?

KEY IDEAS

(1) The system must do positive work because the volume increases. (2) We calculate the work W done by integrating the pressure with respect to the volume (Eq. 18-25).

Calculation: Because here the pressure is constant at 1.01×10^5 Pa, we can take p outside the integral. Thus,

$$W = \int_{V_i}^{V_f} p \, dV = p \int_{V_i}^{V_f} dV = p(V_f - V_i)$$

$$= (1.01 \times 10^5 \text{ Pa})(1.671 \text{ m}^3 - 1.00 \times 10^{-3} \text{ m}^3)$$

$$= 1.69 \times 10^5 \text{ J} = 169 \text{ kJ}. \qquad \text{(Answer)}$$

(b) How much energy is transferred as heat during the process?

KEY IDEA

Because the heat causes only a phase change and not a change in temperature, it is given fully by Eq. 18-16 ($Q = Lm$).

Calculation: Because the change is from liquid to gaseous phase, L is the heat of vaporization L_V, with the value given in Eq. 18-17 and Table 18-4. We find

$$Q = L_V m = (2256 \text{ kJ/kg})(1.00 \text{ kg})$$

$$= 2256 \text{ kJ} \approx 2260 \text{ kJ}. \qquad \text{(Answer)}$$

(c) What is the change in the system's internal energy during the process?

KEY IDEA

The change in the system's internal energy is related to the heat (here, this is energy transferred into the system) and the work (here, this is energy transferred out of the system) by the first law of thermodynamics (Eq. 18-26).

Calculation: We write the first law as

$$\Delta E_{int} = Q - W = 2256 \text{ kJ} - 169 \text{ kJ}$$

$$\approx 2090 \text{ kJ} = 2.09 \text{ MJ}. \qquad \text{(Answer)}$$

This quantity is positive, indicating that the internal energy of the system has increased during the boiling process. This energy goes into separating the H_2O molecules, which strongly attract one another in the liquid state. We see that, when water is boiled, about 7.5% ($= 169$ kJ/2260 kJ) of the heat goes into the work of pushing back the atmosphere. The rest of the heat goes into the system's internal energy.

Fig. 18-17 Water boiling at constant pressure. Energy is transferred from the thermal reservoir as heat until the liquid water has changed completely into steam. Work is done by the expanding gas as it lifts the loaded piston.

 Additional examples, video, and practice available at *WileyPLUS*

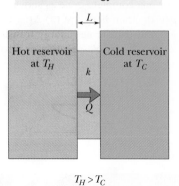

We assume a steady transfer of energy as heat.

$T_H > T_C$

Fig. 18-18 Thermal conduction. Energy is transferred as heat from a reservoir at temperature T_H to a cooler reservoir at temperature T_C through a conducting slab of thickness L and thermal conductivity k.

Table 18-6	
Some Thermal Conductivities	
Substance	k (W/m · K)
Metals	
Stainless steel	14
Lead	35
Iron	67
Brass	109
Aluminum	235
Copper	401
Silver	428
Gases	
Air (dry)	0.026
Helium	0.15
Hydrogen	0.18
Building Materials	
Polyurethane foam	0.024
Rock wool	0.043
Fiberglass	0.048
White pine	0.11
Window glass	1.0

18-12 Heat Transfer Mechanisms

We have discussed the transfer of energy as heat between a system and its environment, but we have not yet described how that transfer takes place. There are three transfer mechanisms: conduction, convection, and radiation.

Conduction

If you leave the end of a metal poker in a fire for enough time, its handle will get hot. Energy is transferred from the fire to the handle by (thermal) **conduction** along the length of the poker. The vibration amplitudes of the atoms and electrons of the metal at the fire end of the poker become relatively large because of the high temperature of their environment. These increased vibrational amplitudes, and thus the associated energy, are passed along the poker, from atom to atom, during collisions between adjacent atoms. In this way, a region of rising temperature extends itself along the poker to the handle.

Consider a slab of face area A and thickness L, whose faces are maintained at temperatures T_H and T_C by a hot reservoir and a cold reservoir, as in Fig. 18-18. Let Q be the energy that is transferred as heat through the slab, from its hot face to its cold face, in time t. Experiment shows that the *conduction rate* P_{cond} (the amount of energy transferred per unit time) is

$$P_{cond} = \frac{Q}{t} = kA\frac{T_H - T_C}{L}, \qquad (18\text{-}32)$$

in which k, called the *thermal conductivity,* is a constant that depends on the material of which the slab is made. A material that readily transfers energy by conduction is a *good thermal conductor* and has a high value of k. Table 18-6 gives the thermal conductivities of some common metals, gases, and building materials.

Thermal Resistance to Conduction (R-Value)

If you are interested in insulating your house or in keeping cola cans cold on a picnic, you are more concerned with poor heat conductors than with good ones. For this reason, the concept of *thermal resistance R* has been introduced into engineering practice. The *R*-value of a slab of thickness L is defined as

$$R = \frac{L}{k}. \qquad (18\text{-}33)$$

The lower the thermal conductivity of the material of which a slab is made, the higher the *R*-value of the slab; so something that has a high *R*-value is a *poor thermal conductor* and thus a *good thermal insulator.*

Note that R is a property attributed to a slab of a specified thickness, not to a material. The commonly used unit for R (which, in the United States at least, is almost never stated) is the square foot–Fahrenheit degree–hour per British thermal unit (ft$^2 \cdot$ F$° \cdot$ h/Btu). (Now you know why the unit is rarely stated.)

Conduction Through a Composite Slab

Figure 18-19 shows a composite slab, consisting of two materials having different thicknesses L_1 and L_2 and different thermal conductivities k_1 and k_2. The temperatures of the outer surfaces of the slab are T_H and T_C. Each face of the slab has area A. Let us derive an expression for the conduction rate through the slab under the assumption that the transfer is a *steady-state* process; that is, the temperatures everywhere in the slab and the rate of energy transfer do not change with time.

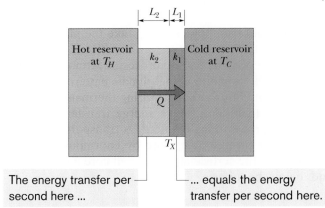

Fig. 18-19 Heat is transferred at a steady rate through a composite slab made up of two different materials with different thicknesses and different thermal conductivities. The steady-state temperature at the interface of the two materials is T_X.

In the steady state, the conduction rates through the two materials must be equal. This is the same as saying that the energy transferred through one material in a certain time must be equal to that transferred through the other material in the same time. If this were not true, temperatures in the slab would be changing and we would not have a steady-state situation. Letting T_X be the temperature of the interface between the two materials, we can now use Eq. 18-32 to write

$$P_{\text{cond}} = \frac{k_2 A(T_H - T_X)}{L_2} = \frac{k_1 A(T_X - T_C)}{L_1}. \tag{18-34}$$

Solving Eq. 18-34 for T_X yields, after a little algebra,

$$T_X = \frac{k_1 L_2 T_C + k_2 L_1 T_H}{k_1 L_2 + k_2 L_1}. \tag{18-35}$$

Substituting this expression for T_X into either equality of Eq. 18-34 yields

$$P_{\text{cond}} = \frac{A(T_H - T_C)}{L_1/k_1 + L_2/k_2}. \tag{18-36}$$

We can extend Eq. 18-36 to apply to any number n of materials making up a slab:

$$P_{\text{cond}} = \frac{A(T_H - T_C)}{\Sigma (L/k)}. \tag{18-37}$$

The summation sign in the denominator tells us to add the values of L/k for all the materials.

CHECKPOINT 7

The figure shows the face and interface temperatures of a composite slab consisting of four materials, of identical thicknesses, through which the heat transfer is steady. Rank the materials according to their thermal conductivities, greatest first.

Convection

When you look at the flame of a candle or a match, you are watching thermal energy being transported upward by **convection.** Such energy transfer occurs when a fluid, such as air or water, comes in contact with an object whose temperature is higher than that of the fluid. The temperature of the part of the fluid that is in contact with the hot object increases, and (in most cases) that fluid expands and thus becomes less dense. Because this expanded fluid is now lighter than the surrounding cooler fluid, buoyant forces cause it to rise. Some of the

Fig. 18-20 A false-color thermogram reveals the rate at which energy is radiated by a cat. The rate is color-coded, with white and red indicating the greatest radiation rate. The nose is cool. *(Edward Kinsman/Photo Researchers)*

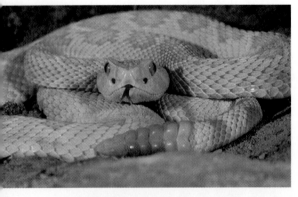

Fig. 18-21 A rattlesnake's face has thermal radiation detectors, allowing the snake to strike at an animal even in complete darkness. *(David A. Northcott/Corbis Images)*

surrounding cooler fluid then flows so as to take the place of the rising warmer fluid, and the process can then continue.

Convection is part of many natural processes. Atmospheric convection plays a fundamental role in determining global climate patterns and daily weather variations. Glider pilots and birds alike seek rising thermals (convection currents of warm air) that keep them aloft. Huge energy transfers take place within the oceans by the same process. Finally, energy is transported to the surface of the Sun from the nuclear furnace at its core by enormous cells of convection, in which hot gas rises to the surface along the cell core and cooler gas around the core descends below the surface.

Radiation

The third method by which an object and its environment can exchange energy as heat is via electromagnetic waves (visible light is one kind of electromagnetic wave). Energy transferred in this way is often called **thermal radiation** to distinguish it from electromagnetic *signals* (as in, say, television broadcasts) and from nuclear radiation (energy and particles emitted by nuclei). (To "radiate" generally means to emit.) When you stand in front of a big fire, you are warmed by absorbing thermal radiation from the fire; that is, your thermal energy increases as the fire's thermal energy decreases. No medium is required for heat transfer via radiation—the radiation can travel through vacuum from, say, the Sun to you.

The rate P_{rad} at which an object emits energy via electromagnetic radiation depends on the object's surface area A and the temperature T of that area in kelvins and is given by

$$P_{rad} = \sigma \varepsilon A T^4. \qquad (18\text{-}38)$$

Here $\sigma = 5.6704 \times 10^{-8}$ W/m$^2 \cdot$ K^4 is called the *Stefan–Boltzmann constant* after Josef Stefan (who discovered Eq. 18-38 experimentally in 1879) and Ludwig Boltzmann (who derived it theoretically soon after). The symbol ε represents the *emissivity* of the object's surface, which has a value between 0 and 1, depending on the composition of the surface. A surface with the maximum emissivity of 1.0 is said to be a *blackbody radiator,* but such a surface is an ideal limit and does not occur in nature. Note again that the temperature in Eq. 18-38 must be in kelvins so that a temperature of absolute zero corresponds to no radiation. Note also that every object whose temperature is above 0 K—including you—emits thermal radiation. (See Fig. 18-20.)

The rate P_{abs} at which an object absorbs energy via thermal radiation from its environment, which we take to be at uniform temperature T_{env} (in kelvins), is

$$P_{abs} = \sigma \varepsilon A T_{env}^4. \qquad (18\text{-}39)$$

The emissivity ε in Eq. 18-39 is the same as that in Eq. 18-38. An idealized blackbody radiator, with $\varepsilon = 1$, will absorb all the radiated energy it intercepts (rather than sending a portion back away from itself through reflection or scattering).

Because an object will radiate energy to the environment while it absorbs energy from the environment, the object's net rate P_{net} of energy exchange due to thermal radiation is

$$P_{net} = P_{abs} - P_{rad} = \sigma \varepsilon A (T_{env}^4 - T^4). \qquad (18\text{-}40)$$

P_{net} is positive if net energy is being absorbed via radiation and negative if it is being lost via radiation.

Thermal radiation is involved in the numerous medical cases of a *dead* rattlesnake striking a hand reaching toward it. Pits between each eye and nostril of a rattlesnake (Fig. 18-21) serve as sensors of thermal radiation. When, say, a mouse moves close to a rattlesnake's head, the thermal radiation from the mouse triggers these sensors, causing a reflex action in which the snake strikes the mouse with its fangs and injects its venom. The thermal radiation from a reaching hand can cause the same reflex action even if the snake has been dead for as long as 30 min because the snake's nervous system continues to function. As one snake expert advised, if you must remove a recently killed rattlesnake, use a long stick rather than your hand.

Sample Problem

Thermal conduction through a layered wall

Figure 18-22 shows the cross section of a wall made of white pine of thickness L_a and brick of thickness L_d ($= 2.0L_a$), sandwiching two layers of unknown material with identical thicknesses and thermal conductivities. The thermal conductivity of the pine is k_a and that of the brick is k_d ($= 5.0k_a$). The face area A of the wall is unknown. Thermal conduction through the wall has reached the steady state; the only known interface temperatures are $T_1 = 25°C$, $T_2 = 20°C$, and $T_5 = -10°C$. What is interface temperature T_4?

KEY IDEAS

(1) Temperature T_4 helps determine the rate P_d at which energy is conducted through the brick, as given by Eq. 18-32. However, we lack enough data to solve Eq. 18-32 for T_4. (2) Because the conduction is steady, the conduction rate P_d through the brick must equal the conduction rate P_a through the pine. That gets us going.

Calculations: From Eq. 18-32 and Fig. 18-22, we can write

$$P_a = k_a A \frac{T_1 - T_2}{L_a} \quad \text{and} \quad P_d = k_d A \frac{T_4 - T_5}{L_d}.$$

Setting $P_a = P_d$ and solving for T_4 yield

$$T_4 = \frac{k_a L_d}{k_d L_a}(T_1 - T_2) + T_5.$$

Letting $L_d = 2.0L_a$ and $k_d = 5.0k_a$, and inserting the known temperatures, we find

$$T_4 = \frac{k_a(2.0L_a)}{(5.0k_a)L_a}(25°C - 20°C) + (-10°C)$$
$$= -8.0°C. \quad \text{(Answer)}$$

Fig. 18-22   Steady-state heat transfer through a wall.

The energy transfer per second is the same in each layer.

 Additional examples, video, and practice available at *WileyPLUS*

REVIEW & SUMMARY

Temperature; Thermometers   Temperature is an SI base quantity related to our sense of hot and cold. It is measured with a thermometer, which contains a working substance with a measurable property, such as length or pressure, that changes in a regular way as the substance becomes hotter or colder.

Zeroth Law of Thermodynamics   When a thermometer and some other object are placed in contact with each other, they eventually reach thermal equilibrium. The reading of the thermometer is then taken to be the temperature of the other object. The process provides consistent and useful temperature measurements because of the **zeroth law of thermodynamics:** If bodies A and B are each in thermal equilibrium with a third body C (the thermometer), then A and B are in thermal equilibrium with each other.

The Kelvin Temperature Scale   In the SI system, temperature is measured on the **Kelvin scale,** which is based on the *triple point* of water (273.16 K). Other temperatures are then defined by use of a *constant-volume gas thermometer,* in which a sample of gas is maintained at constant volume so its pressure is proportional to its temperature. We define the *temperature T* as measured with a gas thermometer to be

$$T = (273.16 \text{ K}) \left(\lim_{\text{gas} \to 0} \frac{p}{p_3} \right). \tag{18-6}$$

Here T is in kelvins, and p_3 and p are the pressures of the gas at 273.16 K and the measured temperature, respectively.

Celsius and Fahrenheit Scales The Celsius temperature scale is defined by

$$T_C = T - 273.15°, \tag{18-7}$$

with T in kelvins. The Fahrenheit temperature scale is defined by

$$T_F = \tfrac{9}{5} T_C + 32°. \tag{18-8}$$

Thermal Expansion All objects change size with changes in temperature. For a temperature change ΔT, a change ΔL in any linear dimension L is given by

$$\Delta L = L \alpha \Delta T, \tag{18-9}$$

in which α is the **coefficient of linear expansion.** The change ΔV in the volume V of a solid or liquid is

$$\Delta V = V \beta \Delta T. \tag{18-10}$$

Here $\beta = 3\alpha$ is the material's **coefficient of volume expansion.**

Heat Heat Q is energy that is transferred between a system and its environment because of a temperature difference between them. It can be measured in **joules** (J), **calories** (cal), **kilocalories** (Cal or kcal), or **British thermal units** (Btu), with

$$1 \text{ cal} = 3.968 \times 10^{-3} \text{ Btu} = 4.1868 \text{ J}. \tag{18-12}$$

Heat Capacity and Specific Heat If heat Q is absorbed by an object, the object's temperature change $T_f - T_i$ is related to Q by

$$Q = C(T_f - T_i), \tag{18-13}$$

in which C is the **heat capacity** of the object. If the object has mass m, then

$$Q = cm(T_f - T_i), \tag{18-14}$$

where c is the **specific heat** of the material making up the object. The **molar specific heat** of a material is the heat capacity per mole, which means per 6.02×10^{23} elementary units of the material.

Heat of Transformation Heat absorbed by a material may change the material's physical state—for example, from solid to liquid or from liquid to gas. The amount of energy required per unit mass to change the state (but not the temperature) of a particular material is its **heat of transformation** L. Thus,

$$Q = Lm. \tag{18-16}$$

The **heat of vaporization** L_V is the amount of energy per unit mass that must be added to vaporize a liquid or that must be removed to condense a gas. The **heat of fusion** L_F is the amount of energy per unit mass that must be added to melt a solid or that must be removed to freeze a liquid.

Work Associated with Volume Change A gas may exchange energy with its surroundings through work. The amount of work W done *by* a gas as it expands or contracts from an initial volume V_i to a final volume V_f is given by

$$W = \int dW = \int_{V_i}^{V_f} p \, dV. \tag{18-25}$$

The integration is necessary because the pressure p may vary during the volume change.

First Law of Thermodynamics The principle of conservation of energy for a thermodynamic process is expressed in the **first law of thermodynamics,** which may assume either of the forms

$$\Delta E_{\text{int}} = E_{\text{int},f} - E_{\text{int},i} = Q - W \quad \text{(first law)} \tag{18-26}$$

or

$$dE_{\text{int}} = dQ - dW \quad \text{(first law)}. \tag{18-27}$$

E_{int} represents the internal energy of the material, which depends only on the material's state (temperature, pressure, and volume). Q represents the energy exchanged as heat between the system and its surroundings; Q is positive if the system absorbs heat and negative if the system loses heat. W is the work done *by* the system; W is positive if the system expands against an external force from the surroundings and negative if the system contracts because of an external force. Q and W are path dependent; ΔE_{int} *is path independent.*

Applications of the First Law The first law of thermodynamics finds application in several special cases:

$$\text{adiabatic processes:} \quad Q = 0, \quad \Delta E_{\text{int}} = -W$$
$$\text{constant-volume processes:} \quad W = 0, \quad \Delta E_{\text{int}} = Q$$
$$\text{cyclical processes:} \quad \Delta E_{\text{int}} = 0, \quad Q = W$$
$$\text{free expansions:} \quad Q = W = \Delta E_{\text{int}} = 0$$

Conduction, Convection, and Radiation The rate P_{cond} at which energy is *conducted* through a slab for which one face is maintained at the higher temperature T_H and the other face is maintained at the lower temperature T_C is

$$P_{\text{cond}} = \frac{Q}{t} = kA \frac{T_H - T_C}{L} \tag{18-32}$$

Here each face of the slab has area A, the length of the slab (the distance between the faces) is L, and k is the thermal conductivity of the material.

Convection occurs when temperature differences cause an energy transfer by motion within a fluid.

Radiation is an energy transfer via the emission of electromagnetic energy. The rate P_{rad} at which an object emits energy via thermal radiation is

$$P_{\text{rad}} = \sigma \varepsilon A T^4, \tag{18-38}$$

where σ ($= 5.6704 \times 10^{-8}$ W/m$^2 \cdot$K^4) is the Stefan–Boltzmann constant, ε is the emissivity of the object's surface, A is its surface area, and T is its surface temperature (in kelvins). The rate P_{abs} at which an object absorbs energy via thermal radiation from its environment, which is at the uniform temperature T_{env} (in kelvins), is

$$P_{\text{abs}} = \sigma \varepsilon A T_{\text{env}}^4. \tag{18-39}$$

QUESTIONS

1 The initial length L, change in temperature ΔT, and change in length ΔL of four rods are given in the following table. Rank the rods according to their coefficients of thermal expansion, greatest first.

Rod	L (m)	ΔT (C°)	ΔL (m)
a	2	10	4×10^{-4}
b	1	20	4×10^{-4}
c	2	10	8×10^{-4}
d	4	5	4×10^{-4}

2 Figure 18-23 shows three linear temperature scales, with the freezing and boiling points of water indicated. Rank the three scales according to the size of one degree on them, greatest first.

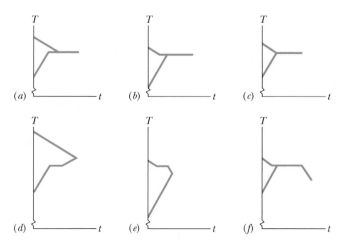

Fig. 18-23 Question 2.

3 Materials A, B, and C are solids that are at their melting temperatures. Material A requires 200 J to melt 4 kg, material B requires 300 J to melt 5 kg, and material C requires 300 J to melt 6 kg. Rank the materials according to their heats of fusion, greatest first.

4 A sample A of liquid water and a sample B of ice, of identical mass, are placed in a thermally insulated container and allowed to come to thermal equilibrium. Figure 18-24a is a sketch of the temperature T of the samples versus time t. (a) Is the equilibrium temperature above, below, or at the freezing point of water? (b) In reaching equilibrium, does the liquid partly freeze, fully freeze, or undergo no freezing? (c) Does the ice partly melt, fully melt, or undergo no melting?

5 Question 4 continued: Graphs b through f of Fig. 18-24 are additional sketches of T versus t, of which one or more are impossible to produce. (a) Which is impossible and why? (b) In the possible ones, is the equilibrium temperature above, below, or at the freezing point of water? (c) As the possible situations reach equilibrium, does the liquid partly freeze, fully freeze, or undergo no freezing? Does the ice partly melt, fully melt, or undergo no melting?

Fig. 18-24 Questions 4 and 5.

6 Figure 18-25 shows three different arrangements of materials 1, 2, and 3 to form a wall. The thermal conductivities are $k_1 > k_2 > k_3$. The left side of the wall is 20 C° higher than

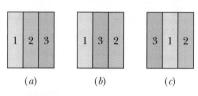

Fig. 18-25 Question 6.

the right side. Rank the arrangements according to (a) the (steady state) rate of energy conduction through the wall and (b) the temperature difference across material 1, greatest first.

7 Figure 18-26 shows two closed cycles on p-V diagrams for a gas. The three parts of cycle 1 are of the same length and shape as those of cycle 2. For each cycle, should the cycle be traversed clockwise or counterclockwise

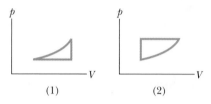

Fig. 18-26 Questions 7 and 8.

if (a) the net work W done by the gas is to be positive and (b) the net energy transferred by the gas as heat Q is to be positive?

8 For which cycle in Fig. 18-26, traversed clockwise, is (a) W greater and (b) Q greater?

9 Three different materials of identical mass are placed one at a time in a special freezer that can extract energy from a material at a certain constant rate. During the cooling process, each material begins in the liquid state and ends in the solid state; Fig. 18-27 shows the tempera-

Fig. 18-27 Question 9.

ture T versus time t. (a) For material 1, is the specific heat for the liquid state greater than or less than that for the solid state? Rank the materials according to (b) freezing-point temperature, (c) specific heat in the liquid state, (d) specific heat in the solid state, and (e) heat of fusion, all greatest first.

10 A solid cube of edge length r, a solid sphere of radius r, and a solid hemisphere of radius r, all made of the same material, are maintained at temperature 300 K in an environment at temperature 350 K. Rank the objects according to the net rate at which thermal radiation is exchanged with the environment, greatest first.

11 A hot object is dropped into a thermally insulated container of water, and the object and water are then allowed to come to thermal equilibrium. The experiment is repeated twice, with different hot objects. All three objects have the same mass and initial temperature, and the mass and initial temperature of the water are the same in the three experiments. For each of the experiments, Fig. 18-28 gives graphs of the temperatures T of the object and the water versus time t. Rank the graphs according to the specific heats of the objects, greatest first.

Fig. 18-28 Question 11.

Tutoring problem available (at instructor's discretion) in *WileyPLUS* and WebAssign

SSM Worked-out solution available in Student Solutions Manual

• – ••• Number of dots indicates level of problem difficulty

WWW Worked-out solution is at

ILW Interactive solution is at http://www.wiley.com/college/halliday

Additional information available in *The Flying Circus of Physics* and at flyingcircusofphysics.com

sec. 18-4 Measuring Temperature

•1 Suppose the temperature of a gas is 373.15 K when it is at the boiling point of water. What then is the limiting value of the ratio of the pressure of the gas at that boiling point to its pressure at the triple point of water? (Assume the volume of the gas is the same at both temperatures.)

•2 Two constant-volume gas thermometers are assembled, one with nitrogen and the other with hydrogen. Both contain enough gas so that $p_3 = 80$ kPa. (a) What is the difference between the pressures in the two thermometers if both bulbs are in boiling water? (*Hint:* See Fig. 18-6.) (b) Which gas is at higher pressure?

•3 A gas thermometer is constructed of two gas-containing bulbs, each in a water bath, as shown in Fig. 18-29. The pressure difference between the two bulbs is measured by a mercury manometer as shown. Appropriate reservoirs, not shown in the diagram, maintain constant gas volume in the two bulbs. There is no difference in pressure when both baths are at the triple point of water. The pressure difference is 120 torr when one bath is at the triple point and the other is at the boiling point of water. It is 90.0 torr when one bath is at the triple point and the other is at an unknown temperature to be measured. What is the unknown temperature?

Fig. 18-29 Problem 3.

sec. 18-5 The Celsius and Fahrenheit Scales

•4 (a) In 1964, the temperature in the Siberian village of Oymyakon reached −71°C. What temperature is this on the Fahrenheit scale? (b) The highest officially recorded temperature in the continental United States was 134°F in Death Valley, California. What is this temperature on the Celsius scale?

•5 At what temperature is the Fahrenheit scale reading equal to (a) twice that of the Celsius scale and (b) half that of the Celsius scale?

••6 On a linear X temperature scale, water freezes at −125.0°X and boils at 375.0°X. On a linear Y temperature scale, water freezes at −70.00°Y and boils at −30.00°Y. A temperature of 50.00°Y corresponds to what temperature on the X scale?

••7 **ILW** Suppose that on a linear temperature scale X, water boils at −53.5°X and freezes at −170°X. What is a temperature of 340 K on the X scale? (Approximate water's boiling point as 373 K.)

sec. 18-6 Thermal Expansion

•8 At 20°C, a brass cube has an edge length of 30 cm. What is the increase in the cube's surface area when it is heated from 20°C to 75°C?

•9 **ILW** A circular hole in an aluminum plate is 2.725 cm in diameter at 0.000°C. What is its diameter when the temperature of the plate is raised to 100.0°C?

•10 An aluminum flagpole is 33 m high. By how much does its length increase as the temperature increases by 15 C°?

•11 What is the volume of a lead ball at 30.00°C if the ball's volume at 60.00°C is 50.00 cm³?

•12 An aluminum-alloy rod has a length of 10.000 cm at 20.000°C and a length of 10.015 cm at the boiling point of water. (a) What is the length of the rod at the freezing point of water? (b) What is the temperature if the length of the rod is 10.009 cm?

•13 **SSM** Find the change in volume of an aluminum sphere with an initial radius of 10 cm when the sphere is heated from 0.0°C to 100°C.

••14 When the temperature of a copper coin is raised by 100 C°, its diameter increases by 0.18%. To two significant figures, give the percent increase in (a) the area of a face, (b) the thickness, (c) the volume, and (d) the mass of the coin. (e) Calculate the coefficient of linear expansion of the coin.

••15 **ILW** A steel rod is 3.000 cm in diameter at 25.00°C. A brass ring has an interior diameter of 2.992 cm at 25.00°C. At what common temperature will the ring just slide onto the rod?

••16 When the temperature of a metal cylinder is raised from 0.0°C to 100°C, its length increases by 0.23%. (a) Find the percent change in density. (b) What is the metal? Use Table 18-2.

••17 **SSM WWW** An aluminum cup of 100 cm³ capacity is completely filled with glycerin at 22°C. How much glycerin, if any, will spill out of the cup if the temperature of both the cup and the glycerin is increased to 28°C? (The coefficient of volume expansion of glycerin is 5.1×10^{-4}/C°.)

••18 At 20°C, a rod is exactly 20.05 cm long on a steel ruler. Both the rod and the ruler are placed in an oven at 270°C, where the rod now measures 20.11 cm on the same ruler. What is the coefficient of linear expansion for the material of which the rod is made?

••19 **GO** A vertical glass tube of length $L = 1.280\,000$ m is half filled with a liquid at 20.000 000°C. How much will the height of the liquid column change when the tube and liquid are heated to 30.000 000°C? Use coefficients $\alpha_{glass} = 1.000\,000 \times 10^{-5}$/K and $\beta_{liquid} = 4.000\,000 \times 10^{-5}$/K.

••20 In a certain experiment, a small radioactive source must move at selected, extremely slow speeds. This motion is accomplished by fastening the source to one end of an aluminum rod and heating the central section of the rod in a controlled way. If the effective heated section of the rod in Fig. 18-30 has length $d = 2.00$ cm, at what constant rate must the temperature of the rod be changed if the source is to move at a constant speed of 100 nm/s?

Fig. 18-30 Problem 20.

•••21 SSM ILW As a result of a temperature rise of 32 C°, a bar with a crack at its center buckles upward (Fig. 18-31). If the fixed distance L_0 is 3.77 m and the coefficient of linear expansion of the bar is $25 \times 10^{-6}/C°$, find the rise x of the center.

Fig. 18-31 Problem 21.

sec. 18-8 The Absorption of Heat by Solids and Liquids

•22 One way to keep the contents of a garage from becoming too cold on a night when a severe subfreezing temperature is forecast is to put a tub of water in the garage. If the mass of the water is 125 kg and its initial temperature is 20°C, (a) how much energy must the water transfer to its surroundings in order to freeze completely and (b) what is the lowest possible temperature of the water and its surroundings until that happens?

•23 SSM A small electric immersion heater is used to heat 100 g of water for a cup of instant coffee. The heater is labeled "200 watts" (it converts electrical energy to thermal energy at this rate). Calculate the time required to bring all this water from 23.0°C to 100°C, ignoring any heat losses.

•24 A certain substance has a mass per mole of 50.0 g/mol. When 314 J is added as heat to a 30.0 g sample, the sample's temperature rises from 25.0°C to 45.0°C. What are the (a) specific heat and (b) molar specific heat of this substance? (c) How many moles are in the sample?

•25 A certain diet doctor encourages people to diet by drinking ice water. His theory is that the body must burn off enough fat to raise the temperature of the water from 0.00°C to the body temperature of 37.0°C. How many liters of ice water would have to be consumed to burn off 454 g (about 1 lb) of fat, assuming that burning this much fat requires 3500 Cal be transferred to the ice water? Why is it not advisable to follow this diet? (One liter = 10^3 cm³. The density of water is 1.00 g/cm³.)

•26 What mass of butter, which has a usable energy content of 6.0 Cal/g (= 6000 cal/g), would be equivalent to the change in gravitational potential energy of a 73.0 kg man who ascends from sea level to the top of Mt. Everest, at elevation 8.84 km? Assume that the average g for the ascent is 9.80 m/s².

•27 SSM Calculate the minimum amount of energy, in joules, required to completely melt 130 g of silver initially at 15.0°C.

•28 How much water remains unfrozen after 50.2 kJ is transferred as heat from 260 g of liquid water initially at its freezing point?

••29 In a solar water heater, energy from the Sun is gathered by water that circulates through tubes in a rooftop collector. The solar radiation enters the collector through a transparent cover and warms the water in the tubes; this water is pumped into a holding tank. Assume that the efficiency of the overall system is 20% (that is, 80% of the incident solar energy is lost from the system). What collector area is necessary to raise the temperature of 200 L of water in the tank from 20°C to 40°C in 1.0 h when the intensity of incident sunlight is 700 W/m²?

••30 A 0.400 kg sample is placed in a cooling apparatus that removes energy as heat at a constant rate. Figure 18-32 gives the temperature T of the sample versus time t; the horizontal scale is set by $t_s = 80.0$ min. The sample freezes during the energy removal. The

specific heat of the sample in its initial liquid phase is 3000 J/kg·K. What are (a) the sample's heat of fusion and (b) its specific heat in the frozen phase?

Fig. 18-32 Problem 30.

••31 ILW What mass of steam at 100°C must be mixed with 150 g of ice at its melting point, in a thermally insulated container, to produce liquid water at 50°C?

••32 The specific heat of a substance varies with temperature according to the function $c = 0.20 + 0.14T + 0.023T^2$, with T in °C and c in cal/g·K. Find the energy required to raise the temperature of 2.0 g of this substance from 5.0°C to 15°C.

••33 *Nonmetric version:* (a) How long does a 2.0×10^5 Btu/h water heater take to raise the temperature of 40 gal of water from 70°F to 100°F? *Metric version:* (b) How long does a 59 kW water heater take to raise the temperature of 150 L of water from 21°C to 38°C?

••34 GO Samples A and B are at different initial temperatures when they are placed in a thermally insulated container and allowed to come to thermal equilibrium. Figure 18-33a gives their temperatures T versus time t. Sample A has a mass of 5.0 kg; sample B has a mass of 1.5 kg. Figure 18-33b is a general plot for the material of sample B. It shows the temperature change ΔT that the material undergoes when energy is transferred to it as heat Q. The change ΔT is plotted versus the energy Q per unit mass of the material, and the scale of the vertical axis is set by $\Delta T_s = 4.0$ C°. What is the specific heat of sample A?

Fig. 18-33 Problem 34.

••35 An insulated Thermos contains 130 cm³ of hot coffee at 80.0°C. You put in a 12.0 g ice cube at its melting point to cool the coffee. By how many degrees has your coffee cooled once the ice has melted and equilibrium is reached? Treat the coffee as though it were pure water and neglect energy exchanges with the environment.

••36 A 150 g copper bowl contains 220 g of water, both at 20.0°C. A very hot 300 g copper cylinder is dropped into the water, causing the water to boil, with 5.00 g being converted to steam. The final temperature of the system is 100°C. Neglect energy transfers with the environment. (a) How much energy (in calories) is transferred to the water as heat? (b) How much to the bowl? (c) What is the original temperature of the cylinder?

••37 A person makes a quantity of iced tea by mixing 500 g of hot tea (essentially water) with an equal mass of ice at its melting point. Assume the mixture has negligible energy exchanges with its environment. If the tea's initial temperature is $T_i = 90°C$, when thermal equilibrium is reached what are (a) the mixture's temperature T_f and (b) the remaining mass m_f of ice? If $T_i = 70°C$, when thermal equilibrium is reached what are (c) T_f and (d) m_f?

••38 A 0.530 kg sample of liquid water and a sample of ice are placed in a thermally insulated container. The container also contains a device that transfers energy as heat from the liquid water to the ice at a constant rate P, until thermal equilibrium is reached. The temperatures T of the liquid water and the ice are given in Fig. 18-34 as functions of time t; the horizontal scale is set by $t_s = 80.0$ min. (a) What is rate P? (b) What is the initial mass of the ice in the container? (c) When thermal equilibrium is reached, what is the mass of the ice produced in this process?

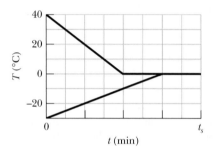

Fig. 18-34 Problem 38.

••39 GO Ethyl alcohol has a boiling point of 78.0°C, a freezing point of −114°C, a heat of vaporization of 879 kJ/kg, a heat of fusion of 109 kJ/kg, and a specific heat of 2.43 kJ/kg·K. How much energy must be removed from 0.510 kg of ethyl alcohol that is initially a gas at 78.0°C so that it becomes a solid at −114°C?

••40 Calculate the specific heat of a metal from the following data. A container made of the metal has a mass of 3.6 kg and contains 14 kg of water. A 1.8 kg piece of the metal initially at a temperature of 180°C is dropped into the water. The container and water initially have a temperature of 16.0°C, and the final temperature of the entire (insulated) system is 18.0°C.

•••41 SSM WWW (a) Two 50 g ice cubes are dropped into 200 g of water in a thermally insulated container. If the water is initially at 25°C, and the ice comes directly from a freezer at −15°C, what is the final temperature at thermal equilibrium? (b) What is the final temperature if only one ice cube is used?

•••42 A 20.0 g copper ring at 0.000°C has an inner diameter of $D = 2.54000$ cm. An aluminum sphere at 100.0°C has a diameter of $d = 2.545\ 08$ cm. The sphere is put on top of the ring (Fig. 18-35), and

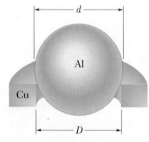

Fig. 18-35 Problem 42.

the two are allowed to come to thermal equilibrium, with no heat lost to the surroundings. The sphere just passes through the ring at the equilibrium temperature. What is the mass of the sphere?

sec. 18-11 Some Special Cases of the First Law of Thermodynamics

•43 In Fig. 18-36, a gas sample expands from V_0 to $4.0V_0$ while its pressure decreases from p_0 to $p_0/4.0$. If $V_0 = 1.0$ m³ and $p_0 = 40$ Pa, how much work is done by the gas if its pressure changes with volume via (a) path A, (b) path B, and (c) path C?

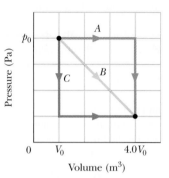

Fig. 18-36 Problem 43.

•44 A thermodynamic system is taken from state A to state B to state C, and then back to A, as shown in the p-V diagram of Fig. 18-37a. The vertical scale is set by $p_s = 40$ Pa, and the horizontal scale is set by $V_s = 4.0$ m³. (a)–(g) Complete the table in Fig. 18-37b by inserting a plus sign, a minus sign, or a zero in each indicated cell. (h) What is the net work done by the system as it moves once through the cycle $ABCA$?

	Q	W	ΔE_{int}
$A \longrightarrow B$	(a)	(b)	+
$B \longrightarrow C$	+	(c)	(d)
$C \longrightarrow A$	(e)	(f)	(g)

Fig. 18-37 Problem 44.

•45 SSM ILW A gas within a closed chamber undergoes the cycle shown in the p-V diagram of Fig. 18-38. The horizontal scale is set by $V_s = 4.0$ m³. Calculate the net energy added to the system as heat during one complete cycle.

Volume (m³)

Fig. 18-38 Problem 45.

•46 Suppose 200 J of work is done on a system and 70.0 cal is extracted from the system as heat. In the sense of the first law of thermodynamics, what are the values (including algebraic signs) of (a) W, (b) Q, and (c) ΔE_{int}?

••47 SSM WWW When a system is taken from state i to state f along path iaf in Fig. 18-39, $Q = 50$ cal and $W = 20$ cal. Along path ibf, $Q = 36$ cal. (a) What is W along path ibf? (b) If $W = -13$ cal for the return path fi, what is Q for this path? (c) If $E_{int,i} = 10$ cal, what is $E_{int,f}$? If $E_{int,b} = 22$ cal, what is Q for (d) path ib and (e) path bf?

Fig. 18-39 Problem 47.

••48 GO Gas held within a chamber passes through the cycle shown in Fig. 18-40. Determine the energy transferred by the system as heat during process CA if the energy added as heat Q_{AB} during process AB is 20.0 J, no energy is transferred as heat during process BC, and the net work done during the cycle is 15.0 J.

••49 GO Figure 18-41 represents a closed cycle for a gas (the figure is not drawn to scale). The change in the internal energy of the gas as it moves from a to c along the path abc is -200 J. As it moves from c to d, 180 J must be transferred to it as heat. An additional transfer of 80 J to it as heat is needed as it moves from d to a. How much work is done on the gas as it moves from c to d?

••50 A lab sample of gas is taken through cycle $abca$ shown in the p-V diagram of Fig. 18-42. The net work done is $+1.2$ J. Along path ab, the change in the internal energy is $+3.0$ J and the magnitude of the work done is 5.0 J. Along path ca, the energy transferred to the gas as heat is $+2.5$ J. How much energy is transferred as heat along (a) path ab and (b) path bc?

sec. 18-12 Heat Transfer Mechanisms

•51 A sphere of radius 0.500 m, temperature 27.0°C, and emissivity 0.850 is located in an environment of temperature 77.0°C. At what rate does the sphere (a) emit and (b) absorb thermal radiation? (c) What is the sphere's net rate of energy exchange?

•52 The ceiling of a single-family dwelling in a cold climate should have an R-value of 30. To give such insulation, how thick would a layer of (a) polyurethane foam and (b) silver have to be?

•53 SSM Consider the slab shown in Fig. 18-18. Suppose that $L = 25.0$ cm, $A = 90.0$ cm^2, and the material is copper. If $T_H = 125$°C, $T_C = 10.0$°C, and a steady state is reached, find the conduction rate through the slab.

Fig. 18-40
Problem 48.

Fig. 18-41
Problem 49.

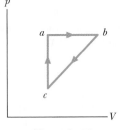

Fig. 18-42
Problem 50.

•54 If you were to walk briefly in space without a spacesuit while far from the Sun (as an astronaut does in the movie *2001, A Space Odyssey*), you would feel the cold of space—while you radiated energy, you would absorb almost none from your environment. (a) At what rate would you lose energy? (b) How much energy would you lose in 30 s? Assume that your emissivity is 0.90, and estimate other data needed in the calculations.

•55 ILW A cylindrical copper rod of length 1.2 m and cross-sectional area 4.8 cm^2 is insulated to prevent heat loss through its surface. The ends are maintained at a temperature difference of 100 C° by having one end in a water–ice mixture and the other in a mixture of boiling water and steam. (a) At what rate is energy conducted along the rod? (b) At what rate does ice melt at the cold end?

••56 The giant hornet *Vespa mandarinia japonica* preys on Japanese bees. However, if one of the hornets attempts to invade a beehive, several hundred of the bees quickly form a compact ball around the hornet to stop it. They don't sting, bite, crush, or suffocate it. Rather they overheat it by quickly raising their body temperatures from the normal 35°C to 47°C or 48°C, which is lethal to the hornet but not to the bees (Fig. 18-43). Assume the following: 500 bees form a ball of radius $R = 2.0$ cm for a time $t = 20$ min, the primary loss of energy by the ball is by thermal radiation, the ball's surface has emissivity $\varepsilon = 0.80$, and the ball has a uniform temperature. On average, how much additional energy must each bee produce during the 20 min to maintain 47°C?

Fig. 18-43 Problem 56.
(© Dr. Masato Ono, Tamagawa University)

••57 (a) What is the rate of energy loss in watts per square meter through a glass window 3.0 mm thick if the outside temperature is -20°F and the inside temperature is $+72$°F? (b) A storm window having the same thickness of glass is installed parallel to the first window, with an air gap of 7.5 cm between the two windows. What now is the rate of energy loss if conduction is the only important energy-loss mechanism?

••58 A solid cylinder of radius $r_1 = 2.5$ cm, length $h_1 = 5.0$ cm, emissivity 0.85, and temperature 30°C is suspended in an environment of temperature 50°C. (a) What is the cylinder's net thermal radiation transfer rate P_1? (b) If the cylinder is stretched until its radius is $r_2 = 0.50$ cm, its net thermal radiation transfer rate becomes P_2. What is the ratio P_2/P_1?

••59 In Fig. 18-44a, two identical rectangular rods of metal are welded end to end, with a temperature of $T_1 = 0$°C on the left side and a temperature of $T_2 = 100$°C on

Fig. 18-44
Problem 59.

the right side. In 2.0 min, 10 J is conducted at a constant rate from the right side to the left side. How much time would be required to conduct 10 J if the rods were welded side to side as in Fig. 18-44b?

••60 **GO** Figure 18-45 shows the cross section of a wall made of three layers. The layer thicknesses are L_1, $L_2 = 0.700L_1$, and $L_3 = 0.350L_1$. The thermal conductivities are k_1, $k_2 = 0.900k_1$, and $k_3 = 0.800k_1$. The temperatures at the left and right sides of the wall are $T_H = 30.0°C$ and $T_C = -15.0°C$, respectively. Thermal conduction is steady. (a) What is the temperature difference ΔT_2 across layer 2 (between the left and right sides of the layer)? If k_2 were, instead, equal to $1.1k_1$, (b) would the rate at which energy is conducted through the wall be greater than, less than, or the same as previously, and (c) what would be the value of ΔT_2?

Fig. 18-45
Problem 60.

••61 **SSM** A tank of water has been outdoors in cold weather, and a slab of ice 5.0 cm thick has formed on its surface (Fig. 18-46). The air above the ice is at $-10°C$. Calculate the rate of ice formation (in centimeters per hour) on the ice slab. Take the thermal conductivity of ice to be 0.0040 cal/s·cm·C° and its density to be 0.92 g/cm³. Assume no energy transfer through the tank walls or bottom.

Fig. 18-46 Problem 61.

••62 Leidenfrost effect. A water drop that is slung onto a skillet with a temperature between 100°C and about 200°C will last about 1 s. However, if the skillet is much hotter, the drop can last several minutes, an effect named after an early investigator. The longer lifetime is due to the support of a thin layer of air and water vapor that separates the drop from the metal (by distance L in Fig. 18-47). Let $L = 0.100$ mm, and assume that the drop is flat with height $h = 1.50$ mm and bottom face area $A = 4.00 \times 10^{-6}$ m². Also assume that the skillet has a constant temperature $T_s = 300°C$ and the drop has a temperature of 100°C. Water has density $\rho = 1000$ kg/m³, and the supporting layer has thermal conductivity $k = 0.026$ W/m·K. (a) At what rate is energy conducted from the skillet to the drop through the drop's bottom surface? (b) If conduction is the primary way energy moves from the skillet to the drop, how long will the drop last?

Fig. 18-47 Problem 62.

••63 **GO** Figure 18-48 shows (in cross section) a wall consisting of

Fig. 18-48 Problem 63.

four layers, with thermal conductivities $k_1 = 0.060$ W/m·K, $k_3 = 0.040$ W/m·K, and $k_4 = 0.12$ W/m·K (k_2 is not known). The layer thicknesses are $L_1 = 1.5$ cm, $L_3 = 2.8$ cm, and $L_4 = 3.5$ cm (L_2 is not known). The known temperatures are $T_1 = 30°C$, $T_{12} = 25°C$, and $T_4 = -10°C$. Energy transfer through the wall is steady. What is interface temperature T_{34}?

••64 Penguin huddling. To withstand the harsh weather of the Antarctic, emperor penguins huddle in groups (Fig. 18-49). Assume that a penguin is a circular cylinder with a top surface area $a = 0.34$ m² and height $h = 1.1$ m. Let P_r be the rate at which an individual penguin radiates energy to the environment (through the top and the sides); thus NP_r is the rate at which N identical, well-separated penguins radiate. If the penguins huddle closely to form a *huddled cylinder* with top surface area Na and height h, the cylinder radiates at the rate P_h. If $N = 1000$, (a) what is the value of the fraction P_h/NP_r and (b) by what percentage does huddling reduce the total radiation loss?

Fig. 18-49 Problem 64.
(Alain Torterotot/Peter Arnold, Inc.)

••65 Ice has formed on a shallow pond, and a steady state has been reached, with the air above the ice at $-5.0°C$ and the bottom of the pond at 4.0°C. If the total depth of *ice + water* is 1.4 m, how thick is the ice? (Assume that the thermal conductivities of ice and water are 0.40 and 0.12 cal/m·C°·s, respectively.)

•••66 **GO** Evaporative cooling of beverages. A cold beverage can be kept cold even on a warm day if it is slipped into a porous ceramic container that has been soaked in water. Assume that energy lost to evaporation matches the net energy gained via the radiation exchange through the top and side surfaces. The container and beverage have temperature $T = 15°C$, the environment has temperature $T_{env} = 32°C$, and the container is a cylinder with radius $r = 2.2$ cm and height 10 cm. Approximate the emissivity as $\varepsilon = 1$, and neglect other energy exchanges. At what rate dm/dt is the container losing water mass?

Additional Problems

67 In the extrusion of cold chocolate from a tube, work is done on the chocolate by the pressure applied by a ram forcing the chocolate through the tube. The work per unit mass of extruded chocolate is equal to p/ρ, where p is the difference between the applied pressure and the pressure where the chocolate emerges from the tube, and ρ is the density of the chocolate.

Rather than increasing the temperature of the chocolate, this work melts cocoa fats in the chocolate. These fats have a heat of fusion of 150 kJ/kg. Assume that all of the work goes into that melting and that these fats make up 30% of the chocolate's mass. What percentage of the fats melt during the extrusion if $p = 5.5$ MPa and $\rho = 1200$ kg/m^3?

68 Icebergs in the North Atlantic present hazards to shipping, causing the lengths of shipping routes to be increased by about 30% during the iceberg season. Attempts to destroy icebergs include planting explosives, bombing, torpedoing, shelling, ramming, and coating with black soot. Suppose that direct melting of the iceberg, by placing heat sources in the ice, is tried. How much energy as heat is required to melt 10% of an iceberg that has a mass of 200 000 metric tons? (Use 1 metric ton = 1000 kg.)

69 Figure 18-50 displays a closed cycle for a gas. The change in internal energy along path ca is -160 J. The energy transferred to the gas as heat is 200 J along path ab, and 40 J along path bc. How much work is done by the gas along (a) path abc and (b) path ab?

Fig. 18-50 Problem 69.

70 In a certain solar house, energy from the Sun is stored in barrels filled with water. In a particular winter stretch of five cloudy days, 1.00×10^6 kcal is needed to maintain the inside of the house at 22.0°C. Assuming that the water in the barrels is at 50.0°C and that the water has a density of 1.00×10^3 kg/m^3, what volume of water is required?

71 A 0.300 kg sample is placed in a cooling apparatus that removes energy as heat at a constant rate of 2.81 W. Figure 18-51 gives the temperature T of the sample versus time t. The temperature scale is set by $T_s = 30$°C and the time scale is set by $t_s = 20$ min. What is the specific heat of the sample?

Fig. 18-51 Problem 71.

72 The average rate at which energy is conducted outward through the ground surface in North America is 54.0 mW/m^2, and the average thermal conductivity of the near-surface rocks is 2.50 W/m · K. Assuming a surface temperature of 10.0°C, find the temperature at a depth of 35.0 km (near the base of the crust). Ignore the heat generated by the presence of radioactive elements.

73 What is the volume increase of an aluminum cube 5.00 cm on an edge when heated from 10.0°C to 60.0°C?

74 In a series of experiments, block B is to be placed in a thermally insulated container with block A, which has the same mass as block B. In each experiment, block B is initially at a certain temperature T_B, but temperature T_A of block A is changed from experiment to experiment. Let T_f represent the final temperature of the two blocks when they reach thermal equilibrium in any of the experiments. Figure 18-52 gives temperature T_f versus the initial temperature T_A for a range of possible values of

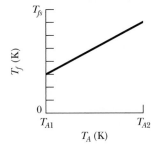

Fig. 18-52 Problem 74.

T_A, from $T_{A1} = 0$ K to $T_{A2} = 500$ K. The vertical axis scale is set by $T_{fs} = 400$ K. What are (a) temperature T_B and (b) the ratio c_B/c_A of the specific heats of the blocks?

75 Figure 18-53 displays a closed cycle for a gas. From c to b, 40 J is transferred from the gas as heat. From b to a, 130 J is transferred from the gas as heat, and the magnitude of the work done by the gas is 80 J. From a to c, 400 J is transferred to the gas as heat. What is the work done by the gas from a to c? (*Hint:* You need to supply the plus and minus signs for the given data.)

Fig. 18-53 Problem 75.

76 Three equal-length straight rods, of aluminum, Invar, and steel, all at 20.0°C, form an equilateral triangle with hinge pins at the vertices. At what temperature will the angle opposite the Invar rod be 59.95°? See Appendix E for needed trigonometric formulas and Table 18-2 for needed data.

77 SSM The temperature of a 0.700 kg cube of ice is decreased to -150°C. Then energy is gradually transferred to the cube as heat while it is otherwise thermally isolated from its environment. The total transfer is 0.6993 MJ. Assume the value of c_{ice} given in Table 18-3 is valid for temperatures from -150°C to 0°C. What is the final temperature of the water?

78 GO *Icicles.* Liquid water coats an active (growing) icicle and extends up a short, narrow tube along the central axis (Fig. 18-54). Because the water–ice interface must have a temperature of 0°C, the water in the tube cannot lose energy through the sides of the icicle or down through the tip because there is no temperature change in those directions. It can lose energy and freeze only by sending energy up (through distance L) to the top of the icicle, where the temperature T_r can be below 0°C. Take $L = 0.12$ m and $T_r = -5$°C. Assume that the central tube and the upward conduction path both have cross-sectional area A. In terms of A, what rate is (a) energy conducted upward and (b) mass converted from liquid to ice at the top of the central tube? (c) At what rate does the top of the tube move downward because of water freezing there? The thermal conductivity of ice is 0.400 W/m·K, and the density of liquid water is 1000 kg/m^3.

Fig. 18-54 Problem 78.

79 SSM A sample of gas expands from an initial pressure and volume of 10 Pa and 1.0 m^3 to a final volume of 2.0 m^3. During the expansion, the pressure and volume are related by the equation $p = aV^2$, where $a = 10$ N/m^8. Determine the work done by the gas during this expansion.

80   Figure 18-55a shows a cylinder containing gas and closed by a movable piston. The cylinder is kept submerged in an ice–water mixture. The piston is *quickly* pushed down from position 1 to position 2 and then held at position 2 until the gas is again at the temperature of the ice–water mixture; it then is *slowly* raised back to position 1. Figure 18-55b is a *p-V* diagram for the process. If 100 g of ice is melted during the cycle, how much work has been done *on* the gas?

Fig. 18-55   Problem 80.

81   **SSM**   A sample of gas undergoes a transition from an initial state *a* to a final state *b* by three different paths (processes), as shown in the *p-V* diagram in Fig. 18-56, where $V_b = 5.00V_i$. The energy transferred to the gas as heat in process 1 is $10p_iV_i$. In terms of p_iV_i, what are (a) the energy transferred to the gas as heat in process 2 and (b) the change in internal energy that the gas undergoes in process 3?

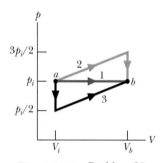

Fig. 18-56   Problem 81.

82   A copper rod, an aluminum rod, and a brass rod, each of 6.00 m length and 1.00 cm diameter, are placed end to end with the aluminum rod between the other two. The free end of the copper rod is maintained at water's boiling point, and the free end of the brass rod is maintained at water's freezing point. What is the steady-state temperature of (a) the copper–aluminum junction and (b) the aluminum–brass junction?

83   **SSM**   The temperature of a Pyrex disk is changed from 10.0°C to 60.0°C. Its initial radius is 8.00 cm; its initial thickness is 0.500 cm. Take these data as being exact. What is the change in the volume of the disk? (See Table 18-2.)

84   (a) Calculate the rate at which body heat is conducted through the clothing of a skier in a steady-state process, given the following data: the body surface area is 1.8 m², and the clothing is 1.0 cm thick; the skin surface temperature is 33°C and the outer surface of the clothing is at 1.0°C; the thermal conductivity of the clothing is 0.040 W/m·K. (b) If, after a fall, the skier's clothes became soaked with water of thermal conductivity 0.60 W/m·K, by how much is the rate of conduction multiplied?

85   **SSM**   A 2.50 kg lump of aluminum is heated to 92.0°C and then dropped into 8.00 kg of water at 5.00°C. Assuming that the lump–water system is thermally isolated, what is the system's equilibrium temperature?

86   A glass window pane is exactly 20 cm by 30 cm at 10°C. By how much has its area increased when its temperature is 40°C, assuming that it can expand freely?

87   A recruit can join the semi-secret "300 F" club at the Amundsen–Scott South Pole Station only when the outside temperature is below −70°C. On such a day, the recruit first basks in a hot sauna and then runs outside wearing only shoes. (This is, of course, extremely dangerous, but the rite is effectively a protest against the constant danger of the cold.)

Assume that upon stepping out of the sauna, the recruit's skin temperature is 102°F and the walls, ceiling, and floor of the sauna room have a temperature of 30°C. Estimate the recruit's surface area, and take the skin emissivity to be 0.80. (a) What is the approximate net rate P_{net} at which the recruit loses energy via thermal radiation exchanges with the room? Next, assume that when outdoors, half the recruit's surface area exchanges thermal radiation with the sky at a temperature of −25°C and the other half exchanges thermal radiation with the snow and ground at a temperature of −80°C. What is the approximate net rate at which the recruit loses energy via thermal radiation exchanges with (b) the sky and (c) the snow and ground?

88   A steel rod at 25.0°C is bolted at both ends and then cooled. At what temperature will it rupture? Use Table 12-1.

89   An athlete needs to lose weight and decides to do it by "pumping iron." (a) How many times must an 80.0 kg weight be lifted a distance of 1.00 m in order to burn off 1.00 lb of fat, assuming that that much fat is equivalent to 3500 Cal? (b) If the weight is lifted once every 2.00 s, how long does the task take?

90   Soon after Earth was formed, heat released by the decay of radioactive elements raised the average internal temperature from 300 to 3000 K, at about which value it remains today. Assuming an average coefficient of volume expansion of 3.0×10^{-5} K^{-1}, by how much has the radius of Earth increased since the planet was formed?

91   It is possible to melt ice by rubbing one block of it against another. How much work, in joules, would you have to do to get 1.00 g of ice to melt?

92   A rectangular plate of glass initially has the dimensions 0.200 m by 0.300 m. The coefficient of linear expansion for the glass is 9.00×10^{-6}/K. What is the change in the plate's area if its temperature is increased by 20.0 K?

93   Suppose that you intercept 5.0×10^{-3} of the energy radiated by a hot sphere that has a radius of 0.020 m, an emissivity of 0.80, and a surface temperature of 500 K. How much energy do you intercept in 2.0 min?

94   A thermometer of mass 0.0550 kg and of specific heat 0.837 kJ/kg·K reads 15.0°C. It is then completely immersed in 0.300 kg of water, and it comes to the same final temperature as the water. If the thermometer then reads 44.4°C, what was the temperature of the water before insertion of the thermometer?

95   A sample of gas expands from $V_1 = 1.0$ m³ and $p_1 = 40$ Pa to $V_2 = 4.0$ m³ and $p_2 = 10$ Pa along path *B* in the *p-V* diagram in Fig. 18-57. It is then compressed back to V_1 along either path *A* or path *C*. Compute the net work done by the gas for the complete cycle along (a) path *BA* and (b) path *BC*.

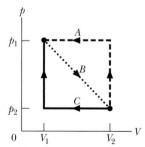

Fig. 18-57   Problem 95.

THE KINETIC THEORY OF GASES

19-1 WHAT IS PHYSICS?

One of the main subjects in thermodynamics is the physics of gases. A gas consists of atoms (either individually or bound together as molecules) that fill their container's volume and exert pressure on the container's walls. We can usually assign a temperature to such a contained gas. These three variables associated with a gas—volume, pressure, and temperature—are all a consequence of the motion of the atoms. The volume is a result of the freedom the atoms have to spread throughout the container, the pressure is a result of the collisions of the atoms with the container's walls, and the temperature has to do with the kinetic energy of the atoms. The **kinetic theory of gases,** the focus of this chapter, relates the motion of the atoms to the volume, pressure, and temperature of the gas.

Applications of the kinetic theory of gases are countless. Automobile engineers are concerned with the combustion of vaporized fuel (a gas) in the automobile engines. Food engineers are concerned with the production rate of the fermentation gas that causes bread to rise as it bakes. Beverage engineers are concerned with how gas can produce the head in a glass of beer or shoot a cork from a champagne bottle. Medical engineers and physiologists are concerned with calculating how long a scuba diver must pause during ascent to eliminate nitrogen gas from the bloodstream (to avoid the *bends*). Environmental scientists are concerned with how heat exchanges between the oceans and the atmosphere can affect weather conditions.

The first step in our discussion of the kinetic theory of gases deals with measuring the amount of a gas present in a sample, for which we use Avogadro's number.

19-2 Avogadro's Number

When our thinking is slanted toward atoms and molecules, it makes sense to measure the sizes of our samples in moles. If we do so, we can be certain that we are comparing samples that contain the same number of atoms or molecules. The *mole* is one of the seven SI base units and is defined as follows:

 One mole is the number of atoms in a 12 g sample of carbon-12.

The obvious question now is: "How many atoms or molecules are there in a mole?" The answer is determined experimentally and, as you saw in Chapter 18, is

$$N_A = 6.02 \times 10^{23}\, \text{mol}^{-1} \quad \text{(Avogadro's number)}, \tag{19-1}$$

where mol^{-1} represents the inverse mole or "per mole," and mol is the abbreviation for mole. The number N_A is called **Avogadro's number** after Italian scientist Amedeo Avogadro (1776–1856), who suggested that all gases occupying the same volume under the same conditions of temperature and pressure contain the same number of atoms or molecules.

The number of moles n contained in a sample of any substance is equal to the ratio of the number of molecules N in the sample to the number of molecules N_A in 1 mol:

$$n = \frac{N}{N_A}. \tag{19-2}$$

(*Caution:* The three symbols in this equation can easily be confused with one another, so you should sort them with their meanings now, before you end in "N-confusion.") We can find the number of moles n in a sample from the mass M_{sam} of the sample and either the *molar mass M* (the mass of 1 mol) or the molecular mass m (the mass of one molecule):

$$n = \frac{M_{sam}}{M} = \frac{M_{sam}}{mN_A}. \tag{19-3}$$

In Eq. 19-3, we used the fact that the mass M of 1 mol is the product of the mass m of one molecule and the number of molecules N_A in 1 mol:

$$M = mN_A. \tag{19-4}$$

19-3 Ideal Gases

Our goal in this chapter is to explain the macroscopic properties of a gas—such as its pressure and its temperature—in terms of the behavior of the molecules that make it up. However, there is an immediate problem: which gas? Should it be hydrogen, oxygen, or methane, or perhaps uranium hexafluoride? They are all different. Experimenters have found, though, that if we confine 1 mol samples of various gases in boxes of identical volume and hold the gases at the same temperature, then their measured pressures are almost the same, and at lower densities the differences tend to disappear. Further experiments show that, at low enough densities, all real gases tend to obey the relation

$$pV = nRT \quad \text{(ideal gas law)}, \tag{19-5}$$

in which p is the absolute (not gauge) pressure, n is the number of moles of gas present, and T is the temperature in kelvins. The symbol R is a constant called the **gas constant** that has the same value for all gases—namely,

$$R = 8.31 \text{ J/mol} \cdot \text{K}. \tag{19-6}$$

Equation 19-5 is called the **ideal gas law.** Provided the gas density is low, this law holds for any single gas or for any mixture of different gases. (For a mixture, n is the total number of moles in the mixture.)

We can rewrite Eq. 19-5 in an alternative form, in terms of a constant called the **Boltzmann constant** k, which is defined as

$$k = \frac{R}{N_A} = \frac{8.31 \text{ J/mol} \cdot \text{K}}{6.02 \times 10^{23} \text{ mol}^{-1}} = 1.38 \times 10^{-23} \text{ J/K}. \tag{19-7}$$

This allows us to write $R = kN_A$. Then, with Eq. 19-2 ($n = N/N_A$), we see that

$$nR = Nk. \tag{19-8}$$

Substituting this into Eq. 19-5 gives a second expression for the ideal gas law:

$$pV = NkT \quad \text{(ideal gas law)}. \tag{19-9}$$

(*Caution:* Note the difference between the two expressions for the ideal gas law—Eq. 19-5 involves the number of moles n, and Eq. 19-9 involves the number of molecules N.)

You may well ask, "What is an *ideal gas,* and what is so 'ideal' about it?" The answer lies in the simplicity of the law (Eqs. 19-5 and 19-9) that governs its macroscopic properties. Using this law—as you will see—we can deduce many properties of the ideal gas in a simple way. Although there is no such thing in nature as a truly ideal gas, *all real* gases approach the ideal state at low enough densities—that is, under conditions in which their molecules are far enough apart that they do not interact with one another. Thus, the ideal gas concept allows us to gain useful insights into the limiting behavior of real gases.

The interior of the railroad tank car in Fig. 19-1 was being cleaned with steam by a crew late one afternoon. When they left for the night, they sealed the car. When they returned the next morning, they discovered that something had crushed the car in spite of its extremely strong steel walls, as if some giant creature from a grade B science fiction movie had stepped on it during a rampage that night.

With Eq. 19-9, we can explain what actually crushed the railroad tank car. When the car was being cleaned, its interior was filled with very hot steam, which is a gas of water molecules. The cleaning crew left the steam inside the car when they closed all the valves on the car at the end of their work shift. At that point the pressure of the gas in the car was equal to atmospheric pressure because the valves had been opened to the atmosphere during the cleaning. As the car cooled during the night, the steam cooled and much of it condensed, which means that the number N of gas molecules and the temperature T of the gas both decreased. Thus, the right side of Eq. 19-9 decreased, and because volume V was constant, the gas pressure p on the left side also decreased. At some point during the night, the gas pressure inside the car reached such a low value that the external atmospheric pressure was able to crush the car's steel walls. The cleaning crew could have prevented this accident by leaving the valves open, so that air could enter the car to keep the internal pressure equal to the external atmospheric pressure.

Fig. 19-1 A railroad tank car crushed overnight. *(Photo courtesy www.Houston.RailFan.net)*

Work Done by an Ideal Gas at Constant Temperature

Suppose we put an ideal gas in a piston–cylinder arrangement like those in Chapter 18. Suppose also that we allow the gas to expand from an initial volume V_i to a final volume V_f while we keep the temperature T of the gas constant. Such a process, at *constant temperature*, is called an **isothermal expansion** (and the reverse is called an **isothermal compression**).

On a p-V diagram, an *isotherm* is a curve that connects points that have the same temperature. Thus, it is a graph of pressure versus volume for a gas whose temperature T is held constant. For n moles of an ideal gas, it is a graph of the equation

$$p = nRT \frac{1}{V} = \text{(a constant)} \frac{1}{V}. \qquad (19\text{-}10)$$

Figure 19-2 shows three isotherms, each corresponding to a different (constant) value of T. (Note that the values of T for the isotherms increase upward to the right.) Superimposed on the middle isotherm is the path followed by a gas during an isothermal expansion from state i to state f at a constant temperature of 310 K.

To find the work done by an ideal gas during an isothermal expansion, we start with Eq. 18-25,

$$W = \int_{V_i}^{V_f} p \, dV. \qquad (19\text{-}11)$$

This is a general expression for the work done during any change in volume of any gas. For an ideal gas, we can use Eq. 19-5 ($pV = nRT$) to substitute for p, obtaining

$$W = \int_{V_i}^{V_f} \frac{nRT}{V} \, dV. \qquad (19\text{-}12)$$

Because we are considering an isothermal expansion, T is constant, so we can

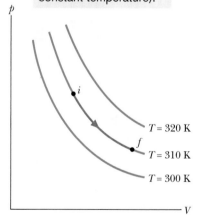

The expansion is along an isotherm (the gas has constant temperature).

Fig. 19-2 Three isotherms on a p-V diagram. The path shown along the middle isotherm represents an isothermal expansion of a gas from an initial state i to a final state f. The path from f to i along the isotherm would represent the reverse process—that is, an isothermal compression.

move it in front of the integral sign to write

$$W = nRT \int_{V_i}^{V_f} \frac{dV}{V} = nRT \left[\ln V \right]_{V_i}^{V_f}. \tag{19-13}$$

By evaluating the expression in brackets at the limits and then using the relationship $\ln a - \ln b = \ln(a/b)$, we find that

$$W = nRT \ln \frac{V_f}{V_i} \qquad \text{(ideal gas, isothermal process).} \tag{19-14}$$

Recall that the symbol ln specifies a *natural* logarithm, which has base *e*.

For an expansion, V_f is greater than V_i, so the ratio V_f/V_i in Eq. 19-14 is greater than unity. The natural logarithm of a quantity greater than unity is positive, and so the work W done by an ideal gas during an isothermal expansion is positive, as we expect. For a compression, V_f is less than V_i, so the ratio of volumes in Eq. 19-14 is less than unity. The natural logarithm in that equation—hence the work W—is negative, again as we expect.

Work Done at Constant Volume and at Constant Pressure

Equation 19-14 does not give the work W done by an ideal gas during *every* thermodynamic process. Instead, it gives the work only for a process in which the temperature is held constant. If the temperature varies, then the symbol T in Eq. 19-12 cannot be moved in front of the integral symbol as in Eq. 19-13, and thus we do not end up with Eq. 19-14.

However, we can always go back to Eq. 19-11 to find the work W done by an ideal gas (or any other gas) during any process, such as a constant-volume process and a constant-pressure process. If the volume of the gas is constant, then Eq. 19-11 yields

$$W = 0 \qquad \text{(constant-volume process).} \tag{19-15}$$

If, instead, the volume changes while the pressure p of the gas is held constant, then Eq. 19-11 becomes

$$W = p(V_f - V_i) = p \, \Delta V \qquad \text{(constant-pressure process).} \tag{19-16}$$

Sample Problem

Ideal gas and changes of temperature, volume, and pressure

A cylinder contains 12 L of oxygen at 20°C and 15 atm. The temperature is raised to 35°C, and the volume is reduced to 8.5 L. What is the final pressure of the gas in atmospheres? Assume that the gas is ideal.

KEY IDEA

Because the gas is ideal, we can use the ideal gas law to relate its parameters, both in the initial state *i* and in the final state *f*.

Calculations: From Eq. 19-5 we can write

$$p_i V_i = nRT_i \quad \text{and} \quad p_f V_f = nRT_f.$$

Dividing the second equation by the first equation and solving for p_f yields

$$p_f = \frac{p_i T_f V_i}{T_i V_f}. \tag{19-17}$$

Note here that if we converted the given initial and final volumes from liters to the proper units of cubic meters, the multiplying conversion factors would cancel out of Eq. 19-17. The same would be true for conversion factors that convert the pressures from atmospheres to the proper pascals. However, to convert the given temperatures to kelvins requires the addition of an amount that would not cancel and thus must be included. Hence, we must write

$$T_i = (273 + 20) \text{ K} = 293 \text{ K}$$

and

$$T_f = (273 + 35) \text{ K} = 308 \text{ K}.$$

Inserting the given data into Eq. 19-17 then yields

$$p_f = \frac{(15 \text{ atm})(308 \text{ K})(12 \text{ L})}{(293 \text{ K})(8.5 \text{ L})} = 22 \text{ atm}. \qquad \text{(Answer)}$$

 Additional examples, video, and practice available at *WileyPLUS*

Work by an ideal gas

One mole of oxygen (assume it to be an ideal gas) expands at a constant temperature T of 310 K from an initial volume V_i of 12 L to a final volume V_f of 19 L. How much work is done by the gas during the expansion?

Fig. 19-3 The shaded area represents the work done by 1 mol of oxygen in expanding from V_i to V_f at a temperature T of 310 K.

Generally we find the work by integrating the gas pressure with respect to the gas volume, using Eq. 19-11. However, because the gas here is ideal and the expansion is isothermal, that integration leads to Eq. 19-14.

Calculation: Therefore, we can write

$$W = nRT \ln \frac{V_f}{V_i}$$

$$= (1 \text{ mol})(8.31 \text{ J/mol} \cdot \text{K})(310 \text{ K}) \ln \frac{19 \text{ L}}{12 \text{ L}}$$

$$= 1180 \text{ J.} \qquad \text{(Answer)}$$

The expansion is graphed in the p-V diagram of Fig. 19-3. The work done by the gas during the expansion is represented by the area beneath the curve if.

You can show that if the expansion is now reversed, with the gas undergoing an isothermal compression from 19 L to 12 L, the work done by the gas will be −1180 J. Thus, an external force would have to do 1180 J of work on the gas to compress it.

 Additional examples, video, and practice available at *WileyPLUS*

19-4 Pressure, Temperature, and RMS Speed

Here is our first kinetic theory problem. Let n moles of an ideal gas be confined in a cubical box of volume V, as in Fig. 19-4. The walls of the box are held at temperature T. What is the connection between the pressure p exerted by the gas on the walls and the speeds of the molecules?

The molecules of gas in the box are moving in all directions and with various speeds, bumping into one another and bouncing from the walls of the box like balls in a racquetball court. We ignore (for the time being) collisions of the molecules with one another and consider only elastic collisions with the walls.

Figure 19-4 shows a typical gas molecule, of mass m and velocity $\vec{v}$, that is about to collide with the shaded wall. Because we assume that any collision of a molecule with a wall is elastic, when this molecule collides with the shaded wall, the only component of its velocity that is changed is the x component, and that component is reversed. This means that the only change in the particle's momentum is along the x axis, and that change is

$$\Delta p_x = (-mv_x) - (mv_x) = -2mv_x.$$

Hence, the momentum Δp_x delivered to the wall by the molecule during the collision is $+2mv_x$. (Because in this book the symbol p represents both momentum and pressure, we must be careful to note that here p represents momentum and is a vector quantity.)

The molecule of Fig. 19-4 will hit the shaded wall repeatedly. The time Δt between collisions is the time the molecule takes to travel to the opposite wall and back again (a distance $2L$) at speed v_x. Thus, Δt is equal to $2L/v_x$. (Note that this result holds even if the molecule bounces off any of the other walls along the way, because those walls are parallel to x and so cannot change v_x.) Therefore, the average

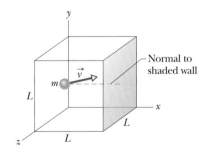

Fig. 19-4 A cubical box of edge length L, containing n moles of an ideal gas. A molecule of mass m and velocity $\vec{v}$ is about to collide with the shaded wall of area L^2. A normal to that wall is shown.

rate at which momentum is delivered to the shaded wall by this single molecule is

$$\frac{\Delta p_x}{\Delta t} = \frac{2mv_x}{2L/v_x} = \frac{mv_x^2}{L}.$$

From Newton's second law ($\vec{F} = d\vec{p}/dt$), the rate at which momentum is delivered to the wall is the force acting on that wall. To find the total force, we must add up the contributions of all the molecules that strike the wall, allowing for the possibility that they all have different speeds. Dividing the magnitude of the total force F_x by the area of the wall ($= L^2$) then gives the pressure p on that wall, where now and in the rest of this discussion, p represents pressure. Thus, using the expression for $\Delta p_x/\Delta t$, we can write this pressure as

$$p = \frac{F_x}{L^2} = \frac{mv_{x1}^2/L + mv_{x2}^2/L + \cdots + mv_{xN}^2/L}{L^2}$$

$$= \left(\frac{m}{L^3}\right)(v_{x1}^2 + v_{x2}^2 + \cdots + v_{xN}^2), \tag{19-18}$$

where N is the number of molecules in the box.

Since $N = nN_A$, there are nN_A terms in the second set of parentheses of Eq. 19-18. We can replace that quantity by $nN_A(v_x^2)_{avg}$, where $(v_x^2)_{avg}$ is the average value of the square of the x components of all the molecular speeds. Equation 19-18 then becomes

$$p = \frac{nmN_A}{L^3}(v_x^2)_{avg}.$$

However, mN_A is the molar mass M of the gas (that is, the mass of 1 mol of the gas). Also, L^3 is the volume of the box, so

$$p = \frac{nM(v_x^2)_{avg}}{V}. \tag{19-19}$$

For any molecule, $v^2 = v_x^2 + v_y^2 + v_z^2$. Because there are many molecules and because they are all moving in random directions, the average values of the squares of their velocity components are equal, so that $v_x^2 = \frac{1}{3}v^2$. Thus, Eq. 19-19 becomes

$$p = \frac{nM(v^2)_{avg}}{3V}. \tag{19-20}$$

The square root of $(v^2)_{avg}$ is a kind of average speed, called the **root-mean-square speed** of the molecules and symbolized by v_{rms}. Its name describes it rather well: You *square* each speed, you find the *mean* (that is, the average) of all these squared speeds, and then you take the square *root* of that mean. With $\sqrt{(v^2)_{avg}} = v_{rms}$, we can then write Eq. 19-20 as

$$p = \frac{nMv_{rms}^2}{3V}. \tag{19-21}$$

This tells us how the pressure of the gas (a purely macroscopic quantity) depends on the speed of the molecules (a purely microscopic quantity).

We can turn Eq. 19-21 around and use it to calculate v_{rms}. Combining Eq. 19-21 with the ideal gas law ($pV = nRT$) leads to

$$v_{rms} = \sqrt{\frac{3RT}{M}}. \tag{19-22}$$

Table 19-1 shows some rms speeds calculated from Eq. 19-22. The speeds are surprisingly high. For hydrogen molecules at room temperature (300 K), the rms speed is 1920 m/s, or 4300 mi/h—faster than a speeding bullet! On the surface of the Sun, where the temperature is 2×10^6 K, the rms speed of hydrogen

Table 19-1

Some RMS Speeds at Room Temperature ($T = 300$ K)[a]

Gas	Molar Mass (10^{-3} kg/mol)	v_{rms} (m/s)
Hydrogen (H_2)	2.02	1920
Helium (He)	4.0	1370
Water vapor (H_2O)	18.0	645
Nitrogen (N_2)	28.0	517
Oxygen (O_2)	32.0	483
Carbon dioxide (CO_2)	44.0	412
Sulfur dioxide (SO_2)	64.1	342

[a]For convenience, we often set room temperature equal to 300 K even though (at 27°C or 81°F) that represents a fairly warm room.

molecules would be 82 times greater than at room temperature were it not for the fact that at such high speeds, the molecules cannot survive collisions among themselves. Remember too that the rms speed is only a kind of average speed; many molecules move much faster than this, and some much slower.

The speed of sound in a gas is closely related to the rms speed of the molecules of that gas. In a sound wave, the disturbance is passed on from molecule to molecule by means of collisions. The wave cannot move any faster than the "average" speed of the molecules. In fact, the speed of sound must be somewhat less than this "average" molecular speed because not all molecules are moving in exactly the same direction as the wave. As examples, at room temperature, the rms speeds of hydrogen and nitrogen molecules are 1920 m/s and 517 m/s, respectively. The speeds of sound in these two gases at this temperature are 1350 m/s and 350 m/s, respectively.

A question often arises: If molecules move so fast, why does it take as long as a minute or so before you can smell perfume when someone opens a bottle across a room? The answer is that, as we shall discuss in Section 19-6, each perfume molecule may have a high speed but it moves away from the bottle only very slowly because its repeated collisions with other molecules prevent it from moving directly across the room to you.

Sample Problem

Average and rms values

Here are five numbers: 5, 11, 32, 67, and 89.

(a) What is the average value n_{avg} of these numbers?

Calculation: We find this from

$$n_{avg} = \frac{5 + 11 + 32 + 67 + 89}{5} = 40.8. \qquad \text{(Answer)}$$

(b) What is the rms value n_{rms} of these numbers?

Calculation: We find this from

$$n_{rms} = \sqrt{\frac{5^2 + 11^2 + 32^2 + 67^2 + 89^2}{5}}$$

$$= 52.1. \qquad \text{(Answer)}$$

The rms value is greater than the average value because the larger numbers—being squared—are relatively more important in forming the rms value.

 Additional examples, video, and practice available at *WileyPLUS*

19-5 Translational Kinetic Energy

We again consider a single molecule of an ideal gas as it moves around in the box of Fig. 19-4, but we now assume that its speed changes when it collides with other molecules. Its translational kinetic energy at any instant is $\frac{1}{2}mv^2$. Its *average* translational kinetic energy over the time that we watch it is

$$K_{avg} = \left(\tfrac{1}{2}mv^2\right)_{avg} = \tfrac{1}{2}m(v^2)_{avg} = \tfrac{1}{2}mv_{rms}^2, \qquad (19\text{-}23)$$

in which we make the assumption that the average speed of the molecule during our observation is the same as the average speed of all the molecules at any given time. (Provided the total energy of the gas is not changing and provided we observe our molecule for long enough, this assumption is appropriate.) Substituting for v_{rms} from Eq. 19-22 leads to

$$K_{avg} = \left(\tfrac{1}{2}m\right)\frac{3RT}{M}.$$

However, M/m, the molar mass divided by the mass of a molecule, is simply Avogadro's number. Thus,

$$K_{avg} = \frac{3RT}{2N_A}.$$

Using Eq. 19-7 ($k = R/N_A$), we can then write

$$K_{avg} = \tfrac{3}{2}kT. \tag{19-24}$$

This equation tells us something unexpected:

> At a given temperature T, all ideal gas molecules—no matter what their mass—have the same average translational kinetic energy—namely, $\tfrac{3}{2}kT$. When we measure the temperature of a gas, we are also measuring the average translational kinetic energy of its molecules.

CHECKPOINT 2

A gas mixture consists of molecules of types 1, 2, and 3, with molecular masses $m_1 > m_2 > m_3$. Rank the three types according to (a) average kinetic energy and (b) rms speed, greatest first.

19-6 Mean Free Path

We continue to examine the motion of molecules in an ideal gas. Figure 19-5 shows the path of a typical molecule as it moves through the gas, changing both speed and direction abruptly as it collides elastically with other molecules. Between collisions, the molecule moves in a straight line at constant speed. Although the figure shows the other molecules as stationary, they are (of course) also moving.

One useful parameter to describe this random motion is the **mean free path** λ of the molecules. As its name implies, λ is the average distance traversed by a molecule between collisions. We expect λ to vary inversely with N/V, the number of molecules per unit volume (or density of molecules). The larger N/V is, the more collisions there should be and the smaller the mean free path. We also expect λ to vary inversely with the size of the molecules—with their diameter d, say. (If the molecules were points, as we have assumed them to be, they would never collide and the mean free path would be infinite.) Thus, the larger the molecules are, the smaller the mean free path. We can even predict that λ should vary (inversely) as the *square* of the molecular diameter because the cross section of a molecule—not its diameter—determines its effective target area.

The expression for the mean free path does, in fact, turn out to be

$$\lambda = \frac{1}{\sqrt{2}\,\pi d^2\, N/V} \qquad \text{(mean free path)}. \tag{19-25}$$

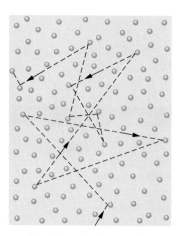

Fig. 19-5 A molecule traveling through a gas, colliding with other gas molecules in its path. Although the other molecules are shown as stationary, they are also moving in a similar fashion.

Fig. 19-6 (a) A collision occurs when the centers of two molecules come within a distance d of each other, d being the molecular diameter. (b) An equivalent but more convenient representation is to think of the moving molecule as having a *radius d* and all other molecules as being points. The condition for a collision is unchanged.

To justify Eq. 19-25, we focus attention on a single molecule and assume—as Fig. 19-5 suggests—that our molecule is traveling with a constant speed v and that all the other molecules are at rest. Later, we shall relax this assumption.

We assume further that the molecules are spheres of diameter d. A collision will then take place if the centers of two molecules come within a distance d of each other, as in Fig. 19-6a. Another, more helpful way to look at the situation is to consider our single molecule to have a *radius* of d and all the other molecules to be *points*, as in Fig. 19-6b. This does not change our criterion for a collision.

As our single molecule zigzags through the gas, it sweeps out a short cylinder of cross-sectional area πd^2 between successive collisions. If we watch this molecule for a time interval Δt, it moves a distance $v\,\Delta t$, where v is its assumed speed. Thus, if we align all the short cylinders swept out in interval Δt, we form a composite cylinder (Fig. 19-7) of length $v\,\Delta t$ and volume $(\pi d^2)(v\,\Delta t)$. The number of collisions that occur in time Δt is then equal to the number of (point) molecules that lie within this cylinder.

Since N/V is the number of molecules per unit volume, the number of molecules in the cylinder is N/V times the volume of the cylinder, or $(N/V)(\pi d^2 v\,\Delta t)$. This is also the number of collisions in time Δt. The mean free path is the length of the path (and of the cylinder) divided by this number:

$$\lambda = \frac{\text{length of path during } \Delta t}{\text{number of collisions in } \Delta t} \approx \frac{v\,\Delta t}{\pi d^2 v\,\Delta t\,N/V}$$

$$= \frac{1}{\pi d^2\,N/V}. \tag{19-26}$$

Fig. 19-7 In time Δt the moving molecule effectively sweeps out a cylinder of length $v\,\Delta t$ and radius d.

This equation is only approximate because it is based on the assumption that all the molecules except one are at rest. In fact, *all* the molecules are moving; when this is taken properly into account, Eq. 19-25 results. Note that it differs from the (approximate) Eq. 19-26 only by a factor of $1/\sqrt{2}$.

The approximation in Eq. 19-26 involves the two v symbols we canceled. The v in the numerator is v_{avg}, the mean speed of the molecules *relative to the container*. The v in the denominator is v_{rel}, the mean speed of our single molecule *relative to the other molecules*, which are moving. It is this latter average speed that determines the number of collisions. A detailed calculation, taking into account the actual speed distribution of the molecules, gives $v_{\text{rel}} = \sqrt{2}\,v_{\text{avg}}$ and thus the factor $\sqrt{2}$.

The mean free path of air molecules at sea level is about 0.1 μm. At an altitude of 100 km, the density of air has dropped to such an extent that the mean free path rises to about 16 cm. At 300 km, the mean free path is about 20 km. A problem faced by those who would study the physics and chemistry of the upper atmosphere in the laboratory is the unavailability of containers large enough to hold gas samples (of Freon, carbon dioxide, and ozone) that simulate upper atmospheric conditions.

✓CHECKPOINT 3

One mole of gas A, with molecular diameter $2d_0$ and average molecular speed v_0, is placed inside a certain container. One mole of gas B, with molecular diameter d_0 and average molecular speed $2v_0$ (the molecules of B are smaller but faster), is placed in an identical container. Which gas has the greater average collision rate within its container?

Sample Problem

Mean free path, average speed, collision frequency

(a) What is the mean free path λ for oxygen molecules at temperature $T = 300$ K and pressure $p = 1.0$ atm? Assume that the molecular diameter is $d = 290$ pm and the gas is ideal.

KEY IDEA

Each oxygen molecule moves among other *moving* oxygen molecules in a zigzag path due to the resulting collisions. Thus, we use Eq. 19-25 for the mean free path.

Calculation: We first need the number of molecules per unit volume, N/V. Because we assume the gas is ideal, we can use the ideal gas law of Eq. 19-9 ($pV = NkT$) to write $N/V = p/kT$. Substituting this into Eq. 19-25, we find

$$\lambda = \frac{1}{\sqrt{2}\,\pi d^2\,N/V} = \frac{kT}{\sqrt{2}\,\pi d^2 p}$$

$$= \frac{(1.38 \times 10^{-23}\,\text{J/K})(300\,\text{K})}{\sqrt{2}\,\pi(2.9 \times 10^{-10}\,\text{m})^2(1.01 \times 10^5\,\text{Pa})}$$

$$= 1.1 \times 10^{-7}\,\text{m}. \qquad \text{(Answer)}$$

This is about 380 molecular diameters.

(b) Assume the average speed of the oxygen molecules is $v = 450$ m/s. What is the average time t between successive

collisions for any given molecule? At what rate does the molecule collide; that is, what is the frequency f of its collisions?

KEY IDEAS

(1) Between collisions, the molecule travels, on average, the mean free path λ at speed v. (2) The average rate or frequency at which the collisions occur is the inverse of the time t between collisions.

Calculations: From the first key idea, the average time between collisions is

$$t = \frac{\text{distance}}{\text{speed}} = \frac{\lambda}{v} = \frac{1.1 \times 10^{-7}\,\text{m}}{450\,\text{m/s}}$$

$$= 2.44 \times 10^{-10}\,\text{s} \approx 0.24\,\text{ns}. \qquad \text{(Answer)}$$

This tells us that, on average, any given oxygen molecule has less than a nanosecond between collisions.

From the second key idea, the collision frequency is

$$f = \frac{1}{t} = \frac{1}{2.44 \times 10^{-10}\,\text{s}} = 4.1 \times 10^9\,\text{s}^{-1}. \qquad \text{(Answer)}$$

This tells us that, on average, any given oxygen molecule makes about 4 billion collisions per second.

 Additional examples, video, and practice available at *WileyPLUS*

19-7 The Distribution of Molecular Speeds

The root-mean-square speed v_{rms} gives us a general idea of molecular speeds in a gas at a given temperature. We often want to know more. For example, what fraction of the molecules have speeds greater than the rms value? What fraction have speeds greater than twice the rms value? To answer such questions, we need to know how the possible values of speed are distributed among the molecules. Figure 19-8a shows this distribution for oxygen molecules at room temperature ($T = 300$ K); Fig. 19-8b compares it with the distribution at $T = 80$ K.

In 1852, Scottish physicist James Clerk Maxwell first solved the problem of finding the speed distribution of gas molecules. His result, known as **Maxwell's speed distribution law,** is

$$P(v) = 4\pi\left(\frac{M}{2\pi RT}\right)^{3/2} v^2 e^{-Mv^2/2RT}. \qquad (19\text{-}27)$$

Here M is the molar mass of the gas, R is the gas constant, T is the gas temperature, and v is the molecular speed. It is this equation that is plotted in Fig. 19-8a, b. The quantity $P(v)$ in Eq. 19-27 and Fig. 19-8 is a *probability distribution function:* For any speed v, the product $P(v)\,dv$ (a dimensionless quantity) is the fraction of molecules with speeds in the interval dv centered on speed v.

As Fig. 19-8a shows, this fraction is equal to the area of a strip with height $P(v)$ and width dv. The total area under the distribution curve corresponds to the fraction of the molecules whose speeds lie between zero and infinity. All molecules fall into this category, so the value of this total area is unity; that is,

$$\int_0^\infty P(v)\,dv = 1. \qquad (19\text{-}28)$$

The fraction (frac) of molecules with speeds in an interval of, say, v_1 to v_2 is then

$$\text{frac} = \int_{v_1}^{v_2} P(v)\,dv. \qquad (19\text{-}29)$$

Average, RMS, and Most Probable Speeds

In principle, we can find the **average speed** v_{avg} of the molecules in a gas with the following procedure: We *weight* each value of v in the distribution; that is, we multiply it by the fraction $P(v)\,dv$ of molecules with speeds in a differential interval dv centered on v. Then we add up all these values of $v\,P(v)\,dv$. The result is v_{avg}. In practice, we do all this by evaluating

$$v_{avg} = \int_0^\infty v\,P(v)\,dv. \qquad (19\text{-}30)$$

Substituting for $P(v)$ from Eq. 19-27 and using generic integral 20 from the list of integrals in Appendix E, we find

$$v_{avg} = \sqrt{\frac{8RT}{\pi M}} \qquad \text{(average speed)}. \qquad (19\text{-}31)$$

Similarly, we can find the average of the square of the speeds $(v^2)_{avg}$ with

$$(v^2)_{avg} = \int_0^\infty v^2\,P(v)\,dv. \qquad (19\text{-}32)$$

Substituting for $P(v)$ from Eq. 19-27 and using generic integral 16 from the list of integrals in Appendix E, we find

$$(v^2)_{avg} = \frac{3RT}{M}. \qquad (19\text{-}33)$$

Fig. 19-8 (a) The Maxwell speed distribution for oxygen molecules at $T = 300$ K. The three characteristic speeds are marked. (b) The curves for 300 K and 80 K. Note that the molecules move more slowly at the lower temperature. Because these are probability distributions, the area under each curve has a numerical value of unity.

The square root of $(v^2)_{avg}$ is the root-mean-square speed v_{rms}. Thus,

$$v_{rms} = \sqrt{\frac{3RT}{M}} \quad \text{(rms speed)}, \quad (19\text{-}34)$$

which agrees with Eq. 19-22.

The **most probable speed** v_P is the speed at which $P(v)$ is maximum (see Fig. 19-8a). To calculate v_P, we set $dP/dv = 0$ (the slope of the curve in Fig. 19-8a is zero at the maximum of the curve) and then solve for v. Doing so, we find

$$v_P = \sqrt{\frac{2RT}{M}} \quad \text{(most probable speed)}. \quad (19\text{-}35)$$

A molecule is more likely to have speed v_P than any other speed, but some molecules will have speeds that are many times v_P. These molecules lie in the *high-speed tail* of a distribution curve like that in Fig. 19-8a. Such higher speed molecules make possible both rain and sunshine (without which we could not exist):

Rain The speed distribution of water molecules in, say, a pond at summertime temperatures can be represented by a curve similar to that of Fig. 19-8a. Most of the molecules do not have nearly enough kinetic energy to escape from the water through its surface. However, small numbers of very fast molecules with speeds far out in the high-speed tail of the curve can do so. It is these water molecules that evaporate, making clouds and rain a possibility.

As the fast water molecules leave the surface, carrying energy with them, the temperature of the remaining water is maintained by heat transfer from the surroundings. Other fast molecules—produced in particularly favorable collisions—quickly take the place of those that have left, and the speed distribution is maintained.

Sunshine Let the distribution curve of Fig. 19-8a now refer to protons in the core of the Sun. The Sun's energy is supplied by a nuclear fusion process that starts with the merging of two protons. However, protons repel each other because of their electrical charges, and protons of average speed do not have enough kinetic energy to overcome the repulsion and get close enough to merge. Very fast protons with speeds in the high-speed tail of the distribution curve can do so, however, and for that reason the Sun can shine.

Sample Problem

Speed distribution in a gas

A container is filled with oxygen gas maintained at room temperature (300 K). What fraction of the molecules have speeds in the interval 599 to 601 m/s? The molar mass M of oxygen is 0.0320 kg/mol.

KEY IDEAS

1. The speeds of the molecules are distributed over a wide range of values, with the distribution $P(v)$ of Eq. 19-27.

2. The fraction of molecules with speeds in a differential interval dv is $P(v)\,dv$.

3. For a larger interval, the fraction is found by integrating $P(v)$ over the interval.

4. However, the interval $\Delta v = 2$ m/s here is small compared to the speed $v = 600$ m/s on which it is centered.

Calculations: Because Δv is small, we can avoid the integration by approximating the fraction as

$$\text{frac} = P(v)\,\Delta v = 4\pi \left(\frac{M}{2\pi RT} \right)^{3/2} v^2 e^{-Mv^2/2RT}\,\Delta v.$$

The function $P(v)$ is plotted in Fig. 19-8a. The total area between the curve and the horizontal axis represents the to-

tal fraction of molecules (unity). The area of the thin gold strip represents the fraction we seek.

To evaluate frac in parts, we can write

$$\text{frac} = (4\pi)(A)(v^2)(e^B)(\Delta v), \qquad (19\text{-}36)$$

where

$$A = \left(\frac{M}{2\pi RT} \right)^{3/2} = \left(\frac{0.0320 \text{ kg/mol}}{(2\pi)(8.31 \text{ J/mol·K})(300 \text{ K})} \right)^{3/2}$$

$$= 2.92 \times 10^{-9} \text{ s}^3/\text{m}^3$$

and $B = -\dfrac{Mv^2}{2RT} = -\dfrac{(0.0320 \text{ kg/mol})(600 \text{ m/s})^2}{(2)(8.31 \text{ J/mol·K})(300 \text{ K})}$

$$= -2.31.$$

Substituting A and B into Eq. 19-36 yields

$$\text{frac} = (4\pi)(A)(v^2)(e^B)(\Delta v)$$

$$= (4\pi)(2.92 \times 10^{-9} \text{ s}^3/\text{m}^3)(600 \text{ m/s})^2(e^{-2.31})(2 \text{ m/s})$$

$$= 2.62 \times 10^{-3}. \qquad \text{(Answer)}$$

Thus, at room temperature, 0.262% of the oxygen molecules will have speeds that lie in the narrow range between 599 and 601 m/s. If the gold strip of Fig. 19-8a were drawn to the scale of this problem, it would be a very thin strip indeed.

Sample Problem

Average speed, rms speed, most probable speed

The molar mass M of oxygen is 0.0320 kg/mol.

(a) What is the average speed v_{avg} of oxygen gas molecules at $T = 300$ K?

KEY IDEA

To find the average speed, we must weight speed v with the distribution function $P(v)$ of Eq. 19-27 and then integrate the resulting expression over the range of possible speeds (from zero to the limit of an infinite speed).

Calculation: We end up with Eq. 19-31, which gives us

$$v_{avg} = \sqrt{\frac{8RT}{\pi M}}$$

$$= \sqrt{\frac{8(8.31 \text{ J/mol·K})(300 \text{ K})}{\pi(0.0320 \text{ kg/mol})}}$$

$$= 445 \text{ m/s.} \qquad \text{(Answer)}$$

This result is plotted in Fig. 19-8a.

(b) What is the root-mean-square speed v_{rms} at 300 K?

KEY IDEA

To find v_{rms}, we must first find $(v^2)_{avg}$ by weighting v^2 with the distribution function $P(v)$ of Eq. 19-27 and then integrating the expression over the range of possible speeds. Then we must take the square root of the result.

Calculation: We end up with Eq. 19-34, which gives us

$$v_{rms} = \sqrt{\frac{3RT}{M}}$$

$$= \sqrt{\frac{3(8.31 \text{ J/mol·K})(300 \text{ K})}{0.0320 \text{ kg/mol}}}$$

$$= 483 \text{ m/s.} \qquad \text{(Answer)}$$

This result, plotted in Fig. 19-8a, is greater than v_{avg} because the greater speed values influence the calculation more when we integrate the v^2 values than when we integrate the v values.

(c) What is the most probable speed v_P at 300 K?

$$v_P = \sqrt{\frac{2RT}{M}}$$

Speed v_P corresponds to the maximum of the distribution function $P(v)$, which we obtain by setting the derivative $dP/dv = 0$ and solving the result for v.

$$= \sqrt{\frac{2(8.31 \text{ J/mol} \cdot \text{K})(300 \text{ K})}{0.0320 \text{ kg/mol}}}$$

Calculation: We end up with Eq. 19-35, which gives us

$= 395$ m/s.   (Answer)

This result is also plotted in Fig. 19-8*a*.

 Additional examples, video, and practice available at *WileyPLUS*

19-8 The Molar Specific Heats of an Ideal Gas

In this section, we want to derive from molecular considerations an expression for the internal energy E_{int} of an ideal gas. In other words, we want an expression for the energy associated with the random motions of the atoms or molecules in the gas. We shall then use that expression to derive the molar specific heats of an ideal gas.

Internal Energy E_{int}

Let us first assume that our ideal gas is a *monatomic gas* (which has individual atoms rather than molecules), such as helium, neon, or argon. Let us also assume that the internal energy E_{int} of our ideal gas is simply the sum of the translational kinetic energies of its atoms. (As explained by quantum theory, individual atoms do not have rotational kinetic energy.)

The average translational kinetic energy of a single atom depends only on the gas temperature and is given by Eq. 19-24 as $K_{avg} = \frac{3}{2}kT$. A sample of n moles of such a gas contains nN_A atoms. The internal energy E_{int} of the sample is then

$$E_{int} = (nN_A)K_{avg} = (nN_A)(\tfrac{3}{2}kT). \qquad (19\text{-}37)$$

Using Eq. 19-7 ($k = R/N_A$), we can rewrite this as

$$E_{int} = \tfrac{3}{2}nRT \qquad \text{(monatomic ideal gas).} \qquad (19\text{-}38)$$

The internal energy E_{int} of an ideal gas is a function of the gas temperature *only*; it does not depend on any other variable.

With Eq. 19-38 in hand, we are now able to derive an expression for the molar specific heat of an ideal gas. Actually, we shall derive two expressions. One is for the case in which the volume of the gas remains constant as energy is transferred to or from it as heat. The other is for the case in which the pressure of the gas remains constant as energy is transferred to or from it as heat. The symbols for these two molar specific heats are C_V and C_p, respectively. (By convention, the capital letter C is used in both cases, even though C_V and C_p represent types of specific heat and not heat capacities.)

Molar Specific Heat at Constant Volume

Figure 19-9*a* shows n moles of an ideal gas at pressure p and temperature T, confined to a cylinder of fixed volume V. This *initial state i* of the gas is marked on the p-V diagram of Fig. 19-9*b*. Suppose now that you add a small amount of

(a)

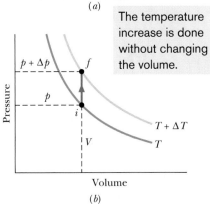

The temperature increase is done without changing the volume.

(b)

Fig. 19-9 (*a*) The temperature of an ideal gas is raised from T to $T + \Delta T$ in a constant-volume process. Heat is added, but no work is done. (*b*) The process on a p-V diagram.

Table 19-2		
Molar Specific Heats at Constant Volume		
Molecule	Example	C_V (J/mol·K)
Monatomic	Ideal	$\frac{3}{2}R = 12.5$
	Real He	12.5
	Ar	12.6
Diatomic	Ideal	$\frac{5}{2}R = 20.8$
	Real N_2	20.7
	O_2	20.8
Polyatomic	Ideal	$3R = 24.9$
	Real NH_4	29.0
	CO_2	29.7

energy to the gas as heat Q by slowly turning up the temperature of the thermal reservoir. The gas temperature rises a small amount to $T + \Delta T$, and its pressure rises to $p + \Delta p$, bringing the gas to *final state f*. In such experiments, we would find that the heat Q is related to the temperature change ΔT by

$$Q = nC_V \Delta T \qquad \text{(constant volume)}, \qquad (19\text{-}39)$$

where C_V is a constant called the **molar specific heat at constant volume.** Substituting this expression for Q into the first law of thermodynamics as given by Eq. 18-26 ($\Delta E_{int} = Q - W$) yields

$$\Delta E_{int} = nC_V \Delta T - W. \qquad (19\text{-}40)$$

With the volume held constant, the gas cannot expand and thus cannot do any work. Therefore, $W = 0$, and Eq. 19-40 gives us

$$C_V = \frac{\Delta E_{int}}{n \, \Delta T}. \qquad (19\text{-}41)$$

From Eq. 19-38, the change in internal energy must be

$$\Delta E_{int} = \tfrac{3}{2}nR \, \Delta T. \qquad (19\text{-}42)$$

Substituting this result into Eq. 19-41 yields

$$C_V = \tfrac{3}{2}R = 12.5 \text{ J/mol·K} \qquad \text{(monatomic gas)}. \qquad (19\text{-}43)$$

As Table 19-2 shows, this prediction of the kinetic theory (for ideal gases) agrees very well with experiment for real monatomic gases, the case that we have assumed. The (predicted and) experimental values of C_V for *diatomic gases* (which have molecules with two atoms) and *polyatomic gases* (which have molecules with more than two atoms) are greater than those for monatomic gases for reasons that will be suggested in Section 19-9.

We can now generalize Eq. 19-38 for the internal energy of any ideal gas by substituting C_V for $\frac{3}{2}R$; we get

$$E_{int} = nC_V T \qquad \text{(any ideal gas)}. \qquad (19\text{-}44)$$

This equation applies not only to an ideal monatomic gas but also to diatomic and polyatomic ideal gases, provided the appropriate value of C_V is used. Just as with Eq. 19-38, we see that the internal energy of a gas depends on the temperature of the gas but not on its pressure or density.

When a confined ideal gas undergoes temperature change ΔT, then from either Eq. 19-41 or Eq. 19-44 the resulting change in its internal energy is

$$\Delta E_{int} = nC_V \Delta T \qquad \text{(ideal gas, any process)}. \qquad (19\text{-}45)$$

This equation tells us:

> A change in the internal energy E_{int} of a confined ideal gas depends on only the change in the temperature, *not* on what type of process produces the change.

The paths are different, but the change in the internal energy is the same.

Fig. 19-10 Three paths representing three different processes that take an ideal gas from an initial state i at temperature T to some final state f at temperature $T + \Delta T$. The change ΔE_{int} in the internal energy of the gas is the same for these three processes and for any others that result in the same change of temperature.

As examples, consider the three paths between the two isotherms in the *p-V* diagram of Fig. 19-10. Path 1 represents a constant-volume process. Path 2 represents a constant-pressure process (that we are about to examine). Path 3 represents a process in which no heat is exchanged with the system's environment (we discuss this in Section 19-11). Although the values of heat Q and work W associated with these three paths differ, as do p_f and V_f, the values of ΔE_{int} associated with the three paths are identical and are all given by Eq. 19-45, because they all involve the same temperature change ΔT. Therefore, no matter what path is actually taken between T and $T + \Delta T$, we can *always* use path 1 and Eq. 19-45 to compute ΔE_{int} easily.

Molar Specific Heat at Constant Pressure

We now assume that the temperature of our ideal gas is increased by the same small amount ΔT as previously but now the necessary energy (heat Q) is added with the gas under constant pressure. An experiment for doing this is shown in Fig. 19-11a; the p-V diagram for the process is plotted in Fig. 19-11b. From such experiments we find that the heat Q is related to the temperature change ΔT by

$$Q = nC_p \Delta T \qquad \text{(constant pressure),} \qquad (19\text{-}46)$$

where C_p is a constant called the **molar specific heat at constant pressure.** This C_p is *greater* than the molar specific heat at constant volume C_V, because energy must now be supplied not only to raise the temperature of the gas but also for the gas to do work—that is, to lift the weighted piston of Fig. 19-11a.

To relate molar specific heats C_p and C_V, we start with the first law of thermodynamics (Eq. 18-26):

$$\Delta E_{\text{int}} = Q - W. \qquad (19\text{-}47)$$

We next replace each term in Eq. 19-47. For ΔE_{int}, we substitute from Eq. 19-45. For Q, we substitute from Eq. 19-46. To replace W, we first note that since the pressure remains constant, Eq. 19-16 tells us that $W = p \, \Delta V$. Then we note that, using the ideal gas equation ($pV = nRT$), we can write

$$W = p \, \Delta V = nR \, \Delta T. \qquad (19\text{-}48)$$

Making these substitutions in Eq. 19-47 and then dividing through by $n \, \Delta T$, we find

$$C_V = C_p - R$$

and then

$$C_p = C_V + R. \qquad (19\text{-}49)$$

This prediction of kinetic theory agrees well with experiment, not only for monatomic gases but also for gases in general, as long as their density is low enough so that we may treat them as ideal.

The left side of Fig. 19-12 shows the relative values of Q for a monatomic gas undergoing either a constant-volume process ($Q = \frac{3}{2}nR \, \Delta T$) or a constant-pressure process ($Q = \frac{5}{2}nR \, \Delta T$). Note that for the latter, the value of Q is higher by the amount W, the work done by the gas in the expansion. Note also that for the constant-volume process, the energy added as Q goes entirely into the change in internal energy ΔE_{int} and for the constant-pressure process, the energy added as Q goes into both ΔE_{int} and the work W.

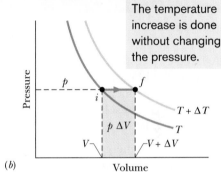

(a) Thermal reservoir

The temperature increase is done without changing the pressure.

(b)

Fig. 19-11 (a) The temperature of an ideal gas is raised from T to $T + \Delta T$ in a constant-pressure process. Heat is added and work is done in lifting the loaded piston. (b) The process on a p-V diagram. The work $p \, \Delta V$ is given by the shaded area.

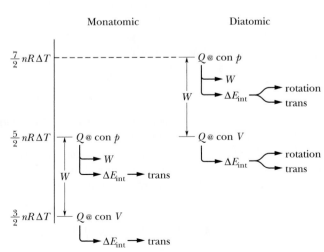

Fig. 19-12 The relative values of Q for a monatomic gas (left side) and a diatomic gas undergoing a constant-volume process (labeled "con V") and a constant-pressure process (labeled "con p"). The transfer of the energy into work W and internal energy (ΔE_{int}) is noted.

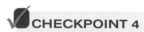

CHECKPOINT 4

The figure here shows five paths traversed by a gas on a *p-V* diagram. Rank the paths according to the change in internal energy of the gas, greatest first.

Sample Problem

Monatomic gas, heat, internal energy, and work

A bubble of 5.00 mol of helium is submerged at a certain depth in liquid water when the water (and thus the helium) undergoes a temperature increase ΔT of 20.0 C° at constant pressure. As a result, the bubble expands. The helium is monatomic and ideal.

(a) How much energy is added to the helium as heat during the increase and expansion?

KEY IDEA

Heat Q is related to the temperature change ΔT by a molar specific heat of the gas.

Calculations: Because the pressure p is held constant during the addition of energy, we use the molar specific heat at constant pressure C_p and Eq. 19-46,

$$Q = nC_p\,\Delta T, \tag{19-50}$$

to find Q. To evaluate C_p we go to Eq. 19-49, which tells us that for any ideal gas, $C_p = C_V + R$. Then from Eq. 19-43, we know that for any *monatomic* gas (like the helium here), $C_V = \frac{3}{2}R$. Thus, Eq. 19-50 gives us

$$Q = n(C_V + R)\,\Delta T = n(\tfrac{3}{2}R + R)\,\Delta T = n(\tfrac{5}{2}R)\,\Delta T$$
$$= (5.00\text{ mol})(2.5)(8.31\text{ J/mol}\cdot\text{K})(20.0\text{ C°})$$
$$= 2077.5\text{ J} \approx 2080\text{ J}. \tag{Answer}$$

(b) What is the change ΔE_{int} in the internal energy of the helium during the temperature increase?

KEY IDEA

Because the bubble expands, this is not a constant-volume process. However, the helium is nonetheless confined (to the bubble). Thus, the change ΔE_{int} is the same as *would occur* in a constant-volume process with the same temperature change ΔT.

Calculation: We can now easily find the constant-volume change ΔE_{int} with Eq. 19-45:

$$\Delta E_{\text{int}} = nC_V\,\Delta T = n(\tfrac{3}{2}R)\,\Delta T$$
$$= (5.00\text{ mol})(1.5)(8.31\text{ J/mol}\cdot\text{K})(20.0\text{ C°})$$
$$= 1246.5\text{ J} \approx 1250\text{ J}. \tag{Answer}$$

(c) How much work W is done by the helium as it expands against the pressure of the surrounding water during the temperature increase?

KEY IDEAS

The work done by *any* gas expanding against the pressure from its environment is given by Eq. 19-11, which tells us to integrate $p\,dV$. When the pressure is constant (as here), we can simplify that to $W = p\,\Delta V$. When the gas is *ideal* (as here), we can use the ideal gas law (Eq. 19-5) to write $p\,\Delta V = nR\,\Delta T$.

Calculation: We end up with

$$W = nR\,\Delta T$$
$$= (5.00\text{ mol})(8.31\text{ J/mol}\cdot\text{K})(20.0\text{ C°})$$
$$= 831\text{ J}. \tag{Answer}$$

Another way: Because we happen to know Q and ΔE_{int}, we can work this problem another way: We can account for the energy changes of the gas with the first law of thermodynamics, writing

$$W = Q - \Delta E_{\text{int}} = 2077.5\text{ J} - 1246.5\text{ J}$$
$$= 831\text{ J}. \tag{Answer}$$

The transfers: Let's follow the energy. Of the 2077.5 J transferred to the helium as heat Q, 831 J goes into the work W required for the expansion and 1246.5 J goes into the internal energy E_{int}, which, for a monatomic gas, is entirely the kinetic energy of the atoms in their translational motion. These several results are suggested on the left side of Fig. 19-12.

 Additional examples, video, and practice available at *WileyPLUS*

19-9 Degrees of Freedom and Molar Specific Heats

As Table 19-2 shows, the prediction that $C_V = \frac{3}{2}R$ agrees with experiment for monatomic gases but fails for diatomic and polyatomic gases. Let us try to explain the discrepancy by considering the possibility that molecules with more than one atom can store internal energy in forms other than translational kinetic energy.

Figure 19-13 shows common models of helium (a *monatomic* molecule, containing a single atom), oxygen (a *diatomic* molecule, containing two atoms), and methane (a *polyatomic* molecule). From such models, we would assume that all three types of molecules can have translational motions (say, moving left–right and up–down) and rotational motions (spinning about an axis like a top). In addition, we would assume that the diatomic and polyatomic molecules can have oscillatory motions, with the atoms oscillating slightly toward and away from one another, as if attached to opposite ends of a spring.

To keep account of the various ways in which energy can be stored in a gas, James Clerk Maxwell introduced the theorem of the **equipartition of energy:**

Every kind of molecule has a certain number f of *degrees of freedom,* which are independent ways in which the molecule can store energy. Each such degree of freedom has associated with it—on average—an energy of $\frac{1}{2}kT$ per molecule (or $\frac{1}{2}RT$ per mole).

Let us apply the theorem to the translational and rotational motions of the molecules in Fig. 19-13. (We discuss oscillatory motion in the next section.) For the translational motion, superimpose an *xyz* coordinate system on any gas. The molecules will, in general, have velocity components along all three axes. Thus, gas molecules of all types have three degrees of translational freedom (three ways to move in translation) and, on average, an associated energy of $3(\frac{1}{2}kT)$ per molecule.

For the rotational motion, imagine the origin of our *xyz* coordinate system at the center of each molecule in Fig. 19-13. In a gas, each molecule should be able to rotate with an angular velocity component along each of the three axes, so each gas should have three degrees of rotational freedom and, on average, an additional energy of $3(\frac{1}{2}kT)$ per molecule. *However,* experiment shows this is true only for the polyatomic molecules. According to *quantum theory,* the physics dealing with the allowed motions and energies of molecules and atoms, a monatomic gas molecule does not rotate and so has no rotational energy (a single atom cannot rotate like a top). A diatomic molecule can rotate like a top only about axes perpendicular to the line connecting the atoms (the axes are shown in Fig. 19-13b) and not about that line itself. Therefore, a diatomic molecule can have only two degrees of rotational freedom and a rotational energy of only $2(\frac{1}{2}kT)$ per molecule.

To extend our analysis of molar specific heats (C_p and C_V, in Section 19-8) to ideal diatomic and polyatomic gases, it is necessary to retrace the derivations of that analysis in detail. First, we replace Eq. 19-38 ($E_{\text{int}} = \frac{3}{2}nRT$) with $E_{\text{int}} = (f/2)nRT$, where f is the number of degrees of freedom listed in Table 19-3. Doing so leads to the prediction

$$C_V = \left(\frac{f}{2}\right)R = 4.16f \text{ J/mol}\cdot\text{K}, \tag{19-51}$$

which agrees—as it must—with Eq. 19-43 for monatomic gases ($f = 3$). As Table 19-2 shows, this prediction also agrees with experiment for diatomic gases ($f = 5$), but it is too low for polyatomic gases ($f = 6$ for molecules comparable to CH_4).

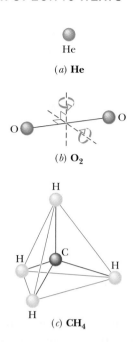

Fig. 19-13 Models of molecules as used in kinetic theory: (*a*) helium, a typical monatomic molecule; (*b*) oxygen, a typical diatomic molecule; and (*c*) methane, a typical polyatomic molecule. The spheres represent atoms, and the lines between them represent bonds. Two rotation axes are shown for the oxygen molecule.

	Table 19-3					
Degrees of Freedom for Various Molecules						
		Degrees of Freedom			Predicted Molar Specific Heats	
Molecule	Example	Translational	Rotational	Total (f)	C_V (Eq. 19-51)	$C_p = C_V + R$
Monatomic	He	3	0	3	$\frac{3}{2}R$	$\frac{5}{2}R$
Diatomic	O_2	3	2	5	$\frac{5}{2}R$	$\frac{7}{2}R$
Polyatomic	CH_4	3	3	6	$3R$	$4R$

Sample Problem

Diatomic gas, heat, temperature, internal energy

We transfer 1000 J as heat Q to a diatomic gas, allowing the gas to expand with the pressure held constant. The gas molecules each rotate around an internal axis but do not oscillate. How much of the 1000 J goes into the increase of the gas's internal energy? Of that amount, how much goes into ΔK_{tran} (the kinetic energy of the translational motion of the molecules) and ΔK_{rot} (the kinetic energy of their rotational motion)?

KEY IDEAS

1. The transfer of energy as heat Q to a gas under constant pressure is related to the resulting temperature increase ΔT via Eq. 19-46 ($Q = nC_p \Delta T$).

2. Because the gas is diatomic with molecules undergoing rotation but not oscillation, the molar specific heat is, from Fig. 19-12 and Table 19-3, $C_p = \frac{7}{2}R$.

3. The increase ΔE_{int} in the internal energy is the same as would occur with a constant-volume process resulting in the same ΔT. Thus, from Eq. 19-45, $\Delta E_{int} = nC_V \Delta T$. From Fig. 19-12 and Table 19-3, we see that $C_V = \frac{5}{2}R$.

4. For the same n and ΔT, ΔE_{int} is greater for a diatomic gas than a monatomic gas because additional energy is required for rotation.

Increase in E_{int}: Let's first get the temperature change ΔT due to the transfer of energy as heat. From Eq. 19-46, substituting $\frac{7}{2}R$ for C_p, we have

$$\Delta T = \frac{Q}{\frac{7}{2}nR}. \qquad (19\text{-}52)$$

We next find ΔE_{int} from Eq. 19-45, substituting the molar specific heat $C_V (= \frac{5}{2}R)$ for a constant-volume process and using the same ΔT. Because we are dealing with a di-

atomic gas, let's call this change $\Delta E_{int,dia}$. Equation 19-45 gives us

$$\Delta E_{int,dia} = nC_V \Delta T = n\frac{5}{2}R\left(\frac{Q}{\frac{7}{2}nR}\right) = \frac{5}{7}Q$$

$$= 0.71428Q = 714.3 \text{ J.} \qquad \text{(Answer)}$$

In words, about 71% of the energy transferred to the gas goes into the internal energy. The rest goes into the work required to increase the volume of the gas, as the gas pushes the walls of its container outward.

Increases in K: If we were to increase the temperature of a *monatomic* gas (with the same value of n) by the amount given in Eq. 19-52, the internal energy would change by a smaller amount, call it $\Delta E_{int, mon}$, because rotational motion is not involved. To calculate that smaller amount, we still use Eq. 19-45 but now we substitute the value of C_V for a monatomic gas—namely, $C_V = \frac{3}{2}R$. So,

$$\Delta E_{int,mon} = n\frac{3}{2}R \Delta T.$$

Substituting for ΔT from Eq. 19-52 leads us to

$$\Delta E_{int,mon} = n\frac{3}{2}R\left(\frac{Q}{n\frac{7}{2}R}\right) = \frac{3}{7}Q$$

$$= 0.42857Q = 428.6 \text{ J.}$$

For the monatomic gas, all this energy would go into the kinetic energy of the translational motion of the atoms. The important point here is that for a diatomic gas with the same values of n and ΔT, the same amount of energy goes into the kinetic energy of the translational motion of the molecules. The rest of $\Delta E_{int, dia}$ (that is, the additional 285.7 J) goes into the rotational motion of the molecules. Thus, for the diatomic gas,

$$\Delta K_{trans} = 428.6 \text{ J} \quad \text{and} \quad \Delta K_{rot} = 285.7 \text{ J.} \qquad \text{(Answer)}$$

 Additional examples, video, and practice available at *WileyPLUS*

19-10 A Hint of Quantum Theory

We can improve the agreement of kinetic theory with experiment by including the oscillations of the atoms in a gas of diatomic or polyatomic molecules. For example, the two atoms in the O_2 molecule of Fig. 19-13b can oscillate toward and away from each other, with the interconnecting bond acting like a spring. However, experiment shows that such oscillations occur only at relatively high temperatures of the gas—the motion is "turned on" only when the gas molecules have relatively large energies. Rotational motion is also subject to such "turning on," but at a lower temperature.

Figure 19-14 is of help in seeing this turning on of rotational motion and oscillatory motion. The ratio C_V/R for diatomic hydrogen gas (H_2) is plotted there against temperature, with the temperature scale logarithmic to cover several orders of magnitude. Below about 80 K, we find that $C_V/R = 1.5$. This result implies that only the three translational degrees of freedom of hydrogen are involved in the specific heat.

As the temperature increases, the value of C_V/R gradually increases to 2.5, implying that two additional degrees of freedom have become involved. Quantum theory shows that these two degrees of freedom are associated with the rotational motion of the hydrogen molecules and that this motion requires a certain minimum amount of energy. At very low temperatures (below 80 K), the molecules do not have enough energy to rotate. As the temperature increases from 80 K, first a few molecules and then more and more of them obtain enough energy to rotate, and the value of C_V/R increases, until all of the molecules are rotating and $C_V/R = 2.5$.

Similarly, quantum theory shows that oscillatory motion of the molecules requires a certain (higher) minimum amount of energy. This minimum amount is not met until the molecules reach a temperature of about 1000 K, as shown in Fig. 19-14. As the temperature increases beyond 1000 K, more and more molecules have enough energy to oscillate and the value of C_V/R increases, until all of the molecules are oscillating and $C_V/R = 3.5$. (In Fig. 19-14, the plotted curve stops at 3200 K because there the atoms of a hydrogen molecule oscillate so much that they overwhelm their bond, and the molecule then *dissociates* into two separate atoms.)

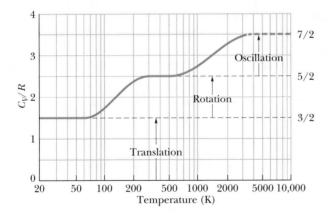

Fig. 19-14 C_V/R versus temperature for (diatomic) hydrogen gas. Because rotational and oscillatory motions begin at certain energies, only translation is possible at very low temperatures. As the temperature increases, rotational motion can begin. At still higher temperatures, oscillatory motion can begin.

19-11 The Adiabatic Expansion of an Ideal Gas

We saw in Section 17-4 that sound waves are propagated through air and other gases as a series of compressions and expansions; these variations in the transmission medium take place so rapidly that there is no time for energy to be transferred from one part of the medium to another as heat. As we saw in Section 18-11, a process for which $Q = 0$ is an *adiabatic process*. We can ensure that $Q = 0$ either by carrying out the process very quickly (as in sound waves) or by doing it (at any rate) in a well-insulated container.

Figure 19-15a shows our usual insulated cylinder, now containing an ideal gas and resting on an insulating stand. By removing mass from the piston, we can allow the gas to expand adiabatically. As the volume increases, both the pressure and the temperature drop. We shall prove next that the relation between the pressure and the volume during such an adiabatic process is

$$pV^{\gamma} = \text{a constant} \qquad \text{(adiabatic process)}, \qquad (19\text{-}53)$$

in which $\gamma = C_p/C_V$, the ratio of the molar specific heats for the gas. On a p-V diagram such as that in Fig. 19-15b, the process occurs along a line (called an *adiabat*) that has the equation $p = (\text{a constant})/V^{\gamma}$. Since the gas goes from an initial state i to a final state f, we can rewrite Eq. 19-53 as

$$p_i V_i^{\gamma} = p_f V_f^{\gamma} \qquad \text{(adiabatic process)}. \qquad (19\text{-}54)$$

To write an equation for an adiabatic process in terms of T and V, we use the ideal gas equation ($pV = nRT$) to eliminate p from Eq. 19-53, finding

$$\left(\frac{nRT}{V}\right) V^{\gamma} = \text{a constant}.$$

Because n and R are constants, we can rewrite this in the alternative form

$$TV^{\gamma-1} = \text{a constant} \qquad \text{(adiabatic process)}, \qquad (19\text{-}55)$$

in which the constant is different from that in Eq. 19-53. When the gas goes from an initial state i to a final state f, we can rewrite Eq. 19-55 as

$$T_i V_i^{\gamma-1} = T_f V_f^{\gamma-1} \qquad \text{(adiabatic process)}. \qquad (19\text{-}56)$$

Understanding adiabatic processes allows you to understand why popping the cork on a cold bottle of champagne or the tab on a cold can of soda causes a slight fog to form at the opening of the container. At the top of any unopened carbonated drink sits a gas of carbon dioxide and water vapor. Because the gas pressure is greater than atmospheric pressure, the gas expands out into the atmosphere

Fig. 19-15 (a) The volume of an ideal gas is increased by removing mass from the piston. The process is adiabatic ($Q = 0$). (b) The process proceeds from i to f along an adiabat on a p-V diagram.

We slowly remove lead shot, allowing an expansion without any heat transfer.

Adiabat ($Q = 0$)

Isotherms:
700 K
500 K
300 K

Insulation

Pressure

Volume

(a)

(b)

when the container is opened. Thus, the gas volume increases, but that means the gas must do work pushing against the atmosphere. Because the expansion is rapid, it is adiabatic, and the only source of energy for the work is the internal energy of the gas. Because the internal energy decreases, the temperature of the gas also decreases, which causes the water vapor in the gas to condense into tiny drops of fog.

Proof of Eq. 19-53

Suppose that you remove some shot from the piston of Fig. 19-15a, allowing the ideal gas to push the piston and the remaining shot upward and thus to increase the volume by a differential amount dV. Since the volume change is tiny, we may assume that the pressure p of the gas on the piston is constant during the change. This assumption allows us to say that the work dW done by the gas during the volume increase is equal to $p\,dV$. From Eq. 18-27, the first law of thermodynamics can then be written as

$$dE_{int} = Q - p\,dV. \qquad (19\text{-}57)$$

Since the gas is thermally insulated (and thus the expansion is adiabatic), we substitute 0 for Q. Then we use Eq. 19-45 to substitute $nC_V\,dT$ for dE_{int}. With these substitutions, and after some rearranging, we have

$$n\,dT = -\left(\frac{p}{C_V}\right)dV. \qquad (19\text{-}58)$$

Now from the ideal gas law ($pV = nRT$) we have

$$p\,dV + V\,dp = nR\,dT. \qquad (19\text{-}59)$$

Replacing R with its equal, $C_p - C_V$, in Eq. 19-59 yields

$$n\,dT = \frac{p\,dV + V\,dp}{C_p - C_V}. \qquad (19\text{-}60)$$

Equating Eqs. 19-58 and 19-60 and rearranging then give

$$\frac{dp}{p} + \left(\frac{C_p}{C_V}\right)\frac{dV}{V} = 0.$$

Replacing the ratio of the molar specific heats with γ and integrating (see integral 5 in Appendix E) yield

$$\ln p + \gamma \ln V = \text{a constant.}$$

Rewriting the left side as $\ln pV^\gamma$ and then taking the antilog of both sides, we find

$$pV^\gamma = \text{a constant.} \qquad (19\text{-}61)$$

Free Expansions

Recall from Section 18-11 that a free expansion of a gas is an adiabatic process with *no* work or change in internal energy. Thus, a free expansion differs from the adiabatic process described by Eqs. 19-53 through 19-61, in which work is done and the internal energy changes. Those equations then do *not* apply to a free expansion, even though such an expansion is adiabatic.

Also recall that in a free expansion, a gas is in equilibrium only at its initial and final points; thus, we can plot only those points, but not the expansion itself, on a p-V diagram. In addition, because $\Delta E_{int} = 0$, the temperature of the final state must be that of the initial state. Thus, the initial and final points on a p-V diagram must be on the same isotherm, and instead of Eq. 19-56 we have

$$T_i = T_f \qquad \text{(free expansion).} \qquad (19\text{-}62)$$

If we next assume that the gas is ideal (so that $pV = nRT$), then because there is no change in temperature, there can be no change in the product pV. Thus, instead of Eq. 19-53 a free expansion involves the relation

$$p_iV_i = p_fV_f \qquad \text{(free expansion).} \qquad (19\text{-}63)$$

Sample Problem

Adiabatic expansion, free expansion

Initially, 1 mol of oxygen (assumed to be an ideal gas) has temperature 310 K and volume 12 L. We will allow it to expand to volume 19 L.

(a) What would be the final temperature if the gas expands adiabatically? Oxygen (O_2) is diatomic and here has rotation but not oscillation.

KEY IDEAS

1. When a gas expands against the pressure of its environment, it must do work.

2. When the process is adiabatic (no energy is transferred as heat), then the energy required for the work can come only from the internal energy of the gas.

3. Because the internal energy decreases, the temperature T must also decrease.

Calculations: We can relate the initial and final temperatures and volumes with Eq. 19-56:

$$T_i V_i^{\gamma-1} = T_f V_f^{\gamma-1}. \qquad (19\text{-}64)$$

Because the molecules are diatomic and have rotation but not oscillation, we can take the molar specific heats from Table 19-3. Thus,

$$\gamma = \frac{C_p}{C_V} = \frac{\frac{7}{2}R}{\frac{5}{2}R} = 1.40.$$

Solving Eq. 19-64 for T_f and inserting known data then yield

$$T_f = \frac{T_i V_i^{\gamma-1}}{V_f^{\gamma-1}} = \frac{(310\text{ K})(12\text{ L})^{1.40-1}}{(19\text{ L})^{1.40-1}}$$

$$= (310\text{ K})(\tfrac{12}{19})^{0.40} = 258\text{ K}. \qquad \text{(Answer)}$$

(b) What would be the final temperature and pressure if, instead, the gas expands freely to the new volume, from an initial pressure of 2.0 Pa?

KEY IDEA

The temperature does not change in a free expansion because there is nothing to change the kinetic energy of the molecules.

Calculation: Thus, the temperature is

$$T_f = T_i = 310\text{ K}. \qquad \text{(Answer)}$$

We find the new pressure using Eq. 19-63, which gives us

$$p_f = p_i \frac{V_i}{V_f} = (2.0\text{ Pa})\frac{12\text{ L}}{19\text{ L}} = 1.3\text{ Pa}. \qquad \text{(Answer)}$$

Problem-Solving Tactics

A Graphical Summary of Four Gas Processes

In this chapter we have discussed four special processes that an ideal gas can undergo. An example of each (for a monatomic ideal gas) is shown in Fig. 19-16, and some associated characteristics are given in Table 19-4, including two process names (isobaric and isochoric) that we have not used but that you might see in other courses.

 CHECKPOINT 5

Rank paths 1, 2, and 3 in Fig. 19-16 according to the energy transfer to the gas as heat, greatest first.

Fig. 19-16 A p-V diagram representing four special processes for an ideal monatomic gas.

Table 19-4

Four Special Processes

Path in Fig. 19-16	Constant Quantity	Process Type	Some Special Results ($\Delta E_{int} = Q - W$ and $\Delta E_{int} = nC_V\Delta T$ for all paths)
1	p	Isobaric	$Q = nC_p\,\Delta T;\ W = p\,\Delta V$
2	T	Isothermal	$Q = W = nRT\ln(V_f/V_i);\ \Delta E_{int} = 0$
3	$pV^\gamma,\ TV^{\gamma-1}$	Adiabatic	$Q = 0;\ W = -\Delta E_{int}$
4	V	Isochoric	$Q = \Delta E_{int} = nC_V\,\Delta T;\ W = 0$

 Additional examples, video, and practice available at *WileyPLUS*

Kinetic Theory of Gases   The *kinetic theory of gases* relates the *macroscopic* properties of gases (for example, pressure and temperature) to the *microscopic* properties of gas molecules (for example, speed and kinetic energy).

Avogadro's Number   One mole of a substance contains N_A (*Avogadro's number*) elementary units (usually atoms or molecules), where N_A is found experimentally to be

$$N_A = 6.02 \times 10^{23} \text{ mol}^{-1} \quad \text{(Avogadro's number).} \quad (19\text{-}1)$$

One molar mass M of any substance is the mass of one mole of the substance. It is related to the mass m of the individual molecules of the substance by

$$M = mN_A. \quad (19\text{-}4)$$

The number of moles n contained in a sample of mass M_{sam}, consisting of N molecules, is given by

$$n = \frac{N}{N_A} = \frac{M_{sam}}{M} = \frac{M_{sam}}{mN_A}. \quad (19\text{-}2, 19\text{-}3)$$

Ideal Gas   An *ideal gas* is one for which the pressure p, volume V, and temperature T are related by

$$pV = nRT \quad \text{(ideal gas law).} \quad (19\text{-}5)$$

Here n is the number of moles of the gas present and R is a constant (8.31 J/mol·K) called the **gas constant.** The ideal gas law can also be written as

$$pV = NkT, \quad (19\text{-}9)$$

where the **Boltzmann constant** k is

$$k = \frac{R}{N_A} = 1.38 \times 10^{-23} \text{ J/K}. \quad (19\text{-}7)$$

Work in an Isothermal Volume Change   The work done *by* an ideal gas during an **isothermal** (constant-temperature) change from volume V_i to volume V_f is

$$W = nRT \ln \frac{V_f}{V_i} \quad \text{(ideal gas, isothermal process).} \quad (19\text{-}14)$$

Pressure, Temperature, and Molecular Speed   The pressure exerted by n moles of an ideal gas, in terms of the speed of its molecules, is

$$p = \frac{nMv_{rms}^2}{3V}, \quad (19\text{-}21)$$

where $v_{rms} = \sqrt{(v^2)_{avg}}$ is the **root-mean-square speed** of the molecules of the gas. With Eq. 19-5 this gives

$$v_{rms} = \sqrt{\frac{3RT}{M}}. \quad (19\text{-}22)$$

Temperature and Kinetic Energy   The average translational kinetic energy K_{avg} per molecule of an ideal gas is

$$K_{avg} = \tfrac{3}{2}kT. \quad (19\text{-}24)$$

Mean Free Path   The *mean free path* λ of a gas molecule is its average path length between collisions and is given by

$$\lambda = \frac{1}{\sqrt{2}\pi d^2 \, N/V}, \quad (19\text{-}25)$$

where N/V is the number of molecules per unit volume and d is the molecular diameter.

Maxwell Speed Distribution   The *Maxwell speed distribution* $P(v)$ is a function such that $P(v) \, dv$ gives the *fraction* of molecules with speeds in the interval dv at speed v:

$$P(v) = 4\pi \left(\frac{M}{2\pi RT}\right)^{3/2} v^2 e^{-Mv^2/2RT}. \quad (19\text{-}27)$$

Three measures of the distribution of speeds among the molecules of a gas are

$$v_{avg} = \sqrt{\frac{8RT}{\pi M}} \quad \text{(average speed),} \quad (19\text{-}31)$$

$$v_P = \sqrt{\frac{2RT}{M}} \quad \text{(most probable speed),} \quad (19\text{-}35)$$

and the rms speed defined above in Eq. 19-22.

Molar Specific Heats   The molar specific heat C_V of a gas at constant volume is defined as

$$C_V = \frac{Q}{n \, \Delta T} = \frac{\Delta E_{int}}{n \, \Delta T}, \quad (19\text{-}39, 19\text{-}41)$$

in which Q is the energy transferred as heat to or from a sample of n moles of the gas, ΔT is the resulting temperature change of the gas, and ΔE_{int} is the resulting change in the internal energy of the gas. For an ideal monatomic gas,

$$C_V = \tfrac{3}{2}R = 12.5 \text{ J/mol·K}. \quad (19\text{-}43)$$

The molar specific heat C_p of a gas at constant pressure is defined to be

$$C_p = \frac{Q}{n \, \Delta T}, \quad (19\text{-}46)$$

in which $Q, n,$ and ΔT are defined as above. C_p is also given by

$$C_p = C_V + R. \quad (19\text{-}49)$$

For n moles of an ideal gas,

$$E_{int} = nC_V T \quad \text{(ideal gas).} \quad (19\text{-}44)$$

If n moles of a confined ideal gas undergo a temperature change ΔT due to *any* process, the change in the internal energy of the gas is

$$\Delta E_{int} = nC_V \, \Delta T \quad \text{(ideal gas, any process).} \quad (19\text{-}45)$$

Degrees of Freedom and C_V   We find C_V by using the *equipartition of energy* theorem, which states that every *degree of freedom* of a molecule (that is, every independent way it can store energy) has associated with it—on average—an energy $\tfrac{1}{2}kT$ per molecule ($= \tfrac{1}{2}RT$ per mole). If f is the number of degrees of freedom, then $E_{int} = (f/2)nRT$ and

$$C_V = \left(\frac{f}{2}\right)R = 4.16f \text{ J/mol·K}. \quad (19\text{-}51)$$

For monatomic gases $f = 3$ (three translational degrees); for diatomic gases $f = 5$ (three translational and two rotational degrees).

Adiabatic Process   When an ideal gas undergoes a slow adiabatic volume change (a change for which $Q = 0$),

$$pV^\gamma = \text{a constant} \quad \text{(adiabatic process),} \quad (19\text{-}53)$$

in which $\gamma \, (= C_p/C_V)$ is the ratio of molar specific heats for the gas. For a free expansion, however, $pV = $ a constant.

QUESTIONS

1 For four situations for an ideal gas, the table gives the energy transferred to or from the gas as heat Q and either the work W done by the gas or the work W_{on} done on the gas, all in joules. Rank the four situations in terms of the temperature change of the gas, most positive first.

	a	b	c	d
Q	−50	+35	−15	+20
W	−50	+35		
W_{on}			−40	+40

2 In the p-V diagram of Fig. 19-17, the gas does 5 J of work when taken along isotherm ab and 4 J when taken along adiabat bc. What is the change in the internal energy of the gas when it is taken along the straight path from a to c?

Fig. 19-17 Question 2.

3 For a temperature increase of ΔT_1, a certain amount of an ideal gas requires 30 J when heated at constant volume and 50 J when heated at constant pressure. How much work is done by the gas in the second situation?

4 The dot in Fig. 19-18a represents the initial state of a gas, and the vertical line through the dot divides the p-V diagram into regions 1 and 2. For the following processes, determine whether the work W done by the gas is positive, negative, or zero: (a) the gas moves up along the vertical line, (b) it moves down along the vertical line, (c) it moves to anywhere in region 1, and (d) it moves to anywhere in region 2.

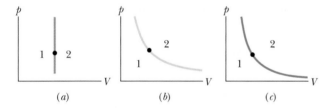

Fig. 19-18 Questions 4, 6, and 8.

5 A certain amount of energy is to be transferred as heat to 1 mol of a monatomic gas (a) at constant pressure and (b) at constant volume, and to 1 mol of a diatomic gas (c) at constant pressure and (d) at constant volume. Figure 19-19 shows four paths from an initial point to four final points on a p-V diagram. Which path goes with which process? (e) Are the molecules of the diatomic gas rotating?

Fig. 19-19 Question 5.

6 The dot in Fig. 19-18b represents the initial state of a gas, and the isotherm through the dot divides the p-V diagram into regions 1 and 2. For the following processes, determine whether the change ΔE_{int} in the internal energy of the gas is positive, negative, or zero: (a) the gas moves up along the isotherm, (b) it moves down along the isotherm, (c) it moves to anywhere in region 1, and (d) it moves to anywhere in region 2.

7 (a) Rank the four paths of Fig. 19-16 according to the work done by the gas, greatest first. (b) Rank paths 1, 2, and 3 according to the change in the internal energy of the gas, most positive first and most negative last.

8 The dot in Fig. 19-18c represents the initial state of a gas, and the adiabat through the dot divides the p-V diagram into regions 1 and 2. For the following processes, determine whether the corresponding heat Q is positive, negative, or zero: (a) the gas moves up along the adiabat, (b) it moves down along the adiabat, (c) it moves to anywhere in region 1, and (d) it moves to anywhere in region 2.

9 An ideal diatomic gas, with molecular rotation but not oscillation, loses energy as heat Q. Is the resulting decrease in the internal energy of the gas greater if the loss occurs in a constant-volume process or in a constant-pressure process?

10 Does the temperature of an ideal gas increase, decrease, or stay the same during (a) an isothermal expansion, (b) an expansion at constant pressure, (c) an adiabatic expansion, and (d) an increase in pressure at constant volume?

PROBLEMS

sec. 19-2 Avogadro's Number

•**1** Find the mass in kilograms of 7.50×10^{24} atoms of arsenic, which has a molar mass of 74.9 g/mol.

•**2** Gold has a molar mass of 197 g/mol. (a) How many moles of gold are in a 2.50 g sample of pure gold? (b) How many atoms are in the sample?

sec. 19-3 Ideal Gases

•**3** **SSM** Oxygen gas having a volume of 1000 cm³ at 40.0°C and 1.01×10^5 Pa expands until its volume is 1500 cm³ and its pressure is

1.06×10^5 Pa. Find (a) the number of moles of oxygen present and (b) the final temperature of the sample.

•**4** A quantity of ideal gas at 10.0°C and 100 kPa occupies a volume of 2.50 m³. (a) How many moles of the gas are present? (b) If the pressure is now raised to 300 kPa and the temperature is raised to 30.0°C, how much volume does the gas occupy? Assume no leaks.

•**5** The best laboratory vacuum has a pressure of about 1.00×10^{-18} atm, or 1.01×10^{-13} Pa. How many gas molecules are there per cubic centimeter in such a vacuum at 293 K?

•6   *Water bottle in a hot car.* In the American Southwest, the temperature in a closed car parked in sunlight during the summer can be high enough to burn flesh. Suppose a bottle of water at a refrigerator temperature of 5.00°C is opened, then closed, and then left in a closed car with an internal temperature of 75.0°C. Neglecting the thermal expansion of the water and the bottle, find the pressure in the air pocket trapped in the bottle. (The pressure can be enough to push the bottle cap past the threads that are intended to keep the bottle closed.)

•7 Suppose 1.80 mol of an ideal gas is taken from a volume of 3.00 m³ to a volume of 1.50 m³ via an isothermal compression at 30°C. (a) How much energy is transferred as heat during the compression, and (b) is the transfer *to* or *from* the gas?

•8 Compute (a) the number of moles and (b) the number of molecules in 1.00 cm³ of an ideal gas at a pressure of 100 Pa and a temperature of 220 K.

•9 An automobile tire has a volume of 1.64×10^{-2} m³ and contains air at a gauge pressure (pressure above atmospheric pressure) of 165 kPa when the temperature is 0.00°C. What is the gauge pressure of the air in the tires when its temperature rises to 27.0°C and its volume increases to 1.67×10^{-2} m³? Assume atmospheric pressure is 1.01×10^5 Pa.

•10 A container encloses 2 mol of an ideal gas that has molar mass M_1 and 0.5 mol of a second ideal gas that has molar mass $M_2 = 3M_1$. What fraction of the total pressure on the container wall is attributable to the second gas? (The kinetic theory explanation of pressure leads to the experimentally discovered law of partial pressures for a mixture of gases that do not react chemically: *The total pressure exerted by the mixture is equal to the sum of the pressures that the several gases would exert separately if each were to occupy the vessel alone.*)

••11 SSM ILW WWW Air that initially occupies 0.140 m³ at a gauge pressure of 103.0 kPa is expanded isothermally to a pressure of 101.3 kPa and then cooled at constant pressure until it reaches its initial volume. Compute the work done by the air. (Gauge pressure is the difference between the actual pressure and atmospheric pressure.)

••12   *Submarine rescue.* When the U. S. submarine *Squalus* became disabled at a depth of 80 m, a cylindrical chamber was lowered from a ship to rescue the crew. The chamber had a radius of 1.00 m and a height of 4.00 m, was open at the bottom, and held two rescuers. It slid along a guide cable that a diver had attached to a hatch on the submarine. Once the chamber reached the hatch and clamped to the hull, the crew could escape into the chamber. During the descent, air was released from tanks to prevent water from flooding the chamber. Assume that the interior air pressure matched the water pressure at depth *h* as given by $p_0 + \rho g h$, where $p_0 = 1.000$ atm is the surface pressure and $\rho = 1024$ kg/m³ is the density of seawater. Assume a surface temperature of 20.0°C and a submerged water temperature of −30.0°C. (a) What is the air volume in the chamber at the surface? (b) If air had not been released from the tanks, what would have been the air volume in the chamber at depth *h* = 80.0 m? (c) How many moles of air were needed to be released to maintain the original air volume in the chamber?

••13 GO A sample of an ideal gas is taken through the cyclic process *abca* shown in Fig. 19-20. The scale of the vertical axis is set by $p_b = 7.5$ kPa and $p_{ac} = 2.5$ kPa. At point *a*, *T* = 200 K. (a) How many moles of gas are in the sample? What are (b) the temperature of the gas at point *b*, (c) the temperature of the gas at point *c*, and (d) the net energy added to the gas as heat during the cycle?

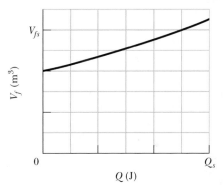

Fig. 19-20   Problem 13.

••14 In the temperature range 310 K to 330 K, the pressure *p* of a certain nonideal gas is related to volume *V* and temperature *T* by

$$p = (24.9 \text{ J/K}) \frac{T}{V} - (0.00662 \text{ J/K}^2) \frac{T^2}{V}.$$

How much work is done by the gas if its temperature is raised from 315 K to 325 K while the pressure is held constant?

••15 Suppose 0.825 mol of an ideal gas undergoes an isothermal expansion as energy is added to it as heat *Q*. If Fig. 19-21 shows the final volume V_f versus *Q*, what is the gas temperature? The scale of the vertical axis is set by $V_{fs} = 0.30$ m³, and the scale of the horizontal axis is set by $Q_s = 1200$ J.

Fig. 19-21   Problem 15.

•••16 An air bubble of volume 20 cm³ is at the bottom of a lake 40 m deep, where the temperature is 4.0°C. The bubble rises to the surface, which is at a temperature of 20°C. Take the temperature of the bubble's air to be the same as that of the surrounding water. Just as the bubble reaches the surface, what is its volume?

•••17 GO Container A in Fig. 19-22 holds an ideal gas at a pressure of 5.0×10^5 Pa and a temperature of 300 K. It is connected by a thin tube (and a closed valve) to container B, with four times the volume of A. Container B holds the same ideal gas at a pressure of 1.0×10^5 Pa and a temperature of 400 K. The valve is opened to allow the pressures to equalize, but the temperature of each container is maintained. What then is the pressure?

Fig. 19-22   Problem 17.

sec. 19-4   Pressure, Temperature, and RMS Speed

•18 The temperature and pressure in the Sun's atmosphere are 2.00×10^6 K and 0.0300 Pa. Calculate the rms speed of free electrons (mass 9.11×10^{-31} kg) there, assuming they are an ideal gas.

•19 (a) Compute the rms speed of a nitrogen molecule at 20.0°C. The molar mass of nitrogen molecules (N_2) is given in Table 19-1. At what temperatures will the rms speed be (b) half that value and (c) twice that value?

•20 Calculate the rms speed of helium atoms at 1000 K. See Appendix F for the molar mass of helium atoms.

•21 SSM The lowest possible temperature in outer space is 2.7 K. What is the rms speed of hydrogen molecules at this temperature? (The molar mass is given in Table 19-1.)

•22 Find the rms speed of argon atoms at 313 K. See Appendix F for the molar mass of argon atoms.

••23 A beam of hydrogen molecules (H_2) is directed toward a wall, at an angle of 55° with the normal to the wall. Each molecule in the beam has a speed of 1.0 km/s and a mass of 3.3×10^{-24} g. The beam strikes the wall over an area of 2.0 cm², at the rate of 10^{23} molecules per second. What is the beam's pressure on the wall?

••24 At 273 K and 1.00×10^{-2} atm, the density of a gas is 1.24×10^{-5} g/cm³. (a) Find v_{rms} for the gas molecules. (b) Find the molar mass of the gas and (c) identify the gas. (*Hint:* The gas is listed in Table 19-1.)

sec. 19-5 Translational Kinetic Energy

•25 Determine the average value of the translational kinetic energy of the molecules of an ideal gas at (a) 0.00°C and (b) 100°C. What is the translational kinetic energy per mole of an ideal gas at (c) 0.00°C and (d) 100°C?

•26 What is the average translational kinetic energy of nitrogen molecules at 1600 K?

••27 Water standing in the open at 32.0°C evaporates because of the escape of some of the surface molecules. The heat of vaporization (539 cal/g) is approximately equal to εn, where ε is the average energy of the escaping molecules and n is the number of molecules per gram. (a) Find ε. (b) What is the ratio of ε to the average kinetic energy of H_2O molecules, assuming the latter is related to temperature in the same way as it is for gases?

sec. 19-6 Mean Free Path

•28 At what frequency would the wavelength of sound in air be equal to the mean free path of oxygen molecules at 1.0 atm pressure and 0.00°C? The molecular diameter is 3.0×10^{-8} cm.

•29 SSM The atmospheric density at an altitude of 2500 km is about 1 molecule/cm³. (a) Assuming the molecular diameter of 2.0×10^{-8} cm, find the mean free path predicted by Eq. 19-25. (b) Explain whether the predicted value is meaningful.

•30 The mean free path of nitrogen molecules at 0.0°C and 1.0 atm is 0.80×10^{-5} cm. At this temperature and pressure there are 2.7×10^{19} molecules/cm³. What is the molecular diameter?

••31 In a certain particle accelerator, protons travel around a circular path of diameter 23.0 m in an evacuated chamber, whose residual gas is at 295 K and 1.00×10^{-6} torr pressure. (a) Calculate the number of gas molecules per cubic centimeter at this pressure. (b) What is the mean free path of the gas molecules if the molecular diameter is 2.00×10^{-8} cm?

••32 At 20°C and 750 torr pressure, the mean free paths for argon gas (Ar) and nitrogen gas (N_2) are $\lambda_{Ar} = 9.9 \times 10^{-6}$ cm and $\lambda_{N_2} = 27.5 \times 10^{-6}$ cm. (a) Find the ratio of the diameter of an Ar atom to that of an N_2 molecule. What is the mean free path of argon at (b) 20°C and 150 torr, and (c) −40°C and 750 torr?

sec. 19-7 The Distribution of Molecular Speeds

•33 SSM The speeds of 10 molecules are 2.0, 3.0, 4.0, . . . , 11 km/s. What are their (a) average speed and (b) rms speed?

•34 The speeds of 22 particles are as follows (N_i represents the number of particles that have speed v_i):

N_i	2	4	6	8	2
v_i (cm/s)	1.0	2.0	3.0	4.0	5.0

What are (a) v_{avg}, (b) v_{rms}, and (c) v_P?

•35 Ten particles are moving with the following speeds: four at 200 m/s, two at 500 m/s, and four at 600 m/s. Calculate their (a) average and (b) rms speeds. (c) Is $v_{rms} > v_{avg}$?

••36 It is found that the most probable speed of molecules in a gas when it has (uniform) temperature T_2 is the same as the rms speed of the molecules in this gas when it has (uniform) temperature T_1. Calculate T_2/T_1.

••37 SSM WWW Figure 19-23 shows a hypothetical speed distribution for a sample of N gas particles (note that $P(v) = 0$ for speed $v > 2v_0$). What are the values of (a) av_0, (b) v_{avg}/v_0, and (c) v_{rms}/v_0? (d) What fraction of the particles has a speed between $1.5v_0$ and $2.0v_0$?

Fig. 19-23 Problem 37.

••38 Figure 19-24 gives the probability distribution for nitrogen gas. The scale of the horizontal axis is set by $v_s = 1200$ m/s. What are the (a) gas temperature and (b) rms speed of the molecules?

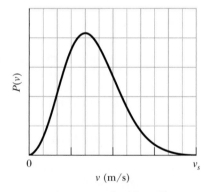

Fig. 19-24 Problem 38.

••39 At what temperature does the rms speed of (a) H_2 (molecular hydrogen) and (b) O_2 (molecular oxygen) equal the escape speed from Earth (Table 13-2)? At what temperature does the rms speed of (c) H_2 and (d) O_2 equal the escape speed from the Moon (where the gravitational acceleration at the surface has magnitude 0.16g)? Considering the answers to parts (a) and (b), should there be much (e) hydrogen and (f) oxygen high in Earth's upper atmosphere, where the temperature is about 1000 K?

••40 Two containers are at the same temperature. The first contains gas with pressure p_1, molecular mass m_1, and rms speed v_{rms1}. The second contains gas with pressure $2.0p_1$, molecular mass m_2, and average speed $v_{avg2} = 2.0v_{rms1}$. Find the mass ratio m_1/m_2.

••41 A hydrogen molecule (diameter 1.0×10^{-8} cm), traveling at the rms speed, escapes from a 4000 K furnace into a chamber containing *cold* argon atoms (diameter 3.0×10^{-8} cm) at a density of 4.0×10^{19} atoms/cm³. (a) What is the speed of the hydrogen molecule? (b) If it collides with an argon atom, what is the closest their centers can be, considering each as spherical? (c) What is the initial number of collisions per second experienced by the hydrogen molecule? (*Hint*: Assume that the argon atoms are stationary. Then the mean free path of the hydrogen molecule is given by Eq. 19-26 and not Eq. 19-25.)

sec. 19-8   The Molar Specific Heats of an Ideal Gas

•42   What is the internal energy of 1.0 mol of an ideal monatomic gas at 273 K?

••43   GO   The temperature of 3.00 mol of an ideal diatomic gas is increased by 40.0 C° without the pressure of the gas changing. The molecules in the gas rotate but do not oscillate. (a) How much energy is transferred to the gas as heat? (b) What is the change in the internal energy of the gas? (c) How much work is done by the gas? (d) By how much does the rotational kinetic energy of the gas increase?

••44   GO   One mole of an ideal diatomic gas goes from *a* to *c* along the diagonal path in Fig. 19-25. The scale of the vertical axis is set by $p_{ab} = 5.0$ kPa and $p_c = 2.0$ kPa, and the scale of the horizontal axis is set by $V_{bc} = 4.0$ m³ and $V_a = 2.0$ m³. During the transition, (a) what is the change in internal energy of the gas, and (b) how much energy is added to the gas as heat? (c) How much heat is required if the gas goes from *a* to *c* along the indirect path *abc*?

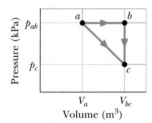

Fig. 19-25   Problem 44.

••45   ILW   The mass of a gas molecule can be computed from its specific heat at constant volume c_V. (Note that this is not C_V.) Take $c_V = 0.075$ cal/g·C° for argon and calculate (a) the mass of an argon atom and (b) the molar mass of argon.

••46   Under constant pressure, the temperature of 2.00 mol of an ideal monatomic gas is raised 15.0 K. What are (a) the work W done by the gas, (b) the energy transferred as heat Q, (c) the change ΔE_{int} in the internal energy of the gas, and (d) the change ΔK in the average kinetic energy per atom?

••47   The temperature of 2.00 mol of an ideal monatomic gas is raised 15.0 K at constant volume. What are (a) the work W done by the gas, (b) the energy transferred as heat Q, (c) the change ΔE_{int} in the internal energy of the gas, and (d) the change ΔK in the average kinetic energy per atom?

••48   When 20.9 J was added as heat to a particular ideal gas, the volume of the gas changed from 50.0 cm³ to 100 cm³ while the pressure remained at 1.00 atm. (a) By how much did the internal energy of the gas change? If the quantity of gas present was 2.00×10^{-3} mol, find (b) C_p and (c) C_V.

••49   SSM   A container holds a mixture of three nonreacting gases: 2.40 mol of gas 1 with $C_{V1} = 12.0$ J/mol·K, 1.50 mol of gas 2 with $C_{V2} = 12.8$ J/mol·K, and 3.20 mol of gas 3 with $C_{V3} = 20.0$ J/mol·K. What is C_V of the mixture?

sec. 19-9   Degrees of Freedom and Molar Specific Heats

•50   We give 70 J as heat to a diatomic gas, which then expands at constant pressure. The gas molecules rotate but do not oscillate. By how much does the internal energy of the gas increase?

•51   ILW   When 1.0 mol of oxygen (O_2) gas is heated at constant pressure starting at 0°C, how much energy must be added to the gas as heat to double its volume? (The molecules rotate but do not oscillate.)

••52   Suppose 12.0 g of oxygen (O_2) gas is heated at constant atmospheric pressure from 25.0°C to 125°C. (a) How many moles of oxygen are present? (See Table 19-1 for the molar mass.) (b) How much energy is transferred to the oxygen as heat? (The molecules rotate but do not oscillate.) (c) What fraction of the heat is used to raise the internal energy of the oxygen?

••53   SSM   WWW   Suppose 4.00 mol of an ideal diatomic gas, with molecular rotation but not oscillation, experienced a temperature increase of 60.0 K under constant-pressure conditions. What are (a) the energy transferred as heat Q, (b) the change ΔE_{int} in internal energy of the gas, (c) the work W done by the gas, and (d) the change ΔK in the total translational kinetic energy of the gas?

sec. 19-11   The Adiabatic Expansion of an Ideal Gas

•54   We know that for an adiabatic process $pV^\gamma = $ a constant. Evaluate "a constant" for an adiabatic process involving exactly 2.0 mol of an ideal gas passing through the state having exactly $p = 1.0$ atm and $T = 300$ K. Assume a diatomic gas whose molecules rotate but do not oscillate.

•55   A certain gas occupies a volume of 4.3 L at a pressure of 1.2 atm and a temperature of 310 K. It is compressed adiabatically to a volume of 0.76 L. Determine (a) the final pressure and (b) the final temperature, assuming the gas to be an ideal gas for which $\gamma = 1.4$.

•56   Suppose 1.00 L of a gas with $\gamma = 1.30$, initially at 273 K and 1.00 atm, is suddenly compressed adiabatically to half its initial volume. Find its final (a) pressure and (b) temperature. (c) If the gas is then cooled to 273 K at constant pressure, what is its final volume?

••57   The volume of an ideal gas is adiabatically reduced from 200 L to 74.3 L. The initial pressure and temperature are 1.00 atm and 300 K. The final pressure is 4.00 atm. (a) Is the gas monatomic, diatomic, or polyatomic? (b) What is the final temperature? (c) How many moles are in the gas?

••58   Opening champagne. In a bottle of champagne, the pocket of gas (primarily carbon dioxide) between the liquid and the cork is at pressure of $p_i = 5.00$ atm. When the cork is pulled from the bottle, the gas undergoes an adiabatic expansion until its pressure matches the ambient air pressure of 1.00 atm. Assume that the ratio of the molar specific heats is $\gamma = \frac{4}{3}$. If the gas has initial temperature $T_i = 5.00$°C, what is its temperature at the end of the adiabatic expansion?

••59   GO   Figure 19-26 shows two paths that may be taken by a gas from an initial point *i* to a final point *f*. Path 1 consists of an isothermal expansion (work is 50 J in magnitude), an adiabatic expansion

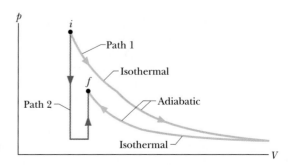

Fig. 19-26   Problem 59.

(work is 40 J in magnitude), an isothermal compression (work is 30 J in magnitude), and then an adiabatic compression (work is 25 J in magnitude). What is the change in the internal energy of the gas if the gas goes from point *i* to point *f* along path 2?

••60 **GO** *Adiabatic wind.* The normal airflow over the Rocky Mountains is west to east. The air loses much of its moisture content and is chilled as it climbs the western side of the mountains. When it descends on the eastern side, the increase in pressure toward lower altitudes causes the temperature to increase. The flow, then called a chinook wind, can rapidly raise the air temperature at the base of the mountains. Assume that the air pressure *p* depends on altitude *y* according to $p = p_0 \exp(-ay)$, where $p_0 = 1.00$ atm and $a = 1.16 \times 10^{-4}$ m^{-1}. Also assume that the ratio of the molar specific heats is $\gamma = \frac{4}{3}$. A parcel of air with an initial temperature of $-5.00°$C descends adiabatically from $y_1 = 4267$ m to $y = 1567$ m. What is its temperature at the end of the descent?

••61 **GO** A gas is to be expanded from initial state *i* to final state *f* along either path 1 or path 2 on a *p-V* diagram. Path 1 consists of three steps: an isothermal expansion (work is 40 J in magnitude), an adiabatic expansion (work is 20 J in magnitude), and another isothermal expansion (work is 30 J in magnitude). Path 2 consists of two steps: a pressure reduction at constant volume and an expansion at constant pressure. What is the change in the internal energy of the gas along path 2?

•••62 **GO** An ideal diatomic gas, with rotation but no oscillation, undergoes an adiabatic compression. Its initial pressure and volume are 1.20 atm and 0.200 m^3. Its final pressure is 2.40 atm. How much work is done by the gas?

•••63 Figure 19-27 shows a cycle undergone by 1.00 mol of an ideal monatomic gas. The temperatures are $T_1 = 300$ K, $T_2 = 600$ K, and $T_3 = 455$ K. For $1 \to 2$, what are (a) heat Q, (b) the change in internal energy ΔE_{int}, and (c) the work done W? For $2 \to 3$, what are (d) Q, (e) ΔE_{int}, and (f) W? For $3 \to 1$, what are (g) Q, (h) ΔE_{int}, and (i) W? For the full cycle, what are (j) Q, (k) ΔE_{int}, and (l) W? The initial pressure at point 1 is 1.00 atm ($= 1.013 \times 10^5$ Pa). What are the (m) volume and (n) pressure at point 2 and the (o) volume and (p) pressure at point 3?

Fig. 19-27 Problem 63.

Additional Problems

64 Calculate the work done by an external agent during an isothermal compression of 1.00 mol of oxygen from a volume of 22.4 L at 0°C and 1.00 atm to a volume of 16.8 L.

65 An ideal gas undergoes an adiabatic compression from $p = 1.0$ atm, $V = 1.0 \times 10^6$ L, $T = 0.0°$C to $p = 1.0 \times 10^5$ atm, $V = 1.0 \times 10^3$ L. (a) Is the gas monatomic, diatomic, or polyatomic? (b) What is its final temperature? (c) How many moles of gas are present? What is the total translational kinetic energy per mole (d) before and (e) after the compression? (f) What is the ratio of the squares of the rms speeds before and after the compression?

66 An ideal gas consists of 1.50 mol of diatomic molecules that rotate but do not oscillate. The molecular diameter is 250 pm. The gas is expanded at a constant pressure of 1.50×10^5 Pa, with a transfer of 200 J as heat. What is the change in the mean free path of the molecules?

67 An ideal monatomic gas initially has a temperature of 330 K and a pressure of 6.00 atm. It is to expand from volume 500 cm^3 to volume 1500 cm^3. If the expansion is isothermal, what are (a) the final pressure and (b) the work done by the gas? If, instead, the expansion is adiabatic, what are (c) the final pressure and (d) the work done by the gas?

68 In an interstellar gas cloud at 50.0 K, the pressure is 1.00×10^{-8} Pa. Assuming that the molecular diameters of the gases in the cloud are all 20.0 nm, what is their mean free path?

69 **SSM** The envelope and basket of a hot-air balloon have a combined weight of 2.45 kN, and the envelope has a capacity (volume) of 2.18×10^3 m^3. When it is fully inflated, what should be the temperature of the enclosed air to give the balloon a *lifting capacity* (force) of 2.67 kN (in addition to the balloon's weight)? Assume that the surrounding air, at 20.0°C, has a weight per unit volume of 11.9 N/m^3 and a molecular mass of 0.028 kg/mol, and is at a pressure of 1.0 atm.

70 An ideal gas, at initial temperature T_1 and initial volume 2.0 m^3, is expanded adiabatically to a volume of 4.0 m^3, then expanded isothermally to a volume of 10 m^3, and then compressed adiabatically back to T_1. What is its final volume?

71 **SSM** The temperature of 2.00 mol of an ideal monatomic gas is raised 15.0 K in an adiabatic process. What are (a) the work W done by the gas, (b) the energy transferred as heat Q, (c) the change ΔE_{int} in internal energy of the gas, and (d) the change ΔK in the average kinetic energy per atom?

72 At what temperature do atoms of helium gas have the same rms speed as molecules of hydrogen gas at 20.0°C? (The molar masses are given in Table 19-1.)

73 **SSM** At what frequency do molecules (diameter 290 pm) collide in (an ideal) oxygen gas (O_2) at temperature 400 K and pressure 2.00 atm?

74 (a) What is the number of molecules per cubic meter in air at 20°C and at a pressure of 1.0 atm ($= 1.01 \times 10^5$ Pa)? (b) What is the mass of 1.0 m^3 of this air? Assume that 75% of the molecules are nitrogen (N_2) and 25% are oxygen (O_2).

75 The temperature of 3.00 mol of a gas with $C_V = 6.00$ cal/mol·K is to be raised 50.0 K. If the process is at *constant volume*, what are (a) the energy transferred as heat Q, (b) the work W done by the gas, (c) the change ΔE_{int} in internal energy of the gas, and (d) the change ΔK in the total translational kinetic energy? If the process is at *constant pressure*, what are (e) Q, (f) W, (g) ΔE_{int}, and (h) ΔK? If the process is *adiabatic*, what are (i) Q, (j) W, (k) ΔE_{int}, and (l) ΔK?

76 During a compression at a constant pressure of 250 Pa, the volume of an ideal gas decreases from 0.80 m^3 to 0.20 m^3. The initial temperature is 360 K, and the gas loses 210 J as heat. What are (a) the change in the internal energy of the gas and (b) the final temperature of the gas?

77 **SSM** Figure 19-28 shows a hypothetical speed distribution

Fig. 19-28 Problem 77.

for particles of a certain gas: $P(v) = Cv^2$ for $0 < v \leq v_0$ and $P(v) = 0$ for $v > v_0$. Find (a) an expression for C in terms of v_0, (b) the average speed of the particles, and (c) their rms speed.

78   (a) An ideal gas initially at pressure p_0 undergoes a free expansion until its volume is 3.00 times its initial volume. What then is the ratio of its pressure to p_0? (b) The gas is next slowly and adiabatically compressed back to its original volume. The pressure after compression is $(3.00)^{1/3}p_0$. Is the gas monatomic, diatomic, or polyatomic? (c) What is the ratio of the average kinetic energy per molecule in this final state to that in the initial state?

79 SSM   An ideal gas undergoes isothermal compression from an initial volume of 4.00 m^3 to a final volume of 3.00 m^3. There is 3.50 mol of the gas, and its temperature is $10.0°\text{C}$. (a) How much work is done by the gas? (b) How much energy is transferred as heat between the gas and its environment?

80   Oxygen (O_2) gas at 273 K and 1.0 atm is confined to a cubical container 10 cm on a side. Calculate $\Delta U_g / K_{avg}$, where ΔU_g is the change in the gravitational potential energy of an oxygen molecule falling the height of the box and K_{avg} is the molecule's average translational kinetic energy.

81   An ideal gas is taken through a complete cycle in three steps: adiabatic expansion with work equal to 125 J, isothermal contraction at 325 K, and increase in pressure at constant volume. (a) Draw a p-V diagram for the three steps. (b) How much energy is transferred as heat in step 3, and (c) is it transferred *to* or *from* the gas?

82   (a) What is the volume occupied by 1.00 mol of an ideal gas at standard conditions—that is, 1.00 atm ($= 1.01 \times 10^5$ Pa) and 273 K? (b) Show that the number of molecules per cubic centimeter (the *Loschmidt number*) at standard conditions is 2.69×10^9.

83 SSM   A sample of ideal gas expands from an initial pressure and volume of 32 atm and 1.0 L to a final volume of 4.0 L. The initial temperature is 300 K. If the gas is monatomic and the expansion isothermal, what are the (a) final pressure p_f, (b) final temperature T_f, and (c) work W done by the gas? If the gas is monatomic and the expansion adiabatic, what are (d) p_f, (e) T_f, and (f) W? If the gas is diatomic and the expansion adiabatic, what are (g) p_f, (h) T_f, and (i) W?

84   An ideal gas with 3.00 mol is initially in state 1 with pressure $p_1 = 20.0$ atm and volume $V_1 = 1500 \text{ cm}^3$. First it is taken to state 2 with pressure $p_2 = 1.50p_1$ and volume $V_2 = 2.00V_1$. Then it is taken to state 3 with pressure $p_3 = 2.00p_1$ and volume $V_3 = 0.500V_1$. What is the temperature of the gas in (a) state 1 and (b) state 2? (c) What is the net change in internal energy from state 1 to state 3?

85   A steel tank contains 300 g of ammonia gas (NH_3) at a pressure of 1.35×10^6 Pa and a temperature of $77°\text{C}$. (a) What is the volume of the tank in liters? (b) Later the temperature is $22°\text{C}$ and the pressure is 8.7×10^5 Pa. How many grams of gas have leaked out of the tank?

86   In an industrial process the volume of 25.0 mol of a monatomic ideal gas is reduced at a uniform rate from 0.616 m^3 to 0.308 m^3 in 2.00 h while its temperature is increased at a uniform rate from $27.0°\text{C}$ to $450°\text{C}$. Throughout the process, the gas passes through thermodynamic equilibrium states. What are (a) the cumulative work done on the gas, (b) the cumulative energy absorbed by the gas as heat, and (c) the molar specific heat for the process? (*Hint:* To evaluate the integral for the work, you might use

$$\int \frac{a + bx}{A + Bx} dx = \frac{bx}{B} + \frac{aB - bA}{B^2} \ln(A + Bx),$$

an indefinite integral.) Suppose the process is replaced with a two-step process that reaches the same final state. In step 1, the gas volume is reduced at constant temperature, and in step 2 the temperature is increased at constant volume. For this process, what are (d) the cumulative work done on the gas, (e) the cumulative energy absorbed by the gas as heat, and (f) the molar specific heat for the process?

87   Figure 19-29 shows a cycle consisting of five paths: AB is isothermal at 300 K, BC is adiabatic with work $= 5.0$ J, CD is at a constant pressure of 5 atm, DE is isothermal, and EA is adiabatic with a change in internal energy of 8.0 J. What is the change in internal energy of the gas along path CD?

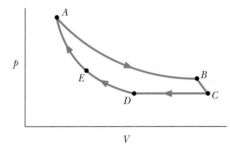

Fig. 19-29   Problem 87.

88   An ideal gas initially at 300 K is compressed at a constant pressure of 25 N/m^2 from a volume of 3.0 m^3 to a volume of 1.8 m^3. In the process, 75 J is lost by the gas as heat. What are (a) the change in internal energy of the gas and (b) the final temperature of the gas?

20

ENTROPY AND THE SECOND LAW OF THERMODYNAMICS

20-1 WHAT IS PHYSICS?

Time has direction, the direction in which we age. We are accustomed to many one-way processes—that is, processes that can occur only in a certain sequence (the right way) and never in the reverse sequence (the wrong way). An egg is dropped onto a floor, a pizza is baked, a car is driven into a lamppost, large waves erode a sandy beach—these one-way processes are **irreversible,** meaning that they cannot be reversed by means of only small changes in their environment.

One goal of physics is to understand why time has direction and why one-way processes are irreversible. Although this physics might seem disconnected from the practical issues of everyday life, it is in fact at the heart of any engine, such as a car engine, because it determines how well an engine can run.

The key to understanding why one-way processes cannot be reversed involves a quantity known as *entropy*.

20-2 Irreversible Processes and Entropy

The one-way character of irreversible processes is so pervasive that we take it for granted. If these processes were to occur *spontaneously* (on their own) in the wrong way, we would be astonished. Yet *none* of these wrong-way events would violate the law of conservation of energy.

For example, if you were to wrap your hands around a cup of hot coffee, you would be astonished if your hands got cooler and the cup got warmer. That is obviously the wrong way for the energy transfer, but the total energy of the closed system (*hands + cup of coffee*) would be the same as the total energy if the process had run in the right way. For another example, if you popped a helium balloon, you would be astonished if, later, all the helium molecules were to gather together in the original shape of the balloon. That is obviously the wrong way for molecules to spread, but the total energy of the closed system (*molecules + room*) would be the same as for the right way.

Thus, changes in energy within a closed system do not set the direction of irreversible processes. Rather, that direction is set by another property that we shall discuss in this chapter—the *change in entropy* ΔS of the system. The change in entropy of a system is defined in the next section, but we can here state its central property, often called the *entropy postulate*:

If an irreversible process occurs in a *closed* system, the entropy S of the system always increases; it never decreases.

Entropy differs from energy in that entropy does *not* obey a conservation law. The *energy* of a closed system is conserved; it always remains constant. For irreversible processes, the *entropy* of a closed system always increases. Because of this property, the change in entropy is sometimes called "the arrow of time." For example, we associate the explosion of a popcorn kernel with the forward direction of time and with an increase in entropy. The backward direction of time (a videotape run backwards) would correspond to the exploded popcorn re-forming the original kernel. Because this backward process would result in an entropy decrease, it never happens.

There are two equivalent ways to define the change in entropy of a system: (1) in terms of the system's temperature and the energy the system gains or loses as heat, and (2) by counting the ways in which the atoms or molecules that make up the system can be arranged. We use the first approach in the next section and the second in Section 20-8.

20-3 Change in Entropy

Let's approach this definition of *change in entropy* by looking again at a process that we described in Sections 18-11 and 19-11: the free expansion of an ideal gas. Figure 20-1a shows the gas in its initial equilibrium state *i*, confined by a closed stopcock to the left half of a thermally insulated container. If we open the stopcock, the gas rushes to fill the entire container, eventually reaching the final equilibrium state *f* shown in Fig. 20-1b. This is an irreversible process; all the molecules of the gas will never return to the left half of the container.

The *p-V* plot of the process, in Fig. 20-2, shows the pressure and volume of the gas in its initial state *i* and final state *f*. Pressure and volume are *state properties*, properties that depend only on the state of the gas and not on how it reached that state. Other state properties are temperature and energy. We now assume that the gas has still another state property—its entropy. Furthermore, we define the **change in entropy** $S_f - S_i$ of a system during a process that takes the system from an initial state *i* to a final state *f* as

$$\Delta S = S_f - S_i = \int_i^f \frac{dQ}{T} \quad \text{(change in entropy defined)}. \quad (20\text{-}1)$$

Here *Q* is the energy transferred as heat to or from the system during the process, and *T* is the temperature of the system in kelvins. Thus, an entropy change depends not only on the energy transferred as heat but also on the temperature at which the transfer takes place. Because *T* is always positive, the sign of ΔS is the same as that of *Q*. We see from Eq. 20-1 that the SI unit for entropy and entropy change is the joule per kelvin.

There is a problem, however, in applying Eq. 20-1 to the free expansion of Fig. 20-1. As the gas rushes to fill the entire container, the pressure, temperature, and volume of the gas fluctuate unpredictably. In other words, they do not have a sequence of well-defined equilibrium values during the intermediate stages of the change from initial state *i* to final state *f*. Thus, we cannot trace a pressure–volume path for the free expansion on the *p-V* plot of Fig. 20-2, and we cannot find a relation between *Q* and *T* that allows us to integrate as Eq. 20-1 requires.

However, if entropy is truly a state property, the difference in entropy between states *i* and *f* must depend *only on those states* and not at all on the way the system went from one state to the other. Suppose, then, that we replace the irreversible free expansion of Fig. 20-1 with a *reversible* process that connects states *i* and *f*. With a reversible process we can trace a pressure–volume path on a *p-V* plot, and we can find a relation between *Q* and *T* that allows us to use Eq. 20-1 to obtain the entropy change.

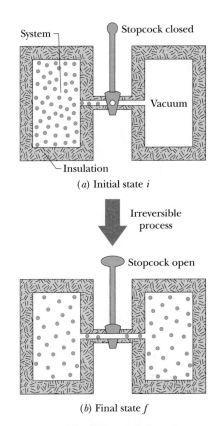

(a) Initial state *i*

Irreversible process

(b) Final state *f*

Fig. 20-1 The free expansion of an ideal gas. (a) The gas is confined to the left half of an insulated container by a closed stopcock. (b) When the stopcock is opened, the gas rushes to fill the entire container. This process is irreversible; that is, it does not occur in reverse, with the gas spontaneously collecting itself in the left half of the container.

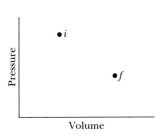

Fig. 20-2 A *p-V* diagram showing the initial state *i* and the final state *f* of the free expansion of Fig. 20-1. The intermediate states of the gas cannot be shown because they are not equilibrium states.

Insulation

Lead shot

Q

Thermal reservoir   Control knob T

(a) Initial state i

Reversible process

Lead shot

T

(b) Final state f

Fig. 20-3 The isothermal expansion of an ideal gas, done in a reversible way. The gas has the same initial state i and same final state f as in the irreversible process of Figs. 20-1 and 20-2.

We saw in Section 19-11 that the temperature of an ideal gas does not change during a free expansion: $T_i = T_f = T$. Thus, points i and f in Fig. 20-2 must be on the same isotherm. A convenient replacement process is then a reversible isothermal expansion from state i to state f, which actually proceeds *along* that isotherm. Furthermore, because T is constant throughout a reversible isothermal expansion, the integral of Eq. 20-1 is greatly simplified.

Figure 20-3 shows how to produce such a reversible isothermal expansion. We confine the gas to an insulated cylinder that rests on a thermal reservoir maintained at the temperature T. We begin by placing just enough lead shot on the movable piston so that the pressure and volume of the gas are those of the initial state i of Fig. 20-1a. We then remove shot slowly (piece by piece) until the pressure and volume of the gas are those of the final state f of Fig. 20-1b. The temperature of the gas does not change because the gas remains in thermal contact with the reservoir throughout the process.

The reversible isothermal expansion of Fig. 20-3 is physically quite different from the irreversible free expansion of Fig. 20-1. However, *both processes have the same initial state and the same final state and thus must have the same change in entropy.* Because we removed the lead shot slowly, the intermediate states of the gas are equilibrium states, so we can plot them on a p-V diagram (Fig. 20-4).

To apply Eq. 20-1 to the isothermal expansion, we take the constant temperature T outside the integral, obtaining

$$\Delta S = S_f - S_i = \frac{1}{T}\int_i^f dQ.$$

Because $\int dQ = Q$, where Q is the total energy transferred as heat during the process, we have

$$\Delta S = S_f - S_i = \frac{Q}{T} \qquad \text{(change in entropy, isothermal process).} \qquad (20\text{-}2)$$

To keep the temperature T of the gas constant during the isothermal expansion of Fig. 20-3, heat Q must have been energy transferred *from* the reservoir *to* the gas. Thus, Q is positive and the entropy of the gas *increases* during the isothermal process and during the free expansion of Fig. 20-1.

To summarize:

> To find the entropy change for an irreversible process occurring in a *closed* system, replace that process with any reversible process that connects the same initial and final states. Calculate the entropy change for this reversible process with Eq. 20-1.

When the temperature change ΔT of a system is small relative to the temperature (in kelvins) before and after the process, the entropy change can be approximated as

$$\Delta S = S_f - S_i \approx \frac{Q}{T_{\text{avg}}}, \qquad (20\text{-}3)$$

where T_{avg} is the average temperature of the system in kelvins during the process.

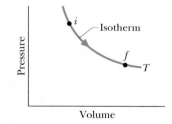

i —Isotherm

Pressure

f T

Volume

Fig. 20-4 A p-V diagram for the reversible isothermal expansion of Fig. 20-3. The intermediate states, which are now equilibrium states, are shown.

CHECKPOINT 1

Water is heated on a stove. Rank the entropy changes of the water as its temperature rises (a) from 20°C to 30°C, (b) from 30°C to 35°C, and (c) from 80°C to 85°C, greatest first.

Entropy as a State Function

We have assumed that entropy, like pressure, energy, and temperature, is a property of the state of a system and is independent of how that state is reached. That entropy is indeed a *state function* (as state properties are usually called) can be deduced only by experiment. However, we can prove it is a state function for the special and important case in which an ideal gas is taken through a reversible process.

To make the process reversible, it is done slowly in a series of small steps, with the gas in an equilibrium state at the end of each step. For each small step, the energy transferred as heat to or from the gas is dQ, the work done by the gas is dW, and the change in internal energy is dE_{int}. These are related by the first law of thermodynamics in differential form (Eq. 18-27):

$$dE_{int} = dQ - dW.$$

Because the steps are reversible, with the gas in equilibrium states, we can use Eq. 18-24 to replace dW with $p\, dV$ and Eq. 19-45 to replace dE_{int} with $nC_V\, dT$. Solving for dQ then leads to

$$dQ = p\, dV + nC_V\, dT.$$

Using the ideal gas law, we replace p in this equation with nRT/V. Then we divide each term in the resulting equation by T, obtaining

$$\frac{dQ}{T} = nR\frac{dV}{V} + nC_V\frac{dT}{T}.$$

Now let us integrate each term of this equation between an arbitrary initial state i and an arbitrary final state f to get

$$\int_i^f \frac{dQ}{T} = \int_i^f nR\frac{dV}{V} + \int_i^f nC_V\frac{dT}{T}.$$

The quantity on the left is the entropy change $\Delta S\ (= S_f - S_i)$ defined by Eq. 20-1. Substituting this and integrating the quantities on the right yield

$$\Delta S = S_f - S_i = nR\ln\frac{V_f}{V_i} + nC_V\ln\frac{T_f}{T_i}. \qquad (20\text{-}4)$$

Note that we did not have to specify a particular reversible process when we integrated. Therefore, the integration must hold for all reversible processes that take the gas from state i to state f. Thus, the change in entropy ΔS between the initial and final states of an ideal gas depends only on properties of the initial state (V_i and T_i) and properties of the final state (V_f and T_f); ΔS does not depend on how the gas changes between the two states.

 CHECKPOINT 2

An ideal gas has temperature T_1 at the initial state i shown in the p-V diagram here. The gas has a higher temperature T_2 at final states a and b, which it can reach along the paths shown. Is the entropy change along the path to state a larger than, smaller than, or the same as that along the path to state b?

Sample Problem

Entropy change of two blocks coming to thermal equilibrium

Figure 20-5a shows two identical copper blocks of mass $m = 1.5$ kg: block L at temperature $T_{iL} = 60°C$ and block R at temperature $T_{iR} = 20°C$. The blocks are in a thermally insulated box and are separated by an insulating shutter. When we lift the shutter, the blocks eventually come to the equilibrium temperature $T_f = 40°C$ (Fig. 20-5b). What is the net entropy change of the two-block system during this irreversible process? The specific heat of copper is 386 J/kg·K.

KEY IDEA

To calculate the entropy change, we must find a reversible process that takes the system from the initial state of Fig. 20-5a to the final state of Fig. 20-5b. We can calculate the net entropy change ΔS_{rev} of the reversible process using Eq. 20-1, and then the entropy change for the irreversible process is equal to ΔS_{rev}.

Calculations: For the reversible process, we need a thermal reservoir whose temperature can be changed slowly (say, by turning a knob). We then take the blocks through the following two steps, illustrated in Fig. 20-6.

Step 1: With the reservoir's temperature set at 60°C, put block L on the reservoir. (Since block and reservoir are at the same temperature, they are already in thermal equilibrium.) Then slowly lower the temperature of the reservoir and the block to 40°C. As the block's temperature changes by each increment dT during this process, energy dQ is transferred as heat *from* the block to the reservoir. Using Eq. 18-14, we can write this transferred energy as $dQ = mc\, dT$, where c is the specific heat of copper. According to Eq. 20-1,

Fig. 20-5 (a) In the initial state, two copper blocks L and R, identical except for their temperatures, are in an insulating box and are separated by an insulating shutter. (b) When the shutter is removed, the blocks exchange energy as heat and come to a final state, both with the same temperature T_f.

Fig. 20-6 The blocks of Fig. 20-5 can proceed from their initial state to their final state in a reversible way if we use a reservoir with a controllable temperature (a) to extract heat reversibly from block L and (b) to add heat reversibly to block R.

the entropy change ΔS_L of block L during the full temperature change from initial temperature T_{iL} ($= 60°C = 333$ K) to final temperature T_f ($= 40°C = 313$ K) is

$$\Delta S_L = \int_i^f \frac{dQ}{T} = \int_{T_{iL}}^{T_f} \frac{mc\,dT}{T} = mc \int_{T_{iL}}^{T_f} \frac{dT}{T}$$
$$= mc \ln \frac{T_f}{T_{iL}}.$$

Inserting the given data yields

$$\Delta S_L = (1.5 \text{ kg})(386 \text{ J/kg·K}) \ln \frac{313 \text{ K}}{333 \text{ K}}$$
$$= -35.86 \text{ J/K}.$$

Step 2: With the reservoir's temperature now set at 20°C, put block R on the reservoir. Then slowly raise the temperature of the reservoir and the block to 40°C. With the same reasoning used to find ΔS_L, you can show that the entropy change ΔS_R of block R during this process is

$$\Delta S_R = (1.5 \text{ kg})(386 \text{ J/kg·K}) \ln \frac{313 \text{ K}}{293 \text{ K}}$$
$$= +38.23 \text{ J/K}.$$

The net entropy change ΔS_{rev} of the two-block system undergoing this two-step reversible process is then

$$\Delta S_{rev} = \Delta S_L + \Delta S_R$$
$$= -35.86 \text{ J/K} + 38.23 \text{ J/K} = 2.4 \text{ J/K}.$$

Thus, the net entropy change ΔS_{irrev} for the two-block system undergoing the actual irreversible process is

$$\Delta S_{irrev} = \Delta S_{rev} = 2.4 \text{ J/K}. \qquad \text{(Answer)}$$

This result is positive, in accordance with the entropy postulate of Section 20-2.

Entropy change of a free expansion of a gas

Suppose 1.0 mol of nitrogen gas is confined to the left side of the container of Fig. 20-1a. You open the stopcock, and the volume of the gas doubles. What is the entropy change of the gas for this irreversible process? Treat the gas as ideal.

(1) We can determine the entropy change for the irreversible process by calculating it for a reversible process that provides the same change in volume. (2) The temperature of the gas does not change in the free expansion. Thus, the reversible process should be an isothermal expansion—namely, the one of Figs. 20-3 and 20-4.

Calculations: From Table 19-4, the energy Q added as heat to the gas as it expands isothermally at temperature T from an initial volume V_i to a final volume V_f is

$$Q = nRT \ln \frac{V_f}{V_i},$$

in which n is the number of moles of gas present. From Eq. 20-2 the entropy change for this reversible process in which the temperature is held constant is

$$\Delta S_{\text{rev}} = \frac{Q}{T} = \frac{nRT \ln(V_f/V_i)}{T} = nR \ln \frac{V_f}{V_i}.$$

Substituting $n = 1.00$ mol and $V_f/V_i = 2$, we find

$$\Delta S_{\text{rev}} = nR \ln \frac{V_f}{V_i} = (1.00 \text{ mol})(8.31 \text{ J/mol} \cdot \text{K})(\ln 2)$$

$$= +5.76 \text{ J/K}.$$

Thus, the entropy change for the free expansion (and for all other processes that connect the initial and final states shown in Fig. 20-2) is

$$\Delta S_{\text{irrev}} = \Delta S_{\text{rev}} = +5.76 \text{ J/K}. \qquad \text{(Answer)}$$

Because ΔS is positive, the entropy increases, in accordance with the entropy postulate of Section 20-2.

 Additional examples, video, and practice available at *WileyPLUS*

20-4 The Second Law of Thermodynamics

Here is a puzzle. If we cause the reversible process of Fig. 20-3 to proceed from (a) to (b) in that figure, the change in entropy of the gas—which we take as our system—is positive. However, because the process is reversible, we can just as easily make it proceed from (b) to (a), simply by slowly adding lead shot to the piston of Fig. 20-3b until the original volume of the gas is restored. In this reverse process, energy must be extracted as heat *from the gas* to keep its temperature from rising. Hence Q is negative and so, from Eq. 20-2, the entropy of the gas must decrease.

Doesn't this decrease in the entropy of the gas violate the entropy postulate of Section 20-2, which states that entropy always increases? No, because that postulate holds only for *irreversible* processes occurring in closed systems. The procedure suggested here does not meet these requirements. The process is *not* irreversible, and (because energy is transferred as heat from the gas to the reservoir) the system—which is the gas alone—is *not* closed.

However, if we include the reservoir, along with the gas, as part of the system, then we do have a closed system. Let's check the change in entropy of the enlarged system *gas + reservoir* for the process that takes it from (b) to (a) in Fig. 20-3. During this reversible process, energy is transferred as heat from the gas to the reservoir—that is, from one part of the enlarged system to another. Let |Q| represent the absolute value (or magnitude) of this heat. With Eq. 20-2, we can then calculate separately the entropy changes for the gas (which loses |Q|) and the reservoir (which gains |Q|). We get

$$\Delta S_{\text{gas}} = -\frac{|Q|}{T}$$

and
$$\Delta S_{res} = +\frac{|Q|}{T}.$$

The entropy change of the closed system is the sum of these two quantities: 0.

With this result, we can modify the entropy postulate of Section 20-2 to include both reversible and irreversible processes:

 If a process occurs in a *closed* system, the entropy of the system increases for irreversible processes and remains constant for reversible processes. It never decreases.

Although entropy may decrease in part of a closed system, there will always be an equal or larger entropy increase in another part of the system, so that the entropy of the system as a whole never decreases. This fact is one form of the **second law of thermodynamics** and can be written as

$$\Delta S \geq 0 \qquad \text{(second law of thermodynamics)}, \qquad (20\text{-}5)$$

where the greater-than sign applies to irreversible processes and the equals sign to reversible processes. Equation 20-5 applies only to closed systems.

In the real world almost all processes are irreversible to some extent because of friction, turbulence, and other factors, so the entropy of real closed systems undergoing real processes always increases. Processes in which the system's entropy remains constant are always idealizations.

Force Due to Entropy

To understand why rubber resists being stretched, let's write the first law of thermodynamics
$$dE = dQ - dW$$

for a rubber band undergoing a small increase in length dx as we stretch it between our hands. The force from the rubber band has magnitude F, is directed inward, and does work $dW = -F\,dx$ during length increase dx. From Eq. 20-2 ($\Delta S = Q/T$), small changes in Q and S at constant temperature are related by $dS = dQ/T$, or $dQ = T\,dS$. So, now we can rewrite the first law as

$$dE = T\,dS + F\,dx. \qquad (20\text{-}6)$$

To good approximation, the change dE in the internal energy of rubber is 0 if the total stretch of the rubber band is not very much. Substituting 0 for dE in Eq. 20-6 leads us to an expression for the force from the rubber band:

$$F = -T\frac{dS}{dx}. \qquad (20\text{-}7)$$

This tells us that F is proportional to the rate dS/dx at which the rubber band's entropy changes during a small change dx in the rubber band's length. Thus, you can *feel* the effect of entropy on your hands as you stretch a rubber band.

To make sense of the relation between force and entropy, let's consider a simple model of the rubber material. Rubber consists of cross-linked polymer chains (long molecules with cross links) that resemble three-dimensional zig-zags (Fig. 20-7). When the rubber band is at its rest length, the polymers are coiled up in a spaghetti-like arrangement. Because of the large disorder of the molecules, this rest state has a high value of entropy. When we stretch a rubber band, we uncoil many of those polymers, aligning them in the direction of stretch. Because the alignment decreases the disorder, the entropy of the stretched rubber band is less. That is, the change dS/dx in Eq. 20-7 is a negative quantity because the entropy decreases with stretching. Thus, the force on our hands from the rubber band is due to the tendency of the polymers to return to their former disordered state and higher value of entropy.

Fig. 20-7   A section of a rubber band (*a*) unstretched and (*b*) stretched, and a polymer within it (*a*) coiled and (*b*) uncoiled.

20-5 Entropy in the Real World: Engines

A **heat engine,** or more simply, an **engine,** is a device that extracts energy from its environment in the form of heat and does useful work. At the heart of every engine is a *working substance.* In a steam engine, the working substance is water, in both its vapor and its liquid form. In an automobile engine the working substance is a gasoline–air mixture. If an engine is to do work on a sustained basis, the working substance must operate in a *cycle;* that is, the working substance must pass through a closed series of thermodynamic processes, called *strokes,* returning again and again to each state in its cycle. Let us see what the laws of thermodynamics can tell us about the operation of engines.

A Carnot Engine

We have seen that we can learn much about real gases by analyzing an ideal gas, which obeys the simple law $pV = nRT$. Although an ideal gas does not exist, any real gas approaches ideal behavior if its density is low enough. Similarly, we can study real engines by analyzing the behavior of an **ideal engine.**

In an ideal engine, all processes are reversible and no wasteful energy transfers occur due to, say, friction and turbulence.

We shall focus on a particular ideal engine called a **Carnot engine** after the French scientist and engineer N. L. Sadi Carnot (pronounced "car-no"), who first proposed the engine's concept in 1824. This ideal engine turns out to be the best (in principle) at using energy as heat to do useful work. Surprisingly, Carnot was able to analyze the performance of this engine before the first law of thermodynamics and the concept of entropy had been discovered.

Figure 20-8 shows schematically the operation of a Carnot engine. During each cycle of the engine, the working substance absorbs energy $|Q_H|$ as heat from a thermal reservoir at constant temperature T_H and discharges energy $|Q_L|$ as heat to a second thermal reservoir at a constant lower temperature T_L.

Fig. 20-8 The elements of a Carnot engine. The two black arrowheads on the central loop suggest the working substance operating in a cycle, as if on a *p-V* plot. Energy $|Q_H|$ is transferred as heat from the high-temperature reservoir at temperature T_H to the working substance. Energy $|Q_L|$ is transferred as heat from the working substance to the low-temperature reservoir at temperature T_L. Work W is done by the engine (actually by the working substance) on something in the environment.

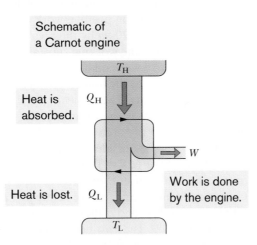

Figure 20-9 shows a *p-V* plot of the *Carnot cycle*—the cycle followed by the working substance. As indicated by the arrows, the cycle is traversed in the clockwise direction. Imagine the working substance to be a gas, confined to an insulating cylinder with a weighted, movable piston. The cylinder may be placed at will on either of the two thermal reservoirs, as in Fig. 20-6, or on an insulating slab. Figure 20-9*a* shows that, if we place the cylinder in contact with the high-temperature reservoir at temperature T_H, heat $|Q_H|$ is transferred *to* the working

Fig. 20-9 A pressure–volume plot of the cycle followed by the working substance of the Carnot engine in Fig. 20-8. The cycle consists of two isothermal (*ab* and *cd*) and two adiabatic processes (*bc* and *da*). The shaded area enclosed by the cycle is equal to the work *W* per cycle done by the Carnot engine.

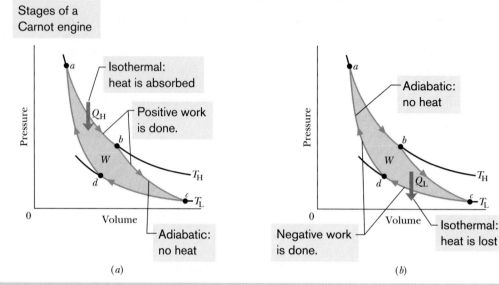

Stages of a Carnot engine

(a) Isothermal: heat is absorbed — Positive work is done. — Adiabatic: no heat

(b) Adiabatic: no heat — Negative work is done. — Isothermal: heat is lost

substance *from* this reservoir as the gas undergoes an isothermal *expansion* from volume V_a to volume V_b. Similarly, with the working substance in contact with the low-temperature reservoir at temperature T_L, heat $|Q_L|$ is transferred *from* the working substance *to* the low-temperature reservoir as the gas undergoes an isothermal *compression* from volume V_c to volume V_d (Fig. 20-9*b*).

In the engine of Fig. 20-8, we assume that heat transfers to or from the working substance can take place *only* during the isothermal processes *ab* and *cd* of Fig. 20-9. Therefore, processes *bc* and *da* in that figure, which connect the two isotherms at temperatures T_H and T_L, must be (reversible) adiabatic processes; that is, they must be processes in which no energy is transferred as heat. To ensure this, during processes *bc* and *da* the cylinder is placed on an insulating slab as the volume of the working substance is changed.

During the processes *ab* and *bc* of Fig. 20-9*a*, the working substance is expanding and thus doing positive work as it raises the weighted piston. This work is represented in Fig. 20-9*a* by the area under curve *abc*. During the processes *cd* and *da* (Fig. 20-9*b*), the working substance is being compressed, which means that it is doing negative work on its environment or, equivalently, that its environment is doing work on it as the loaded piston descends. This work is represented by the area under curve *cda*. The *net work per cycle*, which is represented by *W* in both Figs. 20-8 and 20-9, is the difference between these two areas and is a positive quantity equal to the area enclosed by cycle *abcda* in Fig. 20-9. This work *W* is performed on some outside object, such as a load to be lifted.

Equation 20-1 ($\Delta S = \int dQ/T$) tells us that any energy transfer as heat must involve a change in entropy. To see this for a Carnot engine, we can plot the Carnot cycle on a temperature–entropy (*T-S*) diagram as in Fig. 20-10. The lettered points *a*, *b*, *c*, and *d* there correspond to the lettered points in the *p-V* diagram in Fig. 20-9. The two horizontal lines in Fig. 20-10 correspond to the two isothermal processes of the cycle. Process *ab* is the isothermal expansion of the cycle. As the working substance (reversibly) absorbs energy $|Q_H|$ as heat at constant temperature T_H during the expansion, its entropy increases. Similarly, during the isothermal compression *cd*, the working substance (reversibly) loses energy $|Q_L|$ as heat at constant temperature T_L, and its entropy decreases.

The two vertical lines in Fig. 20-10 correspond to the two adiabatic processes of the Carnot cycle. Because no energy is transferred as heat during the two processes, the entropy of the working substance is constant during them.

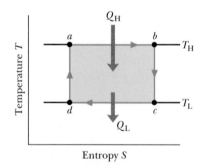

Fig. 20-10 The Carnot cycle of Fig. 20-9 plotted on a temperature–entropy diagram. During processes *ab* and *cd* the temperature remains constant. During processes *bc* and *da* the entropy remains constant.

The Work   To calculate the net work done by a Carnot engine during a cycle, let us apply Eq. 18-26, the first law of thermodynamics ($\Delta E_{int} = Q - W$), to the working substance. That substance must return again and again to any arbitrarily selected state in the cycle. Thus, if X represents any state property of the working substance, such as pressure, temperature, volume, internal energy, or entropy, we must have $\Delta X = 0$ for every cycle. It follows that $\Delta E_{int} = 0$ for a complete cycle of the working substance. Recalling that Q in Eq. 18-26 is the *net* heat transfer per cycle and W is the *net* work, we can write the first law of thermodynamics for the Carnot cycle as

$$W = |Q_H| - |Q_L|. \qquad (20\text{-}8)$$

Entropy Changes   In a Carnot engine, there are *two* (and only two) reversible energy transfers as heat, and thus two changes in the entropy of the working substance—one at temperature T_H and one at T_L. The net entropy change per cycle is then

$$\Delta S = \Delta S_H + \Delta S_L = \frac{|Q_H|}{T_H} - \frac{|Q_L|}{T_L}. \qquad (20\text{-}9)$$

Here ΔS_H is positive because energy $|Q_H|$ is *added to* the working substance as heat (an increase in entropy) and ΔS_L is negative because energy $|Q_L|$ is *removed from* the working substance as heat (a decrease in entropy). Because entropy is a state function, we must have $\Delta S = 0$ for a complete cycle. Putting $\Delta S = 0$ in Eq. 20-9 requires that

$$\frac{|Q_H|}{T_H} = \frac{|Q_L|}{T_L}. \qquad (20\text{-}10)$$

Note that, because $T_H > T_L$, we must have $|Q_H| > |Q_L|$; that is, more energy is extracted as heat from the high-temperature reservoir than is delivered to the low-temperature reservoir.

We shall now derive an expression for the efficiency of a Carnot engine.

Efficiency of a Carnot Engine

The purpose of any engine is to transform as much of the extracted energy Q_H into work as possible. We measure its success in doing so by its **thermal efficiency** ε, defined as the work the engine does per cycle ("energy we get") divided by the energy it absorbs as heat per cycle ("energy we pay for"):

$$\varepsilon = \frac{\text{energy we get}}{\text{energy we pay for}} = \frac{|W|}{|Q_H|} \qquad \text{(efficiency, any engine).} \qquad (20\text{-}11)$$

For a Carnot engine we can substitute for W from Eq. 20-8 to write Eq. 20-11 as

$$\varepsilon_C = \frac{|Q_H| - |Q_L|}{Q_H} = 1 - \frac{|Q_L|}{|Q_H|}. \qquad (20\text{-}12)$$

Using Eq. 20-10 we can write this as

$$\varepsilon_C = 1 - \frac{T_L}{T_H} \qquad \text{(efficiency, Carnot engine),} \qquad (20\text{-}13)$$

where the temperatures T_L and T_H are in kelvins. Because $T_L < T_H$, the Carnot engine necessarily has a thermal efficiency less than unity—that is, less than 100%. This is indicated in Fig. 20-8, which shows that only part of the energy extracted as heat from the high-temperature reservoir is available to do work, and the rest is delivered to the low-temperature reservoir. We shall show in Section 20-7 that no real engine can have a thermal efficiency greater than that calculated from Eq. 20-13.

Fig. 20-11 The elements of a perfect engine—that is, one that converts heat Q_H from a high-temperature reservoir directly to work W with 100% efficiency.

Fig. 20-12 The North Anna nuclear power plant near Charlottesville, Virginia, which generates electric energy at the rate of 900 MW. At the same time, by design, it discards energy into the nearby river at the rate of 2100 MW. This plant and all others like it throw away more energy than they deliver in useful form. They are real counterparts of the ideal engine of Fig. 20-8. *(© Robert Ustinich)*

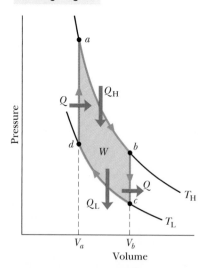

Stages of a Stirling engine

Fig. 20-13 A p-V plot for the working substance of an ideal Stirling engine, with the working substance assumed for convenience to be an ideal gas.

Inventors continually try to improve engine efficiency by reducing the energy $|Q_L|$ that is "thrown away" during each cycle. The inventor's dream is to produce the *perfect engine,* diagrammed in Fig. 20-11, in which $|Q_L|$ is reduced to zero and $|Q_H|$ is converted completely into work. Such an engine on an ocean liner, for example, could extract energy as heat from the water and use it to drive the propellers, with no fuel cost. An automobile fitted with such an engine could extract energy as heat from the surrounding air and use it to drive the car, again with no fuel cost. Alas, a perfect engine is only a dream: Inspection of Eq. 20-13 shows that we can achieve 100% engine efficiency (that is, $\varepsilon = 1$) only if $T_L = 0$ or $T_H \rightarrow \infty$, impossible requirements. Instead, experience gives the following alternative version of the second law of thermodynamics, which says in short, *there are no perfect engines:*

> No series of processes is possible whose sole result is the transfer of energy as heat from a thermal reservoir and the complete conversion of this energy to work.

To summarize: The thermal efficiency given by Eq. 20-13 applies only to Carnot engines. Real engines, in which the processes that form the engine cycle are not reversible, have lower efficiencies. If your car were powered by a Carnot engine, it would have an efficiency of about 55% according to Eq. 20-13; its actual efficiency is probably about 25%. A nuclear power plant (Fig. 20-12), taken in its entirety, is an engine. It extracts energy as heat from a reactor core, does work by means of a turbine, and discharges energy as heat to a nearby river. If the power plant operated as a Carnot engine, its efficiency would be about 40%; its actual efficiency is about 30%. In designing engines of any type, there is simply no way to beat the efficiency limitation imposed by Eq. 20-13.

Stirling Engine

Equation 20-13 applies not to all ideal engines but only to those that can be represented as in Fig. 20-9—that is, to Carnot engines. For example, Fig. 20-13 shows the operating cycle of an ideal **Stirling engine.** Comparison with the Carnot cycle of Fig. 20-9 shows that each engine has isothermal heat transfers at temperatures T_H and T_L. However, the two isotherms of the Stirling engine cycle are connected, not by adiabatic processes as for the Carnot engine but by constant-volume processes. To increase the temperature of a gas at constant volume reversibly from T_L to T_H (process da of Fig. 20-13) requires a transfer of energy as heat to the working substance from a thermal reservoir whose temperature can be varied smoothly between those limits. Also, a reverse transfer is required in process bc. Thus, reversible heat transfers (and corresponding entropy changes) occur in all four of the processes that form the cycle of a Stirling engine, not just two processes as in a Carnot engine. Thus, the derivation that led to Eq. 20-13 does not apply to an ideal Stirling engine. More important, the efficiency of an ideal Stirling engine is lower than that of a Carnot engine operating between the same two temperatures. Real Stirling engines have even lower efficiencies.

The Stirling engine was developed in 1816 by Robert Stirling. This engine, long neglected, is now being developed for use in automobiles and spacecraft. A Stirling engine delivering 5000 hp (3.7 MW) has been built. Because they are quiet, Stirling engines are used on some military submarines.

 CHECKPOINT 3

Three Carnot engines operate between reservoir temperatures of (a) 400 and 500 K, (b) 600 and 800 K, and (c) 400 and 600 K. Rank the engines according to their thermal efficiencies, greatest first.

Sample Problem

Carnot engine, efficiency, power, entropy changes

Imagine a Carnot engine that operates between the temperatures $T_H = 850$ K and $T_L = 300$ K. The engine performs 1200 J of work each cycle, which takes 0.25 s.

(a) What is the efficiency of this engine?

KEY IDEA

The efficiency ε of a Carnot engine depends only on the ratio T_L/T_H of the temperatures (in kelvins) of the thermal reservoirs to which it is connected.

Calculation: Thus, from Eq. 20-13, we have

$$\varepsilon = 1 - \frac{T_L}{T_H} = 1 - \frac{300 \text{ K}}{850 \text{ K}} = 0.647 \approx 65\%. \quad \text{(Answer)}$$

(b) What is the average power of this engine?

KEY IDEA

The average power P of an engine is the ratio of the work W it does per cycle to the time t that each cycle takes.

Calculation: For this Carnot engine, we find

$$P = \frac{W}{t} = \frac{1200 \text{ J}}{0.25 \text{ s}} = 4800 \text{ W} = 4.8 \text{ kW}. \quad \text{(Answer)}$$

(c) How much energy $|Q_H|$ is extracted as heat from the high-temperature reservoir every cycle?

KEY IDEA

The efficiency ε is the ratio of the work W that is done per cycle to the energy $|Q_H|$ that is extracted as heat from the high-temperature reservoir per cycle ($\varepsilon = W/|Q_H|$).

Calculation: Here we have

$$|Q_H| = \frac{W}{\varepsilon} = \frac{1200 \text{ J}}{0.647} = 1855 \text{ J}. \quad \text{(Answer)}$$

(d) How much energy $|Q_L|$ is delivered as heat to the low-temperature reservoir every cycle?

KEY IDEA

For a Carnot engine, the work W done per cycle is equal to the difference in the energy transfers as heat: $|Q_H| - |Q_L|$, as in Eq. 20-8.

Calculation: Thus, we have

$$|Q_L| = |Q_H| - W$$
$$= 1855 \text{ J} - 1200 \text{ J} = 655 \text{ J}. \quad \text{(Answer)}$$

(e) By how much does the entropy of the working substance change as a result of the energy transferred to it from the high-temperature reservoir? From it to the low-temperature reservoir?

KEY IDEA

The entropy change ΔS during a transfer of energy as heat Q at constant temperature T is given by Eq. 20-2 ($\Delta S = Q/T$).

Calculations: Thus, for the *positive* transfer of energy Q_H from the high-temperature reservoir at T_H, the change in the entropy of the working substance is

$$\Delta S_H = \frac{Q_H}{T_H} = \frac{1855 \text{ J}}{850 \text{ K}} = +2.18 \text{ J/K}. \quad \text{(Answer)}$$

Similarly, for the *negative* transfer of energy Q_L to the low-temperature reservoir at T_L, we have

$$\Delta S_L = \frac{Q_L}{T_L} = \frac{-655 \text{ J}}{300 \text{ K}} = -2.18 \text{ J/K}. \quad \text{(Answer)}$$

Note that the net entropy change of the working substance for one cycle is zero, as we discussed in deriving Eq. 20-10.

 Additional examples, video, and practice available at *WileyPLUS*

Sample Problem

Impossibly efficient engine

An inventor claims to have constructed an engine that has an efficiency of 75% when operated between the boiling and freezing points of water. Is this possible?

KEY IDEA

The efficiency of a real engine must be less than the efficiency of a Carnot engine operating between the same two temperatures.

Calculation: From Eq. 20-13, we find that the efficiency of a Carnot engine operating between the boiling and freezing points of water is

$$\varepsilon = 1 - \frac{T_L}{T_H} = 1 - \frac{(0 + 273)\text{ K}}{(100 + 273)\text{ K}} = 0.268 \approx 27\%.$$

Thus, for the given temperatures, the claimed efficiency of 75% for a real engine (with its irreversible processes and wasteful energy transfers) is impossible.

WILEY PLUS   Additional examples, video, and practice available at *WileyPLUS*

20-6 Entropy in the Real World: Refrigerators

A **refrigerator** is a device that uses work in order to transfer energy from a low-temperature reservoir to a high-temperature reservoir as the device continuously repeats a set series of thermodynamic processes. In a household refrigerator, for example, work is done by an electrical compressor to transfer energy from the food storage compartment (a low-temperature reservoir) to the room (a high-temperature reservoir).

Air conditioners and heat pumps are also refrigerators. The differences are only in the nature of the high- and low-temperature reservoirs. For an air conditioner, the low-temperature reservoir is the room that is to be cooled and the high-temperature reservoir is the (presumably warmer) outdoors. A heat pump is an air conditioner that can be operated in reverse to heat a room; the room is the high-temperature reservoir, and heat is transferred to it from the (presumably cooler) outdoors.

Let us consider an *ideal refrigerator:*

> In an ideal refrigerator, all processes are reversible and no wasteful energy transfers occur as a result of, say, friction and turbulence.

Figure 20-14 shows the basic elements of an ideal refrigerator. Note that its operation is the reverse of how the Carnot engine of Fig. 20-8 operates. In other words, all the energy transfers, as either heat or work, are reversed from those of a Carnot engine. We can call such an ideal refrigerator a **Carnot refrigerator.**

The designer of a refrigerator would like to extract as much energy $|Q_L|$ as possible from the low-temperature reservoir (what we want) for the least amount of work $|W|$ (what we pay for). A measure of the efficiency of a refrigerator, then, is

$$K = \frac{\text{what we want}}{\text{what we pay for}} = \frac{|Q_L|}{|W|} \qquad \begin{array}{l}\text{(coefficient of performance,} \\ \text{any refrigerator),}\end{array} \qquad (20\text{-}14)$$

where K is called the *coefficient of performance.* For a Carnot refrigerator, the first law of thermodynamics gives $|W| = |Q_H| - |Q_L|$, where $|Q_H|$ is the magnitude of the energy transferred as heat to the high-temperature reservoir. Equation 20-14 then becomes

$$K_C = \frac{|Q_L|}{|Q_H| - |Q_L|}. \qquad (20\text{-}15)$$

Because a Carnot refrigerator is a Carnot engine operating in reverse, we can

Schematic of a refrigerator

Fig. 20-14 The elements of a refrigerator. The two black arrowheads on the central loop suggest the working substance operating in a cycle, as if on a *p-V* plot. Energy is transferred as heat Q_L to the working substance from the low-temperature reservoir. Energy is transferred as heat Q_H to the high-temperature reservoir from the working substance. Work W is done on the refrigerator (on the working substance) by something in the environment.

combine Eq. 20-10 with Eq. 20-15; after some algebra we find

$$K_C = \frac{T_L}{T_H - T_L} \quad \text{(coefficient of performance, Carnot refrigerator).} \quad (20\text{-}16)$$

For typical room air conditioners, $K \approx 2.5$. For household refrigerators, $K \approx 5$. Perversely, the value of K is higher the closer the temperatures of the two reservoirs are to each other. That is why heat pumps are more effective in temperate climates than in climates where the outside temperature is much lower than the desired inside temperature.

It would be nice to own a refrigerator that did not require some input of work — that is, one that would run without being plugged in. Figure 20-15 represents another "inventor's dream," a *perfect refrigerator* that transfers energy as heat Q from a cold reservoir to a warm reservoir without the need for work. Because the unit operates in cycles, the entropy of the working substance does not change during a complete cycle. The entropies of the two reservoirs, however, do change: The entropy change for the cold reservoir is $-|Q|/T_L$, and that for the warm reservoir is $+|Q|/T_H$. Thus, the net entropy change for the entire system is

$$\Delta S = -\frac{|Q|}{T_L} + \frac{|Q|}{T_H}.$$

Because $T_H > T_L$, the right side of this equation is negative and thus the net change in entropy per cycle for the closed system *refrigerator + reservoirs* is also negative. Because such a decrease in entropy violates the second law of thermodynamics (Eq. 20-5), a perfect refrigerator does not exist. (If you want your refrigerator to operate, you must plug it in.)

This result leads us to another (equivalent) formulation of the second law of thermodynamics:

 No series of processes is possible whose sole result is the transfer of energy as heat from a reservoir at a given temperature to a reservoir at a higher temperature.

In short, *there are no perfect refrigerators.*

CHECKPOINT 4

You wish to increase the coefficient of performance of an ideal refrigerator. You can do so by (a) running the cold chamber at a slightly higher temperature, (b) running the cold chamber at a slightly lower temperature, (c) moving the unit to a slightly warmer room, or (d) moving it to a slightly cooler room. The magnitudes of the temperature changes are to be the same in all four cases. List the changes according to the resulting coefficients of performance, greatest first.

Perfect refrigerator: total transfer of heat from cold to hot without any work

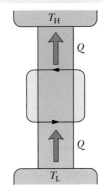

Fig. 20-15 The elements of a perfect refrigerator — that is, one that transfers energy from a low-temperature reservoir to a high-temperature reservoir without any input of work.

20-7 The Efficiencies of Real Engines

Let ε_C be the efficiency of a Carnot engine operating between two given temperatures. In this section we prove that no real engine operating between those temperatures can have an efficiency greater than ε_C. If it could, the engine would violate the second law of thermodynamics.

Let us assume that an inventor, working in her garage, has constructed an engine X, which she claims has an efficiency ε_X that is greater than ε_C:

$$\varepsilon_X > \varepsilon_C \quad \text{(a claim).} \quad (20\text{-}17)$$

Let us couple engine X to a Carnot refrigerator, as in Fig. 20-16a. We adjust the strokes of the Carnot refrigerator so that the work it requires per cycle is just equal to that provided by engine X. Thus, no (external) work is performed on or by the combination *engine + refrigerator* of Fig. 20-16a, which we take as our system.

If Eq. 20-17 is true, from the definition of efficiency (Eq. 20-11), we must have

$$\frac{|W|}{|Q'_H|} > \frac{|W|}{|Q_H|},$$

where the prime refers to engine X and the right side of the inequality is the efficiency of the Carnot refrigerator when it operates as an engine. This inequality requires that

$$|Q_H| > |Q'_H|. \tag{20-18}$$

Because the work done by engine X is equal to the work done on the Carnot refrigerator, we have, from the first law of thermodynamics as given by Eq. 20-8,

$$|Q_H| - |Q_L| = |Q'_H| - |Q'_L|,$$

which we can write as

$$|Q_H| - |Q'_H| = |Q_L| - |Q'_L| = Q. \tag{20-19}$$

Because of Eq. 20-18, the quantity Q in Eq. 20-19 must be positive.

Comparison of Eq. 20-19 with Fig. 20-16 shows that the net effect of engine X and the Carnot refrigerator working in combination is to transfer energy Q as heat from a low-temperature reservoir to a high-temperature reservoir without the requirement of work. Thus, the combination acts like the perfect refrigerator of Fig. 20-15, whose existence is a violation of the second law of thermodynamics.

Something must be wrong with one or more of our assumptions, and it can only be Eq. 20-17. We conclude that *no real engine can have an efficiency greater than that of a Carnot engine when both engines work between the same two temperatures.* At most, the real engine can have an efficiency equal to that of a Carnot engine. In that case, the real engine *is* a Carnot engine.

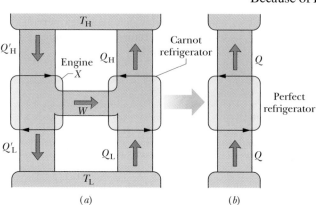

Fig. 20-16 (*a*) Engine X drives a Carnot refrigerator. (*b*) If, as claimed, engine X is more efficient than a Carnot engine, then the combination shown in (*a*) is equivalent to the perfect refrigerator shown here. This violates the second law of thermodynamics, so we conclude that engine X *cannot* be more efficient than a Carnot engine.

20-8 A Statistical View of Entropy

In Chapter 19 we saw that the macroscopic properties of gases can be explained in terms of their microscopic, or molecular, behavior. For one example, recall that we were able to account for the pressure exerted by a gas on the walls of its container in terms of the momentum transferred to those walls by rebounding gas molecules. Such explanations are part of a study called **statistical mechanics.**

Here we shall focus our attention on a single problem, one involving the distribution of gas molecules between the two halves of an insulated box. This problem is reasonably simple to analyze, and it allows us to use statistical mechanics to calculate the entropy change for the free expansion of an ideal gas. You will see that statistical mechanics leads to the same entropy change as we would find using thermodynamics.

Figure 20-17 shows a box that contains six identical (and thus indistinguishable) molecules of a gas. At any instant, a given molecule will be in either the left or the right half of the box; because the two halves have equal volumes, the molecule has the same likelihood, or probability, of being in either half.

Table 20-1 shows the seven possible *configurations* of the six molecules, each configuration labeled with a Roman numeral. For example, in configuration I, all six molecules are in the left half of the box ($n_1 = 6$) and none are in the right half ($n_2 = 0$). We see that, in general, a given configuration can be achieved in a number of different ways. We call these different arrangements of the molecules *microstates*. Let us see how to calculate the number of microstates that correspond to a given configuration.

	Table 20-1			

Six Molecules in a Box

Configuration			Multiplicity W	Calculation of W	Entropy 10^{-23} J/K
Label	n_1	n_2	(number of microstates)	(Eq. 20-20)	(Eq. 20-21)
I	6	0	1	$6!/(6!\ 0!) = 1$	0
II	5	1	6	$6!/(5!\ 1!) = 6$	2.47
III	4	2	15	$6!/(4!\ 2!) = 15$	3.74
IV	3	3	20	$6!/(3!\ 3!) = 20$	4.13
V	2	4	15	$6!/(2!\ 4!) = 15$	3.74
VI	1	5	6	$6!/(1!\ 5!) = 6$	2.47
VII	0	6	1	$6!/(0!\ 6!) = 1$	0
			Total = 64		

Suppose we have N molecules, distributed with n_1 molecules in one half of the box and n_2 in the other. (Thus $n_1 + n_2 = N$.) Let us imagine that we distribute the molecules "by hand," one at a time. If $N = 6$, we can select the first molecule in six independent ways; that is, we can pick any one of the six molecules. We can pick the second molecule in five ways, by picking any one of the remaining five molecules; and so on. The total number of ways in which we can select all six molecules is the product of these independent ways, or $6 \times 5 \times 4 \times 3 \times 2 \times 1 = 720$. In mathematical shorthand we write this product as $6! = 720$, where $6!$ is pronounced "six factorial." Your hand calculator can probably calculate factorials. For later use you will need to know that $0! = 1$. (Check this on your calculator.)

However, because the molecules are indistinguishable, these 720 arrangements are not all different. In the case that $n_1 = 4$ and $n_2 = 2$ (which is configuration III in Table 20-1), for example, the order in which you put four molecules in one half of the box does not matter, because after you have put all four in, there is no way that you can tell the order in which you did so. The number of ways in which you can order the four molecules is $4! = 24$. Similarly, the number of ways in which you can order two molecules for the other half of the box is simply $2! = 2$. To get the number of *different* arrangements that lead to the $(4, 2)$ split of configuration III, we must divide 720 by 24 and also by 2. We call the resulting quantity, which is the number of microstates that correspond to a given configuration, the *multiplicity W* of that configuration. Thus, for configuration III,

$$W_{\text{III}} = \frac{6!}{4!\ 2!} = \frac{720}{24 \times 2} = 15.$$

Thus, Table 20-1 tells us there are 15 independent microstates that correspond to configuration III. Note that, as the table also tells us, the total number of microstates for six molecules distributed over the seven configurations is 64.

Extrapolating from six molecules to the general case of N molecules, we have

$$W = \frac{N!}{n_1!\ n_2!} \qquad \text{(multiplicity of configuration).} \qquad (20\text{-}20)$$

You should verify that Eq. 20-20 gives the multiplicities for all the configurations listed in Table 20-1.

The basic assumption of statistical mechanics is

 All microstates are equally probable.

In other words, if we were to take a great many snapshots of the six molecules as

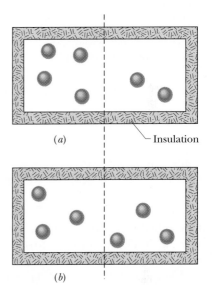

Fig. 20-17 An insulated box contains six gas molecules. Each molecule has the same probability of being in the left half of the box as in the right half. The arrangement in (*a*) corresponds to configuration III in Table 20-1, and that in (*b*) corresponds to configuration IV.

Fig. 20-18 For a *large* number of molecules in a box, a plot of the number of microstates that require various percentages of the molecules to be in the left half of the box. Nearly all the microstates correspond to an approximately equal sharing of the molecules between the two halves of the box; those microstates form the *central configuration peak* on the plot. For $N \approx 10^{22}$, the central configuration peak is much too narrow to be drawn on this plot.

they jostle around in the box of Fig. 20-17 and then count the number of times each microstate occurred, we would find that all 64 microstates would occur equally often. Thus the system will spend, on average, the same amount of time in each of the 64 microstates.

Because all microstates are equally probable but different configurations have different numbers of microstates, the configurations are *not* all equally probable. In Table 20-1 configuration IV, with 20 microstates, is the *most probable configuration,* with a probability of 20/64 = 0.313. This result means that the system is in configuration IV 31.3% of the time. Configurations I and VII, in which all the molecules are in one half of the box, are the least probable, each with a probability of 1/64 = 0.016 or 1.6%. It is not surprising that the most probable configuration is the one in which the molecules are evenly divided between the two halves of the box, because that is what we expect at thermal equilibrium. However, it *is* surprising that there is *any* probability, however small, of finding all six molecules clustered in half of the box, with the other half empty.

For large values of N there are extremely large numbers of microstates, but nearly all the microstates belong to the configuration in which the molecules are divided equally between the two halves of the box, as Fig. 20-18 indicates. Even though the measured temperature and pressure of the gas remain constant, the gas is churning away endlessly as its molecules "visit" all probable microstates with equal probability. However, because so few microstates lie outside the very narrow central configuration peak of Fig. 20-18, we might as well assume that the gas molecules are always divided equally between the two halves of the box. As we shall see, this is the configuration with the greatest entropy.

Sample Problem

Microstates and multiplicity

Suppose that there are 100 indistinguishable molecules in the box of Fig. 20-17. How many microstates are associated with the configuration $n_1 = 50$ and $n_2 = 50$, and with the configuration $n_1 = 100$ and $n_2 = 0$? Interpret the results in terms of the relative probabilities of the two configurations.

KEY IDEA

The multiplicity W of a configuration of indistinguishable molecules in a closed box is the number of independent microstates with that configuration, as given by Eq. 20-20.

Calculations: Thus, for the (n_1, n_2) configuration $(50, 50)$,

$$W = \frac{N!}{n_1! \, n_2!} = \frac{100!}{50! \, 50!}$$

$$= \frac{9.33 \times 10^{157}}{(3.04 \times 10^{64})(3.04 \times 10^{64})}$$

$$= 1.01 \times 10^{29}. \qquad \text{(Answer)}$$

Similarly, for the configuration $(100, 0)$, we have

$$W = \frac{N!}{n_1! \, n_2!} = \frac{100!}{100! \, 0!} = \frac{1}{0!} = \frac{1}{1} = 1. \quad \text{(Answer)}$$

The meaning: Thus, a 50–50 distribution is more likely than a 100–0 distribution by the enormous factor of about 1×10^{29}. If you could count, at one per nanosecond, the number of microstates that correspond to the 50–50 distribution, it would take you about 3×10^{12} years, which is about 200 times longer than the age of the universe. Keep in mind that the 100 molecules used in this sample problem is a very small number. Imagine what these calculated probabilities would be like for a mole of molecules, say about $N = 10^{24}$. Thus, you need never worry about suddenly finding all the air molecules clustering in one corner of your room, with you gasping for air in another corner. So, you can breathe easy because of the physics of entropy.

 Additional examples, video, and practice available at *WileyPLUS*

Probability and Entropy

In 1877, Austrian physicist Ludwig Boltzmann (the Boltzmann of Boltzmann's constant k) derived a relationship between the entropy S of a configuration of a gas and the multiplicity W of that configuration. That relationship is

$$S = k \ln W \qquad \text{(Boltzmann's entropy equation).} \qquad (20\text{-}21)$$

This famous formula is engraved on Boltzmann's tombstone.

It is natural that S and W should be related by a logarithmic function. The total entropy of two systems is the *sum* of their separate entropies. The probability of occurrence of two independent systems is the *product* of their separate probabilities. Because $\ln ab = \ln a + \ln b$, the logarithm seems the logical way to connect these quantities.

Table 20-1 displays the entropies of the configurations of the six-molecule system of Fig. 20-17, computed using Eq. 20-21. Configuration IV, which has the greatest multiplicity, also has the greatest entropy.

When you use Eq. 20-20 to calculate W, your calculator may signal "OVER-FLOW" if you try to find the factorial of a number greater than a few hundred. Instead, you can use **Stirling's approximation** for $\ln N!$:

$$\ln N! \approx N(\ln N) - N \qquad \text{(Stirling's approximation).} \qquad (20\text{-}22)$$

The Stirling of this approximation was an English mathematician and not the Robert Stirling of engine fame.

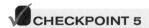 CHECKPOINT 5

A box contains 1 mol of a gas. Consider two configurations: (a) each half of the box contains half the molecules and (b) each third of the box contains one-third of the molecules. Which configuration has more microstates?

Entropy change of free expansion using microstates

In the first sample problem of this chapter, we showed that when n moles of an ideal gas doubles its volume in a free expansion, the entropy increase from the initial state i to the final state f is $S_f - S_i = nR \ln 2$. Derive this result with statistical mechanics.

KEY IDEA

We can relate the entropy S of any given configuration of the molecules in the gas to the multiplicity W of microstates for that configuration, using Eq. 20-21 ($S = k \ln W$).

Calculations: We are interested in two configurations: the final configuration f (with the molecules occupying the full volume of their container in Fig. 20-1b) and the initial configuration i (with the molecules occupying the left half of the container). Because the molecules are in a closed container, we can calculate the multiplicity W of their microstates with Eq. 20-20. Here we have N molecules in the n moles of the gas. Initially, with the molecules all in the left half of the container, their (n_1, n_2) configuration is $(N, 0)$. Then, Eq. 20-20 gives their multiplicity as

$$W_i = \frac{N!}{N! \, 0!} = 1.$$

Finally, with the molecules spread through the full volume, their (n_1, n_2) configuration is $(N/2, N/2)$. Then, Eq. 20-20 gives their multiplicity as

$$W_f = \frac{N!}{(N/2)! \, (N/2)!}.$$

From Eq. 20-21, the initial and final entropies are

$$S_i = k \ln W_i = k \ln 1 = 0$$

and

$$S_f = k \ln W_f = k \ln(N!) - 2k \ln[(N/2)!]. \qquad (20\text{-}23)$$

In writing Eq. 20-23, we have used the relation

$$\ln \frac{a}{b^2} = \ln a - 2 \ln b.$$

Now, applying Eq. 20-22 to evaluate Eq. 20-23, we find that

$$\begin{aligned}
S_f &= k \ln(N!) - 2k \ln[(N/2)!] \\
&= k[N(\ln N) - N] - 2k[(N/2) \ln(N/2) - (N/2)] \\
&= k[N(\ln N) - N - N \ln(N/2) + N] \\
&= k[N(\ln N) - N(\ln N - \ln 2)] = Nk \ln 2. \qquad (20\text{-}24)
\end{aligned}$$

From Eq. 19-8 we can substitute nR for Nk, where R is the universal gas constant. Equation 20-24 then becomes

$$S_f = nR \ln 2.$$

The change in entropy from the initial state to the final is thus

$$\begin{aligned}
S_f - S_i &= nR \ln 2 - 0 \\
&= nR \ln 2, \qquad \text{(Answer)}
\end{aligned}$$

which is what we set out to show. In the first sample problem of this chapter we calculated this entropy increase for a free expansion with thermodynamics by finding an equivalent reversible process and calculating the entropy change for *that* process in terms of temperature and heat transfer. In this sample problem, we calculate the same increase in entropy with statistical mechanics using the fact that the system consists of molecules. In short, the two, very different approaches give the same answer.

 Additional examples, video, and practice available at *WileyPLUS*

REVIEW & SUMMARY

One-Way Processes An **irreversible process** is one that cannot be reversed by means of small changes in the environment. The direction in which an irreversible process proceeds is set by the *change in entropy* ΔS of the system undergoing the process. Entropy S is a *state property* (or *state function*) of the system; that is, it depends only on the state of the system and not on the way in which the system reached that state. The *entropy postulate* states (in part): *If an irreversible process occurs in a closed system, the entropy of the system always increases.*

Calculating Entropy Change The **entropy change** ΔS for an irreversible process that takes a system from an initial state i to a final state f is exactly equal to the entropy change ΔS for *any reversible process* that takes the system between those same two states. We can compute the latter (but not the former) with

$$\Delta S = S_f - S_i = \int_i^f \frac{dQ}{T}. \tag{20-1}$$

Here Q is the energy transferred as heat to or from the system during the process, and T is the temperature of the system in kelvins during the process.

For a reversible isothermal process, Eq. 20-1 reduces to

$$\Delta S = S_f - S_i = \frac{Q}{T}. \tag{20-2}$$

When the temperature change ΔT of a system is small relative to the temperature (in kelvins) before and after the process, the entropy change can be approximated as

$$\Delta S = S_f - S_i \approx \frac{Q}{T_{\text{avg}}}, \tag{20-3}$$

where T_{avg} is the system's average temperature during the process.

When an ideal gas changes reversibly from an initial state with temperature T_i and volume V_i to a final state with temperature T_f and volume V_f, the change ΔS in the entropy of the gas is

$$\Delta S = S_f - S_i = nR \ln \frac{V_f}{V_i} + nC_V \ln \frac{T_f}{T_i}. \tag{20-4}$$

The Second Law of Thermodynamics This law, which is an extension of the entropy postulate, states: *If a process occurs in a closed system, the entropy of the system increases for irreversible processes and remains constant for reversible processes. It never decreases.* In equation form,

$$\Delta S \geq 0. \tag{20-5}$$

Engines An **engine** is a device that, operating in a cycle, extracts energy as heat $|Q_H|$ from a high-temperature reservoir and does a cer-

tain amount of work $|W|$. The *efficiency* ε of any engine is defined as

$$\varepsilon = \frac{\text{energy we get}}{\text{energy we pay for}} = \frac{|W|}{|Q_H|}. \tag{20-11}$$

In an **ideal engine,** all processes are reversible and no wasteful energy transfers occur due to, say, friction and turbulence. A **Carnot engine** is an ideal engine that follows the cycle of Fig. 20-9. Its efficiency is

$$\varepsilon_C = 1 - \frac{|Q_L|}{|Q_H|} = 1 - \frac{T_L}{T_H}, \tag{20-12, 20-13}$$

in which T_H and T_L are the temperatures of the high- and low-temperature reservoirs, respectively. Real engines always have an efficiency lower than that given by Eq. 20-13. Ideal engines that are not Carnot engines also have lower efficiencies.

A *perfect engine* is an imaginary engine in which energy extracted as heat from the high-temperature reservoir is converted completely to work. Such an engine would violate the second law of thermodynamics, which can be restated as follows: No series of processes is possible whose sole result is the absorption of energy as heat from a thermal reservoir and the complete conversion of this energy to work.

Refrigerators A refrigerator is a device that, operating in a cycle, has work W done on it as it extracts energy $|Q_L|$ as heat from a low-temperature reservoir. The coefficient of performance K of a refrigerator is defined as

$$K = \frac{\text{what we want}}{\text{what we pay for}} = \frac{|Q_L|}{|W|}. \tag{20-14}$$

A **Carnot refrigerator** is a Carnot engine operating in reverse. For a Carnot refrigerator, Eq. 20-14 becomes

$$K_C = \frac{|Q_L|}{|Q_H| - |Q_L|} = \frac{T_L}{T_H - T_L}. \tag{20-15, 20-16}$$

A *perfect refrigerator* is an imaginary refrigerator in which energy extracted as heat from the low-temperature reservoir is converted completely to heat discharged to the high-temperature reservoir, without any need for work. Such a refrigerator would violate the second law of thermodynamics, which can be restated as follows: No series of processes is possible whose sole result is the transfer of energy as heat from a reservoir at a given temperature to a reservoir at a higher temperature.

Entropy from a Statistical View The entropy of a system can be defined in terms of the possible distributions of its molecules. For identical molecules, each possible distribution of molecules is called a **microstate** of the system. All equivalent microstates are grouped into a **configuration** of the system. The number of microstates in a configuration is the **multiplicity** W of the configuration.

For a system of N molecules that may be distributed between the two halves of a box, the multiplicity is given by

$$W = \frac{N!}{n_1! \, n_2!}, \qquad (20\text{-}20)$$

in which n_1 is the number of molecules in one half of the box and n_2 is the number in the other half. A basic assumption of **statistical mechanics** is that all the microstates are equally probable. Thus, configurations with a large multiplicity occur most often. When N is very large (say, $N = 10^{22}$ molecules or more), the molecules are nearly always in the configuration in which $n_1 = n_2$.

The multiplicity W of a configuration of a system and the entropy S of the system in that configuration are related by Boltzmann's entropy equation:

$$S = k \ln W, \qquad (20\text{-}21)$$

where $k = 1.38 \times 10^{-23}$ J/K is the Boltzmann constant.

When N is very large (the usual case), we can approximate $\ln N!$ with *Stirling's approximation*:

$$\ln N! \approx N(\ln N) - N. \qquad (20\text{-}22)$$

QUESTIONS

1 Point i in Fig. 20-19 represents the initial state of an ideal gas at temperature T. Taking algebraic signs into account, rank the entropy changes that the gas undergoes as it moves, successively and reversibly, from point i to points a, b, c, and d, greatest first.

Fig. 20-19   Question 1.

2 In four experiments, blocks A and B, starting at different initial temperatures, were brought together in an insulating box and allowed to reach a common final temperature. The entropy changes for the blocks in the four experiments had the following values (in joules per kelvin), but not necessarily in the order given. Determine which values for A go with which values for B.

Block	Values			
A	8	5	3	9
B	−3	−8	−5	−2

3 A gas, confined to an insulated cylinder, is compressed adiabatically to half its volume. Does the entropy of the gas increase, decrease, or remain unchanged during this process?

4 An ideal monatomic gas at initial temperature T_0 (in kelvins) expands from initial volume V_0 to volume $2V_0$ by each of the five processes indicated in the T-V diagram of Fig. 20-20. In which process is the expansion (a) isothermal, (b) isobaric (constant pressure), and (c) adiabatic? Explain your answers. (d) In which processes does the entropy of the gas decrease?

5 In four experiments, 2.5 mol of hydrogen gas undergoes reversible isothermal expansions, starting from the same volume but at different temperatures. The corresponding p-V plots are shown in Fig. 20-21. Rank the situations according to the change in the entropy of the gas, greatest first.

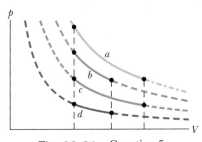

Fig. 20-21   Question 5.

6 A box contains 100 atoms in a configuration that has 50 atoms in each half of the box. Suppose that you could count the different microstates associated with this configuration at the rate of 100 billion states per second, using a supercomputer. Without written calculation, guess how much computing time you would need: a day, a year, or much more than a year.

7 Does the entropy per cycle increase, decrease, or remain the same for (a) a Carnot engine, (b) a real engine, and (c) a perfect engine (which is, of course, impossible to build)?

8 Three Carnot engines operate between temperature limits of (a) 400 and 500 K, (b) 500 and 600 K, and (c) 400 and 600 K. Each engine extracts the same amount of energy per cycle from the high-temperature reservoir. Rank the magnitudes of the work done by the engines per cycle, greatest first.

9 An inventor claims to have invented four engines, each of which operates between constant-temperature reservoirs at 400 and 300 K. Data on each engine, per cycle of operation, are: engine A, $Q_H = 200$ J, $Q_L = -175$ J, and $W = 40$ J; engine B, $Q_H = 500$ J, $Q_L = -200$ J, and $W = 400$ J; engine C, $Q_H = 600$ J, $Q_L = -200$ J, and $W = 400$ J; engine D, $Q_H = 100$ J, $Q_L = -90$ J, and $W = 10$ J. Of the first and second laws of thermodynamics, which (if either) does each engine violate?

10 Does the entropy per cycle increase, decrease, or remain the same for (a) a Carnot refrigerator, (b) a real refrigerator, and (c) a perfect refrigerator (which is, of course, impossible to build)?

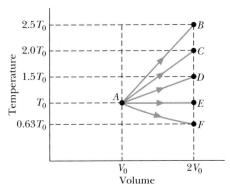

Fig. 20-20   Question 4.

PROBLEMS

sec. 20-3 Change in Entropy

•1 SSM Suppose 4.00 mol of an ideal gas undergoes a reversible isothermal expansion from volume V_1 to volume $V_2 = 2.00V_1$ at temperature $T = 400$ K. Find (a) the work done by the gas and (b) the entropy change of the gas. (c) If the expansion is reversible and adiabatic instead of isothermal, what is the entropy change of the gas?

•2 An ideal gas undergoes a reversible isothermal expansion at 77.0°C, increasing its volume from 1.30 L to 3.40 L. The entropy change of the gas is 22.0 J/K. How many moles of gas are present?

•3 ILW A 2.50 mol sample of an ideal gas expands reversibly and isothermally at 360 K until its volume is doubled. What is the increase in entropy of the gas?

•4 How much energy must be transferred as heat for a reversible isothermal expansion of an ideal gas at 132°C if the entropy of the gas increases by 46.0 J/K?

•5 ILW Find (a) the energy absorbed as heat and (b) the change in entropy of a 2.00 kg block of copper whose temperature is increased reversibly from 25.0°C to 100°C. The specific heat of copper is 386 J/kg·K.

•6 (a) What is the entropy change of a 12.0 g ice cube that melts completely in a bucket of water whose temperature is just above the freezing point of water? (b) What is the entropy change of a 5.00 g spoonful of water that evaporates completely on a hot plate whose temperature is slightly above the boiling point of water?

••7 ILW A 50.0 g block of copper whose temperature is 400 K is placed in an insulating box with a 100 g block of lead whose temperature is 200 K. (a) What is the equilibrium temperature of the two-block system? (b) What is the change in the internal energy of the system between the initial state and the equilibrium state? (c) What is the change in the entropy of the system? (See Table 18-3.)

••8 At very low temperatures, the molar specific heat C_V of many solids is approximately $C_V = AT^3$, where A depends on the particular substance. For aluminum, $A = 3.15 \times 10^{-5}$ J/mol·K⁴. Find the entropy change for 4.00 mol of aluminum when its temperature is raised from 5.00 K to 10.0 K.

••9 A 10 g ice cube at −10°C is placed in a lake whose temperature is 15°C. Calculate the change in entropy of the cube–lake system as the ice cube comes to thermal equilibrium with the lake. The specific heat of ice is 2220 J/kg·K. (*Hint:* Will the ice cube affect the lake temperature?)

••10 A 364 g block is put in contact with a thermal reservoir. The block is initially at a lower temperature than the reservoir. Assume that the consequent transfer of energy as heat from the reservoir to the block is reversible. Figure 20-22 gives the change

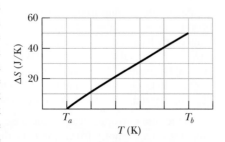

Fig. 20-22 Problem 10.

in entropy ΔS of the block until thermal equilibrium is reached. The scale of the horizontal axis is set by $T_a = 280$ K and $T_b = 380$ K. What is the specific heat of the block?

••11 SSM WWW In an experiment, 200 g of aluminum (with a specific heat of 900 J/kg·K) at 100°C is mixed with 50.0 g of water at 20.0°C, with the mixture thermally isolated. (a) What is the equilibrium temperature? What are the entropy changes of (b) the aluminum, (c) the water, and (d) the aluminum–water system?

••12 A gas sample undergoes a reversible isothermal expansion. Figure 20-23 gives the change ΔS in entropy of the gas versus the final volume V_f of the gas. The scale of the vertical axis is set by $\Delta S_s = 64$ J/K. How many moles are in the sample?

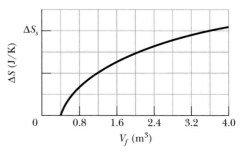

Fig. 20-23 Problem 12.

••13 In the irreversible process of Fig. 20-5, let the initial temperatures of the identical blocks L and R be 305.5 and 294.5 K, respectively, and let 215 J be the energy that must be transferred between the blocks in order to reach equilibrium. For the reversible processes of Fig. 20-6, what is ΔS for (a) block L, (b) its reservoir, (c) block R, (d) its reservoir, (e) the two-block system, and (f) the system of the two blocks and the two reservoirs?

••14 (a) For 1.0 mol of a monatomic ideal gas taken through the cycle in Fig. 20-24, where $V_1 = 4.00V_0$, what is W/p_0V_0 as the gas goes from state a to state c along path abc? What is $\Delta E_{int}/p_0V_0$ in going (b) from b to c and (c) through one full cycle? What is ΔS in going (d) from b to c and (e) through one full cycle?

Fig. 20-24 Problem 14.

••15 A mixture of 1773 g of water and 227 g of ice is in an initial equilibrium state at 0.000°C. The mixture is then, in a reversible process, brought to a second equilibrium state where the water–ice ratio, by mass, is 1.00:1.00 at 0.000°C. (a) Calculate the entropy change of the system during this process. (The heat of fusion for water is 333 kJ/kg.) (b) The system is then returned to the initial equilibrium state in an irreversible process (say, by using a Bunsen burner). Calculate the entropy change of the system during this process. (c) Are your answers consistent with the second law of thermodynamics?

••16 **GO** An 8.0 g ice cube at $-10°C$ is put into a Thermos flask containing 100 cm³ of water at 20°C. By how much has the entropy of the cube–water system changed when equilibrium is reached? The specific heat of ice is 2220 J/kg·K.

••17 In Fig. 20-25, where $V_{23} = 3.00V_1, n$ moles of a diatomic ideal gas are taken through the cycle with the molecules rotating but not oscillating. What are (a) p_2/p_1, (b) p_3/p_1, and (c) T_3/T_1? For path $1 \rightarrow 2$, what are (d) W/nRT_1, (e) Q/nRT_1, (f) $\Delta E_{int}/nRT_1$, and (g) $\Delta S/nR$? For path $2 \rightarrow 3$, what are (h) W/nRT_1, (i) Q/nRT_1, (j) $\Delta E_{int}/nRT_1$, (k) $\Delta S/nR$? For path $3 \rightarrow 1$, what are (l) W/nRT_1, (m) Q/nRT_1, (n) $\Delta E_{int}/nRT_1$, and (o) $\Delta S/nR$?

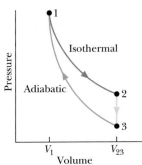

Fig. 20-25 Problem 17.

••18 **GO** A 2.0 mol sample of an ideal monatomic gas undergoes the reversible process shown in Fig. 20-26. The scale of the vertical axis is set by $T_s = 400.0$ K and the scale of the horizontal axis is set by $S_s = 20.0$ J/K. (a) How much energy is absorbed as heat by the gas? (b) What is the change in the internal energy of the gas? (c) How much work is done by the gas?

Fig. 20-26 Problem 18.

•••19 Suppose 1.00 mol of a monatomic ideal gas is taken from initial pressure p_1 and volume V_1 through two steps: (1) an isothermal expansion to volume $2.00V_1$ and (2) a pressure increase to $2.00p_1$ at constant volume. What is Q/p_1V_1 for (a) step 1 and (b) step 2? What is W/p_1V_1 for (c) step 1 and (d) step 2? For the full process, what are (e) $\Delta E_{int}/p_1V_1$ and (f) ΔS? The gas is returned to its initial state and again taken to the same final state but now through these two steps: (1) an isothermal compression to pressure $2.00p_1$ and (2) a volume increase to $2.00V_1$ at constant pressure. What is Q/p_1V_1 for (g) step 1 and (h) step 2? What is W/p_1V_1 for (i) step 1 and (j) step 2? For the full process, what are (k) $\Delta E_{int}/p_1V_1$ and (l) ΔS?

•••20 Expand 1.00 mol of an monatomic gas initially at 5.00 kPa and 600 K from initial volume $V_i = 1.00$ m³ to final volume $V_f = 2.00$ m³. At any instant during the expansion, the pressure p and volume V of the gas are related by $p = 5.00 \exp[(V_i - V)/a]$, with p in kilopascals, V_i and V in cubic meters, and $a = 1.00$ m³. What are the final (a) pressure and (b) temperature of the gas? (c) How much work is done by the gas during the expansion? (d) What is ΔS for the expansion? (*Hint:* Use two simple reversible processes to find ΔS.)

•••21 **GO** Energy can be removed from water as heat at and even below the normal freezing point (0.0°C at atmospheric pressure) without causing the water to freeze; the water is then said to be *supercooled*. Suppose a 1.00 g water drop is supercooled until its temperature is that of the surrounding air, which is at $-5.00°C$. The drop then suddenly and irreversibly freezes, transferring energy to the air as heat. What is the entropy change for the drop? (*Hint:* Use a three-step reversible process as if the water were taken through the normal freezing point.) The specific heat of ice is 2220 J/kg·K.

•••22 **GO** An insulated Thermos contains 130 g of water at 80.0°C. You put in a 12.0 g ice cube at 0°C to form a system of *ice + original water*. (a) What is the equilibrium temperature of the system? What are the entropy changes of the water that was originally the ice cube (b) as it melts and (c) as it warms to the equilibrium temperature? (d) What is the entropy change of the original water as it cools to the equilibrium temperature? (e) What is the net entropy change of the *ice + original water* system as it reaches the equilibrium temperature?

sec. 20-5   Entropy in the Real World: Engines

•23 A Carnot engine whose low-temperature reservoir is at 17°C has an efficiency of 40%. By how much should the temperature of the high-temperature reservoir be increased to increase the efficiency to 50%?

•24 A Carnot engine absorbs 52 kJ as heat and exhausts 36 kJ as heat in each cycle. Calculate (a) the engine's efficiency and (b) the work done per cycle in kilojoules.

•25 A Carnot engine has an efficiency of 22.0%. It operates between constant-temperature reservoirs differing in temperature by 75.0 C°. What is the temperature of the (a) lower-temperature and (b) higher-temperature reservoir?

•26 In a hypothetical nuclear fusion reactor, the fuel is deuterium gas at a temperature of 7×10^8 K. If this gas could be used to operate a Carnot engine with $T_L = 100°C$, what would be the engine's efficiency? Take both temperatures to be exact and report your answer to seven significant figures.

•27 **SSM** **WWW** A Carnot engine operates between 235°C and 115°C, absorbing 6.30×10^4 J per cycle at the higher temperature. (a) What is the efficiency of the engine? (b) How much work per cycle is this engine capable of performing?

••28 In the first stage of a two-stage Carnot engine, energy is absorbed as heat Q_1 at temperature T_1, work W_1 is done, and energy is expelled as heat Q_2 at a lower temperature T_2. The second stage absorbs that energy as heat Q_2, does work W_2, and expels energy as heat Q_3 at a still lower temperature T_3. Prove that the efficiency of the engine is $(T_1 - T_3)/T_1$.

••29 **GO** Figure 20-27 shows a reversible cycle through which 1.00 mol of a monatomic ideal gas is taken. Assume that $p = 2p_0$, $V = 2V_0$, $p_0 = 1.01 \times 10^5$ Pa, and $V_0 = 0.0225$ m³. Calculate (a) the work done during the cycle, (b) the energy added as heat during stroke *abc*, and (c) the efficiency of the cycle. (d) What is the efficiency of a Carnot engine operating between

Fig. 20-27 Problem 29.

the highest and lowest temperatures that occur in the cycle? (e) Is this greater than or less than the efficiency calculated in (c)?

••30 A 500 W Carnot engine operates between constant-temperature reservoirs at 100°C and 60.0°C. What is the rate at which energy is (a) taken in by the engine as heat and (b) exhausted by the engine as heat?

••31 The efficiency of a particular car engine is 25% when the engine does 8.2 kJ of work per cycle. Assume the process is reversible. What are (a) the energy the engine gains per cycle as heat Q_{gain} from the fuel combustion and (b) the energy the engine loses per cycle as heat Q_{lost}. If a tune-up increases the efficiency to 31%, what are (c) Q_{gain} and (d) Q_{lost} at the same work value?

••**32** 🔵 A Carnot engine is set up to produce a certain work W per cycle. In each cycle, energy in the form of heat Q_H is transferred to the working substance of

Fig. 20-28   Problem 32.

the engine from the higher-temperature thermal reservoir, which is at an adjustable temperature T_H. The lower-temperature thermal reservoir is maintained at temperature $T_L = 250$ K. Figure 20-28 gives Q_H for a range of T_H. The scale of the vertical axis is set by $Q_{Hs} = 6.0$ kJ. If T_H is set at 550 K, what is Q_H?

••**33** SSM ILW Figure 20-29 shows a reversible cycle through which 1.00 mol of a monatomic ideal gas is taken. Volume $V_c = 8.00V_b$. Process bc is an adiabatic expansion, with $p_b = 10.0$ atm and $V_b = 1.00 \times 10^{-3}$ m³. For the cycle, find (a) the energy added to the gas as heat, (b) the energy leaving the gas as heat, (c) the net work done by the gas, and (d) the efficiency of the cycle.

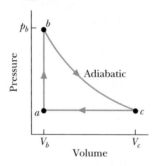

Fig. 20-29   Problem 33.

••**34** An ideal gas (1.0 mol) is the working substance in an engine that operates on the cycle shown in Fig. 20-30. Processes BC and DA are reversible and adiabatic. (a) Is the gas monatomic, diatomic, or polyatomic? (b) What is the engine efficiency?

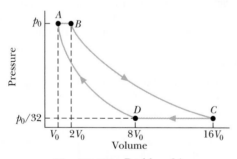

Fig. 20-30   Problem 34.

•••**35** The cycle in Fig. 20-31 represents the operation of a gaso-

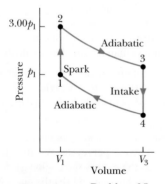

Fig. 20-31   Problem 35.

line internal combustion engine. Volume $V_3 = 4.00V_1$. Assume the gasoline–air intake mixture is an ideal gas with $\gamma = 1.30$. What are the ratios (a) T_2/T_1, (b) T_3/T_1, (c) T_4/T_1, (d) p_3/p_1, and (e) p_4/p_1? (f) What is the engine efficiency?

sec. 20-6 Entropy in the Real World: Refrigerators

•**36** How much work must be done by a Carnot refrigerator to transfer 1.0 J as heat (a) from a reservoir at 7.0°C to one at 27°C, (b) from a reservoir at −73°C to one at 27°C, (c) from a reservoir at −173°C to one at 27°C, and (d) from a reservoir at −223°C to one at 27°C?

•**37** SSM A heat pump is used to heat a building. The outside temperature is 25.0°C, and the temperature inside the building is to be maintained at 22°C. The pump's coefficient of performance is 3.8, and the heat pump delivers 7.54 MJ as heat to the building each hour. If the heat pump is a Carnot engine working in reverse, at what rate must work be done to run it?

•**38** The electric motor of a heat pump transfers energy as heat from the outdoors, which is at −5.0°C, to a room that is at 17°C. If the heat pump were a Carnot heat pump (a Carnot engine working in reverse), how much energy would be transferred as heat to the room for each joule of electric energy consumed?

•**39** SSM A Carnot air conditioner takes energy from the thermal energy of a room at 70°F and transfers it as heat to the outdoors, which is at 96°F. For each joule of electric energy required to operate the air conditioner, how many joules are removed from the room?

•**40** To make ice, a freezer that is a reverse Carnot engine extracts 42 kJ as heat at −15°C during each cycle, with coefficient of performance 5.7. The room temperature is 30.3°C. How much (a) energy per cycle is delivered as heat to the room and (b) work per cycle is required to run the freezer?

••**41** ILW An air conditioner operating between 93°F and 70°F is rated at 4000 Btu/h cooling capacity. Its coefficient of performance is 27% of that of a Carnot refrigerator operating between the same two temperatures. What horsepower is required of the air conditioner motor?

••**42** The motor in a refrigerator has a power of 200 W. If the freezing compartment is at 270 K and the outside air is at 300 K, and assuming the efficiency of a Carnot refrigerator, what is the maximum amount of energy that can be extracted as heat from the freezing compartment in 10.0 min?

••**43** 🔵 Figure 20-32 represents a Carnot engine that works between temperatures $T_1 = 400$ K and $T_2 = 150$ K and drives a

Fig. 20-32   Problem 43.

Carnot refrigerator that works between temperatures $T_3 = 325$ K and $T_4 = 225$ K. What is the ratio Q_3/Q_1?

••44 (a) During each cycle, a Carnot engine absorbs 750 J as heat from a high-temperature reservoir at 360 K, with the low-temperature reservoir at 280 K. How much work is done per cycle? (b) The engine is then made to work in reverse to function as a Carnot refrigerator between those same two reservoirs. During each cycle, how much work is required to remove 1200 J as heat from the low-temperature reservoir?

sec. 20-8   A Statistical View of Entropy

•45 Construct a table like Table 20-1 for eight molecules.

••46 A box contains N identical gas molecules equally divided between its two halves. For $N = 50$, what are (a) the multiplicity W of the central configuration, (b) the total number of microstates, and (c) the percentage of the time the system spends in the central configuration? For $N = 100$, what are (d) W of the central configuration, (e) the total number of microstates, and (f) the percentage of the time the system spends in the central configuration? For $N = 200$, what are (g) W of the central configuration, (h) the total number of microstates, and (i) the percentage of the time the system spends in the central configuration? (j) Does the time spent in the central configuration increase or decrease with an increase in N?

•••47 SSM WWW A box contains N gas molecules. Consider the box to be divided into three equal parts. (a) By extension of Eq. 20-20, write a formula for the multiplicity of any given configuration. (b) Consider two configurations: configuration A with equal numbers of molecules in all three thirds of the box, and configuration B with equal numbers of molecules in each half of the box divided into two equal parts rather than three. What is the ratio W_A/W_B of the multiplicity of configuration A to that of configuration B? (c) Evaluate W_A/W_B for $N = 100$. (Because 100 is not evenly divisible by 3, put 34 molecules into one of the three box parts of configuration A and 33 in each of the other two parts.)

Additional Problems

48 Four particles are in the insulated box of Fig. 20-17. What are (a) the least multiplicity, (b) the greatest multiplicity, (c) the least entropy, and (d) the greatest entropy of the four-particle system?

49 A cylindrical copper rod of length 1.50 m and radius 2.00 cm is insulated to prevent heat loss through its curved surface. One end is attached to a thermal reservoir fixed at 300°C; the other is attached to a thermal reservoir fixed at 30.0°C. What is the rate at which entropy increases for the rod–reservoirs system?

50 Suppose 0.550 mol of an ideal gas is isothermally and reversibly expanded in the four situations given below. What is the change in the entropy of the gas for each situation?

Situation	(a)	(b)	(c)	(d)
Temperature (K)	250	350	400	450
Initial volume (cm³)	0.200	0.200	0.300	0.300
Final volume (cm³)	0.800	0.800	1.20	1.20

51 SSM As a sample of nitrogen gas (N_2) undergoes a temperature increase at constant volume, the distribution of molecular speeds increases. That is, the probability distribution function $P(v)$ for the molecules spreads to higher speed values, as sug-

gested in Fig. 19-8b. One way to report the spread in $P(v)$ is to measure the difference Δv between the most probable speed v_P and the rms speed v_{rms}. When $P(v)$ spreads to higher speeds, Δv increases. Assume that the gas is ideal and the N_2 molecules rotate but do not oscillate. For 1.5 mol, an initial temperature of 250 K, and a final temperature of 500 K, what are (a) the initial difference Δv_i, (b) the final difference Δv_f, and (c) the entropy change ΔS for the gas?

52 Suppose 1.0 mol of a monatomic ideal gas initially at 10 L and 300 K is heated at constant volume to 600 K, allowed to expand isothermally to its initial pressure, and finally compressed at constant pressure to its original volume, pressure, and temperature. During the cycle, what are (a) the net energy entering the system (the gas) as heat and (b) the net work done by the gas? (c) What is the efficiency of the cycle?

53 Suppose that a deep shaft were drilled in Earth's crust near one of the poles, where the surface temperature is $-40°C$, to a depth where the temperature is 800°C. (a) What is the theoretical limit to the efficiency of an engine operating between these temperatures? (b) If all the energy released as heat into the low-temperature reservoir were used to melt ice that was initially at $-40°C$, at what rate could liquid water at 0°C be produced by a 100 MW power plant (treat it as an engine)? The specific heat of ice is 2220 J/kg·K; water's heat of fusion is 333 kJ/kg. (Note that the engine can operate only between 0°C and 800°C in this case. Energy exhausted at $-40°C$ cannot warm anything above $-40°C$.)

54 What is the entropy change for 3.20 mol of an ideal monatomic gas undergoing a reversible increase in temperature from 380 K to 425 K at constant volume?

55 A 600 g lump of copper at 80.0°C is placed in 70.0 g of water at 10.0°C in an insulated container. (See Table 18-3 for specific heats.) (a) What is the equilibrium temperature of the copper–water system? What entropy changes do (b) the copper, (c) the water, and (d) the copper–water system undergo in reaching the equilibrium temperature?

56 ~~✂~~ Figure 20-33 gives the force magnitude F versus stretch distance x for a rubber band, with the scale of the F axis set by $F_s = 1.50$ N and the scale of the x axis set by $x_s = 3.50$ cm. The temperature is 2.00°C. When the rubber band is stretched by $x = 1.70$ cm, at what rate does the entropy of the rubber band change during a small additional stretch?

Fig. 20-33
Problem 56.

57 The temperature of 1.00 mol of a monatomic ideal gas is raised reversibly from 300 K to 400 K, with its volume kept constant. What is the entropy change of the gas?

58 Repeat Problem 57, with the pressure now kept constant.

59 SSM A 0.600 kg sample of water is initially ice at temperature $-20°C$. What is the sample's entropy change if its temperature is increased to 40°C?

60 A three-step cycle is undergone by 3.4 mol of an ideal diatomic gas: (1) the temperature of the gas is increased from 200 K to 500 K at constant volume; (2) the gas is then isothermally expanded to its original pressure; (3) the gas is then contracted at constant pressure back to its original volume. Throughout the cycle, the molecules rotate but do not oscillate. What is the efficiency of the cycle?

61 An inventor has built an engine X and claims that its efficiency ε_X is greater than the efficiency ε of an ideal engine operating between the same two temperatures. Suppose you couple engine X to an ideal refrigerator (Fig. 20-34a) and adjust the cycle of engine X so that the work per cycle it provides equals the work per cycle required by the ideal refrigerator. Treat this combination as a single unit and show that if the inventor's claim were true (if $\varepsilon_X >$ ε), the combined unit would act as a perfect refrigerator (Fig. 20-34b), transferring energy as heat from the low-temperature reservoir to the high-temperature reservoir without the need for work.

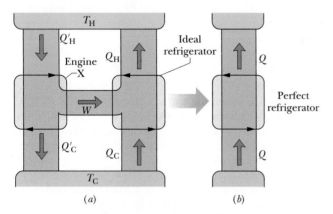

Fig. 20-34 Problem 61.

62 Suppose 2.00 mol of a diatomic gas is taken reversibly around the cycle shown in the T-S diagram of Fig. 20-35, where $S_1 = 6.00$ J/K and $S_2 = 8.00$ J/K. The molecules do not rotate or oscillate. What is the energy transferred as heat Q for (a) path $1 \rightarrow 2$, (b) path $2 \rightarrow 3$, and (c) the full cycle? (d) What is the work W for the isothermal process? The volume V_1 in state 1 is 0.200 m³. What is the volume in (e) state 2 and (f) state 3?

Fig. 20-35 Problem 62.

What is the change ΔE_{int} for (g) path $1 \rightarrow 2$, (h) path $2 \rightarrow 3$, and (i) the full cycle? (*Hint:* (h) can be done with one or two lines of calculation using Section 19-8 or with a page of calculation using Section 19-11.) (j) What is the work W for the adiabatic process?

63 A three-step cycle is undergone reversibly by 4.00 mol of an ideal gas: (1) an adiabatic expansion that gives the gas 2.00 times its initial volume, (2) a constant-volume process, (3) an isothermal compression back to the initial state of the gas. We do not know whether the gas is monatomic or diatomic; if it is diatomic, we do not know whether the molecules are rotating or oscillating. What are the entropy changes for (a) the cycle, (b) process 1, (c) process 3, and (d) process 2?

64 (a) A Carnot engine operates between a hot reservoir at 320 K and a cold one at 260 K. If the engine absorbs 500 J as heat per cycle at the hot reservoir, how much work per cycle does it deliver? (b) If the engine working in reverse functions as a refrigerator between the same two reservoirs, how much work per cycle must be supplied to remove 1000 J as heat from the cold reservoir?

65 A 2.00 mol diatomic gas initially at 300 K undergoes this cycle: It is (1) heated at constant volume to 800 K, (2) then allowed to expand isothermally to its initial pressure, (3) then compressed at constant pressure to its initial state. Assuming the gas molecules neither rotate nor oscillate, find (a) the net energy transferred as heat to the gas, (b) the net work done by the gas, and (c) the efficiency of the cycle.

66 An ideal refrigerator does 150 J of work to remove 560 J as heat from its cold compartment. (a) What is the refrigerator's coefficient of performance? (b) How much heat per cycle is exhausted to the kitchen?

67 Suppose that 260 J is conducted from a constant-temperature reservoir at 400 K to one at (a) 100 K, (b) 200 K, (c) 300 K, and (d) 360 K. What is the net change in entropy ΔS_{net} of the reservoirs in each case? (e) As the temperature difference of the two reservoirs decreases, does ΔS_{net} increase, decrease, or remain the same?

68 An apparatus that liquefies helium is in a room maintained at 300 K. If the helium in the apparatus is at 4.0 K, what is the minimum ratio Q_{to}/Q_{from}, where Q_{to} is the energy delivered as heat to the room and Q_{from} is the energy removed as heat from the helium?

69 GO A brass rod is in thermal contact with a constant-temperature reservoir at 130°C at one end and a constant-temperature reservoir at 24.0°C at the other end. (a) Compute the total change in entropy of the rod–reservoirs system when 5030 J of energy is conducted through the rod, from one reservoir to the other. (b) Does the entropy of the rod change?

70 A 45.0 g block of tungsten at 30.0°C and a 25.0 g block of silver at −120°C are placed together in an insulated container. (See Table 18-3 for specific heats.) (a) What is the equilibrium temperature? What entropy changes do (b) the tungsten, (c) the silver, and (d) the tungsten–silver system undergo in reaching the equilibrium temperature?

71 A box contains N molecules. Consider two configurations: configuration A with an equal division of the molecules between the two halves of the box, and configuration B with 60.0% of the molecules in the left half of the box and 40.0% in the right half. For $N = 50$, what are (a) the multiplicity W_A of configuration A, (b) the multiplicity W_B of configuration B, and (c) the ratio $f_{B/A}$ of the time the system spends in configuration B to the time it spends in configuration A? For $N = 100$, what are (d) W_A, (e) W_B, and (f) $f_{B/A}$? For $N = 200$, what are (g) W_A, (h) W_B? and (i) $f_{B/A}$? (j) With increasing N, does f increase, decrease, or remain the same?

72 Calculate the efficiency of a fossil-fuel power plant that consumes 380 metric tons of coal each hour to produce useful work at the rate of 750 MW. The heat of combustion of coal (the heat due to burning it) is 28 MJ/kg.

73 SSM A Carnot refrigerator extracts 35.0 kJ as heat during each cycle, operating with a coefficient of performance of 4.60. What are (a) the energy per cycle transferred as heat to the room and (b) the work done per cycle?

74 A Carnot engine whose high-temperature reservoir is at 400 K has an efficiency of 30.0%. By how much should the temperature of the low-temperature reservoir be changed to increase the efficiency to 40.0%?

75 SSM System A of three particles and system B of five particles are in insulated boxes like that in Fig. 20-17. What is the least multiplicity W of (a) system A and (b) system B? What is the greatest multiplicity W of (c) A and (d) B? What is the greatest entropy of (e) A and (f) B?

The International System of Units (SI)*

TABLE 1

The SI Base Units

Quantity	Name	Symbol	Definition
length	meter	m	"... the length of the path traveled by light in vacuum in 1/299,792,458 of a second." (1983)
mass	kilogram	kg	"... this prototype [a certain platinum–iridium cylinder] shall henceforth be considered to be the unit of mass." (1889)
time	second	s	"... the duration of 9,192,631,770 periods of the radiation corresponding to the transition between the two hyperfine levels of the ground state of the cesium-133 atom." (1967)
electric current	ampere	A	"... that constant current which, if maintained in two straight parallel conductors of infinite length, of negligible circular cross section, and placed 1 meter apart in vacuum, would produce between these conductors a force equal to 2×10^{-7} newton per meter of length." (1946)
thermodynamic temperature	kelvin	K	"... the fraction 1/273.16 of the thermodynamic temperature of the triple point of water." (1967)
amount of substance	mole	mol	"... the amount of substance of a system which contains as many elementary entities as there are atoms in 0.012 kilogram of carbon-12." (1971)
luminous intensity	candela	cd	"... the luminous intensity, in a given direction, of a source that emits monochromatic radiation of frequency 540×10^{12} hertz and that has a radiant intensity in that direction of 1/683 watt per steradian." (1979)

*Adapted from "The International System of Units (SI)," National Bureau of Standards Special Publication 330, 1972 edition. The definitions above were adopted by the General Conference of Weights and Measures, an international body, on the dates shown. In this book we do not use the candela.

TABLE 2

Some SI Derived Units

Quantity	Name of Unit	Symbol	
area	square meter	m^2	
volume	cubic meter	m^3	
frequency	hertz	Hz	s^{-1}
mass density (density)	kilogram per cubic meter	kg/m^3	
speed, velocity	meter per second	m/s	
angular velocity	radian per second	rad/s	
acceleration	meter per second per second	m/s^2	
angular acceleration	radian per second per second	rad/s^2	
force	newton	N	$kg \cdot m/s^2$
pressure	pascal	Pa	N/m^2
work, energy, quantity of heat	joule	J	$N \cdot m$
power	watt	W	J/s
quantity of electric charge	coulomb	C	$A \cdot s$
potential difference, electromotive force	volt	V	W/A
electric field strength	volt per meter (or newton per coulomb)	V/m	N/C
electric resistance	ohm	Ω	V/A
capacitance	farad	F	$A \cdot s/V$
magnetic flux	weber	Wb	$V \cdot s$
inductance	henry	H	$V \cdot s/A$
magnetic flux density	tesla	T	Wb/m^2
magnetic field strength	ampere per meter	A/m	
entropy	joule per kelvin	J/K	
specific heat	joule per kilogram kelvin	$J/(kg \cdot K)$	
thermal conductivity	watt per meter kelvin	$W/(m \cdot K)$	
radiant intensity	watt per steradian	W/sr	

TABLE 3

The SI Supplementary Units

Quantity	Name of Unit	Symbol
plane angle	radian	rad
solid angle	steradian	sr

Some Fundamental Constants of Physics*

Constant	Symbol	Computational Value	Best (1998) Value Value[a]	Best (1998) Value Uncertainty[b]
Speed of light in a vacuum	c	3.00×10^8 m/s	2.997 924 58	exact
Elementary charge	e	1.60×10^{-19} C	1.602 176 487	0.025
Gravitational constant	G	6.67×10^{-11} m³/s²·kg	6.674 28	100
Universal gas constant	R	8.31 J/mol·K	8.314 472	1.7
Avogadro constant	N_A	6.02×10^{23} mol⁻¹	6.022 141 79	0.050
Boltzmann constant	k	1.38×10^{-23} J/K	1.380 650 4	1.7
Stefan–Boltzmann constant	σ	5.67×10^{-8} W/m²·K⁴	5.670 400	7.0
Molar volume of ideal gas at STP[d]	V_m	2.27×10^{-2} m³/mol	2.271 098 1	1.7
Permittivity constant	ϵ_0	8.85×10^{-12} F/m	8.854 187 817 62	exact
Permeability constant	μ_0	1.26×10^{-6} H/m	1.256 637 061 43	exact
Planck constant	h	6.63×10^{-34} J·s	6.626 068 96	0.050
Electron mass[c]	m_e	9.11×10^{-31} kg	9.109 382 15	0.050
		5.49×10^{-4} u	5.485 799 094 3	4.2×10^{-4}
Proton mass[c]	m_p	1.67×10^{-27} kg	1.672 621 637	0.050
		1.0073 u	1.007 276 466 77	1.0×10^{-4}
Ratio of proton mass to electron mass	m_p/m_e	1840	1836.152 672 47	4.3×10^{-4}
Electron charge-to-mass ratio	e/m_e	1.76×10^{11} C/kg	1.758 820 150	0.025
Neutron mass[c]	m_n	1.68×10^{-27} kg	1.674 927 211	0.050
		1.0087 u	1.008 664 915 97	4.3×10^{-4}
Hydrogen atom mass[c]	m_{1_H}	1.0078 u	1.007 825 031 6	0.0005
Deuterium atom mass[c]	m_{2_H}	2.0136 u	2.013 553 212 724	3.9×10^{-5}
Helium atom mass[c]	$m_{4_{He}}$	4.0026 u	4.002 603 2	0.067
Muon mass	m_μ	1.88×10^{-28} kg	1.883 531 30	0.056
Electron magnetic moment	μ_e	9.28×10^{-24} J/T	9.284 763 77	0.025
Proton magnetic moment	μ_p	1.41×10^{-26} J/T	1.410 606 662	0.026
Bohr magneton	μ_B	9.27×10^{-24} J/T	9.274 009 15	0.025
Nuclear magneton	μ_N	5.05×10^{-27} J/T	5.050 783 24	0.025
Bohr radius	a	5.29×10^{-11} m	5.291 772 085 9	6.8×10^{-4}
Rydberg constant	R	1.10×10^7 m⁻¹	1.097 373 156 852 7	6.6×10^{-6}
Electron Compton wavelength	λ_C	2.43×10^{-12} m	2.426 310 217 5	0.0014

[a]Values given in this column should be given the same unit and power of 10 as the computational value.

[b]Parts per million.

[c]Masses given in u are in unified atomic mass units, where 1 u = 1.660 538 782 $\times 10^{-27}$ kg.

[d]STP means standard temperature and pressure: 0°C and 1.0 atm (0.1 MPa).

*The values in this table were selected from the 1998 CODATA recommended values (www.physics.nist.gov).

Some Astronomical Data

Some Distances from Earth

To the Moon*	3.82×10^8 m	To the center of our galaxy	2.2×10^{20} m
To the Sun*	1.50×10^{11} m	To the Andromeda Galaxy	2.1×10^{22} m
To the nearest star (Proxima Centauri)	4.04×10^{16} m	To the edge of the observable universe	$\sim 10^{26}$ m

*Mean distance.

The Sun, Earth, and the Moon

Property	Unit	Sun	Earth	Moon
Mass	kg	1.99×10^{30}	5.98×10^{24}	7.36×10^{22}
Mean radius	m	6.96×10^8	6.37×10^6	1.74×10^6
Mean density	kg/m³	1410	5520	3340
Free-fall acceleration at the surface	m/s²	274	9.81	1.67
Escape velocity	km/s	618	11.2	2.38
Period of rotation[a]	—	37 d at poles[b] 26 d at equator[b]	23 h 56 min	27.3 d
Radiation power[c]	W	3.90×10^{26}		

[a]Measured with respect to the distant stars.

[b]The Sun, a ball of gas, does not rotate as a rigid body.

[c]Just outside Earth's atmosphere solar energy is received, assuming normal incidence, at the rate of 1340 W/m².

Some Properties of the Planets

	Mercury	Venus	Earth	Mars	Jupiter	Saturn	Uranus	Neptune	Pluto
Mean distance from Sun, 10^6 km	57.9	108	150	228	778	1430	2870	4500	5900
Period of revolution, y	0.241	0.615	1.00	1.88	11.9	29.5	84.0	165	248
Period of rotation,[a] d	58.7	−243[b]	0.997	1.03	0.409	0.426	−0.451[b]	0.658	6.39
Orbital speed, km/s	47.9	35.0	29.8	24.1	13.1	9.64	6.81	5.43	4.74
Inclination of axis to orbit	<28°	≈3°	23.4°	25.0°	3.08°	26.7°	97.9°	29.6°	57.5°
Inclination of orbit to Earth's orbit	7.00°	3.39°		1.85°	1.30°	2.49°	0.77°	1.77°	17.2°
Eccentricity of orbit	0.206	0.0068	0.0167	0.0934	0.0485	0.0556	0.0472	0.0086	0.250
Equatorial diameter, km	4880	12 100	12 800	6790	143 000	120 000	51 800	49 500	2300
Mass (Earth = 1)	0.0558	0.815	1.000	0.107	318	95.1	14.5	17.2	0.002
Density (water = 1)	5.60	5.20	5.52	3.95	1.31	0.704	1.21	1.67	2.03
Surface value of g,[c] m/s²	3.78	8.60	9.78	3.72	22.9	9.05	7.77	11.0	0.5
Escape velocity,[c] km/s	4.3	10.3	11.2	5.0	59.5	35.6	21.2	23.6	1.3
Known satellites	0	0	1	2	63 + ring	60 + rings	27 + rings	13 + rings	3

[a]Measured with respect to the distant stars.

[b]Venus and Uranus rotate opposite their orbital motion.

[c]Gravitational acceleration measured at the planet's equator.

Conversion Factors

Conversion factors may be read directly from these tables. For example, 1 degree = 2.778 × 10^{-3} revolutions, so 16.7° = 16.7 × 2.778 × 10^{-3} rev. The SI units are fully capitalized. Adapted in part from G. Shortley and D. Williams, *Elements of Physics*, 1971, Prentice-Hall, Englewood Cliffs, NJ.

Plane Angle

	°	′	″	RADIAN	rev
1 degree =	1	60	3600	1.745×10^{-2}	2.778×10^{-3}
1 minute =	1.667×10^{-2}	1	60	2.909×10^{-4}	4.630×10^{-5}
1 second =	2.778×10^{-4}	1.667×10^{-2}	1	4.848×10^{-6}	7.716×10^{-7}
1 RADIAN =	57.30	3438	2.063×10^{5}	1	0.1592
1 revolution =	360	2.16×10^{4}	1.296×10^{6}	6.283	1

Solid Angle

1 sphere = 4π steradians = 12.57 steradians

Length

	cm	METER	km	in.	ft	mi
1 centimeter =	1	10^{-2}	10^{-5}	0.3937	3.281×10^{-2}	6.214×10^{-6}
1 METER =	100	1	10^{-3}	39.37	3.281	6.214×10^{-4}
1 kilometer =	10^{5}	1000	1	3.937×10^{4}	3281	0.6214
1 inch =	2.540	2.540×10^{-2}	2.540×10^{-5}	1	8.333×10^{-2}	1.578×10^{-5}
1 foot =	30.48	0.3048	3.048×10^{-4}	12	1	1.894×10^{-4}
1 mile =	1.609×10^{5}	1609	1.609	6.336×10^{4}	5280	1

1 angström = 10^{-10} m
1 nautical mile = 1852 m
 = 1.151 miles = 6076 ft

1 fermi = 10^{-15} m
1 light-year = 9.461×10^{12} km
1 parsec = 3.084×10^{13} km

1 fathom = 6 ft
1 Bohr radius = 5.292×10^{-11} m
1 yard = 3 ft

1 rod = 16.5 ft
1 mil = 10^{-3} in.
1 nm = 10^{-9} m

Area

	METER^2	cm^2	ft^2	in.^{2}
1 SQUARE METER =	1	10^{4}	10.76	1550
1 square centimeter =	10^{-4}	1	1.076×10^{-3}	0.1550
1 square foot =	9.290×10^{-2}	929.0	1	144
1 square inch =	6.452×10^{-4}	6.452	6.944×10^{-3}	1

1 square mile = 2.788×10^{7} ft^2 = 640 acres
1 barn = 10^{-28} m^2

1 acre = 43 560 ft^2
1 hectare = 10^{4} m^2 = 2.471 acres

Volume

	METER^3	cm^3	L	ft^3	in.^{3}
1 CUBIC METER = 1		10^6	1000	35.31	6.102×10^4
1 cubic centimeter = 10^{-6}		1	1.000×10^{-3}	3.531×10^{-5}	6.102×10^{-2}
1 liter = 1.000×10^{-3}		1000	1	3.531×10^{-2}	61.02
1 cubic foot = 2.832×10^{-2}		2.832×10^4	28.32	1	1728
1 cubic inch = 1.639×10^{-5}		16.39	1.639×10^{-2}	5.787×10^{-4}	1

1 U.S. fluid gallon = 4 U.S. fluid quarts = 8 U.S. pints = 128 U.S. fluid ounces = 231 in.^{3}
1 British imperial gallon = 277.4 in.^{3} = 1.201 U.S. fluid gallons

Mass

Quantities in the colored areas are not mass units but are often used as such. For example, when we write 1 kg "=" 2.205 lb, this means that a kilogram is a *mass* that *weighs* 2.205 pounds at a location where g has the standard value of 9.80665 m/s^2.

	g	KILOGRAM	slug	u	oz	lb	ton
1 gram = 1		0.001	6.852×10^{-5}	6.022×10^{23}	3.527×10^{-2}	2.205×10^{-3}	1.102×10^{-6}
1 KILOGRAM = 1000		1	6.852×10^{-2}	6.022×10^{26}	35.27	2.205	1.102×10^{-3}
1 slug = 1.459×10^4		14.59	1	8.786×10^{27}	514.8	32.17	1.609×10^{-2}
1 atomic mass unit = 1.661×10^{-24}		1.661×10^{-27}	1.138×10^{-28}	1	5.857×10^{-26}	3.662×10^{-27}	1.830×10^{-30}
1 ounce = 28.35		2.835×10^{-2}	1.943×10^{-3}	1.718×10^{25}	1	6.250×10^{-2}	3.125×10^{-5}
1 pound = 453.6		0.4536	3.108×10^{-2}	2.732×10^{26}	16	1	0.0005
1 ton = 9.072×10^5		907.2	62.16	5.463×10^{29}	3.2×10^4	2000	1

1 metric ton = 1000 kg

Density

Quantities in the colored areas are weight densities and, as such, are dimensionally different from mass densities. See the note for the mass table.

	slug/ft^3	KILOGRAM/ METER^3	g/cm^3	lb/ft^3	lb/in.^{3}
1 slug per foot^3 = 1		515.4	0.5154	32.17	1.862×10^{-2}
1 KILOGRAM per METER^3 = 1.940×10^{-3}		1	0.001	6.243×10^{-2}	3.613×10^{-5}
1 gram per centimeter^3 = 1.940		1000	1	62.43	3.613×10^{-2}
1 pound per foot^3 = 3.108×10^{-2}		16.02	16.02×10^{-2}	1	5.787×10^{-4}
1 pound per inch^3 = 53.71		2.768×10^4	27.68	1728	1

Time

	y	d	h	min	SECOND
1 year = 1		365.25	8.766×10^3	5.259×10^5	3.156×10^7
1 day = 2.738×10^{-3}		1	24	1440	8.640×10^4
1 hour = 1.141×10^{-4}		4.167×10^{-2}	1	60	3600
1 minute = 1.901×10^{-6}		6.944×10^{-4}	1.667×10^{-2}	1	60
1 SECOND = 3.169×10^{-8}		1.157×10^{-5}	2.778×10^{-4}	1.667×10^{-2}	1

Speed

	ft/s	km/h	METER/SECOND	mi/h	cm/s
1 foot per second = 1	1.097	0.3048	0.6818	30.48
1 kilometer per hour = 0.9113	1	0.2778	0.6214	27.78
1 METER per SECOND = 3.281	3.6	1	2.237	100
1 mile per hour = 1.467	1.609	0.4470	1	44.70
1 centimeter per second = 3.281×10^{-2}	3.6×10^{-2}	0.01	2.237×10^{-2}	1

1 knot = 1 nautical mi/h = 1.688 ft/s   1 mi/min = 88.00 ft/s = 60.00 mi/h

Force

Force units in the colored areas are now little used. To clarify: 1 gram-force (= 1 gf) is the force of gravity that would act on an object whose mass is 1 gram at a location where g has the standard value of 9.80665 m/s^2.

	dyne	NEWTON	lb	pdl	gf	kgf
1 dyne = 1	10^{-5}	2.248×10^{-6}	7.233×10^{-5}	1.020×10^{-3}	1.020×10^{-6}
1 NEWTON = 10^5	1	0.2248	7.233	102.0	0.1020
1 pound = 4.448×10^5	4.448	1	32.17	453.6	0.4536
1 poundal = 1.383×10^4	0.1383	3.108×10^{-2}	1	14.10	1.410×10^2
1 gram-force = 980.7	9.807×10^{-3}	2.205×10^{-3}	7.093×10^{-2}	1	0.001
1 kilogram-force = 9.807×10^5	9.807	2.205	70.93	1000	1

1 ton = 2000 lb

Pressure

	atm	dyne/cm^2	inch of water	cm Hg	PASCAL	lb/in.^{2}	lb/ft^2
1 atmosphere = 1	1.013×10^6	406.8	76	1.013×10^5	14.70	2116
1 dyne per centimeter^2 = 9.869×10^{-7}	1	4.015×10^{-4}	7.501×10^{-5}	0.1	1.405×10^{-5}	2.089×10^{-3}
1 inch of water^a at 4°C = 2.458×10^{-3}	2491	1	0.1868	249.1	3.613×10^{-2}	5.202
1 centimeter of mercury^a at 0°C = 1.316×10^{-2}	1.333×10^4	5.353	1	1333	0.1934	27.85
1 PASCAL = 9.869×10^{-6}	10	4.015×10^{-3}	7.501×10^{-4}	1	1.450×10^{-4}	2.089×10^{-2}
1 pound per inch^2 = 6.805×10^{-2}	6.895×10^4	27.68	5.171	6.895×10^3	1	144
1 pound per foot^2 = 4.725×10^{-4}	478.8	0.1922	3.591×10^{-2}	47.88	6.944×10^{-3}	1

^{a}Where the acceleration of gravity has the standard value of 9.80665 m/s^2.

1 bar = 10^6 dyne/cm^2 = 0.1 MPa   1 millibar = 10^3 dyne/cm^2 = 10^2 Pa   1 torr = 1 mm Hg

Energy, Work, Heat

Quantities in the colored areas are not energy units but are included for convenience. They arise from the relativistic mass–energy equivalence formula $E = mc^2$ and represent the energy released if a kilogram or unified atomic mass unit (u) is completely converted to energy (bottom two rows) or the mass that would be completely converted to one unit of energy (rightmost two columns).

	Btu	erg	ft·lb	hp·h	JOULE	cal	kW·h	eV	MeV	kg	u
1 British thermal unit =	1	1.055×10^{10}	777.9	3.929×10^{-4}	1055	252.0	2.930×10^{-4}	6.585×10^{21}	6.585×10^{15}	1.174×10^{-14}	7.070×10^{12}
1 erg =	9.481×10^{-11}	1	7.376×10^{-8}	3.725×10^{-14}	10^{-7}	2.389×10^{-8}	2.778×10^{-14}	6.242×10^{11}	6.242×10^{5}	1.113×10^{-24}	670.2
1 foot-pound =	1.285×10^{-3}	1.356×10^{7}	1	5.051×10^{-7}	1.356	0.3238	3.766×10^{-7}	8.464×10^{18}	8.464×10^{12}	1.509×10^{-17}	9.037×10^{9}
1 horsepower-hour =	2545	2.685×10^{13}	1.980×10^{6}	1	2.685×10^{6}	6.413×10^{5}	0.7457	1.676×10^{25}	1.676×10^{19}	2.988×10^{-11}	1.799×10^{16}
1 JOULE =	9.481×10^{-4}	10^{7}	0.7376	3.725×10^{-7}	1	0.2389	2.778×10^{-7}	6.242×10^{18}	6.242×10^{12}	1.113×10^{-17}	6.702×10^{9}
1 calorie =	3.968×10^{-3}	4.1868×10^{7}	3.088	1.560×10^{-6}	4.1868	1	1.163×10^{-6}	2.613×10^{19}	2.613×10^{13}	4.660×10^{-17}	2.806×10^{10}
1 kilowatt-hour =	3413	3.600×10^{13}	2.655×10^{6}	1.341	3.600×10^{6}	8.600×10^{5}	1	2.247×10^{25}	2.247×10^{19}	4.007×10^{-11}	2.413×10^{16}
1 electron-volt =	1.519×10^{-22}	1.602×10^{-12}	1.182×10^{-19}	5.967×10^{-26}	1.602×10^{-19}	3.827×10^{-20}	4.450×10^{-26}	1	10^{-6}	1.783×10^{-36}	1.074×10^{-9}
1 million electron-volts =	1.519×10^{-16}	1.602×10^{-6}	1.182×10^{-13}	5.967×10^{-20}	1.602×10^{-13}	3.827×10^{-14}	4.450×10^{-20}	10^{-6}	1	1.783×10^{-30}	1.074×10^{-3}
1 kilogram =	8.521×10^{13}	8.987×10^{23}	6.629×10^{16}	3.348×10^{10}	8.987×10^{16}	2.146×10^{16}	2.497×10^{10}	5.610×10^{35}	5.610×10^{29}	1	6.022×10^{26}
1 unified atomic mass unit =	1.415×10^{-13}	1.492×10^{-3}	1.101×10^{-10}	5.559×10^{-17}	1.492×10^{-10}	3.564×10^{-11}	4.146×10^{-17}	9.320×10^{8}	932.0	1.661×10^{-27}	1

Power

	Btu/h	ft·lb/s	hp	cal/s	kW	WATT
1 British thermal unit per hour =	1	0.2161	3.929×10^{-4}	6.998×10^{-2}	2.930×10^{-4}	0.2930
1 foot-pound per second =	4.628	1	1.818×10^{-3}	0.3239	1.356×10^{-3}	1.356
1 horsepower =	2545	550	1	178.1	0.7457	745.7
1 calorie per second =	14.29	3.088	5.615×10^{-3}	1	4.186×10^{-3}	4.186
1 kilowatt =	3413	737.6	1.341	238.9	1	1000
1 WATT =	3.413	0.7376	1.341×10^{-3}	0.2389	0.001	1

Magnetic Field

	gauss	TESLA	milligauss
1 gauss =	1	10^{-4}	1000
1 TESLA =	10^{4}	1	10^{7}
1 milligauss =	0.001	10^{-7}	1

Magnetic Flux

	maxwell	WEBER
1 maxwell =	1	10^{-8}
1 WEBER =	10^{8}	1

1 tesla = 1 weber/meter^2

Mathematical Formulas

Geometry

Circle of radius r: circumference $= 2\pi r$; area $= \pi r^2$.

Sphere of radius r: area $= 4\pi r^2$; volume $= \frac{4}{3}\pi r^3$.

Right circular cylinder of radius r and height h:
area $= 2\pi r^2 + 2\pi rh$; volume $= \pi r^2 h$.

Triangle of base a and altitude h: area $= \frac{1}{2}ah$.

Quadratic Formula

If $ax^2 + bx + c = 0$, then $x = \dfrac{-b \pm \sqrt{b^2 - 4ac}}{2a}$.

Trigonometric Functions of Angle θ

$$\sin\theta = \frac{y}{r} \qquad \cos\theta = \frac{x}{r}$$

$$\tan\theta = \frac{y}{x} \qquad \cot\theta = \frac{x}{y}$$

$$\sec\theta = \frac{r}{x} \qquad \csc\theta = \frac{r}{y}$$

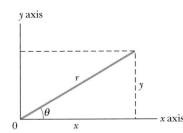

Pythagorean Theorem

In this right triangle,
$$a^2 + b^2 = c^2$$

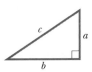

Triangles

Angles are A, B, C

Opposite sides are a, b, c

Angles $A + B + C = 180°$

$$\frac{\sin A}{a} = \frac{\sin B}{b} = \frac{\sin C}{c}$$

$$c^2 = a^2 + b^2 - 2ab\cos C$$

Exterior angle $D = A + C$

Mathematical Signs and Symbols

$=$ equals

$\approx$ equals approximately

$\sim$ is the order of magnitude of

$\neq$ is not equal to

$\equiv$ is identical to, is defined as

$>$ is greater than ($\gg$ is much greater than)

$<$ is less than ($\ll$ is much less than)

$\geq$ is greater than or equal to (or, is no less than)

$\leq$ is less than or equal to (or, is no more than)

$\pm$ plus or minus

$\propto$ is proportional to

Σ the sum of

x_{avg} the average value of x

Trigonometric Identities

$$\sin(90° - \theta) = \cos\theta$$

$$\cos(90° - \theta) = \sin\theta$$

$$\sin\theta/\cos\theta = \tan\theta$$

$$\sin^2\theta + \cos^2\theta = 1$$

$$\sec^2\theta - \tan^2\theta = 1$$

$$\csc^2\theta - \cot^2\theta = 1$$

$$\sin 2\theta = 2\sin\theta\cos\theta$$

$$\cos 2\theta = \cos^2\theta - \sin^2\theta = 2\cos^2\theta - 1 = 1 - 2\sin^2\theta$$

$$\sin(\alpha \pm \beta) = \sin\alpha\cos\beta \pm \cos\alpha\sin\beta$$

$$\cos(\alpha \pm \beta) = \cos\alpha\cos\beta \mp \sin\alpha\sin\beta$$

$$\tan(\alpha \pm \beta) = \frac{\tan\alpha \pm \tan\beta}{1 \mp \tan\alpha\tan\beta}$$

$$\sin\alpha \pm \sin\beta = 2\sin\tfrac{1}{2}(\alpha \pm \beta)\cos\tfrac{1}{2}(\alpha \mp \beta)$$

$$\cos\alpha + \cos\beta = 2\cos\tfrac{1}{2}(\alpha + \beta)\cos\tfrac{1}{2}(\alpha - \beta)$$

$$\cos\alpha - \cos\beta = -2\sin\tfrac{1}{2}(\alpha + \beta)\sin\tfrac{1}{2}(\alpha - \beta)$$

Binomial Theorem

$$(1 + x)^n = 1 + \frac{nx}{1!} + \frac{n(n-1)x^2}{2!} + \cdots \qquad (x^2 < 1)$$

Exponential Expansion

$$e^x = 1 + x + \frac{x^2}{2!} + \frac{x^3}{3!} + \cdots$$

Logarithmic Expansion

$$\ln(1 + x) = x - \tfrac{1}{2}x^2 + \tfrac{1}{3}x^3 - \cdots \qquad (|x| < 1)$$

Trigonometric Expansions
(θ in radians)

$$\sin \theta = \theta - \frac{\theta^3}{3!} + \frac{\theta^5}{5!} - \cdots$$

$$\cos \theta = 1 - \frac{\theta^2}{2!} + \frac{\theta^4}{4!} - \cdots$$

$$\tan \theta = \theta + \frac{\theta^3}{3} + \frac{2\theta^5}{15} + \cdots$$

Cramer's Rule

Two simultaneous equations in unknowns x and y,

$$a_1x + b_1y = c_1 \quad \text{and} \quad a_2x + b_2y = c_2,$$

have the solutions

$$x = \frac{\begin{vmatrix} c_1 & b_1 \\ c_2 & b_2 \end{vmatrix}}{\begin{vmatrix} a_1 & b_1 \\ a_2 & b_2 \end{vmatrix}} = \frac{c_1b_2 - c_2b_1}{a_1b_2 - a_2b_1}$$

and

$$y = \frac{\begin{vmatrix} a_1 & c_1 \\ a_2 & c_2 \end{vmatrix}}{\begin{vmatrix} a_1 & b_1 \\ a_2 & b_2 \end{vmatrix}} = \frac{a_1c_2 - a_2c_1}{a_1b_2 - a_2b_1}.$$

Products of Vectors

Let $\hat{i}$, $\hat{j}$, and $\hat{k}$ be unit vectors in the x, y, and z directions. Then

$$\hat{i}\cdot\hat{i} = \hat{j}\cdot\hat{j} = \hat{k}\cdot\hat{k} = 1, \quad \hat{i}\cdot\hat{j} = \hat{j}\cdot\hat{k} = \hat{k}\cdot\hat{i} = 0,$$

$$\hat{i} \times \hat{i} = \hat{j} \times \hat{j} = \hat{k} \times \hat{k} = 0,$$

$$\hat{i} \times \hat{j} = \hat{k}, \quad \hat{j} \times \hat{k} = \hat{i}, \quad \hat{k} \times \hat{i} = \hat{j}$$

Any vector $\vec{a}$ with components a_x, a_y, and a_z along the x, y, and z axes can be written as

$$\vec{a} = a_x\hat{i} + a_y\hat{j} + a_z\hat{k}.$$

Let $\vec{a}$, $\vec{b}$, and $\vec{c}$ be arbitrary vectors with magnitudes a, b, and c. Then

$$\vec{a} \times (\vec{b} + \vec{c}) = (\vec{a} \times \vec{b}) + (\vec{a} \times \vec{c})$$

$$(s\vec{a}) \times \vec{b} = \vec{a} \times (s\vec{b}) = s(\vec{a} \times \vec{b}) \qquad (s = \text{a scalar}).$$

Let θ be the smaller of the two angles between $\vec{a}$ and $\vec{b}$. Then

$$\vec{a}\cdot\vec{b} = \vec{b}\cdot\vec{a} = a_xb_x + a_yb_y + a_zb_z = ab \cos \theta$$

$$\vec{a} \times \vec{b} = -\vec{b} \times \vec{a} = \begin{vmatrix} \hat{i} & \hat{j} & \hat{k} \\ a_x & a_y & a_z \\ b_x & b_y & b_z \end{vmatrix}$$

$$= \hat{i}\begin{vmatrix} a_y & a_z \\ b_y & b_z \end{vmatrix} - \hat{j}\begin{vmatrix} a_x & a_z \\ b_x & b_z \end{vmatrix} + \hat{k}\begin{vmatrix} a_x & a_y \\ b_x & b_y \end{vmatrix}$$

$$= (a_yb_z - b_ya_z)\hat{i} + (a_zb_x - b_za_x)\hat{j}$$

$$+ (a_xb_y - b_xa_y)\hat{k}$$

$$|\vec{a} \times \vec{b}| = ab \sin \theta$$

$$\vec{a}\cdot(\vec{b} \times \vec{c}) = \vec{b}\cdot(\vec{c} \times \vec{a}) = \vec{c}\cdot(\vec{a} \times \vec{b})$$

$$\vec{a} \times (\vec{b} \times \vec{c}) = (\vec{a}\cdot\vec{c})\vec{b} - (\vec{a}\cdot\vec{b})\vec{c}$$

Derivatives and Integrals

In what follows, the letters u and v stand for any functions of x, and a and m are constants. To each of the indefinite integrals should be added an arbitrary constant of integration. The *Handbook of Chemistry and Physics* (CRC Press Inc.) gives a more extensive tabulation.

1. $\dfrac{dx}{dx} = 1$

2. $\dfrac{d}{dx}(au) = a\dfrac{du}{dx}$

3. $\dfrac{d}{dx}(u + v) = \dfrac{du}{dx} + \dfrac{dv}{dx}$

4. $\dfrac{d}{dx}x^m = mx^{m-1}$

5. $\dfrac{d}{dx}\ln x = \dfrac{1}{x}$

6. $\dfrac{d}{dx}(uv) = u\dfrac{dv}{dx} + v\dfrac{du}{dx}$

7. $\dfrac{d}{dx}e^x = e^x$

8. $\dfrac{d}{dx}\sin x = \cos x$

9. $\dfrac{d}{dx}\cos x = -\sin x$

10. $\dfrac{d}{dx}\tan x = \sec^2 x$

11. $\dfrac{d}{dx}\cot x = -\csc^2 x$

12. $\dfrac{d}{dx}\sec x = \tan x \sec x$

13. $\dfrac{d}{dx}\csc x = -\cot x \csc x$

14. $\dfrac{d}{dx}e^u = e^u\dfrac{du}{dx}$

15. $\dfrac{d}{dx}\sin u = \cos u\dfrac{du}{dx}$

16. $\dfrac{d}{dx}\cos u = -\sin u\dfrac{du}{dx}$

1. $\displaystyle\int dx = x$

2. $\displaystyle\int au\,dx = a\int u\,dx$

3. $\displaystyle\int (u + v)\,dx = \int u\,dx + \int v\,dx$

4. $\displaystyle\int x^m\,dx = \dfrac{x^{m+1}}{m+1}\ (m \neq -1)$

5. $\displaystyle\int \dfrac{dx}{x} = \ln |x|$

6. $\displaystyle\int u\dfrac{dv}{dx}\,dx = uv - \int v\dfrac{du}{dx}\,dx$

7. $\displaystyle\int e^x\,dx = e^x$

8. $\displaystyle\int \sin x\,dx = -\cos x$

9. $\displaystyle\int \cos x\,dx = \sin x$

10. $\displaystyle\int \tan x\,dx = \ln |\sec x|$

11. $\displaystyle\int \sin^2 x\,dx = \tfrac{1}{2}x - \tfrac{1}{4}\sin 2x$

12. $\displaystyle\int e^{-ax}\,dx = -\dfrac{1}{a}e^{-ax}$

13. $\displaystyle\int xe^{-ax}\,dx = -\dfrac{1}{a^2}(ax + 1)e^{-ax}$

14. $\displaystyle\int x^2 e^{-ax}\,dx = -\dfrac{1}{a^3}(a^2x^2 + 2ax + 2)e^{-ax}$

15. $\displaystyle\int_0^\infty x^n e^{-ax}\,dx = \dfrac{n!}{a^{n+1}}$

16. $\displaystyle\int_0^\infty x^{2n} e^{-ax^2}\,dx = \dfrac{1 \cdot 3 \cdot 5 \cdots (2n-1)}{2^{n+1}a^n}\sqrt{\dfrac{\pi}{a}}$

17. $\displaystyle\int \dfrac{dx}{\sqrt{x^2 + a^2}} = \ln(x + \sqrt{x^2 + a^2})$

18. $\displaystyle\int \dfrac{x\,dx}{(x^2 + a^2)^{3/2}} = -\dfrac{1}{(x^2 + a^2)^{1/2}}$

19. $\displaystyle\int \dfrac{dx}{(x^2 + a^2)^{3/2}} = \dfrac{x}{a^2(x^2 + a^2)^{1/2}}$

20. $\displaystyle\int_0^\infty x^{2n+1} e^{-ax^2}\,dx = \dfrac{n!}{2a^{n+1}}\ (a > 0)$

21. $\displaystyle\int \dfrac{x\,dx}{x + d} = x - d\ln(x + d)$

Properties of the Elements

All physical properties are for a pressure of 1 atm unless otherwise specified.

Element	Symbol	Atomic Number Z	Molar Mass, g/mol	Density, g/cm³ at 20°C	Melting Point, °C	Boiling Point, °C	Specific Heat, J/(g·°C) at 25°C
Actinium	Ac	89	(227)	10.06	1323	(3473)	0.092
Aluminum	Al	13	26.9815	2.699	660	2450	0.900
Americium	Am	95	(243)	13.67	1541	—	—
Antimony	Sb	51	121.75	6.691	630.5	1380	0.205
Argon	Ar	18	39.948	1.6626×10^{-3}	−189.4	−185.8	0.523
Arsenic	As	33	74.9216	5.78	817 (28 atm)	613	0.331
Astatine	At	85	(210)	—	(302)	—	—
Barium	Ba	56	137.34	3.594	729	1640	0.205
Berkelium	Bk	97	(247)	14.79	—	—	—
Beryllium	Be	4	9.0122	1.848	1287	2770	1.83
Bismuth	Bi	83	208.980	9.747	271.37	1560	0.122
Bohrium	Bh	107	262.12	—	—	—	—
Boron	B	5	10.811	2.34	2030	—	1.11
Bromine	Br	35	79.909	3.12 (liquid)	−7.2	58	0.293
Cadmium	Cd	48	112.40	8.65	321.03	765	0.226
Calcium	Ca	20	40.08	1.55	838	1440	0.624
Californium	Cf	98	(251)	—	—	—	—
Carbon	C	6	12.01115	2.26	3727	4830	0.691
Cerium	Ce	58	140.12	6.768	804	3470	0.188
Cesium	Cs	55	132.905	1.873	28.40	690	0.243
Chlorine	Cl	17	35.453	3.214×10^{-3} (0°C)	−101	−34.7	0.486
Chromium	Cr	24	51.996	7.19	1857	2665	0.448
Cobalt	Co	27	58.9332	8.85	1495	2900	0.423
Copernicium	Cp	112	(285)	—	—	—	—
Copper	Cu	29	63.54	8.96	1083.40	2595	0.385
Curium	Cm	96	(247)	13.3	—	—	—
Darmstadtium	Ds	110	(271)	—	—	—	—
Dubnium	Db	105	262.114	—	—	—	—
Dysprosium	Dy	66	162.50	8.55	1409	2330	0.172
Einsteinium	Es	99	(254)	—	—	—	—
Erbium	Er	68	167.26	9.15	1522	2630	0.167
Europium	Eu	63	151.96	5.243	817	1490	0.163
Fermium	Fm	100	(237)	—	—	—	—
Fluorine	F	9	18.9984	1.696×10^{-3} (0°C)	−219.6	−188.2	0.753
Francium	Fr	87	(223)	—	(27)	—	—
Gadolinium	Gd	64	157.25	7.90	1312	2730	0.234
Gallium	Ga	31	69.72	5.907	29.75	2237	0.377
Germanium	Ge	32	72.59	5.323	937.25	2830	0.322
Gold	Au	79	196.967	19.32	1064.43	2970	0.131

Element	Symbol	Atomic Number Z	Molar Mass, g/mol	Density, g/cm³ at 20°C	Melting Point, °C	Boiling Point, °C	Specific Heat, J/(g·°C) at 25°C
Hafnium	Hf	72	178.49	13.31	2227	5400	0.144
Hassium	Hs	108	(265)	—	—	—	—
Helium	He	2	4.0026	0.1664×10^{-3}	−269.7	−268.9	5.23
Holmium	Ho	67	164.930	8.79	1470	2330	0.165
Hydrogen	H	1	1.00797	0.08375×10^{-3}	−259.19	−252.7	14.4
Indium	In	49	114.82	7.31	156.634	2000	0.233
Iodine	I	53	126.9044	4.93	113.7	183	0.218
Iridium	Ir	77	192.2	22.5	2447	(5300)	0.130
Iron	Fe	26	55.847	7.874	1536.5	3000	0.447
Krypton	Kr	36	83.80	3.488×10^{-3}	−157.37	−152	0.247
Lanthanum	La	57	138.91	6.189	920	3470	0.195
Lawrencium	Lr	103	(257)	—	—	—	—
Lead	Pb	82	207.19	11.35	327.45	1725	0.129
Lithium	Li	3	6.939	0.534	180.55	1300	3.58
Lutetium	Lu	71	174.97	9.849	1663	1930	0.155
Magnesium	Mg	12	24.312	1.738	650	1107	1.03
Manganese	Mn	25	54.9380	7.44	1244	2150	0.481
Meitnerium	Mt	109	(266)	—	—	—	—
Mendelevium	Md	101	(256)	—	—	—	—
Mercury	Hg	80	200.59	13.55	−38.87	357	0.138
Molybdenum	Mo	42	95.94	10.22	2617	5560	0.251
Neodymium	Nd	60	144.24	7.007	1016	3180	0.188
Neon	Ne	10	20.183	0.8387×10^{-3}	−248.597	−246.0	1.03
Neptunium	Np	93	(237)	20.25	637	—	1.26
Nickel	Ni	28	58.71	8.902	1453	2730	0.444
Niobium	Nb	41	92.906	8.57	2468	4927	0.264
Nitrogen	N	7	14.0067	1.1649×10^{-3}	−210	−195.8	1.03
Nobelium	No	102	(255)	—	—	—	—
Osmium	Os	76	190.2	22.59	3027	5500	0.130
Oxygen	O	8	15.9994	1.3318×10^{-3}	−218.80	−183.0	0.913
Palladium	Pd	46	106.4	12.02	1552	3980	0.243
Phosphorus	P	15	30.9738	1.83	44.25	280	0.741
Platinum	Pt	78	195.09	21.45	1769	4530	0.134
Plutonium	Pu	94	(244)	19.8	640	3235	0.130
Polonium	Po	84	(210)	9.32	254	—	—
Potassium	K	19	39.102	0.862	63.20	760	0.758
Praseodymium	Pr	59	140.907	6.773	931	3020	0.197
Promethium	Pm	61	(145)	7.22	(1027)	—	—
Protactinium	Pa	91	(231)	15.37 (estimated)	(1230)	—	—
Radium	Ra	88	(226)	5.0	700	—	—
Radon	Rn	86	(222)	9.96×10^{-3} (0°C)	(−71)	−61.8	0.092
Rhenium	Re	75	186.2	21.02	3180	5900	0.134
Rhodium	Rh	45	102.905	12.41	1963	4500	0.243
Roentgenium	Rg	111	(280)	—	—	—	—
Rubidium	Rb	37	85.47	1.532	39.49	688	0.364
Ruthenium	Ru	44	101.107	12.37	2250	4900	0.239
Rutherfordium	Rf	104	261.11	—	—	—	—

Element	Symbol	Atomic Number Z	Molar Mass, g/mol	Density, g/cm^3 at 20°C	Melting Point, °C	Boiling Point, °C	Specific Heat, J/(g · °C) at 25°C
Samarium	Sm	62	150.35	7.52	1072	1630	0.197
Scandium	Sc	21	44.956	2.99	1539	2730	0.569
Seaborgium	Sg	106	263.118	—	—	—	—
Selenium	Se	34	78.96	4.79	221	685	0.318
Silicon	Si	14	28.086	2.33	1412	2680	0.712
Silver	Ag	47	107.870	10.49	960.8	2210	0.234
Sodium	Na	11	22.9898	0.9712	97.85	892	1.23
Strontium	Sr	38	87.62	2.54	768	1380	0.737
Sulfur	S	16	32.064	2.07	119.0	444.6	0.707
Tantalum	Ta	73	180.948	16.6	3014	5425	0.138
Technetium	Tc	43	(99)	11.46	2200	—	0.209
Tellurium	Te	52	127.60	6.24	449.5	990	0.201
Terbium	Tb	65	158.924	8.229	1357	2530	0.180
Thallium	Tl	81	204.37	11.85	304	1457	0.130
Thorium	Th	90	(232)	11.72	1755	(3850)	0.117
Thulium	Tm	69	168.934	9.32	1545	1720	0.159
Tin	Sn	50	118.69	7.2984	231.868	2270	0.226
Titanium	Ti	22	47.90	4.54	1670	3260	0.523
Tungsten	W	74	183.85	19.3	3380	5930	0.134
Unnamed	Uut	113	(284)	—	—	—	—
Unnamed	Unq	114	(289)	—	—	—	—
Unnamed	Uup	115	(288)	—	—	—	—
Unnamed	Uuh	116	(293)	—	—	—	—
Unnamed	Uus	117	—	—	—	—	—
Unnamed	Uuo	118	(294)	—	—	—	—
Uranium	U	92	(238)	18.95	1132	3818	0.117
Vanadium	V	23	50.942	6.11	1902	3400	0.490
Xenon	Xe	54	131.30	5.495×10^{-3}	−111.79	−108	0.159
Ytterbium	Yb	70	173.04	6.965	824	1530	0.155
Yttrium	Y	39	88.905	4.469	1526	3030	0.297
Zinc	Zn	30	65.37	7.133	419.58	906	0.389
Zirconium	Zr	40	91.22	6.506	1852	3580	0.276

The values in parentheses in the column of molar masses are the mass numbers of the longest-lived isotopes of those elements that are radioactive. Melting points and boiling points in parentheses are uncertain.

The data for gases are valid only when these are in their usual molecular state, such as H_2, He, O_2, Ne, etc. The specific heats of the gases are the values at constant pressure.

Source: Adapted from J. Emsley, *The Elements,* 3rd ed., 1998, Clarendon Press, Oxford. See also www.webelements.com for the latest values and newest elements.

Periodic Table of the Elements

	Metals
	Metalloids
	Nonmetals

Period	IA	IIA	IIIB	IVB	VB	VIB	VIIB	VIIIB			IB	IIB	IIIA	IVA	VA	VIA	VIIA	0
1	1 H																	2 He
2	3 Li	4 Be											5 B	6 C	7 N	8 O	9 F	10 Ne
3	11 Na	12 Mg											13 Al	14 Si	15 P	16 S	17 Cl	18 Ar
4	19 K	20 Ca	21 Sc	22 Ti	23 V	24 Cr	25 Mn	26 Fe	27 Co	28 Ni	29 Cu	30 Zn	31 Ga	32 Ge	33 As	34 Se	35 Br	36 Kr
5	37 Rb	38 Sr	39 Y	40 Zr	41 Nb	42 Mo	43 Tc	44 Ru	45 Rh	46 Pd	47 Ag	48 Cd	49 In	50 Sn	51 Sb	52 Te	53 I	54 Xe
6	55 Cs	56 Ba	57-71 *	72 Hf	73 Ta	74 W	75 Re	76 Os	77 Ir	78 Pt	79 Au	80 Hg	81 Tl	82 Pb	83 Bi	84 Po	85 At	86 Rn
7	87 Fr	88 Ra	89-103 †	104 Rf	105 Db	106 Sg	107 Bh	108 Hs	109 Mt	110 Ds	111 Rg	112 Cp	113	114	115	116	117	118

THE HORIZONTAL PERIODS

Alkali metals IA

Noble gases 0

Transition metals

Inner transition metals

Lanthanide series *	57 La	58 Ce	59 Pr	60 Nd	61 Pm	62 Sm	63 Eu	64 Gd	65 Tb	66 Dy	67 Ho	68 Er	69 Tm	70 Yb	71 Lu
Actinide series †	89 Ac	90 Th	91 Pa	92 U	93 Np	94 Pu	95 Am	96 Cm	97 Bk	98 Cf	99 Es	100 Fm	101 Md	102 No	103 Lr

Evidence for the discovery of elements 113 through 118 has been reported. See www.webelements.com for the latest information and newest elements.

CHAPTER 12

CP **1.** c, e, f **2.** (a) no; (b) at site of $\vec{F}_1$, perpendicular to plane of figure; (c) 45 N **3.** d
Q **1.** (a) 1 and 3 tie, then 2; (b) all tie; (c) 1 and 3 tie, then 2 (zero) **3.** a and c (forces and torques balance) **5.** (a) 12 kg; (b) 3 kg; (c) 1 kg **7.** (a) at C (to eliminate forces there from a torque equation); (b) plus; (c) minus; (d) equal **9.** increase
P **1.** (a) 1.00 m; (b) 2.00 m; (c) 0.987 m; (d) 1.97 m **3.** (a) 9.4 N; (b) 4.4 N **5.** 7.92 kN **7.** (a) 2.8×10^2 N; (b) 8.8×10^2 N; (c) $71°$
9. 74.4 g **11.** (a) 1.2 kN; (b) down; (c) 1.7 kN; (d) up; (e) left; (f) right **13.** (a) 2.7 kN; (b) up; (c) 3.6 kN; (d) down **15.** (a) 5.0 N; (b) 30 N; (c) 1.3 m **17.** (a) 0.64 m; (b) increased **19.** 8.7 N
21. (a) 6.63 kN; (b) 5.74 kN; (c) 5.96 kN **23.** (a) 192 N; (b) 96.1 N; (c) 55.5 N **25.** 13.6 N **27.** (a) 1.9 kN; (b) up; (c) 2.1 kN; (d) down **29.** (a) $(-80 \text{ N})\hat{i} + (1.3 \times 10^2 \text{ N})\hat{j}$; (b) $(80 \text{ N})\hat{i} + (1.3 \times 10^2 \text{ N})\hat{j}$
31. 2.20 m **33.** (a) $60.0°$; (b) 300 N **35.** (a) 445 N; (b) 0.50; (c) 315 N **37.** 0.34 **39.** (a) 211 N; (b) 534 N; (c) 320 N **41.** (a) slides; (b) $31°$; (c) tips; (d) $34°$ **43.** (a) 6.5×10^6 N/m^2; (b) 1.1×10^{-5} m
45. (a) 0.80; (b) 0.20; (c) 0.25 **47.** (a) 1.4×10^9 N; (b) 75
49. (a) 866 N; (b) 143 N; (c) 0.165 **51.** (a) 1.2×10^2 N; (b) 68 N
53. (a) 1.8×10^7 N; (b) 1.4×10^7 N; (c) 16 **55.** 0.29 **57.** 76 N
59. (a) 8.01 kN; (b) 3.65 kN; (c) 5.66 kN **61.** 71.7 N **63.** (a) $L/2$; (b) $L/4$; (c) $L/6$; (d) $L/8$; (e) $25L/24$ **65.** (a) 88 N; (b) $(30\hat{i} + 97\hat{j})$ N
67. 2.4×10^9 N/m^2 **69.** $60°$ **71.** (a) $\mu < 0.57$; (b) $\mu > 0.57$
73. (a) $(35\hat{i} + 200\hat{j})$ N; (b) $(-45\hat{i} + 200\hat{j})$ N; (c) 1.9×10^2 N
75. (a) BC, CD, DA; (b) 535 N; (c) 757 N **77.** (a) 1.38 kN; (b) 180 N
79. (a) $a_1 = L/2, a_2 = 5L/8, h = 9L/8$; (b) $b_1 = 2L/3, b_2 = L/2$, $h = 7L/6$ **81.** $L/4$ **83.** (a) 106 N; (b) $64.0°$

CHAPTER 13

CP **1.** all tie **2.** (a) 1, tie of 2 and 4, then 3; (b) line d
3. (a) increase; (b) negative **4.** (a) 2; (b) 1 **5.** (a) path 1 (decreased E (more negative) gives decreased a); (b) less (decreased a gives decreased T)
Q **1.** $3GM^2/d^2$, leftward **3.** Gm^2/r^2, upward **5.** b and c tie, then a (zero) **7.** 1, tie of 2 and 4, then 3 **9.** (a) positive y; (b) yes, rotates counterclockwise until it points toward particle B **11.** b, d, and f all tie, then e, c, a
P **1.** $\frac{1}{2}$ **3.** 19 m **5.** 0.8 m **7.** $-5.00d$ **9.** 2.60×10^5 km
11. (a) $M = m$; (b) 0 **13.** 8.31×10^{-9} N **15.** (a) $-1.88d$; (b) $-3.90d$; (c) $0.489d$ **17.** (a) 17 N; (b) 2.4 **19.** 2.6×10^6 m
21. 5×10^{24} kg **23.** (a) 7.6 m/s^2; (b) 4.2 m/s^2 **25.** (a) $(3.0 \times 10^{-7}$ N/kg$)m$; (b) $(3.3 \times 10^{-7}$ N/kg$)m$; (c) $(6.7 \times 10^{-7}$ N/kg·m$)mr$
27. (a) 9.83 m/s^2; (b) 9.84 m/s^2; (c) 9.79 m/s^2 **29.** 5.0×10^9 J
31. (a) 0.74; (b) 3.8 m/s^2; (c) 5.0 km/s **33.** (a) 0.0451; (b) 28.5
35. -4.82×10^{-13} J **37.** (a) 0.50 pJ; (b) -0.50 pJ **39.** (a) 1.7 km/s; (b) 2.5×10^5 m; (c) 1.4 km/s **41.** (a) 82 km/s; (b) 1.8×10^4 km/s
43. (a) 7.82 km/s; (b) 87.5 min **45.** 6.5×10^{23} kg **47.** 5×10^{10} stars
49. (a) 1.9×10^{13} m; (b) $3.6R_P$ **51.** (a) 6.64×10^3 km; (b) 0.0136
53. 5.8×10^6 m **57.** 0.71 y **59.** $(GM/L)^{0.5}$ **61.** (a) 3.19×10^3 km; (b) lifting **63.** (a) 2.8 y; (b) 1.0×10^{-4} **65.** (a) $r^{1.5}$; (b) r^{-1}; (c) $r^{0.5}$; (d) $r^{-0.5}$ **67.** (a) 7.5 km/s; (b) 97 min; (c) 4.1×10^2 km; (d) 7.7 km/s; (e) 93 min; (f) 3.2×10^{-3} N; (g) no; (h) yes **69.** 1.1 s
71. (a) $GMmx(x^2 + R^2)^{-3/2}$; (b) $[2GM(R^{-1} - (R^2 + x^2)^{-1/2})]^{1/2}$
73. (a) 1.0×10^3 kg; (b) 1.5 km/s **75.** 3.2×10^{-7} N **77.** $.037\hat{j}$ μN
79. $2\pi r^{1.5}G^{-0.5}(M + m/4)^{-0.5}$ **81.** (a) 2.2×10^{-7} rad/s; (b) 89 km/s
83. (a) 2.15×10^4 s; (b) 12.3 km/s; (c) 12.0 km/s; (d) 2.17×10^{11} J; (e) -4.53×10^{11} J; (f) -2.35×10^{11} J; (g) 4.04×10^7 m; (h) $1.22 \times$

10^3 s; (i) elliptical **85.** 2.5×10^4 km **87.** (a) 1.4×10^6 m/s; (b) 3×10^6 m/s^2 **89.** (a) 0; (b) 1.8×10^{32} J; (c) 1.8×10^{32} J; (d) 0.99 km/s
91. (a) Gm^2/R_i; (b) $Gm^2/2R_i$; (c) $(Gm/R_i)^{0.5}$; (d) $2(Gm/R_i)^{0.5}$; (e) Gm^2/R_i; (f) $(2Gm/R_i)^{0.5}$; (g) The center-of-mass frame is an inertial frame, and in it the principle of conservation of energy may be written as in Chapter 8; the reference frame attached to body A is noninertial, and the principle cannot be written as in Chapter 8. Answer (d) is correct. **93.** 2.4×10^4 m/s **95.** $-0.044\hat{j}$ μN
97. $GM_Em/12R_E$

CHAPTER 14

CP **1.** all tie **2.** (a) all tie (the gravitational force on the penguin is the same); (b) $0.95\rho_0, \rho_0, 1.1\rho_0$ **3.** 13 cm^3/s, outward
4. (a) all tie; (b) 1, then 2 and 3 tie, 4 (wider means slower); (c) 4, 3, 2, 1 (wider and lower mean more pressure)
Q **1.** (a) moves downward; (b) moves downward **3.** (a) downward; (b) downward; (c) same **5.** b, then a and d tie (zero), then c
7. (a) 1 and 4; (b) 2; (c) 3 **9.** B, C, A
P **1.** 0.074 **3.** 1.1×10^5 Pa **5.** 2.9×10^4 N **7.** (b) 26 kN
9. (a) 1.0×10^3 torr; (b) 1.7×10^3 torr **11.** (a) 94 torr; (b) 4.1×10^2 torr; (c) 3.1×10^2 torr **13.** 1.08×10^3 atm **15.** -2.6×10^4 Pa
17. 7.2×10^5 N **19.** 4.69×10^5 N **21.** 0.635 J **23.** 44 km
25. 739.26 torr **27.** (a) 7.9 km; (b) 16 km **29.** 8.50 kg **31.** (a) 6.7×10^2 kg/m^3; (b) 7.4×10^2 kg/m^3 **33.** (a) 2.04×10^{-2} m^3; (b) 1.57 kN **35.** five **37.** 57.3 cm **39.** (a) 1.2 kg; (b) 1.3×10^3 kg/m^3 **41.** (a) 0.10; (b) 0.083 **43.** (a) 637.8 cm^3; (b) 5.102 m^3; (c) 5.102×10^3 kg **45.** 0.126 m^3 **47.** (a) 1.80 m^3; (b) 4.75 m^3
49. (a) 3.0 m/s; (b) 2.8 m/s **51.** 8.1 m/s **53.** 66 W **55.** 1.4×10^5 J
57. (a) 1.6×10^{-3} m^3/s; (b) 0.90 m **59.** (a) 2.5 m/s; (b) 2.6×10^5 Pa
61. (a) 3.9 m/s; (b) 88 kPa **63.** 1.1×10^2 m/s **65.** (a) 2.0×10^{-2} m^3/s **67.** (a) 74 N; (b) 1.5×10^2 m^3 **69.** (a) 0.0776 m^3/s; (b) 69.8 kg/s **71.** (a) 35 cm; (b) 30 cm; (c) 20 cm **73.** 1.5 g/cm^3 **75.** 5.11×10^{-7} kg **77.** 44.2 g **79.** 6.0×10^2 kg/m^3 **81.** 45.3 cm^3
83. (a) 3.2 m/s; (b) 9.2×10^4 Pa; (c) 10.3 m **85.** 1.07×10^3 g

CHAPTER 15

CP **1.** (sketch x versus t) (a) $-x_m$; (b) $+x_m$; (c) 0 **2.** a (F must have the form of Eq. 15-10) **3.** (a) 5 J; (b) 2 J; (c) 5 J **4.** all tie (in Eq. 15-29, m is included in I) **5.** 1, 2, 3 (the ratio m/b matters; k does not)
Q **1.** a and b **3.** (a) 2; (b) positive; (c) between 0 and $+x_m$
5. (a) between D and E; (b) between $3\pi/2$ rad and 2π rad
7. (a) all tie; (b) 3, then 1 and 2 tie; (c) 1, 2, 3 (zero); (d) 1, 2, 3 (zero); (e) 1, 3, 2 **9.** b (infinite period, does not oscillate), c, a
11. (a) greater; (b) same; (c) same; (d) greater; (e) greater
P **1.** (a) 0.50 s; (b) 2.0 Hz; (c) 18 cm **3.** 37.8 m/s^2 **5.** (a) 1.0 mm; (b) 0.75 m/s; (c) 5.7×10^2 m/s^2 **7.** (a) 498 Hz; (b) greater
9. (a) 3.0 m; (b) -49 m/s; (c) -2.7×10^2 m/s^2; (d) 20 rad; (e) 1.5 Hz; (f) 0.67 s **11.** 39.6 Hz **13.** (a) 0.500 s; (b) 2.00 Hz; (c) 12.6 rad/s; (d) 79.0 N/m; (e) 4.40 m/s; (f) 27.6 N **15.** (a) $0.18A$; (b) same direction **17.** (a) 5.58 Hz; (b) 0.325 kg; (c) 0.400 m **19.** (a) 25 cm; (b) 2.2 Hz **21.** 54 Hz **23.** 3.1 cm **25.** (a) 0.525 m; (b) 0.686 s
27. (a) 0.75; (b) 0.25; (c) $2^{-0.5}x_m$ **29.** 37 mJ **31.** (a) 2.25 Hz; (b) 125 J; (c) 250 J; (d) 86.6 cm **33.** (a) 1.1 m/s; (b) 3.3 cm
35. (a) 3.1 ms; (b) 4.0 m/s; (c) 0.080 J; (d) 80 N; (e) 40 N
37. (a) 2.2 Hz; (b) 56 cm/s; (c) 0.10 kg; (d) 20.0 cm **39.** (a) 39.5 rad/s; (b) 34.2 rad/s; (c) 124 rad/s^2 **41.** (a) 0.205 kg·m^2; (b) 47.7 cm; (c) 1.50 s **43.** (a) 1.64 s; (b) equal **45.** 8.77 s **47.** 0.366 s

49. (a) 0.845 rad; (b) 0.0602 rad   **51.** (a) 0.53 m; (b) 2.1 s
53. 0.0653 s   **55.** (a) 2.26 s; (b) increases; (c) same   **57.** 6.0%
59. (a) 14.3 s; (b) 5.27   **61.** (a) $F_m/b\omega$; (b) F_m/b **63.** 5.0 cm
65. (a) 2.8×10^3 rad/s; (b) 2.1 m/s; (c) 5.7 km/s^2 **67.** (a) 1.1 Hz;
(b) 5.0 cm   **69.** 7.2 m/s   **71.** (a) 7.90 N/m; (b) 1.19 cm; (c) 2.00 Hz
73. (a) 1.3×10^2 N/m; (b) 0.62 s; (c) 1.6 Hz; (d) 5.0 cm; (e) 0.51 m/s
75. (a) 16.6 cm; (b) 1.23%   **77.** (a) 1.2 J; (b) 50   **79.** 1.53 m
81. (a) 0.30 m; (b) 0.28 s; (c) 1.5×10^2 m/s^2; (d) 11 J   **83.** (a) 1.23
kN/m; (b) 76.0 N   **85.** 1.6 kg   **87.** (a) 0.735 kg · m^2; (b) 0.0240 N · m;
(c) 0.181 rad/s   **89.** (a) 3.5 m; (b) 0.75 s   **91.** (a) 0.35 Hz; (b) 0.39
Hz; (c) 0 (no oscillation)   **93.** (a) 245 N/m; (b) 0.284 s
95. 0.079 kg · m^2 **97.** (a) 8.11×10^{-5} kg · m^2; (b) 3.14 rad/s
99. 14.0°   **101.** (a) 3.2 Hz; (b) 0.26 m; (c) $x =$
$(0.26$ m$) \cos(20t - \pi/2)$, with t in seconds   **103.** (a) 0.44 s; (b) 0.18 m
105. (a) 0.45 s; (b) 0.10 m above and 0.20 m below; (c) 0.15 m;
(d) 2.3 J   **107.** 7×10^2 N/m

CHAPTER 16

CP   **1.** a, 2; b, 3; c, 1 (compare with the phase in Eq. 16-2, then see
Eq. 16-5)   **2.** (a) 2, 3, 1 (see Eq. 16-12); (b) 3, then 1 and 2 tie (find
amplitude of dy/dt)   **3.** (a) same (independent of f); (b) decrease
($\lambda = v/f$); (c) increase; (d) increase   **4.** 0.20 and 0.80 tie, then 0.60,
0.45   **5.** (a) 1; (b) 3; (c) 2   **6.** (a) 75 Hz; (b) 525 Hz
Q   **1.** (a) 1, 4, 2, 3; (b) 1, 4, 2, 3   **3.** a, upward; b, upward; c, downward; d, downward; e, downward; f, downward; g, upward; h, upward
5. intermediate (closer to fully destructive)   **7.** (a) 0, 0.2 wavelength, 0.5 wavelength (zero); (b) $4P_{\text{avg,1}}$ **9.** d **11.** c, a, b
P   **1.** 1.1 ms   **3.** (a) 3.49 m^{-1}; (b) 31.5 m/s   **5.** (a) 0.680 s; (b) 1.47
Hz; (c) 2.06 m/s   **7.** (a) 64 Hz; (b) 1.3 m; (c) 4.0 cm; (d) 5.0 m^{-1};
(e) 4.0×10^2 s^{-1}; (f) $\pi/2$ rad; (g) minus   **9.** (a) 3.0 mm; (b) 16 m^{-1};
(c) 2.4×10^2 s^{-1}; (d) minus   **11.** (a) negative; (b) 4.0 cm; (c) 0.31
cm^{-1}; (d) 0.63 s^{-1}; (e) π rad; (f) minus; (g) 2.0 cm/s; (h) -2.5 cm/s
13. (a) 11.7 cm; (b) π rad   **15.** (a) 0.12 mm; (b) 141 m^{-1}; (c) 628 s^{-1};
(d) plus   **17.** (a) 15 m/s; (b) 0.036 N   **19.** 129 m/s   **21.** 2.63 m
23. (a) 5.0 cm; (b) 40 cm; (c) 12 m/s; (d) 0.033 s; (e) 9.4 m/s;
(f) 16 m^{-1}; (g) 1.9×10^2 s^{-1}; (h) 0.93 rad; (i) plus   **27.** 3.2 mm
29. 0.20 m/s   **31.** $1.41y_m$ **33.** (a) 9.0 mm; (b) 16 m^{-1}; (c) 1.1×10^3
s^{-1}; (d) 2.7 rad; (e) plus   **35.** 5.0 cm   **37.** (a) 3.29 mm; (b) 1.55 rad;
(c) 1.55 rad   **39.** 84°   **41.** (a) 82.0 m/s; (b) 16.8 m; (c) 4.88 Hz
43. (a) 7.91 Hz; (b) 15.8 Hz; (c) 23.7 Hz   **45.** (a) 105 Hz; (b) 158 m/s
47. 260 Hz   **49.** (a) 144 m/s; (b) 60.0 cm; (c) 241 Hz   **51.** (a) 0.50 cm;
(b) 3.1 m^{-1}; (c) 3.1×10^2 s^{-1}; (d) minus   **53.** (a) 0.25 cm; (b) 1.2×10^2
cm/s; (c) 3.0 cm; (d) 0   **55.** 0.25 m   **57.** (a) 2.00 Hz; (b) 2.00 m; (c) 4.00
m/s; (d) 50.0 cm; (e) 150 cm; (f) 250 cm; (g) 0; (h) 100 cm; (i) 200 cm
59. (a) 324 Hz; (b) eight   **61.** 36 N   **63.** (a) 75 Hz; (b) 13 ms
65. (a) 2.0 mm; (b) 95 Hz; (c) $+30$ m/s; (d) 31 cm; (e) 1.2 m/s
67. (a) 0.31 m; (b) 1.64 rad; (c) 2.2 mm   **69.** (a) $0.83y_1$; (b) 37°
71. (a) 3.77 m/s; (b) 12.3 N; (c) 0; (d) 46.4 W; (e) 0; (f) 0; (g) $\pm$ 0.50 cm
73. 1.2 rad   **75.** (a) 300 m/s; (b) no   **77.** (a) $[k \Delta\ell(\ell + \Delta\ell)/m]^{0.5}$
79. (a) 144 m/s; (b) 3.00 m; (c) 1.50 m; (d) 48.0 Hz; (e) 96.0 Hz
81. (a) 1.00 cm; (b) 3.46×10^3 s^{-1}; (c) 10.5 m^{-1}; (d) plus   **83.** (a)
$2\pi y_m/\lambda$; (b) no   **85.** (a) 240 cm; (b) 120 cm; (c) 80 cm   **87.** (a) 1.33
m/s; (b) 1.88 m/s; (c) 16.7 m/s^2; (d) 23.7 m/s^2 **89.** (a) 0.52 m; (b) 40
m/s; (c) 0.40 m   **91.** (a) 0.16 m; (b) 2.4×10^2 N; (c) $y(x, t) =$
$(0.16$ m$) \sin[(1.57$ m$^{-1})x] \sin[(31.4$ s$^{-1})t]$ **93.** (c) 2.0 m/s; (d) $-x$

CHAPTER 17

CP   **1.** beginning to decrease (example: mentally move the curves
of Fig. 17-6 rightward past the point at $x = 42$ cm)   **2.** (a) 1 and 2
tie, then 3 (see Eq. 17-28); (b) 3, then 1 and 2 tie (see Eq. 17-26)
3. second (see Eqs. 17-39 and 17-41)   **4.** a, greater; b, less; c, can't
tell; d, can't tell; e, greater; f, less

Q   **1.** (a) 0, 0.2 wavelength, 0.5 wavelength (zero); (b) $4P_{\text{avg,1}}$
3. C, then A and B tie   **5.** E, A, D, C, B **7.** 1, 4, 3, 2   **9.** 150 Hz
and 450 Hz

P   **1.** (a) 79 m; (b) 41 m; (c) 89 m   **3.** (a) 2.6 km; (b) 2.0×10^2
5. 1.9×10^3 km   **7.** 40.7 m   **9.** 0.23 ms   **11.** (a) 76.2 μm; (b) 0.333 mm
13. 960 Hz   **15.** (a) 2.3×10^2 Hz; (b) higher   **17.** (a) 143 Hz; (b) 3;
(c) 5; (d) 286 Hz; (e) 2; (f) 3   **19.** (a) 14; (b) 14   **21.** (a) 343 Hz;
(b) 3; (c) 5; (d) 686 Hz; (e) 2; (f) 3   **23.** (a) 0; (b) fully constructive;
(c) increase; (d) 128 m; (e) 63.0 m; (f) 41.2 m   **25.** 36.8 nm
27. (a) 1.0×10^3; (b) 32   **29.** 15.0 mW   **31.** 2 μW   **33.** 0.76 μm
35. (a) 5.97×10^{-5} W/m^2; (b) 4.48 nW   **37.** (a) 0.34 nW; (b) 0.68 nW;
(c) 1.4 nW; (d) 0.88 nW; (e) 0   **39.** (a) 405 m/s; (b) 596 N; (c) 44.0
cm; (d) 37.3 cm   **41.** (a) 833 Hz; (b) 0.418 m   **43.** (a) 3; (b) 1129 Hz;
(c) 1506 Hz   **45.** (a) 2; (b) 1   **47.** 12.4 m   **49.** 45.3 N   **51.** 2.25 ms
53. 0.020   **55.** (a) 526 Hz; (b) 555 Hz   **57.** 0   **59.** (a) 1.022 kHz;
(b) 1.045 kHz   **61.** 41 kHz   **63.** 155 Hz   **65.** (a) 2.0 kHz; (b) 2.0
kHz   **67.** (a) 485.8 Hz; (b) 500.0 Hz; (c) 486.2 Hz; (d) 500.0 Hz
69. (a) 42°; (b) 11 s   **71.** 1 cm   **73.** 2.1 m   **75.** (a) 39.7 μW/m^2;
(b) 171 nm; (c) 0.893 Pa   **77.** 0.25   **79.** (a) 2.10 m; (b) 1.47 m
81. (a) 59.7; (b) 2.81×10^{-4} **83.** (a) rightward; (b) 0.90 m/s; (c) less
85. (a) 11 ms; (b) 3.8 m   **87.** (a) 9.7×10^2 Hz; (b) 1.0 kHz; (c) 60 Hz,
no   **89.** (a) 21 nm; (b) 35 cm; (c) 24 nm; (d) 35 cm   **91.** (a) 7.70 Hz;
(b) 7.70 Hz   **93.** (a) 5.2 kHz; (b) 2   **95.** (a) 10 W; (b) 0.032 W/m^2;
(c) 99 dB   **97.** (a) 0; (b) 0.572 m; (c) 1.14 m   **99.** 171 m   **101.** (a)
3.6×10^2 m/s; (b) 150 Hz   **103.** 400 Hz   **105.** (a) 14; (b) 12

CHAPTER 18

CP   **1.** (a) all tie; (b) 50°X, 50°Y, 50°W   **2.** (a) 2 and 3 tie, then 1,
then 4; (b) 3, 2, then 1 and 4 tie (from Eqs. 18-9 and 18-10, assume
that change in area is proportional to initial area)   **3.** A (see
Eq. 18-14)   **4.** c and e (maximize area enclosed by a clockwise
cycle)   **5.** (a) all tie (ΔE_{int} depends on i and f, not on path); (b) 4, 3,
2, 1 (compare areas under curves); (c) 4, 3, 2, 1 (see Eq. 18-26)
6. (a) zero (closed cycle); (b) negative (W_{net} is negative; see
Eq. 18-26)   **7.** b and d tie, then a, c (P_{cond} identical; see Eq. 18-32)
Q   **1.** c, then the rest tie   **3.** B, then A and C tie   **5.** (a) f, because
ice temperature will not rise to freezing point and then drop; (b) b
and c at freezing point, d above, e below; (c) in b liquid partly
freezes and no ice melts; in c no liquid freezes and no ice melts; in d
no liquid freezes and ice fully melts; in e liquid fully freezes and no
ice melts   **7.** (a) both clockwise; (b) both clockwise   **9.** (a) greater;
(b) 1, 2, 3; (c) 1, 3, 2; (d) 1, 2, 3; (e) 2, 3, 1   **11.** c, b, a
P   **1.** 1.366   **3.** 348 K   **5.** (a) 320°F; (b) -12.3°F   **7.** -92.1°X
9. 2.731 cm   **11.** 49.87 cm^3 **13.** 29 cm^3 **15.** 360°C   **17.** 0.26 cm^3
19. 0.13 mm   **21.** 7.5 cm   **23.** 160 s   **25.** 94.6 L   **27.** 42.7 kJ
29. 33 m^2 **31.** 33 g   **33.** 3.0 min   **35.** 13.5 C°   **37.** (a) 5.3°C; (b) 0;
(c) 0°C; (d) 60 g   **39.** 742 kJ   **41.** (a) 0°C; (b) 2.5°C   **43.** (a) 1.2 $\times$
10^2 J; (b) 75 J; (c) 30 J   **45.** -30 J   **47.** (a) 6.0 cal; (b) -43 cal;
(c) 40 cal; (d) 18 cal; (e) 18 cal   **49.** 60 J   **51.** (a) 1.23 kW; (b)
2.28 kW; (c) 1.05 kW   **53.** 1.66 kJ/s   **55.** (a) 16 J/s; (b) 0.048 g/s
57. (a) 1.7×10^4 W/m^2; (b) 18 W/m^2 **59.** 0.50 min   **61.** 0.40 cm/h
63. -4.2°C   **65.** 1.1 m   **67.** 10%   **69.** (a) 80 J; (b) 80 J   **71.** 4.5 $\times$
10^2 J/kg·K   **73.** 0.432 cm^3 **75.** 3.1×10^2 J   **77.** 79.5°C   **79.** 23 J
81. (a) $11p_1V_1$; (b) $6p_1V_1$ **83.** 4.83×10^{-2} cm^3 **85.** 10.5°C
87. (a) 90 W; (b) 2.3×10^2 W; (c) 3.3×10^2 W   **89.** (a) 1.87×10^4;
(b) 10.4 h   **91.** 333 J   **93.** 8.6 J   **95.** (a) -45 J; (b) $+45$ J

CHAPTER 19

CP   **1.** all but c **2.** (a) all tie; (b) 3, 2, 1   **3.** gas A **4.** 5 (greatest
change in T), then tie of 1, 2, 3, and 4   **5.** 1, 2, 3 ($Q_3 = 0$, Q_2 goes
into work W_2, but Q_1 goes into greater work W_1 and increases
gas temperature)

Q **1.** d, then a and b tie, then c **3.** 20 J **5.** (a) 3; (b) 1; (c) 4; (d) 2; (e) yes **7.** (a) 1, 2, 3, 4; (b) 1, 2, 3 **9.** constant-volume process
P **1.** 0.933 kg **3.** (a) 0.0388 mol; (b) 220°C **5.** 25 molecules/cm³
7. (a) 3.14×10^3 J; (b) from **9.** 186 kPa **11.** 5.60 kJ
13. (a) 1.5 mol; (b) 1.8×10^3 K; (c) 6.0×10^2 K; (d) 5.0 kJ
15. 360 K **17.** 2.0×10^5 Pa **19.** (a) 511 m/s; (b) −200°C; (c) 899°C
21. 1.8×10^2 m/s **23.** 1.9 kPa **25.** (a) 5.65×10^{-21} J; (b) 7.72×10^{-21} J; (c) 3.40 kJ; (d) 4.65 kJ **27.** (a) 6.76×10^{-20} J; (b) 10.7
29. (a) 6×10^9 km **31.** (a) 3.27×10^{10} molecules/cm³; (b) 172 m
33. (a) 6.5 km/s; (b) 7.1 km/s **35.** (a) 420 m/s; (b) 458 m/s; (c) yes
37. (a) 0.67; (b) 1.2; (c) 1.3; (d) 0.33 **39.** (a) 1.0×10^4 K; (b) 1.6×10^5 K; (c) 4.4×10^2 K; (d) 7.0×10^3 K; (e) no; (f) yes **41.** (a) 7.0 km/s; (b) 2.0×10^{-8} cm; (c) 3.5×10^{10} collisions/s **43.** (a) 3.49 kJ; (b) 2.49 kJ; (c) 997 J; (d) 1.00 kJ **45.** (a) 6.6×10^{-26} kg; (b) 40 g/mol **47.** (a) 0; (b) +374 J; (c) +374 J; (d) $+3.11 \times 10^{-22}$ J
49. 15.8 J/mol·K **51.** 8.0 kJ **53.** (a) 6.98 kJ; (b) 4.99 kJ; (c) 1.99 kJ; (d) 2.99 kJ **55.** (a) 14 atm; (b) 6.2×10^2 K **57.** (a) diatomic; (b) 446 K; (c) 8.10 mol **59.** −15 J **61.** −20 J **63.** (a) 3.74 kJ; (b) 3.74 kJ; (c) 0; (d) 0; (e) −1.81 kJ; (f) 1.81 kJ; (g) −3.22 kJ; (h) −1.93 kJ; (i) −1.29 kJ; (j) 520 J; (k) 0; (l) 520 J; (m) 0.0246 m³; (n) 2.00 atm; (o) 0.0373 m³; (p) 1.00 atm **65.** (a) monatomic; (b) 2.7×10^4 K; (c) 4.5×10^4 mol; (d) 3.4 kJ; (e) 3.4×10^2 kJ; (f) 0.010 **67.** (a) 2.00 atm; (b) 333 J; (c) 0.961 atm; (d) 236 J
69. 349 K **71.** (a) −374 J; (b) 0; (c) +374 J; (d) $+3.11 \times 10^{-22}$ J
73. 7.03×10^9 s⁻¹ **75.** (a) 900 cal; (b) 0; (c) 900 cal; (d) 450 cal; (e) 1200 cal; (f) 300 cal; (g) 900 cal; (h) 450 cal; (i) 0; (j) −900 cal; (k) 900 cal; (l) 450 cal **77.** (a) $3/v_0^3$; (b) $0.750v_0$; (c) $0.775v_0$
79. (a) −2.37 kJ; (b) 2.37 kJ **81.** (b) 125 J; (c) to **83.** (a) 8.0 atm; (b) 300 K; (c) 4.4 kJ; (d) 3.2 atm; (e) 120 K; (f) 2.9 kJ; (g) 4.6 atm; (h) 170 K; (i) 3.4 kJ **85.** (a) 38 L; (b) 71 g **87.** −3.0 J

CHAPTER 20

CP **1.** a, b, c **2.** smaller (Q is smaller) **3.** c, b, a **4.** a, d, c, b **5.** b
Q **1.** b, a, c, d **3.** unchanged **5.** a and c tie, then b and d tie
7. (a) same; (b) increase; (c) decrease **9.** A, first; B, first and second; C, second; D, neither
P **1.** (a) 9.22 kJ; (b) 23.1 J/K; (c) 0 **3.** 14.4 J/K **5.** (a) 5.79×10^4 J; (b) 173 J/K **7.** (a) 320 K; (b) 0; (c) +1.72 J/K **9.** +0.76 J/K
11. (a) 57.0°C; (b) −22.1 J/K; (c) +24.9 J/K; (d) +2.8 J/K
13. (a) −710 mJ/K; (b) +710 mJ/K; (c) +723 mJ/K; (d) −723 mJ/K; (e) +13 mJ/K; (f) 0 **15.** (a) −943 J/K; (b) +943 J/K; (c) yes
17. (a) 0.333; (b) 0.215; (c) 0.644; (d) 1.10; (e) 1.10; (f) 0; (g) 1.10; (h) 0; (i) −0.889; (j) −0.889; (k) −1.10; (l) −0.889; (m) 0; (n) 0.889; (o) 0 **19.** (a) 0.693; (b) 4.50; (c) 0.693; (d) 0; (e) 4.50; (f) 23.0 J/K; (g) −0.693; (h) 7.50; (i) −0.693; (j) 3.00; (k) 4.50; (l) 23.0 J/K
21. −1.18 J/K **23.** 97 K **25.** (a) 266 K; (b) 341 K **27.** (a) 23.6%; (b) 1.49×10^4 J **29.** (a) 2.27 kJ; (b) 14.8 kJ; (c) 15.4%; (d) 75.0%; (e) greater **31.** (a) 33 kJ; (b) 25 kJ; (c) 26 kJ; (d) 18 kJ
33. (a) 1.47 kJ; (b) 554 J; (c) 918 J; (d) 62.4% **35.** (a) 3.00; (b) 1.98; (c) 0.660; (d) 0.495; (e) 0.165; (f) 34.0% **37.** 440 W **39.** 20 J
41. 0.25 hp **43.** 2.03 **47.** (a) $W = N!/(n_1!\, n_2!\, n_3!)$; (b) $[(N/2)!\, (N/2)!]/[(N/3)!\, (N/3)!\, (N/3)!]$; (c) 4.2×10^{16} **49.** 0.141 J/K·s
51. (a) 87 m/s; (b) 1.2×10^2 m/s; (c) 22 J/K **53.** (a) 78%; (b) 82 kg/s
55. (a) 40.9°C; (b) −27.1 J/K; (c) 30.3 J/K; (d) 3.18 J/K **57.** +3.59 J/K
59. 1.18×10^3 J/K **63.** (a) 0; (b) 0; (c) −23.0 J/K; (d) 23.0 J/K
65. (a) 25.5 kJ; (b) 4.73 kJ; (c) 18.5% **67.** (a) 1.95 J/K; (b) 0.650 J/K; (c) 0.217 J/K; (d) 0.072 J/K; (e) decrease **69.** (a) 4.45 J/K; (b) no
71. (a) 1.26×10^{14}; (b) 4.71×10^{13}; (c) 0.37; (d) 1.01×10^{29}; (e) 1.37×10^{28}; (f) 0.14; (g) 9.05×10^{58}; (h) 1.64×10^{57}; (i) 0.018; (j) decrease **73.** (a) 42.6 kJ; (b) 7.61 kJ **75.** (a) 1; (b) 1; (c) 3; (d) 10; (e) 1.5×10^{-23} J/K; (f) 3.2×10^{-23} J/K

I N D E X

Figures are noted by page numbers in *italics*, tables are indicated by t following the page number.